Identification of Feathers
of the Birds of Western Europe

Cloé Fraigneau

HELM

HELM
Bloomsbury Publishing Plc
50 Bedford Square, London, WC1B 3DP, UK
29 Earlsfort Terrace, Dublin 2, Ireland

BLOOMSBURY, HELM and the Helm logo are trademarks of Bloomsbury Publishing Plc

First published in the United Kingdom by Bloomsbury Publishing 2021

This edition published by arrangement with Delachaux et Niestlé, Paris, France

First published in 2017 in France as *Identifier les plumes des oiseaux d'Europe occidentale* by Delachaux et
Niestlé, Paris. © Delachaux et Niestlé, Paris, 2017
Translation © Tony Williams, 2021

A catalogue record for this book is available from the British Library

ISBN: HB: 978-1-4729-7172-2; ePub: 978-1-4729-7170-8;
ePDF: 978-1-4729-7171-5

2 4 6 8 10 9 7 5 3 1

Layout for this edition by Rod Teasdale.
Printed and bound in China by RR Donnelley Asia Printing Solutions Limited Company.

To find out more about our authors and books visit www.bloomsbury.com
and sign up for our newsletters

Cover photographs
Upper right: © Luis Castaneda/AGF foto/Biosphoto – upper left: © Fabrice Chanson/Biosphoto – middle left and
rear cover: Cloé Fraigneau – bottom left: © Serge Hänzi/Biosphoto – bottom right: © Bruno Mathieu/Biosphoto.

Disclaimer
The author would like to point out that no birds were mistreated or killed in the preparation of this book.
The feathers shown here come from natural moults, plucked carcasses, or birds that died in a care centre or were
found dead in the wild (as a result of predation, hunting, collision, etc.). The origin of the feathers presented in the
'Species Description' section is shown in an appendix (Table of illustration credits and origin of feathers).
European regulations have listed a large number of protected species, for which destruction, mutilation, capture,
taking from the natural environment, intentional disturbance, detention, transport, naturalisation, and commercial
or non-commercial use of specimens are prohibited The term specimen includes all or part of individuals,
including eggs. This therefore includes the possession and transport of feathers, as it cannot be proven
that the bird was not disturbed, injured or killed in order to obtain them.

Contents

Abbreviations used in this guide

Please note that abbreviations are sometimes used in combination, e.g. Pi = inner primary, TP = total length of a primary, NP = number of primaries.

Abbreviation	
♀	Female (used on the plates)
♂	Male (used on the plates)
-	Inapplicable information (e.g. the feather does not occur in the species concerned)
<z	Only the maximum measurement is known (z: size in cm)
>y	Only the minimum measurement is known (y: size in cm)
ad	Adult; definitive plumage (but might change according to season). Used on the plates after feather type.
Al	Alula
axil	Axillary (underwing-covert of the axilla), 'armpit'.
C	Calamus proportion as a % of total feather length T, from the base to the position of the first barbs (see 'feather measurements' p.18)
c	Central, used for rectrices, e.g. **Rc**: one or two pairs of feathers in the centre of the tail
cm	Centimetres
Em	Position of emargination, as a % of the total length of the primary feather concerned (see p. 19) (biometrics table p. 374)
f	Female (used in the tables)
FW	Folded wing, measured in cm (biometrics table)
GC	Greater coverts (of the secondaries)
i	Inner: towards the body or the centre of a bird (median axis) in feather descriptions (e.g. **Pi** or **Si**: primary inner, secondary inner)
imm	Immature (first-year for certain species): plumage not yet that of an adult; likely to change over the years. Indicated on the plates after the feather type.
inc	Incomplete; for bars
ind	An individual
irid	Iridescence
irr	Irregular
iv	Inner vane
juv	Juvenile: the first plumage, of which part may last until the following year (most often flight or tail feathers)
kg	kilogram
L	Length: size of a bird in cm (biometrics table p. 374)
LC	Lesser coverts (of the secondaries)
m	Median/middle (in feather descriptions, e.g. **Pm**, **Sm** or **Rm**: primary median, secondary median, rectrix median)
m	Male (used in the tables)
M	Mass in grams in the biometrics table.
mm	millimetres
MC	Median coverts (of the secondaries)

Abbreviation	
N	Number of (followed by the type of flight or tail feather), .g. **NP**, **NS**, **NR** (number of primaries, etc).
n.d.	No data: no information available on size or characteristics
Not	Position of the notch, as a % of the total length of the primary feather concerned (see 'feather measurements' p.18) (biometrics table p. 374)
o	Outer: at the far end from the body or the median axis (in describing feathers, e.g. **Po**, **So** or **Ro**: primary outer, secondary outer, rectrix outer)
ov	Outer vane
P / Ps	Primary / Primaries. Where individual primaries are identified, these are numbered as **P1**, **P2** etc.
PC	Primary coverts
Pmin and Pmax	Size of the shortest primary and of the longest primary
R / Rs	Rectrix / Rectrices: tail feathers. Where individual rectrices are identified, these are numbered **R1**, **R2** etc.
Rmin and Rmax	Size of shortest rectrix and of the longest rectrix
s	Subterminal: for a bar or a mark (not quite at feather tip)
S / Ss	Secondary / Secondaries. Where individual secondaries are identified, these are numbered as **S1**, **S2** etc.
scap	Scapulars
SC	Secondary coverts (Greater, unless indicated)
Smin and Smax	Size of the shortest secondary and of the longest secondary
sp.	Species (plural: spp.)
SS	Small reference sample (≤ 3 individuals) indicated for the description, or size, or both (some samples that have not been measured have however been used in descriptions, and certain measurements are given even though a usable image was not available).
t	Terminal: for a bar or a spot (at the feather tip)
T	Total length of the feather. Also used in conjunction with feather abbreviations, e.g. **TP**, **TS**, **TR** (total length of primary, etc).
Ter	Tertial
undt-cov	Undertail-covert (on the underside of the base of tail feathers)
undw-cov	Underwing-covert of primary or secondary (at the base of flight feathers)
uppt-cov	Uppertail-covert (on the upperside of the base of tail feathers)
uppw-cov	Upperwing-covert of primary or secondary (at base of flight feathers)
W	Wingspan in cm (biometrics table)
x	A character present on at least one category of the feathers concerned; (x) indicates that the character is rarely seen or is insignificant.
y	Young (1st-year plumage): plumage of an immature in certain cases. By default describes a juvenile (having recently left the nest) and other non-adult individuals (immatures).

Introduction

The aim of this guide is to enable the most accurate identification possible of the 'large' feathers (primaries, secondaries, tertials and tail feathers) and certain coverts of European birds. The principle is that of the usual identification key for naturalists, adapted to the particular case of feathers (any species carries dozens of feathers of different shapes and colours). Through a succession of keys and comparative tables, the observer is led to an increasingly small group of species, and at best to a single species from which the studied feather originates. The illustrations can then be used to confirm or disprove the identification.

This guide should allow the identification of feathers found in isolation (naturally moulted), found in the various natural environments of Western Europe. In consequence it uses different features to those applicable to live birds or parts of birds.

Similar to a species identification guide, this book brings together the feathers of different species from the same family. Indeed, despite sometimes significant variations in size and/or colour, the shape and structure of feathers are generally fairly similar within the same family. Shape and structure are criteria more or less easy to ascertain, but in any case are much more reliable than size and colour, which may vary within a species according to sex, age or geographical origin of the individual. Certain characteristic adaptations often point to a particular family or group from which the feather originated.

On the other hand, the sequence in which species are described, within families or orders, does not follow the usual systematic order. An attempt has been made here to describe groups that may be confused in close proximity, either because of their similar morphological characteristics or because they occupy the same habitats (so that their feathers may be found in the same places). A list of observable features to be seen on the flight and tail feathers, or coverts, is presented before the detailed presentation of the groups. Species displaying these features are listed, and descriptions of groups and species are given for reference and for further clarification.

To present the different feathers to the reader in the most practical way possible, a large format has been chosen for this guide and this may limit the practicality of taking it into the field. However, unlike live birds, feathers and other remains can be easily photographed or even brought home where this guide can be consulted at leisure, with reference to the specimens collected in the field.

1. Collecting methods, conservation and identification

FINDING FEATHERS

THE SEASON: for the most part, European species moult after they have reproduced. Naturally moulted feathers are therefore found more often in summer, from July to September. Large species, and certain other groups, moult gradually throughout the year, which explains why feathers are also moulted outside this season. In addition, birds suffering accidents or predation leave scattered feathers all year round.

▷ Feathers from seabirds are often found among the debris washed-up on the high-tide line.

WHERE: moulted feathers usually fall when a bird is preening, so look for resting sites (a sheltered lake bank, a large tree, and at the foot of a roosting site, etc.). Shorelines, especially seashores, are also sites to visit, especially after high tides and gales that bring in feathers that have been lost offshore. Bird colonies, visited after breeding, are also good places for feather collecting, although the feathers found are more frequently soiled.

GATHERING THE FIRST CLUES

As with birdwatching, discovery conditions can give important clues to identification. Therefore, take note of the location as precisely as possible (at least county and country), the date (at least the month) and the habitat (wood, forest edge, embankment, urban, etc.), all are useful clues to later identification. It should be noted whether feathers are from a plucked or dead bird or found as isolated feathers. Details can be quickly written on the bag or envelope containing the feathers. If a single feather is collected, it is prudent to attach a label to its calamus, and to indicate location and habitat, the date, and any other remarks, etc.

MAKING FEATHERS IDENTIFIABLE

As soon as a feather is collected, or after drying (or even later), it can be cleaned when dry and smoothed out. Remove debris with a small paint brush or toothbrush (for rigid feathers) and then smooth out the barbs with your fingers, without pressing them, from base to tip. This usually gives the feather its correct form and shape and is useful for identification. Rigid or stained feathers can be washed in soap and rinsed with clear water, but it is advisable to limit the use of water (small feathers, and those of owls, for example, could be damaged by such treatment).

PRESERVING FEATHERS

In order to prevent natural feather consumers, notably certain insects, from destroying your collection, insecticides (anti-mites, camphor, etc.) should be placed with the feathers and their condition should be checked regularly to prevent attacks. Keeping newly found feathers apart is also a good way to avoid contaminating an already disinfected collection. Once feathers have been dried and the bulk of any dirt removed, they can be kept for decades without decay.

Identified feathers can be placed in plastic pouches or files, paper envelopes or boxes. Binders with stiff pouches can be used to hold sheets of paper on which to fix finds and are very practical; notes can be written on the paper around the feathers. Adhesive tape or weak glue can be used without damaging the feather, provided it is fixed only by the calamus, without sticking down the barbs.

Feathers can of course be collected for their aesthetic appeal (and why not use them in 'collages'), but the curious naturalist generally seeks to identify their origin. Once the feather has been identified to the species it came from, it is best to group feathers belonging to the same species. Closely related species can then be placed near to each other, as in identification guides, to allow for easier comparison.

Various possibilities exist, the most common being to adopt the systematic order presented in field guides. For a trip abroad, a separate feather album may be preferable to mixing them with your general collection.

IDENTIFYING FEATHERS

First, it should be noted that this is rarely simple, is often arduous, but is always enriching. Some species of common or emblematic birds have easily recognisable feathers; but most have body feathers that remain unidentified by the finder when they are retrieved alone.

If all the leaves of a tree are more or less similar to each other and are relatively similar within a species, the same is not true for the feathers of a bird – far from it!

A more or less marked variability in feather size and colour is visible between individuals of a species, but above all each individual bird has thousands of feathers, some being unique in shape or design, and most being different depending on their location on the body. Therefore, when you study an isolated feather, it is like finding the location of a unique piece from one of the thousands of pieces of a jig-saw puzzle, among hundreds of different puzzles.

It is therefore understandable that adopting a strategy based on concrete elements will soon become necessary, if not essential. Of course, a good knowledge of birds, their habits and their morphology, will be invaluable. However, a patient and methodical amateur will do better than a seasoned ornithologist who is in a hurry in this particular inquiry – the identification of a feather found in the field. In reality only a tiny fraction of the feathers lost by birds are identifiable, at least without the help of genetic analysis. Indeed, out of the several thousand feathers on a bird, it is usually only a hundred that will be recognisable. However, because large feathers are more easily found (because they are more visible and degrade less rapidly after shedding) than smaller ones, the majority of feathers found will be useful and hopefully identifiable.

However, it will be necessary to admit from time to time that one cannot name the species from which the feather originated and postpone identification. Even more trying, the 'featherologist' will sometimes have to question his or her previous identifications as their knowledge and deductive abilities improve. Thus, the feather of a rare species once identified with confidence may prove a few years later to be that of a common species that hitherto wasn't well known.

Noting doubts during identification can be very useful, for example with one or two question marks on the label associated with the feather. Also, looking again at doubtful feathers after some time can be a good way to progress, based on previously ignored or unrecognised features.

The identification process can be frustrating in cases where the feather can only be assigned to a large group of species.

However, unanswered questions also provide a stimulus for improving one's knowledge of birds. The amateur who looks diligently for the answer will be a better observer of wild birds in the field by frequenting natural history museums or consulting books and websites to find the answer. And one day perhaps, the solution to the problem will appear at the turn of a page, or from a hide!

• AN IDENTIFICATION STRATEGY

Depending on their knowledge or motivation, the feather-finder may seek to identify very precisely the feather found, or conversely will be content with a minimum of knowledge. Everyone can therefore adopt a personalised strategy to their work. General guidelines are given below to help the investigator.

• NOTING ASSOCIATED CLUES

When a feather is found, information noted at the time can be valuable: place (or at least the county), region, country, etc.: corresponding to a species' distribution; environment: woodland, sea or lake shore, steppe, cliff base, urban park, etc.: link to a species' way of life; date (at least the month): link to a species' status in the region concerned (breeding, overwintering); whether the feather belonged to a plucked bird or other more precise information (roadside, under a high-voltage line, etc.): link to predators or the risks encountered by a species.

• SIZE AND COLOUR

These are the most commonly used features. However, they should be used with caution. Here are some pitfalls to consider.

COLOUR: This is the most obvious criterion, but attention should be paid to the objectivity of the observer, to the descriptive terms used (an issue even within this book, etc.), and to variations in vocabulary. Beware of variability within a species (depending on age, sex, origin, etc.), the degradation of colours with feather wear, and possible pigment aberrations, etc.

Abnormalities of coloration and shape also occur: some individuals have 'defects' (aberrant, albinism, leucism, melanism, etc.) but these are very rare in nature, and such feathers are therefore only found exceptionally. On the other hand, certain 'anomalies'

are regularly observed: growth bars, occasional pigmentation defects (corvids), wear of the feather tip and/or white areas of the feather (thus abnormal shape), added colouration (small gulls, etc.).

SIZE : the most objective criterion. Very useful if the feather is in good condition (but not if the base or tip has been cut), but is very variable depending on the position of the feather on a bird. The subjectivity of the observer is less involved (feathers are measured more or less flat).

• STRUCTURE AND SILHOUETTE

While feathers have the same overall structure (see p. 10) , some details are characteristic of the position of the observed feather or the group to which it belongs, such as down density, distribution of connected/free barbs, shape of hyporachis, etc.

The shape is primarily an indicator of the location of the feather on a bird's body. But for a given place, there are specifics according to the feather groups: proportion of the calamus, emargination, width of the vanes, thickness of the barbs, curvature, etc.

Less obvious than colour and size for the inexperienced, these two features are, however, much more reliable. They are consistent within a species (regardless of age, sex or origin of the individual) and are often shared by closely related species. For example, the rigid, lanceolate tail feathers of woodpeckers occur in all members of this family, which use their tails as a support. The 'hairy' appearance of the feather surface of owls is also shared by all of the species in this order.

We sometimes see convergence in form between distantly related groups with the same lifestyle, or that are subject to the same constraints. Therefore, we can note the elongated calamus of the primaries in birds with a high wing-loading, or the fingers at the wing-tips of gliding specialists.

• A STRATEGY FOR IDENTIFICATION

In some cases ('little brown jobs' in particular), species cannot be readily distinguished from each other, so that the found feather cannot be identified to species level. However, a genus or family can still be assigned in some cases. But sometimes even that will not be possible. Not being able to identify a feather is just part of the game! The feather can

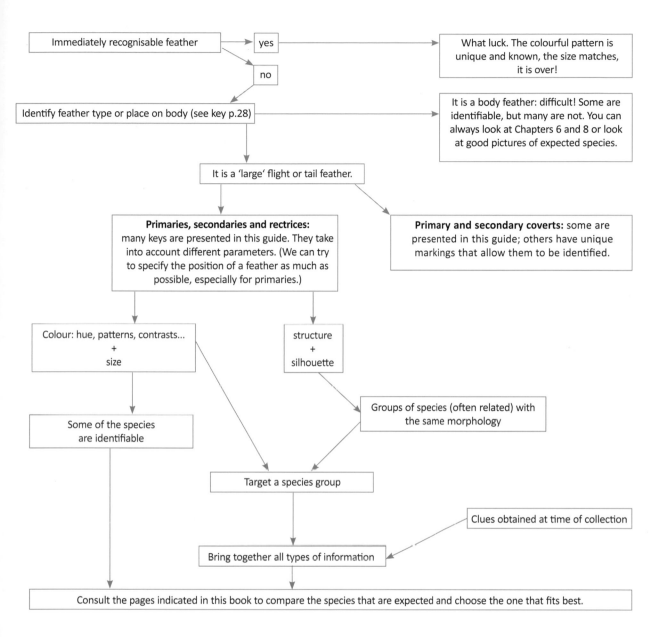

always be re-examined later, when the investigator has acquired more experience, etc.

SUMMARY OF IDENTIFICATION TECHNIQUES

Decide on the feather's original **position**: flight feather, tail feather, body feather, coverts, etc.

See if a **three-dimensional structural criterion** is visible: 'hairs', tegmen, hyporachis, curved profile, etc.

See if a **silhouette criterion** is visible: notch, long calamus, curvature, narrowness, etc.

See if a **colour criterion** is visible: white feather, barring, coloured notches, bright colour, spots, etc.

Consult the lists of species with the appropriate features and **compare** with the illustrations. Take into account the clues recorded during collection to check if this correlates with the ecology of species under consideration. If no particular features are visible, **consult the beginning of the group descriptions**. They indicate the general appearance of the feathers. The related illustrations often allow one to target or eliminate groups of species.

2. Factors useful for identification

This chapter presents the terminology used in this guide, as well as the various methods available for the identification of feathers: names of the parts of the feather, names of the different feathers, particular vocabulary for colours and patterns, measurement methods and size ranges, graphs on relative sizes, and identification tables.

THE FEATHER

• IDENTITY

Feathers are epidermal growths composed of keratins; complex sets of proteins insoluble in water. Like hair or nails they are 'dead' structures, not irrigated by blood vessels and without nerves once their growth is completed. They are regularly renewed during the life of a bird during a complex, usually seasonal phenomenon, called moult.

The feather grows in a sheath produced by the skin; it nourishes the feather and dries out once growth is complete. The location of the follicular cells within the sheath determines the size and shape of the feather and its finer structure, as well as its pigmentation.

• STRUCTURE

The main axis of the feather is termed the rachis or shaft. Parallel blades, the barbs, grow from the rachis. On these are the barbules, a kind of miniature barb, which have hooks and notches allowing them to attach to each other and thus allow the barbs to interlock. The assemblage of barbs and barbules on one side of the rachis is termed a vane (or web). Usually the inner vane, located towards the central axis of a bird, is distinguished from the outer vane, located towards the outside of a bird's body; the outer vane is often stiffer and narrower.

Depending on the feather's function, the barbs can be: all linked, providing a certain rigidity (flight feathers);

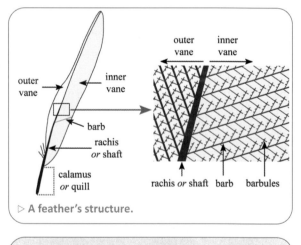

▷ **A feather's structure.**

▷ **Compare the extent of free barbs with that of linked barbs.**
Blackbird feathers, with from left to right: down from the belly, body feather from the back with barbs linked only towards the tip, and a wing-covert with nearly all of the barbs linked.

– partially linked – the tip with linked barbs making it watertight, while at the base of the feather they are not (body feathers);
– all of the barbs are free – the feather is 'fluffy' and serves as a thermal insulating layer (mainly on down feathers, hidden under the other feathers).

• DISTINGUISH THE TWO SIDES

While forming, the feather is at first tube-shaped. During its growth, it splits and spreads. The exterior of the tube becomes the upperside and the interior the underside. The base does not split and remains as a hollow tube, the calamus. It is largely embedded in the skin which allows a bird to orient the feather by changing the tension of the skin. The rachis, the central axis of the feather, is often flat or domed above, and concave, or forming two ridges below. Colours are usually brighter on the top of the feather, but not always (Columbidae, Psittacidae, etc.). On the underside, the two vanes meet at the top of the calamus. At this point, the feather forms two parts, the smallest being called the 'hyporachis'. In modern birds, the hyporachis is always fluffy and usually without any colour patterns.

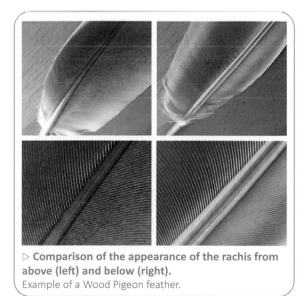

▷ **Comparison of the appearance of the rachis from above (left) and below (right).**
Example of a Wood Pigeon feather.

• DISTINGUISHING THE TWO VANES

When a feather forms part of the wing or tail, it is generally asymmetrical (the two vanes are dissimilar), this difference diminishing the closer the feather is to the bird's body or towards the middle of the tail. However, even on these almost symmetrical feathers, the two vanes can be distinguished by their curvature when viewed in profile. Feathers closer to a bird's central axis cover those towards the exterior when the wing or tail is folded. Looking at the feather in profile, from above, the edge of the inner vane tends to turn upwards and that of the outer vane downwards. This provides maximum contact between neighbouring feathers so that the plumage forms a continuous surface. When a feather serves as a watertight or insulating cover, it does not necessarily show such a difference in curvature between the two vanes: both curve downwards to maximise contact with the feather below (such as in the majority of 'contour' feathers).

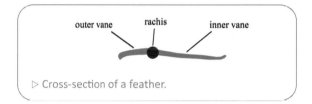

outer vane   rachis   inner vane

▷ Cross-section of a feather.

THE NAMES OF THE FEATHERS ON A BIRD

• THE WINGS

The remiges: these are the large feathers of the wings; they are implanted on the hand (primaries), on the forearm (secondaries) and possibly also on the upper arm in certain species (tertials). In birds whose arms have well-developed feathers (boobies, albatrosses, pelicans, etc.), the wings appear to be folded in three and not in two as in other species. In cases where the upper arm does not carry well-developed flight feathers, the term 'tertials' may apply to the few innermost 'secondary' feathers, attached near the elbow (usually three in number).

Primaries: (on the hand) are mainly used to propel a bird and are therefore quite rigid – these are the largest feathers of the wing. The shorter and wider secondaries (on the forearm) form the surface providing lift – they are also used for propulsion but to a lesser extent. The tertials, when well-developed, play the same role as the secondaries and resemble them closely in shape. When reduced in size, they have an intermediate role between that of the secondaries and scapulars.

The wing-coverts are implanted at the base of the flight feathers, in several superimposed rows, each row filling the space between the two feathers of the lower row as much as possible. They are found on the upperwing, as well as the underwing, where they are much thinner and softer (underwing-coverts). On the hand there is a row of **primary coverts**, the bases of which are covered with small feathers forming the leading edge of the wing. On the forearm, the **secondary coverts** are more clearly divided into successive layers. The **greater coverts** cover the secondaries, the **median coverts** cover the greater coverts, and finally several rows of **lesser coverts** perfect the aerodynamics of the wing right up to the leading edge.

Most European birds have 10 or 11 primaries and at least nine secondaries (including tertials) of which there may be more than 20 per wing. In certain groups there is an additional, intermediate flight feather fixed to the wrist termed the 'carpal primary'; it has the shape and colour of an inner primary or outer secondary but is clearly smaller. Where there are well-developed tertials, there are often between 10 and 20 secondaries (including tertials). There are the same numbers of primary and greater coverts as there are of the corresponding flight feathers.

The alula is a small tuft of feathers (three to seven on each wing) attached to the first digit (thumb) of a bird. Seemingly of little importance, the alula is actually a fantastic airstream deflector, copied by aircraft designers, which prevents a bird stalling at low speed, allowing it to slow down as much as possible without falling, for example before landing. The action of the alula also helps a bird to stabilise itself in flight during turbulence, by increasing uplift above the wing.

The hand provides the main thrust when the wings are flapped. Its feathers are therefore particularly rigid and adapted to penetrating the air. The outer webs of the primaries and secondaries are narrow and thick, and their calamus is quite long. The primary coverts also have a very long, bent calamus, and press strongly onto the primaries.

The forearm (or ulna) is more devoted to lifting a bird in the air, and its feathers are wider than those of the hand, more flexible, and with vanes of similar or identical width. Towards the base of the wing, the remiges and greater coverts change shape, either to become rounder and shorter, or conversely longer and more lanceolate, depending on the species.

• THE TAIL

The rectrices are the large tail feathers; they are implanted above the pygostyle, the tip of a bird's spine. The term 'rectrices', as well as the word 'direction' comes from the Latin *directio*, meaning 'who directs'. These feathers are used not only to perform changes in direction, but also to stabilise a bird in flight. They help in braking when a bird spreads them to increase the surface area of its tail, especially before landing. The tail can also serve as a counterbalance, especially when it is long (magpies, wagtails, etc.), or increase the aerodynamic area in gliding birds. Also, some birds also use it for purposes other than flight, such as for moving on tree trunks (woodpeckers, treecreepers, etc.) or in the water (gannets, cormorants, etc.).

As in the wing feathers, the base of the rectrices is protected with coverts, logically called **uppertail-coverts** above the tail and **undertail-coverts** below. The latter usually have a very characteristic spoon-shape and press strongly onto the base of the rectrices. The uppertail-coverts are flatter and have some characteristics in common with the rectrices (see Chapter 3 and Chapter 6, p. 26 and 100). The so-called rump feathers in turn cover the base of the uppertail-coverts; those of the vent are superimposed on the undertail-coverts.

• THE BODY

Outer feathers

Most body feathers are placed in a similar fashion to tiles on a house roof, protecting a bird from both adverse weather (rain, wind, etc.) and physical blows. The bases of the body feathers are fluffy (free barbs) and without any colour pattern (usually white, grey or black). Placed against the body, the body plumage forms a soft layer that insulates and absorbs shocks. On the other hand, this layer can easily absorb water, particularly when the plumage is poorly cared for (birds that are sick or too weak to preen their plumage). The tips of the body feathers have linked barbs forming a waterproof layer resistant to physical shocks, and thus constitute a kind of shell around a bird. The colour of the tip of the body feathers forms the majority of the plumage pattern, at least when a bird is at rest. Thus, these body feathers often have spots, streaks or coloured borders that may be characteristic of the species. The body feathers can therefore be very useful in camouflage and communication between individuals, because all or some of them show colour related to the age and/or sex of a bird.

Insulating feathers
The down itself is totally made up of feathers with free barbs which when entwined trap air and form a more or less dense envelope around the body. By fluffing up its plumage, a bird traps a layer of air around itself, isolating it from the outside with its tightly packed body plumage. On the other hand, by flattening its

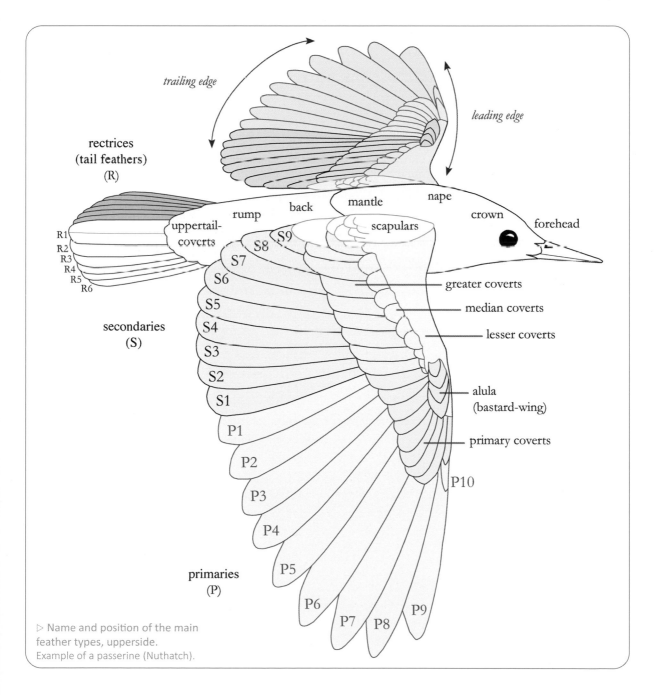

▷ Name and position of the main feather types, upperside.
Example of a passerine (Nuthatch).

plumage, the air is expelled. It can also separate the body feathers to let cool air circulate around its body (the 'venetian blind' system). The thick layer of down also protects the body from shocks, for example in the event of an accident, capture by a predator or a fight with a congener.

Feather sensitivity and care
Since all feathers are embedded in the skin and connected to nerves, a bird knows their position by tactile sensation, as in the fur of a mammal. A few feathers, filoplumes and vibrissae also have a special role. The last are found around the bill. Long and narrow, without barbs, they allow the position of food to be sensed in relation to the corners of the bill. In certain species, they are particularly developed and also play a role in capturing food. Examples include nightjars and swifts, which have a very large bill opening, even if it appears small when closed. The vibrissae of the mouth corners spread vertically when the bill opens, forming two parallel grids. When the bill is wide open horizontally, the whole forms a net

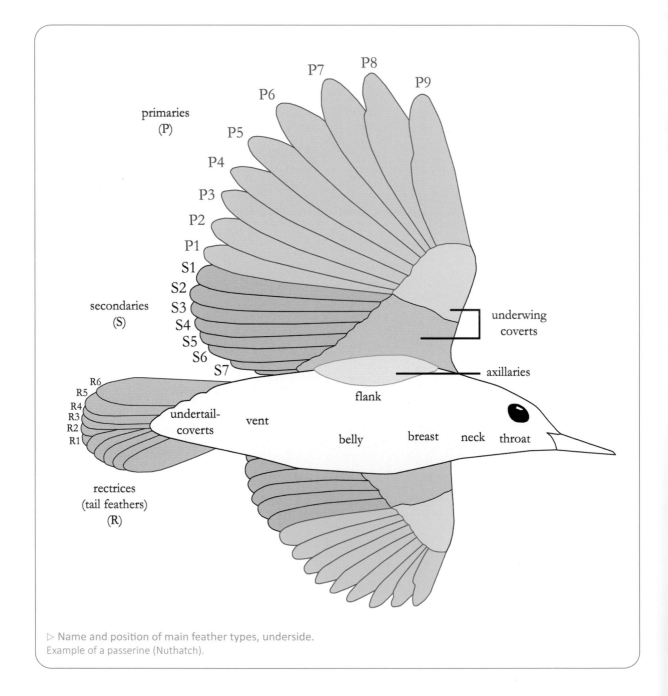

▷ Name and position of main feather types, underside. Example of a passerine (Nuthatch).

for trawling for aerial plankton. Of course, birds do not normally fly with their beaks open permanently, but only when approaching prey. This gigantic funnel makes it much easier to catch small, mobile prey such as insects. The filoplumes, on the other hand, have a function related to plumage care. The tactile sensations perceived by a bird during grooming inform it of the quality and positioning of its plumage. Finally, there is a particular tuft of feathers near the uropygial gland, in the form of a very short little brush. By pressing this gland, a bird releases secretions useful in plumage care (they have antibacterial properties and help slow feather wear). All of these feathers are rarely found as a result of moulting but can be seen on a whole bird or a living bird held in the hand.

This book deals mainly with the characteristics of the remiges (wing) and rectrices (tail), collectively called 'flight feathers', and in a less detailed way with the more visible body (or 'contour') feathers.

• THE NUMBERING OF FLIGHT FEATHERS

The rectrices (R) are always numbered from the centre of the tail. The two central feathers are numbered R1, the following R2, etc. For example, in corvids there are 12 rectrices, numbered from R1 to R6 on each side. On the other hand, the numbering of remiges (primaries in particular) varies according to usage, country and author. The choice was made here to number primaries (P) and secondaries (S) starting from the wrist. Thus, the nearer the secondaries are to the body, the larger their number (S1 is on the wrist, then S2 inwards, etc.). But for primaries, the opposite is true: the nearer a primary is to the wrist, the lower its number (with the highest number being nearest to the wing-tip). This choice of numbering is due to the highly variable size of the outer primaries in different species groups. In some species the outer primary is similar in

▷ Outer primaries of a Blackbird: primary P10 above and primary P9 below.

size to the next primary, in others it is much shorter and functions with the alula in its role as a flight deflector. It is then called the 'remicle'. Due to its small size, it is rarely found on the ground and is difficult to identify. So, in order to correctly number the primaries from the outside inwards, we have to know the species that is being examined and refer to specialised guides (see References, p. 400) to find out if there is a very small primary or not, and to count it as P1. The first large outer primary would therefore be P1 or P2 depending on the group, and the numbering moved from 1 in case it isn't included if P1 is actually very small and has been omitted in the numbering. However, by numbering the primaries from the wrist outwards, the problem related to the presence or absence of a small outer primary no longer arises. The primary of the wrist is P1, and then the numbering continues outwards: P5 towards the middle, P9 towards the tip of the wing, etc. In the case of nine large primaries, numbering goes from P1 to P9, and if there is a small outer primary, it will be numbered P10, and its possible presence does not affect the numbering of the other primaries.

DESCRIPTIVE VOCABULARY

• COLOURED MARKS ON A FEATHER

Depending on the position, shape and outline of the marks observed, a more or less precise vocabulary is used. For example, if an edge is thin and becomes a border when it is wider. A bar is said to be terminal when it touches the tip of the feather, subterminal if the tip is of another colour. Very thin, zigzag and sometimes discontinuous bars are called vermiculations, while incomplete bars, often rounded or triangular, are notches. The diagram on p. 16 illustrates some of the most commonly encountered forms of marks.

• FEATHER SHAPE

The so-called 'classic' forms do not have obvious features but may be more or less broad or narrow, rounded or pointed, etc., depending on the species groups. In some cases, a general silhouette or

Arrangement of primaries, secondaries, rectrices and coverts (upperside)

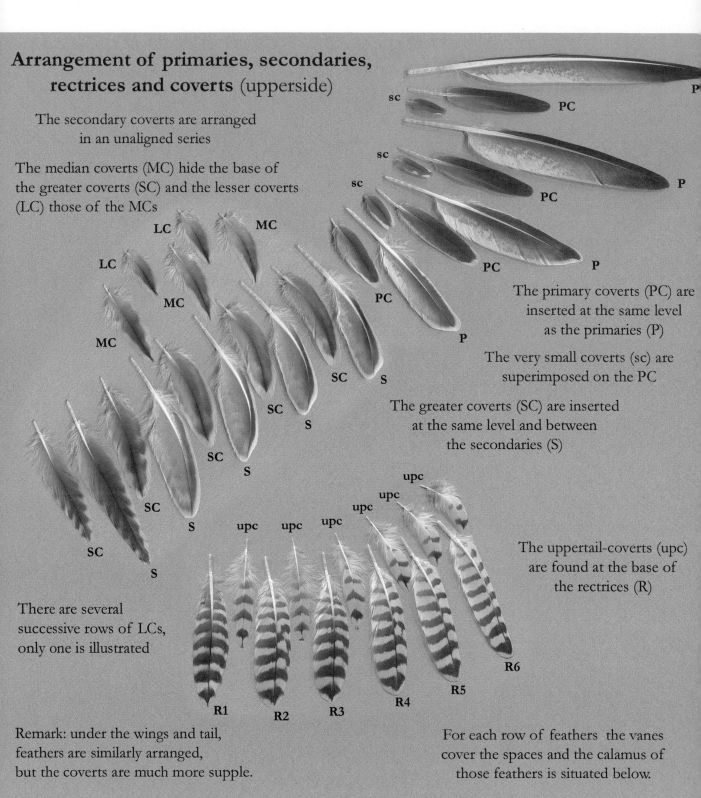

The secondary coverts are arranged in an unaligned series

The median coverts (MC) hide the base of the greater coverts (SC) and the lesser coverts (LC) those of the MCs

The primary coverts (PC) are inserted at the same level as the primaries (P)

The very small coverts (sc) are superimposed on the PC

The greater coverts (SC) are inserted at the same level and between the secondaries (S)

The uppertail-coverts (upc) are found at the base of the rectrices (R)

There are several successive rows of LCs, only one is illustrated

Remark: under the wings and tail, feathers are similarly arranged, but the coverts are much more supple.

Note: only a selection of the wing feathers is presented.

For each row of feathers the vanes cover the spaces and the calamus of those feathers is situated below.

Example: right-side feathers of a godwit

shape detail may be described that will help the observer to compare species. It is especially useful to look at the general outline of the feather (parallel edges, narrowed tip, lateral curvature, etc.), the shape of the tip (rounded, rectangular, wedge-shaped, pointed, tapered, etc.) or the silhouette (wide, stubby, elongated, etc.). Some

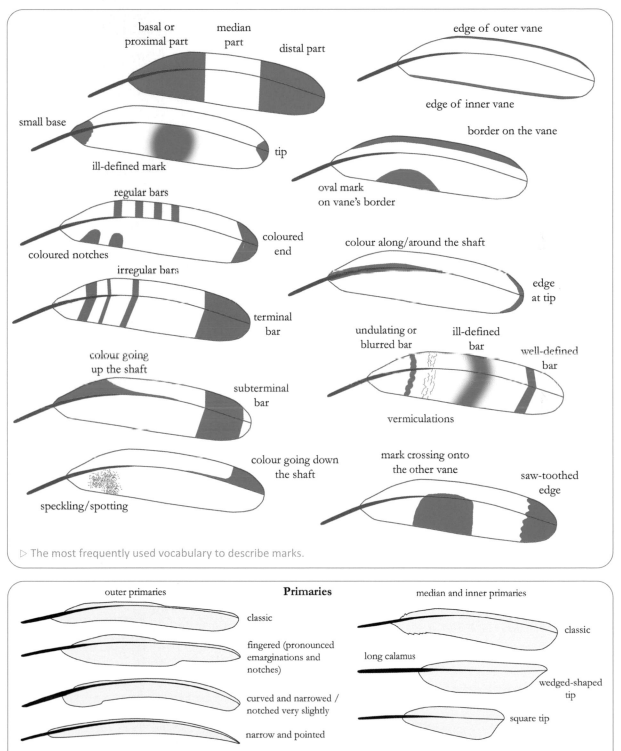

▷ The most frequently used vocabulary to describe marks.

▷ Vocabulary used to describe feather shape.

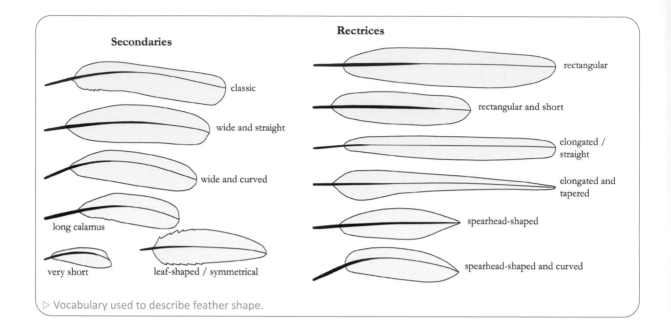

Secondaries

classic

wide and straight

wide and curved

long calamus

very short   leaf-shaped / symmetrical

Rectrices

rectangular

rectangular and short

elongated / straight

elongated and tapered

spearhead-shaped

spearhead-shaped and curved

▷ Vocabulary used to describe feather shape.

examples of feather shapes are illustrated on the plates (pp. 17 and 18).

• TAIL SHAPE

Bird's tails have many shapes, which can be roughly categorised according to the relative proportions of the different feathers. With increasingly large central rectrices, they are described as: with streamers, forked, notched, square or rectangular, rounded, wedge-shaped or stepped and pointed. The diagrams on p. 39 show the relative sizes of the rectrices and the silhouette of the outer and central rectrices for each of these tail shapes.

FEATHER MEASUREMENTS

• FEATHER SIZE

Specialised guides (ringer's manuals, collection data, etc.) usually have precise size measurements to the nearest millimetre. Due to the variation found in the sources consulted and the variability within a species,

we felt it was irrelevant to maintain this precision. As a result we have decided to give measurements to the nearest centimetre, or half-centimetre for the smallest feathers.

To measure the total length of a feather (T), the most practical way is to use a stopped ruler, pushing the end of the calamus towards 0, or to use a calliper. The feather may be flattened or not depending on each author's measurement practices. Since a rigid feather cannot be completely flattened, it seemed more logical to keep the feathers' natural curvature when the measurements of samples were taken for this guide. There are also two types of measurements given in the literature (Brown *et al.*, for example): the total length (moulted feather) or its length in place on a bird (corpse or live bird). The latter, called M, is necessarily less than T. M can be estimated for an entire feather by noting the position of a small ring of skin surrounding the calamus (if the feather is not too damaged). It indicates how far the feather was embedded in the skin, and so M can be found by measuring the feather

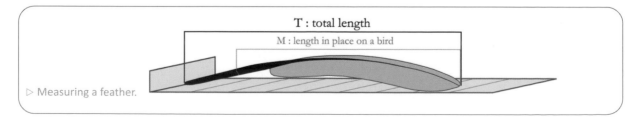

T : total length

M : length in place on a bird

▷ Measuring a feather.

from this ring to its tip. This then allows comparison of the measurement of the collected feather with those given in the literature. In this book, it is always the total length T that is indicated.

Note: obviously measuring growing feathers has no relevance here. We simply know that their final size will be larger than measured. Similarly, broken feathers will be shorter than the measurements given.

• THE POSITION OF EMARGINATIONS AND NOTCHES

Depending on the species and the position of the feather, the measurement of emarginations (outer vane) and/or notches (outer vane) can provide valuable information. This narrowing of the vanes is positioned differently on the feathers and varies in length depending on the type of flight performed by the species in question. In general, emarginations improve manoeuvrability in flight. A bird can quickly vary the shape of the wing tip with small movements of the 'arm'. Thus, their number often increases in forest species or gliding specialists, but decreases in species of open habitats, and they are often absent in certain groups (gulls and waders, swifts and swallows, etc.). When useful, their measurements have therefore been incorporated into the descriptions of the groups. Measurements are given in proportion to the total length of the feather (not the length of the vanes).

In the next figure, Not shows the measurement of the notch and Em the measurement of the emargination. For either Not (notch) or Em (emargination), with T as the total length of the feather, the value of Not or Em as a % = Not or Em (cm) / T(cm) x 100. Not or Em must be measured from the tip of the calamus to the narrowest part of the vane, which is easier to locate precisely than the beginning of the narrowing of the vane.

They are termed low-positioned emarginations/notches when Em/Not has a low value and high-positioned when Em/Not has a high value.

• THE PROPORTIONS OF THE CALAMUS

The relative length of the calamus is a good indicator of the pressure placed on the feather. The greater the pressure, the longer the calamus to withstand the lever effect of the arm and provide greater fixation in the wing.

Thus, on the same bird, feathers which must withstand greater pressure will have a longer calamus, such as the primaries or primary coverts. Conversely, the calamus will be reduced in feathers experiencing low pressure, such as the tertials and secondary coverts.

Comparing species, a longer calamus on the primaries indicates a higher wing-loading, and/or a more sustained flapping flight. On the tail, it is an indication of active use of the tail, for example in diving or climbing a tree trunk.

The following figures show how to measure the calamus. Starting from the feather's base, the entire length must be taken into account until the first barbs appear on one of the two vanes (often the outer web). It was decided to measure the calamus in this way as it is sometimes very difficult to spot the small hole on the underside of the spine that in theory delimits the end of the calamus

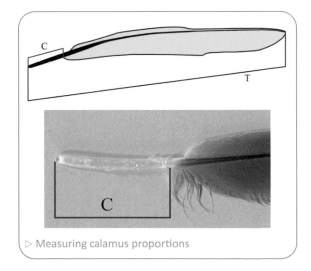

▷ Measuring calamus proportions

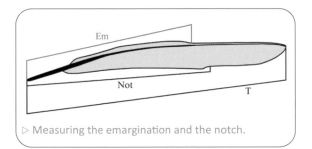

▷ Measuring the emargination and the notch.

on an old feather. When useful, the proportions of the calamus are indicated in the descriptions. To calculate the calamus proportion, measure C (the length of the calamus) and T (the total length of the feather); C in % = C(cm) / T(cm) x 100.

FEATHER SIZE VARIATION

The pairs of measures given here for primaries, secondaries and rectrices take into account all of the data collected by the author (from the literature and direct measurements). Other than the constraints of larger or smaller datasets, they are also subject to different measuring techniques used by their authors (feathers flattened or not). However, the differences between measures related to the amount of curvature is generally less than natural variability within any species and can therefore be considered negligible in most cases.

Only total T lengths are given in the tables and text. They are accurate to one centimetre for the majority of species and to half a centimetre for the smallest, and only occasionally to the nearest millimetre for passerines in particular.

The tables show maximum and minimum feather sizes. For example, Rmin is the measure of the smallest feather of the smallest individual and Rmax is the measure of the largest feather of the largest individual.

Note that in the rest of the book: Rmin and Rmax are for rectrices, Pmin and Pmax for primaries, and Smin and Smax for secondaries.

For the rectrices (tail feathers), the position of Rmax and Rmin depends on tail shape (see graphs for each group). In the rare cases in which the outer rectrices are much smaller than the others, their size is indicated separately.

For the primaries, P1 (at the wrist) is normally the smallest of the primaries. If the outer primary is very small, it is shown separately in the table or indicated only in the biometric data synthesis table at the end of the guide (see p. 374). The position of the longest primary (Pmax) varies depending on the shape of the wing: it is the outer primary for pointed wings but more median for rounded or rectangular wings (see diagram p. 22). This position is clear on the graphs of groups, since it is the primary that reaches 100%.

For the secondaries, size is much more constant throughout the forearm. However, it varies more towards the body and elbow. Some species have three inner secondaries that rapidly decrease in size and are smaller than the previous ones (the tertials: majority of passerines, woodpeckers, etc.). Other species have inner secondaries larger than the middle ones (Anatidae, waders, etc.). Thus, to allow for comparison, the tables indicate the sizes of the outer and median secondaries only. When available, the minimum and/or maximum sizes of the inner secondaries and tertials are given separately, or in the table at the end of the guide (see p. 374). These flight feathers are often distinguished by their shape and sometimes by their colour. Thus, when a median secondary differs from the others, it is considered to be the first of the inner secondaries or tertials, often S7 (sometimes S8) in passerines, but more often S10 when there are more secondaries. For example, on the schematic bird on p. 13, measurements are shown for primaries P1 to P9 (P10 is not considered), secondaries S1 to S6 (S7 to S9 not taken into account) and for all rectrices (tail feathers).

To facilitate comparisons and calculations, different scales of feather representation were chosen for the plates. The details of the use of these scales are shown on p. 115.

GRAPHS OF RELATIVE FEATHER SIZES

These only concern primaries, secondaries and tail feathers and indicate their sizes relative to the size of the largest feather of the wing or tail.

• MAKING OF THE GRAPHS

These indicate the size of the wing or tail feathers as a percentage of the largest wing (Pmax) or tail (Rmax) feather, which measures 100%. To obtain them, it suffices to measure all the remiges or rectrices of a species. All of these actual measurements are then expressed as a percentage of this largest feather. Graphs can thus be obtained for all sets of P, S and R feathers that can be measured.

Theoretically, the exact size of the other feathers can then be calculated by their size proportional to the measurements given in the tables.

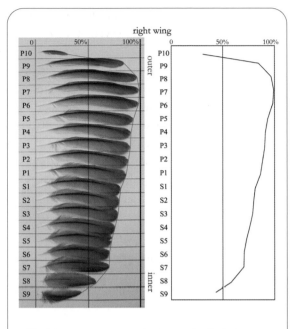

▷ Obtaining measurements for the graph of relative feather sizes. Left : In sequence flight feathers
Right: graph obtained from feather measurements relative to that of the longest feather, here P7 (7.2 cm long).

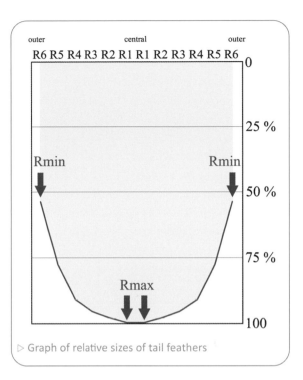

▷ Graph of relative sizes of tail feathers

x 12 = 18.5 cm. The sizes of all other remiges can then be deduced. The same is true for the rectrices.

• USING THE GRAPHS FOR IDENTIFICATION

These graphs show the average relative sizes observed within a group of species when they have homogeneous curves, or the minimum and maximum noted within the group. However, slight variations may be noted outside of these measurements, these graphs being just one of several aids to identification.

The numerical values used for the graphs are given in an appendix (see p. 388).
The graphs allow:
– a visual representation of the evolution of the size of remiges or rectrices for a homogeneous species or group of species. These curves are relatively similar within a species (+/− 2%);
– a comparison between species of different size;
– an estimate of the size of the other large feathers based on a single measured feather from a known position (one collected, or a measurement taken from a published description);
– confirmation or otherwise of an identification in the case of a plucked bird (even incomplete) using the shape of the wing or tail reconstructed from the measured feathers.

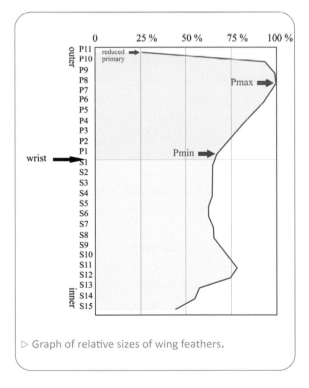

▷ Graph of relative sizes of wing feathers.

For example, if the longest P (Pmax) = 10 cm, it can be deduced that the others measure: e.g. for P1 = 65% so that T(P1) = 6.5 cm. Inversely, if T(P1) = 12 cm for 65%, them Pmax = 100/65 x T(1) = 100/65

Visual indications

Even if the actual shape of a wing depends in part on the length of the bones and the angles between them, the relative sizes of the remiges provides useful information. Similarly, the variable spread of the tail and the angled implantation of the rectrices does not hide the relationship between the relative sizes of the rectrices and the shape of the tail.

With a little training, a quick look of the outlines allows the observer to know immediately what wing shape (and therefore what type of flight) and what tail shape is being dealt with.

Calculating measurements

For practical reasons, not all feather measurements are indicated for each species described. A large sample can be found in some guides (see References, p. 400). Only the extreme ranges of size are noted.

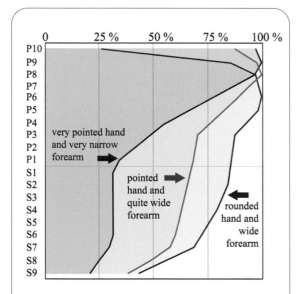

▷ Comparison of a long-handed, straight wing (brown) with a triangular-handed, pointed wing (green) and a rounded, wide wing (blue).
- The quicker the primaries decrease in size towards the wrist (the steeper the curve slope), the longer and/or more pointed the hand.
- The smaller the secondaries compared to Pmax, the narrower the forearm.
- The closer the Pmax is towards the outside of the wing, the more pointed the hand. In this case, the size of the primaries often decreases quite quickly towards the wrist.
- The closer Pmax is to the inside of the wing, the wider and more rounded the hand. In general, primaries close to Pmax have fairly similar measurements.

These tables of precise measurements of the size of each feather are valid only for the individual measured, and do not account for variability within species (often +/− 10%).

However, the relative proportions of the feathers of the same individual are much more consistent.

But be careful, because due to the variability in size between individuals of the same species, there is a corresponding variation in the range of sizes between small and large individuals for the same feather.

Using a simple calculation, taking the largest measurement shown in the table for the group of feathers considered (the largest feather of the largest individual) and assigning it to the maximum value on the graph (100%), or taking the smallest measurement (the smallest feather of the smallest individual) to correspond to the minimum value, we can find two tables and therefore the two curves indicating the minimum and maximum size for each feather (see the two examples below).

Using the same technique, by measuring just one of the feathers (always providing its precise location is known), it is possible to deduce the size of all the other feathers of the wing or tail for the individual in question by using the graph.

The sizes of feathers of the same type can therefore be found, either according to the measurements given in the tables or using the measurement of a found feather. However, it is best to rely on measurements of primary or outer secondary feathers when looking for the sizes of the other remiges.

Calculating feather size from the range of sizes given in the tables

In the tables the minimum (Tmin) and maximum (Tmax) measurements of feathers in cm. are indicated for a species Y. The range in size for each feather can therefore be found.

For example, for the rectrices: in the table or the text it states Rmin =15 cm and Rmax = 40 cm.

The graph indicates that for R6, Rmin is 50% and that Rmax (100%) applies to R1 (indicating a highly rounded tail). In the case of a 'large' individual, Rmax = 40 cm, so R1 measures 40 cm and R6: 50/100 x 40 = 0·5 x 40 = 20 cm. The rectrices therefore measure between 20 and 40 cm for a large individual. For a 'small' individual,

Rmin = 15 cm, so R6 measures 15 cm and R1: 100/50 x 15 = 2 x 15 = 30 cm. Rectrices measure between 15 and 30 cm for a small individual. By reading the proportions of the other rectrices on the graph, it is possible to trace two curves, one for the smallest the other for the largest individual measured.

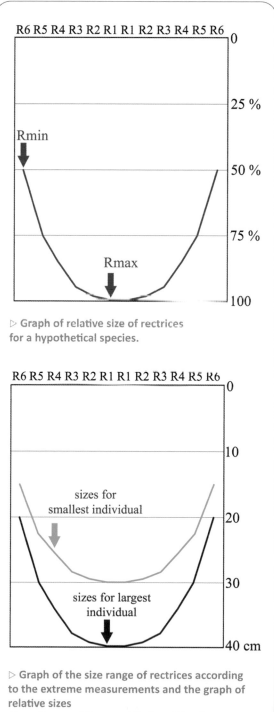

▷ **Graph of relative size of rectrices for a hypothetical species.**

▷ **Graph of the size range of rectrices according to the extreme measurements and the graph of relative sizes**
(here Rmax = 40 cm and Rmin = 15 cm).

So, a size range for each feather can be determined depending on its position. For example, if R4 measures 85% of R1 on the graph, it will measure between 85% of the R1 of the largest (85% x 40 = 34 cm) and 85% of the R1 of the smallest (85% x 30 = 25.5 cm) individual. If it is thought that an R4 of this species has been found, it should normally be in this size range.

Calculating feather measurements from the size of a found feather.

When an isolated large feather is found, it is worth verifying whether its size is consistent with that of the supposed species.

For example, a primary measuring 15 cm is found. From its shape, we guess that it is an inner primary, apparently between P4 and P1.

The supposed species X has a Pmin measurement of 12 cm for P1 (small individual) and Pmax of 24 cm for P8 (large individual).

The graph indicates that P1 measures 70% of P8 and that P4 measures 80% of P8.

To obtain a size range of primaries from P1 to P4 for this species, we take as a reference the smallest of the two for a small individual (here P1) and the

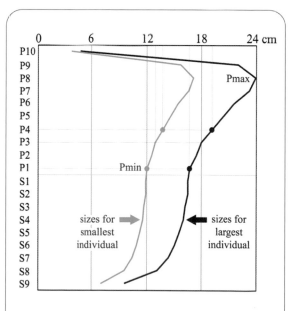

▷ **Graph of species X, indicating the theoretical measurements of the flight feathers of extreme individuals**

largest of the two for a large individual (here P4). For example, a 'small' P1 measures 12 cm (Pmin shown in the table) and a 'large' P4 measures 80% Pmax, or 0·8 x 24 = 19·2 cm.

Assuming that the feather is probably a P3 or P4, we have a 'large' P4 that measures 19.2 cm. It is now necessary to find the size of a 'small' P3. The graph indicates that P3 measures 75% of P8. With a P1 that measures 70% of P8, we obtain a 'small' P3 measuring 75/70 x Pmin, as Pmin is for a 'small' P1. Thus, we have T (small P3) = 75/70 x 12 = 12·9 cm. So, the theoretical size range for P3-P4 is 12·9-19·2 cm. It is sufficient to check whether the feather found matches these measurements.

If the feather's position is known precisely, it is possible, from its size and the graph data, to find all of the measurements of the other feathers of the same type using the same logic.

To calculate the theoretical size of a secondary, the extreme primary measurements can also be used, since the relative sizes of secondaries is expressed as a percentage of the size of the largest primary. Of course, the Smin and Smax values shown in the tables can be used, but an additional calculation will be needed (by bringing the proportional size of the largest secondary to 100%). In addition, because the size of secondaries is less variable for the same individual, we obtain more reliable measurements using primary values as references.

KEY TABLES

How to use them. These tables do not pretend to provide a solution to all enigmas. However, in most cases they will allow potential answers to be restricted to a small number of species. Examination and comparison of illustrations or accurate descriptions will be made easier by being reduced in number, and thus identification becomes easier.

After having determined the approximate placement of the feather and, if possible, to have targeted a group of likely species it is necessary to look at what features of structure, shape or colour are present on the feather. Depending on the exact position of the feather and individual variability (age, sex, geographical origin, etc.), either all or some of the criteria indicated will be visible.

1. Choose the main feather category (primary, secondary, rectrice), a size class, etc. If you are not sure about feather category, try the corresponding keys one after the other and note the potential species indicated (there will be more species than necessary, but this way you are less likely to eliminate the correct species).

Example: rectrice more than 20 cm long, barred, not downy ? diurnal raptor.

2. For each criterion, look at which species shows them and list the species. If several features are present, cross-reference the information to eliminate as many non-compatible species as possible.

Example: regular, well-defined bars + terminal bar ⇨ eliminate adult Honey Buzzard.

3. Several potential species may sometimes remain. In this case, a more detailed study is needed: consult the corresponding descriptions and illustrations (see Index p. 396) and possibly a complementary guide (see References p. 400) or try to better target the likely species by using information not brought into your investigation so far: feather structure, habitat where found, probability of encountering the species, etc.

The list of abbreviations used in the key tables is presented on p. 4.

Remarks:

Sizes are shown in centimetres (cm), with an accuracy of 0.5 or 1 cm. Except as noted in particular cases, this is the measurement of the entire feather, including the calamus. The cited measurements represent the range of sizes of all feathers in each category (P, S or R) and of several individuals. However, some small variability outside this range may be encountered.

For the primaries: some species have a very small outer primary and the size of this outermost primary is not taken into account in the general size range. This feather is generally difficult to identify. When measurements are available, the size of this outer primary is shown in the general table at the end of the book.

Example: in the Blackbird, primaries measure between 9 and 12 cm long, except for the outer one, which is about 3 cm long. The table indicates 9 − 12 cm.

For the secondaries: in general, the measurements indicated do not take into account the innermost secondaries (tertials), in order to simplify comparisons. These feathers are recognisable by their lanceolate shape, very similar to that of scapulars, with which they also often share the same colour. In most passerines, the two innermost secondaries are significantly shorter. In waders and ducks it is the opposite and some are significantly longer. Sometimes, however, the description of the secondaries may also apply to these inner secondaries. When measurements are available, the sizes of these feathers are shown in the table of biometric data at the end of the book (p. 374).

For the rectrices (tail feathers): all are measured and given, with the exception of a very few specific cases (indicated).

The number of feathers: the general total number of feathers per family or species is indicated in the table of biometric data (p. 374) and at the beginning of the descriptions.

3. Determining the location of a feather on a bird

This chapter allows a feather's position on a bird to be located (wing, tail, body). The various categories in Chapter 2 (see p. 10 and after) will be called 'feather type'. Determining the type of feather is the first step in establishing which species it belongs to. Over time, this stage will appear obvious, or at least be quickly achieved. Identifying the more or less precise position of a feather on a bird is key to arriving at the identification of the species concerned.

HOW TO USE THE IDENTIFICATION KEY

This key leads the observer step by step towards a type of feather and towards the location of a feather on a bird's body. For experienced observers, it may be useful to recall certain differences between the categories of feathers.

The key distinguishes between: body feathers, scapulars, underwing-coverts, secondary coverts, coverts, primaries, secondaries, and alula feathers.

The principle of the key is to choose, at each numbered step, the description that best corresponds to the feather among all those proposed bearing the same numeral (for example, choose between 3a, 3b and 3c). The determination continues each time following the number shown at the end of the correct choice, until a feather type is established. In theory, the different criteria set out in each choice will all be visible on the feather, but some are more obvious than others and there are special cases. It is possible that a feather does not correspond to any description exactly. It should then be compared to other types of feathers identified correctly. The illustrations in Chapters 6, 7 and 8 can be used as a basis for comparison, especially for large feathers.

Some feathers are particularly difficult to position, especially those located towards the central axis of the body and the base of the wings. The observer may fail to determine the location of a feather discovered in isolation, but this is inevitable. Experience helps, so do revisit any outstanding problems later on, when your chance of success will have improved.

REMARKS ON THE FEATHER DESCRIPTIONS AND THE VOCABULARY USED

FROM WHICH SIDE DOES THIS FEATHER COME?

Only the feathers on the right side of a bird are shown in this book, to facilitate comparison. Obviously, each feather has its mirror copy on the left side.

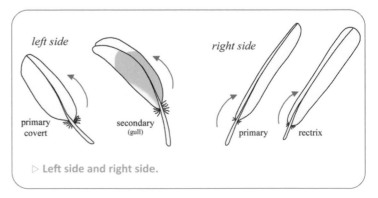

left side

primary covert

secondary (gull)

right side

primary   rectrix

▷ **Left side and right side.**

First, determine which side of a bird the feather comes from. This seemingly unimportant detail will prove useful later, especially to distinguish the inner vane from the outer vane, which is generally narrower.

When the observer looks at a large feather or covert from above, the base towards the observer, if it 'turns left', it comes from the left side of a bird (conversely for the right side). This is true for the majority of feathers, but less obvious in body feathers.

Note: where the curvature of the calamus differs from the rest of the shaft (primary covert, for example), only take the part with barbs into account.

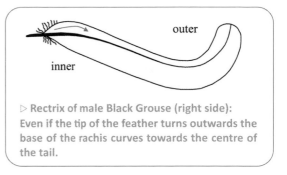

▷ Rectrix of male Black Grouse (right side): Even if the tip of the feather turns outwards the base of the rachis curves towards the centre of the tail.

Remark: for a rectrice, when it 'turns to the left' with the calamus towards the observer, it comes from the left half of the tail. Of course, in its natural position, the calamus is obviously 'away from the observer', towards the front of the bird.

Special case: in birds whose tails are used as ornaments (Lyrebird, Black Grouse, etc.), the rectrices turn outwards from the tail. However, the inner vane will be wider than the outer vane and the curvature will be towards the inside of the tail at the base of the feather. This is generally applicable to scapulars and secondaries, which may also have a double curvature.

Beware, the inner remiges on the arm may have an inverted curvature and then appear to come from the other wing of a bird (e.g. grebes): looking at the curve of the vane in profile can often help (see p. 11).

'DOWN'

True down feathers (with free, uncoloured barbs) are not described as they are rarely identifiable. The term 'down' is used for convenience to refer to soft, barb-free feather bases. These barbs are often grey or white and are hidden in the plumage. Their abundance at the base of the feather can be a useful feature.

'SMALL' AND 'LARGE'

The terms 'small' and 'large' are of course relative. A large sparrow feather is tiny compared to a small eagle feather. 'Large' in a key refers to fairly rigid feathers (and is used as a general term for primaries, secondaries and the large tail feathers). A 'small' feather can be identified by its flexibility. To test the rigidity of a feather, hold the base of the shaft with one hand (1) and press the surface with barbs on the other palm (2). A 'small' feather offers little or no

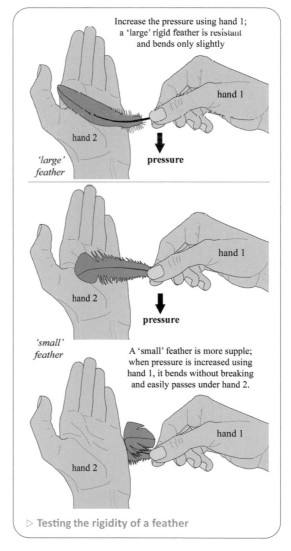

Increase the pressure using hand 1; a 'large' rigid feather is resistant and bends only slightly

hand 1

hand 2

'large' feather

pressure

hand 1

hand 2

'small' feather

A 'small' feather is more supple; when pressure is increased using hand 1, it bends without breaking and easily passes under hand 2.

hand 1

hand 2

▷ Testing the rigidity of a feather

resistance; hand 1 can be lowered without forcing or breaking the feather, which passes under hand 2 easily.

The length of the calamus is also an indicator of the pressure exerted on the feather during flight. A short calamus often means that the feather is not used to provide power in flight: body feathers, secondaries, underwing-coverts, alula, tertials and scapulars.

The tail feathers often have a proportionately shorter calamus than the flight feathers of the same bird.

KEY FOR DETERMINING FEATHER TYPE

• **1a**

Feather soft or rigid, with little down at the base ⇨ **2.**

• **1b**

Feather very soft, with a lot of down at the base and barbs at the tip connected together. ⇨ **BODY FEATHER.**

Large 'body' feathers on the wing (coverts) are described later in the key because they can be distinguished (see step 2 and following). It is difficult to discover exactly which part of the body feathers come from, and especially to which species it belonged, unless it has a distinctive shape or pattern, or a brightly coloured mark. The species to which it may belong should be queried directly by consulting field guides or photographic guides. Some identifiable body feathers are presented in Chapter 6, p. 100

Here is a brief description of body feather shapes based on their location on a bird.

▷ **1b Body feathers.**

Obviously concave. Often quite elongated: they may come from the belly (Be). They are almost always paler than those of the back (Ba) of the same species, the latter being much flatter, often with more connected barbs. Sometimes with a wide tip, for example in the Anatidae (A).

Lanceolate shape. Possibly scapulars, sometimes also ornamental feathers of the neck or mantle. Large scapulars (lower) have little down at the base and resemble inner remiges.

▷ **1b Scapulars.**

▷ **1b Undertail-coverts, sketch in profile and photos from below.**

▷ Greater, median and lesser coverts.

▷ Flank feather.

Spoon-shaped, with the distal part concave. These are usually undertail-coverts, the ends of which press strongly on the tail feather bases. Their coloured part is almost always the same colour as the vent. The uppertail-coverts have this same shape, although less pronounced; their tips are the same colour as the base of the upperside of the rectrices, except in cases of a contrasting rump (frequent in passerines). The calamus may be slightly upturned, as can the rectrices (see 6b).

Quite oval, with reduced amount of down. Possibly lesser or median coverts (on the forearm). Those on the leading edge are clearly curved, sometimes two-tone (above/below the wing).

Fluffy appearance. All the barbs are unattached, the tip is coloured: probably from the flank (F), an area that is little exposed being partially hidden under the wing at rest.

Very short. Probably from the head or near head (H). As wide as they are long on the throat; more elongated on the chest and neck as they near the shoulders. Those on the cap are sometimes quite tapered. The shorter the feather, the less of the body it covers.

- **2a**
'Small', supple feather ⇨**3.**
- **2b**
(Very) rigid feather ⇨**4.**
- **3a**
Oval form, often concave, short calamus:
⇨ **SECONDARY COVERT.**

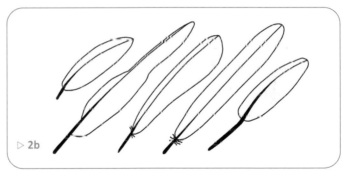

▷ 2b

For a given species, the size and stiffness decrease when moving from large to small (and the amount of 'down' at the base increases).
Field identification guides can often be used to find the species concerned, as these feathers are partly visible on the closed wing, and therefore illustrated in these guides.

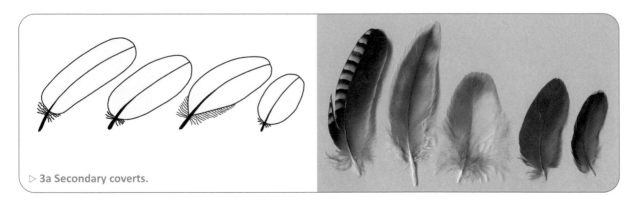

▷ 3a Secondary coverts.

• 3b

Generally, appear slender and 'flat'; fragile in form. Calamus in profile short or very short, not thick and flat, and always thin.

Feather sometimes striped, often pale with dark base. Tip usually the same colour as the flanks. More or less wide depending on the species; those of the forearm are wider. Variant with more compact form: axillaries (almond-shaped).

⇨ **UNDERWING-COVERTS (or AXILLARIES).**

• 3c

Lanceolate in form and rachis often straight (or slightly curved) with a quite short calamus.

⇨ **SCAPULAR.**

Note: colour and size similar to that of the inner secondaries of the same species, but with thinner rachis, more flexible feather and with more down at the base (see 7a).

▷ 3b Underwing-coverts.

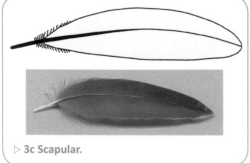

▷ 3c Scapular.

• 4a

Viewed from above, the rachis has a kink (usually) where the barbs begin:

– the feather is very fragile for its size;

– the calamus is long compared to the rest of the feather (as much as a half of the total length);

– the tip is rounded (or pointed on the outer feathers that are shorter and narrower).

⇨ **PRIMARY COVERT.**

• 4b

No kink where the calamus meets the rest of rachis ⇨ **5.**

▷ 4a Primary coverts.

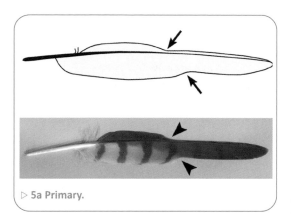

▷ 5a Primary.

• 5a

Presence of an abrupt change in the edge of one or both vanes: the vane narrows abruptly (emargination/notch), one vane much narrower than the other:

⇨ **PRIMARY.**

Check the presence of the other features in 9a; see also woodpecker rectrices in 6b and description of the Picidae p. 204 .

• 5b

No notch, vanes with regular edges ⇨ **6.**

• 6a

The calamus, viewed in profile, follows the curve of the rachis: see figures that follow for comparison with 6b (example with remiges) ⇨ **7.**

• 6b

The calamus, viewed in profile, is slightly upturned (sometimes not very visible but very reliable when detected): consult figures on next page to compare with 6a. The calamus is also more or less clearly upturned on the uppertail- and undertail-coverts.

– calamus quite short compared to the rest of the feather;

– silhouette of the tip of the feather quite symmetrical and wide, rounded or square;

– rachis almost straight on the distal ¾ of the feather;

– vanes of quite uniform width along their whole length (parallel-sided):

⇨ **RECTRICE.**

Exceptions to normal shapes of rectrices

Among the most frequently encountered are:

– curved edges and rachis: Anatidae and a few other species (see figure on following page and key p. 96 . R5 curved, pointed, or rounded);

▷ 6b Rectrices.

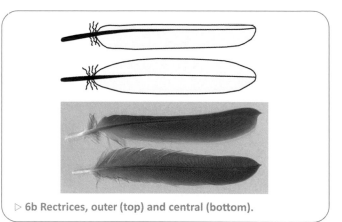

▷ 6b Rectrices, outer (top) and central (bottom).

x

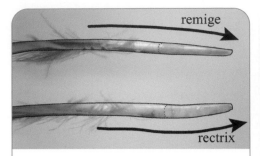

▷ **6a and 6b Comparison of the curve of the calamus between rectrices and remiges.**
Curve of the calamus (viewed in profile): remige (upper), rectrix (lower).
6a: the curve of the calamus follows that of the rachis: remiges and other wing feathers.
6b: the calamus is upturned: rectrix.

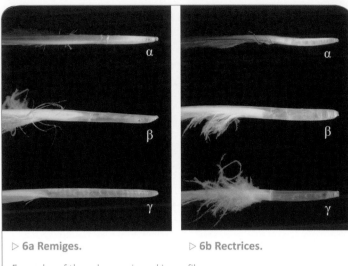

▷ **6a Remiges.**   ▷ **6b Rectrices.**

Examples of the calamus viewed in profile
α: Magpie / β: Common Buzzard / γ: Mallard (not to scale)

▷ **6b The rectrices of a woodpecker (upper) and of Anatidae (lower).**

▷ **The curve in the calamus of rectrices (viewed in profile).** Upturned calamus, usual appearance, example: Jay
Straight calamus:
– diving birds, example: cormorant.
– climbing birds, example: woodpecker.

– gannets, cormorants, woodpeckers and treecreepers: in profile the calamus is straight or only slightly upturned and quite long, and is more or less pointed (pointed and tapered in woodpeckers and treecreepers, see p. 204 woodpeckers and p. 148 treecreepers). **Beware not to mistake** the narrowing of the outline of the rectrices of woodpeckers with the emarginations/notches of the primaries (see 5a), and curved rectrices with the secondaries (8b).

• **7a**
Frequently, the rachis is (almost) straight OR the rachis is consistently bent, in which case the curvature gradually changes direction towards the tip of the feather;
– the tip is rounded or pointed, with both vanes approximately the same width;
– leaf-shaped outline (similar in shape to a plant leaf).
  ⇨ **INNER SECONDARY**
    OR TERTIAL (closer to the body).

It is sometimes difficult to determine which side these feathers come from. On remiges, observing the vane's curvature in profile should help (see p. 11).

Inner remiges can be confused with scapulars, especially for species larger than a Blackbird. In these species, the difference in stiffness is not very obvious between large scapulars and the innermost secondaries (that are in theory more robust). Generally, the diameter of the calamus is slightly larger on the remiges than on the scapulars.

▷ **7a Tertials (or inner secondaries).**

▷ **7b Tertial.**

• **7b**

Rachis clearly bent in the same direction along its entire length where there are barbs:

⇨ **8.**

• **8a**

One vane is (much) narrower than the other and ends more abruptly; – rachis slightly curved:

⇨ **9.**

• **8b**

The tip of the feather is rounded or square (sometimes with a longer inner vane, or split tip); – the rachis is (very) curved towards its base on the side with the widest (inner) vane;

⇨ **SECONDARIES.**

A TIP: To compare the curvature of the calamus with that of a primary (see in 9a), overlay the calamus of a primary with that of a secondary of equivalent size that comes from the same side of the bird. When the calami are aligned, the main part of the rachis has a sharper curve. This criterion is sometimes the only way to distinguish an inner primary from an outer secondary (adjacent on the wing). Secondaries also often have more unattached barbs at the base compared to the primaries.

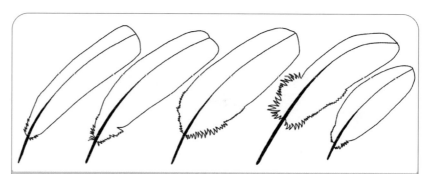

▷ **8a Secondaries.**

▷ **Comparison of rachis curvature between flight feathers.**

primary

secondary

• 9a

Feather quite elongated and 'classic' in shape: one vane is (much) narrower than the other;

– the tip of the feather is finger-shaped (one of the vanes narrows more sharply than the other);

– the rachis is (very) slightly and regularly curved (but more clearly curved in Galliformes, small Rallidae, etc.);

⇨ **PRIMARY.**

• 9b

The feather very much resembles a small primary, but is very stiff, pointed at the tip, quite wide and short, with a short and thick calamus.

⇨ **AN ALULA.**

▷ **9b Alula feathers.**

▷ **9a Primaries.**

SUMMARY OF THE MAIN FEATURES OF FEATHERS IN RESPECT TO THEIR LOCATION ON A BIRD						
Feather type	General shape	Tip	Vanes	Rachis	Calamus	Exceptions
Primary	elongated, asymmetrical, possibly a notch on its edge (outer P)	finger-shaped, pointed, wedge-shaped or rounded	outer obviously narrower than inner	slightly curved inwards	quite long	rachis quite inwardly curved: small Galliformes, Rallidae, swifts, storm petrels, etc.
Secondary	wide and curved	usually rounded or square	little difference in width, or even equal	clearly curved inwards	not long	Columbidae: median S almost straight. Gulls/waders, Anatidae: long calamus
Tertial (or inner secondary)	symmetrical, leaf-shaped	rounded but narrower than the base, pointed	the same width	straight or doubly curved	rather short	some are curved outwards (grebes…)
Rectrix (Tail feather)	symmetrical, with parallel sides	rounded or square, sometimes tapered	outer vane very narrow in outer part of tail; vanes of equal width on central feathers	quite straight, except at base	quite short, upturned	curved in the Anatidae, etc. rachis long, straight and not upturned in woodpeckers, cormorants, gannets. Curves outwards in Black Grouse
Alula	short	often pointed	of different widths	rigid, slightly curved inwards	quite short but strong	tip sometimes rounder (cuckoos…) or slightly tapering (swans…)
Primary covert	slightly elongated	rounded or pointed	of different widths	quite rigid; generally bends where barbs begin	long and thick	more or less pointed and rigid depending on species
Greater covert	oval, possibly slightly curved	wide and rounded	little difference in width, even the same	supple	short	more or less elongated depending on species
Other wing and body feathers	variable, with much down at base	variable	little difference in width, or even equal	very supple	short	very variable

Note: flight feathers (primaries, secondaries and rectrices): as a general rule, the primaries have a fairly straight rachis viewed from above and the secondaries have a rounded or squared tip. The rectrices have both characters.

The approximate location of the isolated feather: hand, forearm, tail, body, etc., should now have been found. If this is not the case, try the key from the beginning again or compare the problem feather with other feathers from a known location.

Before searching for the species the feather has come from, it is possible to further clarify its position on the body. The following paragraphs provide guidance with this in mind.

• TIPS FOR MORE PRECISE DETERMINATION OF FEATHER LOCATION

Before looking at which species the feather may have come from, it is possible to further clarify its position on a bird's body. The following provides guidance for this purpose.

Once feather type is established, it is possible to try to be a little more precise as to its location; this will help in species determination.

Unless the researcher has a reference bird or a complete skin, it will generally be difficult to assign a precise location to collected feathers. However, terms such as outer, median and inner may often be used.

When looking at a 'large' feather, the narrower vane is the outer one in relation to the body (tail or folded wing), and leading (open wing). This vane is narrow but thick, which improves its robustness when going into the wind (in flight). The inner vane is supple. As the inner feathers cover the outer ones, the inner vane of a feather pushes against the upper feather during flight, improving the bearing surface of the wing or tail and cushioning air turbulence (see diagram p. 11).

For the remiges

The narrower and stiffer the outer vane, the more the feather is located outwards/forwards on the wing (facing the wind). If there is at least one emargination, the closer the emargination reaches the base of the feather, the more that feather is located towards the outside of the wing.

For the rectrices

The narrower the outer vane, the more the feather is located towards the outside of the tail. Conversely, the more similar the vanes in width the more the feather is located towards the centre; the central rectrices are almost symmetrical. The straighter the rachis (especially at its base), the more the feather is located towards the centre of the tail.

Using these clues, it becomes possible to 'reconstruct' a wing or tail with feathers found in isolation or, better, with a plucked corpse. Feathers placed side by side, like the pieces of a jigsaw, can reveal the pattern and shape of the wings or tail.

In the case of a plucked bird, once arranged, the feathers can give valuable information about the shape of the wing or tail, and thus point the finder to a group of species. An example of the reconstruction of a plucked bird can be found in the guide *Tracks & Signs of the Birds of Britain and Europe* (see References p. 400).

The number of 'large' feathers per species is shown in the biometric data table on p. 374 and in the group descriptions.

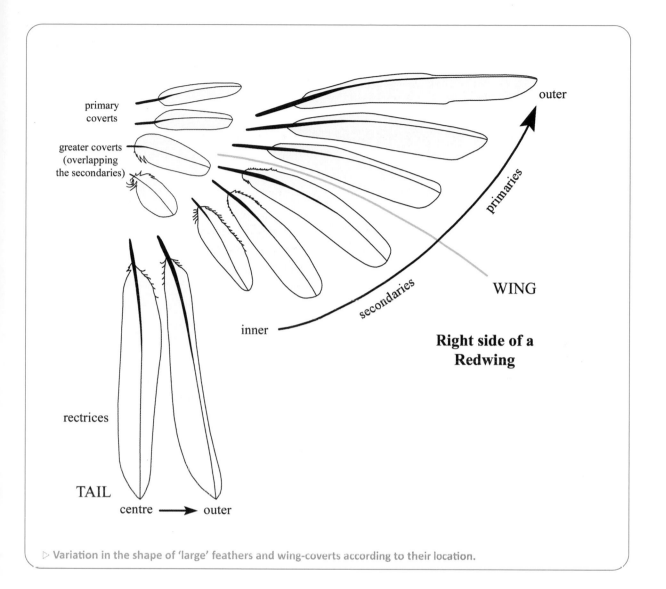

primary
coverts

greater coverts
(overlapping
the secondaries)

primaries

outer

WING

secondaries

inner

**Right side of a
Redwing**

rectrices

TAIL

centre ⟶ outer

▷ Variation in the shape of 'large' feathers and wing-coverts according to their location.

4. Flight and tail feather shapes and adaptations

RELATIONSHIP BETWEEN FEATHER SHAPE AND THE FORM OF THE WINGS AND TAIL

The varied lifestyles of birds have led to the evolution of many forms of wings and tails, adapted to the type of flight, movement, or diet of each species. Wing shape is closely related to the type of flight used most frequently by a species, and each mode of flight has its own corresponding form of tail.

• TAIL FORMS

These are largely a consequence of the relative lengths of the rectrices, which can be the same size for the entire tail, vary slightly, or show large variations between the central and outer feathers. The shape and length of the tail affect the balance of a bird in flight and also assist as a counterbalance when walking, or as a rudder when diving. Species that require high precision in flight have forked tails, and in the extreme have streamers (very elongated outer rectrices). Forked tails facilitate changes in direction and occur in many species. Square, rectangular, or rounded tails have no special function, other than providing the possibility of additional uplift when spread in flight. Pointed, wedge-shaped tails that are quite elongated, like long rectangular tails, can be useful when walking, aiding balance in a similar way to the pole of a tight-rope walker. In addition, the shape and size of the tail (e.g. the length of the streamers or central feathers) can play a role in mate selection (often of males by females) during displaying and courtship.

The following diagrams show the different categories found in the lengths of birds' rectrices, and examples of the shape of the outer and central rectrices for each category.

When an isolated rectrix is found, it is sometimes difficult to make the connection with the corresponding tail shape. However, by examining the proportions of the feather, its outline, and the shape of the tip it is possible to come to some conclusions.

Here are three examples:

– This feather is quite rectangular in silhouette with a symmetrical, rounded or square tip.

– This rectrix is probably from a rounded or rectangular tail, possibly notched. If the tip is clearly asymmetrical, the tail is more likely to be notched or rounded as the tip of the rectrix follows the general outline of the tail. If the rectrix is squat, it is probably from a square or short tail.

– This rachis seems thin and the feather appears quite narrow. This rectrix obviously comes from an elongated tail, either in its entirety, in the centre (a pointed tail), or at the exterior (forked tail). Whether it is centrally or externally positioned

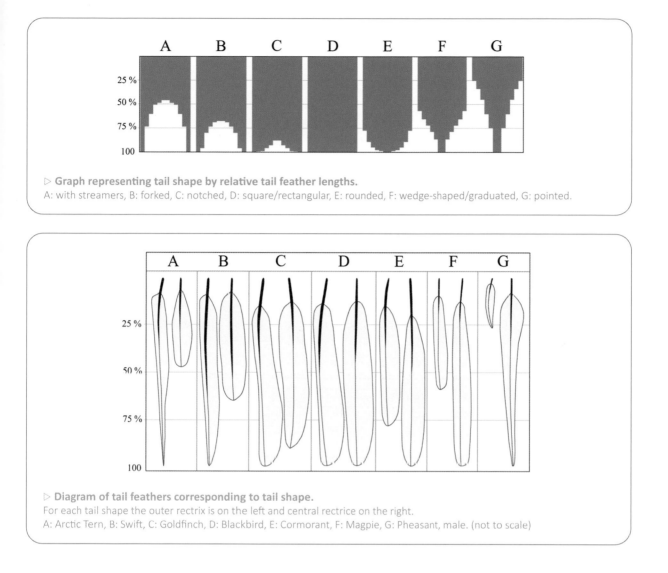

▷ **Graph representing tail shape by relative tail feather lengths.**
A: with streamers, B: forked, C: notched, D: square/rectangular, E: rounded, F: wedge-shaped/graduated, G: pointed.

▷ **Diagram of tail feathers corresponding to tail shape.**
For each tail shape the outer rectrix is on the left and central rectrice on the right.
A: Arctic Tern, B: Swift, C: Goldfinch, D: Blackbird, E: Cormorant, F: Magpie, G: Pheasant, male. (not to scale)

can decide between a pointed or forked tail shape, respectively.

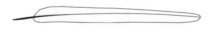

• WING SHAPES

Wing shapes arise from the relative lengths of each bone in the wing and the angles between them (wings being longer or shorter), but especially from the relative sizes and shapes of the primaries and secondaries.

There are three main types of flight and each species uses one preferentially, although all use flapping flight on take-off, and gliding on landing. Each type of flight has its own requirements and the different wing shapes are adapted to suit the preferred flight-type of a particular species. Certain species are very specialised and essentially have only one type of flight, but others adapt to the surrounding environment (topography, weather, vegetation, etc.) or modify their flight according to their needs (gain in altitude, speed, hunting, etc.).

Gliding flight: also practiced by other groups of vertebrates (reptiles, mammals, fish), it mainly uses the weight of the body for forward movement and expends little energy. Breaking from a dive is a common example which is clearly visible on landing. Specialist species, mainly marine, take advantage of the lift provided just above the waves to propel themselves forwards. Horizontal air currents also circulate along cliff faces and allow birds to arrive at their nests effortlessly.

Gliding is therefore used by many large pelagic seabirds, such as albatrosses, gannets, fulmars, and some of the larger gulls and terns.

Their wings are narrow and elongated, the hand is triangular, and the forearm is long. The arm can also sometimes carry remiges, as in gannets and albatrosses. The size of primaries decreases rapidly, while secondaries are numerous and of similar size (quite short). The tip of the primaries is clearly wedge-shaped, although in some cases it can be rounded (fulmars). The tail is often short to reduce drag.

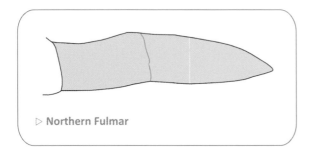

▷ **Northern Fulmar**

Soaring: birds that soar, often called 'soaring birds', are very light for their size and their wings and tail have a large surface area when spread. Many diurnal raptors soar, as well as corvids, storks, etc., and other migratory birds.

Air currents in the environment again provide the energy necessary for soaring flight. However, in this case it is vertical (or almost) currents that are the driving force, pushing a bird upwards, along cliffs or in 'thermals' produced by rising warm air. The trajectory is less direct than for gliding flight, since a bird must remain in the rising column of air by spiralling. Thermals do not form over the sea, which often forces detours for long-distance travel.

On the other hand, it is a very economical mode of flight, since the finger-like emargination of the wingtips provides forward suction, while the curvature of the wings helps to lift a bird upwards. The trailing edge of the wing is also often 'fingered', very visible in some diurnal birds of prey: this helps to reduce turbulence during flight.

The ideal wing for gliding is wide, with a rectangular or rounded hand, and a forearm almost equal to the hand. The outer primaries (often 5, 6 and 7) narrow towards the tip and the secondaries are wide and strong.

A large tail is also an asset as it increases the aerodynamic surface. Some raptors, such as harriers,

hunt low over the ground and therefore have narrow, elongated wings, somewhat reminiscent of those of the gliding flight specialists.

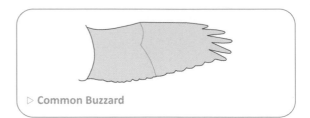

▷ **Common Buzzard**

Flapping flight: practiced by the majority of species, some of which are specialists and use no other mode of flight. All birds use it to take-off or when weather conditions do not allow them to use air currents. It requires more energy than soaring or gliding but allows birds to move in all weathers and in all environments. Heavy birds are obliged to flap continuously, such as: swans, geese and ducks, pigeons, grebes, coots, and cormorants. For lighter species, flapping flight is sometimes interrupted by glides, resulting in an undulating flight. The hand is classically triangular, with a fairly short but wide forearm. Depending on the habitat and lifestyle of the species, multiple variations are possible.

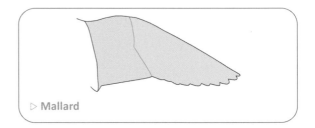

▷ **Mallard**

It is possible to differentiate between several subtypes of flapping flight, which result in more or less clear changes in the shape of the wing and of the remiges.

Direct flapping flight: may involve heavy species, primaries are often quite rigid, with a long calamus. The secondaries are much shorter, also quite rigid; sometimes with longer innermost feathers that improve cohesion between the wing and the body.

Undulating flapping flight: wing flapping alternates with glides, with the wings more or less closed. The more a bird weighs, the longer the periods of

Evolution of the shape of wing and tail feathers for three types of flight and three tail shapes.

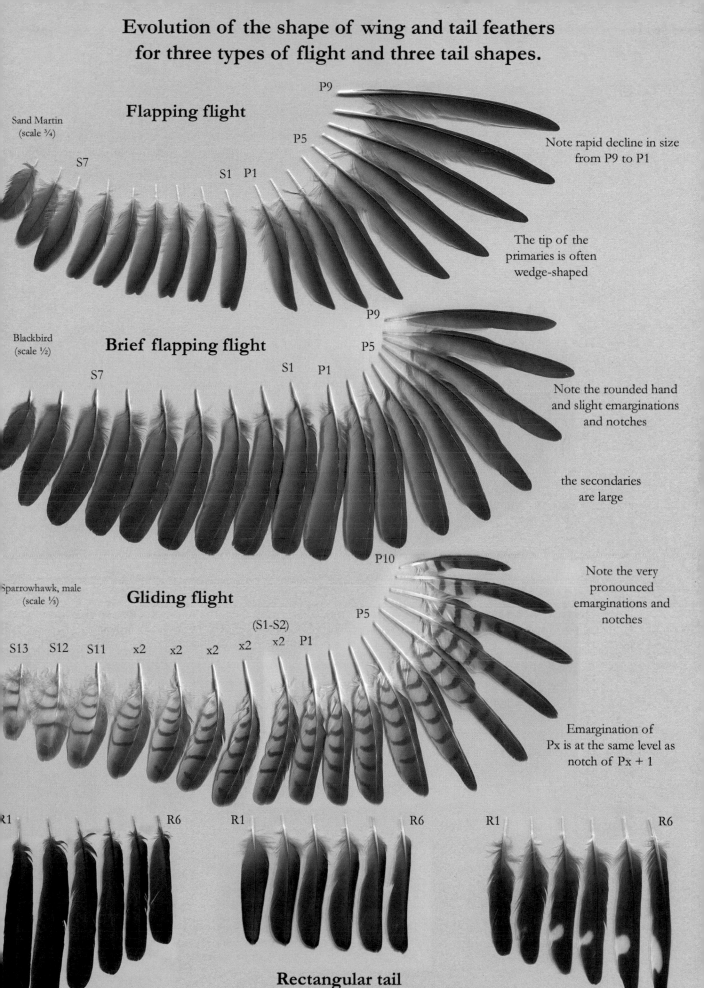

Flapping flight

Sand Martin (scale ¾)

S7 S1 P1 P5 P9

Note rapid decline in size from P9 to P1

The tip of the primaries is often wedge-shaped

Brief flapping flight

Blackbird (scale ½)

S7 S1 P1 P5 P9

Note the rounded hand and slight emarginations and notches

the secondaries are large

Gliding flight

Sparrowhawk, male (scale ⅓)

S13 S12 S11 x2 x2 x2 x2 (S1-S2) x2 P1 P5 P10

Note the very pronounced emarginations and notches

Emargination of Px is at the same level as notch of Px + 1

R1 R6 R1 R6 R1 R6

Graduated tail

Magpie (scale ¼)

Rectangular tail

Starling (scale ½)

Forked tail

Barn swallow, young (scale ¾)

gliding, and the flight more undulating. In this case the hand is a little shorter, either triangular (species of open and semi-open environments) or rounded (e.g. woodpeckers).

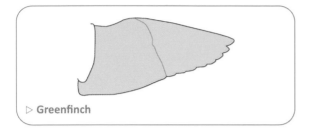

▷ **Greenfinch**

Brief flapping flight: except when on migration, these birds fly by continuously flapping their wings, but perch or land often (mainly forest birds). The hand is rounded and wide, which helps with manoeuvrability in flight (e.g. Blackbird, owls).

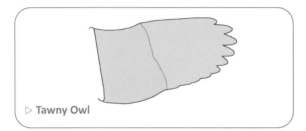

▷ **Tawny Owl**

Accelerating flapping flight: very powerful, it allows for the rapid take-off of relatively heavy birds (Galliformes notably), thanks to narrowed primaries. The wing is rounded and wide, and the remiges are often rigid to compensate for the heavy load. This type of flight is also found in some forest raptors (goshawks, sparrowhawks) and allows them to hunt in relatively thick vegetation with pursuits between obstacles.

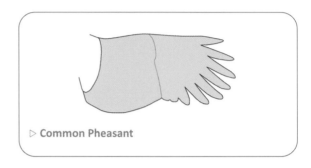

▷ **Common Pheasant**

If we look only at the relative sizes of the remiges, we can distinguish these types of flight by the shape of the curves. More details are given on the shape of the wings and tail in relation to a species' behaviour in the guide *Tracks & Signs of the Birds of Britain and Europe* (see References p. 400)

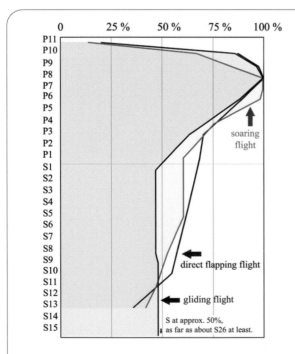

▷ **Graph of relative remige sizes for the three main types of flight: gliding (blue), soaring (green) and direct flapping flight (brown).**

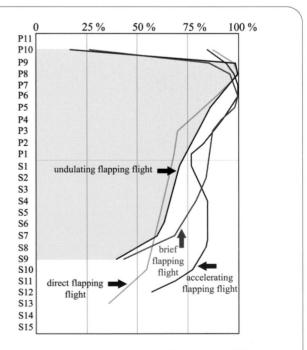

▷ **Graph of the relative remige sizes for different types of flapping flight: direct (brown), undulating (blue), short (green) and accelerating (pink).**

▷ The lines represent a stylised species for each type of flight.

ADAPTATIONS AND CONVERGENCIES

• CONTRASTING COLOURS IN WINGS

Many species that live in groups and/or in open environments, especially in winter, have contrasting coloured patches on their wings. In fact, contrasting areas in motion are easier to spot in flight than uniform-coloured wings (e.g. against a backdrop of sky, meadow, or sea) and make it easier to see other members of the group. The wing markings also help to break up a bird's outline and can confuse predators, especially when birds are flying in flocks. Groups with such markings include Anatidae, waders, finches, etc. Mostly, these wing markings include a white (or pale) wing bar, placed in the middle of the wing, towards the base of the remiges and the tips of the coverts. Sometimes this bar is highlighted with bright colours, as in the Anatidae. This very brightly coloured area is then called the speculum. The term mirror is used for the white marks on the wing-tips of gulls. The white on the wings 'blinks' in flight, whether the weather is clear, cloudy, or even very dull. This visual signal can therefore be seen from afar and is especially useful when birds are migrating; like the flashes of light sent in a Morse code signal.

Colourful areas may also be present on other parts of the body, especially in areas visible in flight (back, rump, etc.). For example, white (or light) markings are often present on the outer rectrices (many finches, pipits, etc.) and sometimes on the rump or back (some 'shanks and sandpipers, wheatears, Jay, Bullfinch, etc.). When taking-off, these pale markings can also serve as a warning for nearby birds, similar to the spread white tail of frightened deer.

Other examples of similar colour markings can be found on p. 48 and beyond.

▷ **Examples of wing markings visible in flight.**
Above: Teal; right: Dunlin (above), Hawfinch (below).

• FEATHER STRENGTH

Depending on lifestyle and evolutionary relationships, one or more solutions to the issue of feather strength have been adopted by birds. Primaries are the most modified feathers in relation to flight requirements, as they provide most of the thrust during wing flapping. They also show characteristics related to wing-loading.

This can be appreciated by looking at illustrations of birds in flight, comparing wing surface area with the known weight of the species. For the same bird weight, the smaller the wings, the higher the wing-loading. Similarly, for the same wing surface area the greater the weight, the higher the wing-loading. In general, the following links are therefore apparent.

The greater the wing-loading:
– the longer the calamus (it offers better leverage);
– the wider the calamus (which increases its strength);
– the more the barbs are reinforced, especially around the rachis (tegmen and other thickenings);
– the more curved the feather when viewed in profile (being concave partly compensates for the pressure exerted on the feather).

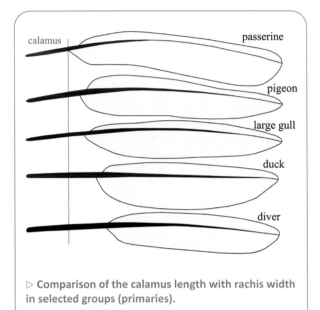

▷ Comparison of the calamus length with rachis width in selected groups (primaries).

• CERTAIN CHARACTERISTICS COMMON TO GROUPS OF SPECIES

Caution: the species mentioned may or may not be taxonomically related. The characteristics described may lead to confusion between certain species when identifying large feathers.

Swimming aquatic species. e.g.: ducks, gulls, auks, divers, coots, etc.

Their body feathers are very insulating and numerous. The down is abundant and fine, and the base of the body feathers is very fluffy. The feathers are highly curved, especially the ventral body feathers. Everything helps to trap as much air as possible between feathers and skin, in order to avoid the body cooling. These birds may remain on cold water for hours, days or months, so perfect insulation is essential.

Swimming and diving aquatic species. e.g.: divers, grebes, auks, coots, diving ducks, etc.

Note: cormorants are a group apart, see the description of this group (p. 285).

The wing-loading is usually high. Their wings are often narrow (and short), and these species are heavy, allowing them to dive more easily. They are forced to run on the water to gain speed before they can take-off (most swimming birds that do not dive can take-off directly). They also show certain general features that increase the robustness of their large feathers. Some cases are quite extreme: the divers have a long calamus, reaching 30 to 35% of the total length of the primaries. The primaries of grebes are highly curved in profile. The Anatidae (whether or not they dive) have highly reinforced barbs around the calamus (tegmen).

Heavy birds that fly little. e.g.: Galliformes, bustards, etc.

The primaries, and to a lesser extent the secondaries, are heavily curved when viewed in profile. These species have powerful and noisy flight but are not very good at gliding. Their wing-loading is at a maximum.

Birds that use their tail when swimming (as a rudder) **or climbing** (as a third support). e.g.: cormorants, gannets, woodpeckers, treecreepers, etc.

Their rectrices show certain adaptations that contribute to their robustness:
– calamus not upcurved and quite long;
– extremity very rigid but breaks easily or wears rapidly;
– rachis and barbs more rigid than most of the rectrices;
– difference in width of the two vanes less marked than usual.

Birds adapted for gliding. e.g.: buzzards, eagles, vultures, the all-black corvids, and storks.

Their hand is described as 'fingered', with the primaries narrowed towards their tips (up to seven primaries). The vanes are narrower towards the tip, to such an extent that the ends of the feathers are separated which allows vortexes to form at the wing-tips, ensuring extra lift without using energy.

▷ Ventral body feather of an aquatic bird that swims, viewed in profile (Mute Swan).

Birds that fly little (excluding possible migratory flights), requiring good camouflage, and **nocturnal and crepuscular species**, that stay motionless during the day.

Their body feathers are often brown (more or less pale) and marked with dark spots imitating vegetation (dry grasses, reeds, bark, dead leaves, etc.).

The larger feathers may also be cryptically coloured. Examples include: pheasants, Grey Partridge, Common Quail, Woodcock, Bittern, nightjars, owls.

5. Keys to flight and tail feather determination

Body feathers are described separately in Chapter 6 or occasionally illustrated in the plates.
The identification keys and tables in this guide focus on the larger feathers (primaries, secondaries and tail feathers). However, various body feathers are illustrated for many species and some are specially treated in the following chapter.

Three main categories of criteria are covered in the keys:
– **colour criteria:** generally, fairly easy to use; however, colours can vary within a species depending on the sex, age, and geographic origin of a bird. Colour therefore provides essential clues, but should be used with care;
– **criteria of three-dimensional structure:** sometimes easily visible and are valid for all individuals of a species or group of species. Structural features can be used when manipulating the feather while observing it from different angles, so that texture and rigidity can be assessed, etc.

These features are difficult to show in illustrations, because different viewing angles would have to be represented for the same feather. The descriptions describe the features not visible in the illustrations. These can, even without taking into account colour or size, lead directly to a small group of species. They should therefore be the first point of reference used when the feather in hand is in good condition.
– **criteria of form and silhouette:** require closer observation but become easier to discriminate as the observer gains experience and compares different feathers one with another.

They may therefore seem arduous to use at first, but quickly become an indispensable element of identification.

• WHAT FEATURES APPEAR TO DIFFER BETWEEN A FEATHER SEEN ON A BIRD AND ONE HELD IN THE HAND?

The colour and shape of a feather found in isolation can confuse the observer, even an experienced ornithologist. In fact, the colours of a single feather are only a tiny part of the array of features that give clues to the appearance of a bird's plumage – but they are the most obvious clues. On the contrary, the shapes of feathers are difficult to observe on a bird, as the feathers are packed together. However, examination of shape may reveal important adaptations in relation to a bird's way of life.

The base of a feather is hidden by those that cover it (particularly the wing and tail coverts) and may have colour that is invisible on the living bird. The inner vane of large feathers is also largely hidden on a bird, even when it spreads its wings or tail. Only the lower face is visible from below, but it is generally pale and not indicative. However, it is coloured more intensely in certain groups (Columbidae, Psittacidae, Laridae, etc.).

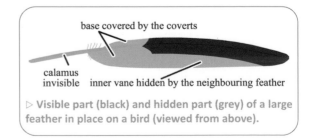

base covered by the coverts

calamus invisible   inner vane hidden by the neighbouring feather

▷ **Visible part (black) and hidden part (grey) of a large feather in place on a bird (viewed from above).**

Considering shape, the difference in an isolated feather is mainly due to the calamus being visible; it is totally hidden when the feather is in place, being buried in the skin and under the coverts. And yet its thickness and especially its length provide, in some cases, good indications as to which species the feather came from.

▷ **Deterioration of colour pigment with time.**
Tip of a primary of an auk: the colour of the tip has become very faded due to environmental conditions (sun, saltwater, friction, etc.), whereas the rest has been protected by the feathers covering it.

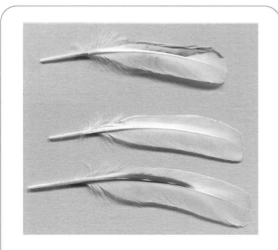

▷ **Secondary of a Great Crested Grebe coloured by environmental contamination (top).**
Top: The feather was white after moulting but has been in contact with iron oxide or another dye (especially on the outer vane). Fortuitously, the pattern thus created resembles that found in grebes. Without detecting this anomaly, we could have identified it to this group anyway. Below: two unstained secondaries.

• CONFUSION CAUSED BY WEAR

Colour: naturally diminishes in intensity over time. Some colours disappear when the feather is damaged (see p. 48 'Colour criteria').

So, when identifying a feather, also look at the criteria for darker shades than those observed. Details on the margin or tip of the feather can also disappear with abrasion. Thus, the descriptions often indicate the presence of a pale edge on the outer vane or a tip of a different colour, but on worn feathers these features may have disappeared.

After shedding, feathers are sometimes tinted by natural or artificial elements with which they come into contact (algae, decomposing material, rusty metal, etc.).

It is therefore necessary to be careful and check the distribution and homogeneity of the colours. Sometimes, washing the feather with soap allows the removal of some of the staining and helps to detect accidental colouring.

Shape: first, it is necessary to note any wear at the feather's tip, particularly in primaries, and sometimes also in the rectrices.

Such wear can alter the original outline of the feather and remove certain features. Damaged feathers, however, sometimes show a suspicious curvature, not necessarily natural, either horizontally or vertically. Viewed from above, be careful of a feather's curvature if you observe V-slits on one of the vanes and superimposed barbs at the same level on the other vane: the feather has been forced into its curvature. It is possible to try and straighten the feather by pulling strongly but delicately in the opposite direction.

Viewed in profile, the curvature of a feather may be an accident if irregularities are observed in its curvature and if there is a shiny bar on the shaft where it bends.

The spine is probably broken, so just push the feather upwards to check if it bends at the suspected place.

Artificial curvatures are frequently observed on feathers from plucked prey, where the predator has pulled heavily on the feather to extract it. Sometimes it is even the bird itself that has damaged the feather by pulling to get rid of it.

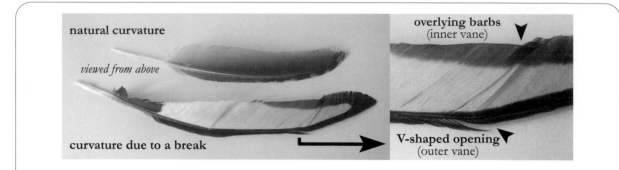

▷ **Unnatural curvature viewed from above.**
Above: Primary of a pheasant, naturally curved. Below and Right: Curvature due to a break is detected by a V-shaped opening of a vane (often the outer vane) and overlying barbs in the same position but on the other vane.

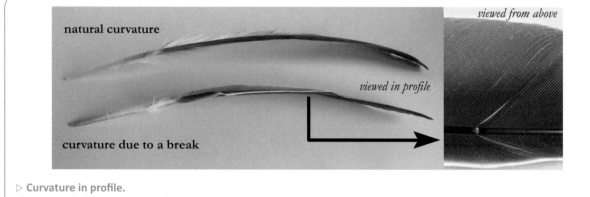

▷ **Curvature in profile.**
Above: a grebe's primary, naturally curved. Below and Right : Curvature due to a break is detected by the curvature being non-regular with bent parts visible on the shaft.

Colour and shape are therefore subject to wear and contact with the environment, which can give an unusual look to a feather. Here is a final example:

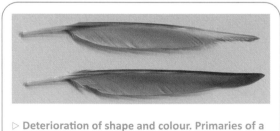

▷ **Deterioration of shape and colour. Primaries of a duck (tegmen present).**
Top: a feather with a very worn tip (pointed shape). It is also faded and has then been coloured (at the same time?) by environmental factors after moulting.
Bottom: this feather is almost intact (*Aythya* sp.).

COLOUR CRITERIA

• FEATHER COLORATION

Feather colour is due to three main phenomena, which vary in their resistance to time and wear.

The coloration of a feather is mainly due to incorporation of pigments within the keratins during growth. These pigments are of different type and density: melanins (black, brown to varying degrees), and carotenoids and associated pigments (red, pink, yellow, etc.). These pigments fade more or less quickly over time depending on a bird's lifestyle (especially as a result of the action of the sun) so that a feather that is black when just grown may appear brown after a few months.

Another type of coloration is produced in dark pigments by the super-imposition of a particular keratin structure. Reflected light will appear as different shades depending on the angle of view and produce sheen and iridescence (blue, green, purple, etc.). These colours can be altered by adjusting the angle of a feather and be slightly improved with careful preening of the barbs.

Finally, some birds can add pigments produced by their uropygial gland (preen gland), or taken from the environment, to the exterior of their feathers. These pigments may disappear with washing and after a while in the field. To keep the colour, birds preen their feathers regularly, and so these colours disappear quite quickly on moulted feathers.

• COLOUR DIVERSITY WITHIN A SPECIES

As a general rule, young birds (juveniles and early immature stages), more frequently have pale edges to their large feathers, coverts and body feathers. Adult colours are often less vivid in young birds or have darker markings in order to improve camouflage during the early stages of their lives. Therefore, we can find feathers very similar to those of adults, but which differ to a greater or lesser degree in coloration. For example, while age-related differences are very subtle on the tail feathers of owls, they are much more obvious on the larger feathers of gulls.

In many species there are also more or less marked differences in colouring between the sexes in all or part of the plumage. For example, in adult Pheasants, all the male's feathers are of different colour to those of the female. In many dabbling ducks, the primaries and secondaries are identical in both sexes but their rectrices, secondary coverts and body feathers are differently coloured.

Wherever possible, the descriptions and illustrations in the text indicate if they concern immatures, or adult males or females.

• THE POSITION OF COLOURS ON THE FEATHER

Feather colouring is not due to chance but to a combination of environmental factors which favour or inhibit certain colours or patterns in the species concerned. For effective camouflage, colours close to those of the species' habitat will be selected for the visible parts of a bird at rest. Conversely, particularly in open environments, gregarious species reveal contrasting colours in flight to maintain visual contact between individuals during flight. Visual disruption of a bird's silhouette confuses predators, so many species also show significant contrasts between different parts of the body, which sometimes makes it difficult to assign isolated feathers to the same individual. It is also interesting to note that certain pigments increase feather strength and are concentrated in areas of the feather that receive the most stress (for example, black is common at the end of large feathers). It should also be noted that females or males sometimes have preferences for certain attributes of the other sex that lead to similarities between disparate groups.

We can therefore observe colour similarities in species that are not closely related. Similar environmental constraints have led to convergences that can make identification from colour relatively difficult (for examples of convergence in feather shape, see pages 43 and 81).

For all of these reasons, we must stress the importance of cross-referencing colour information with other criteria such as shape, structure or size in order to best target the suspected species or group of species.

• ATYPICAL COLORATION

Due to a total or partial pigmentation defect, some individuals show coloration atypical of the species. More often than not this results in paler or even white areas in the plumage. Sometimes it is the brown colours that become more reddish or pinkish. Unusual patterns may also be noted, for example in the number or extent of spots or bars.

Such feathers are relatively rare, but when found, colouring and pattern can obviously be confusing. Comparison with a reference collection and verification of structural and silhouette criteria are useful in such cases.

▷ **A few colour aberrations (from top to bottom).**
Blackbird: some individuals are partially lacking in pigment, either with all-white feathers or with areas less rich in melanins. Here there is a pale, diffuse band across a secondary, which is normally uniform.

Carrion Crow: many birds appear to be 'discoloured', with a more or less well marked and extensive black pigmentation defect. Here it is also accompanied by an irregularity in the growth of the primary, which has resulted in bands of different thickness.

Jay: generally the tail feathers of young birds have slightly fewer bars than those of adults, but this one (still growing) has grey barring along almost its entire length.

Magpie: this secondary has an unusual white edge on the outer vane. There is also a small defect in the shape at the end of the outer vane (the inner vane is damaged).

LIST OF THE COLOUR CRITERIA

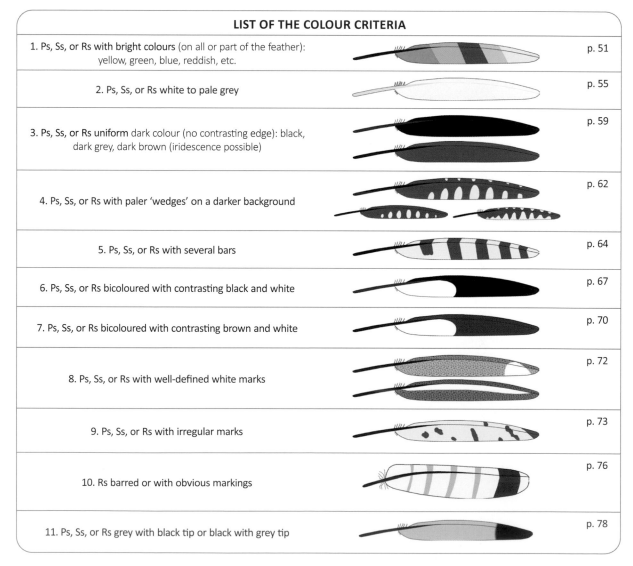

1. Ps, Ss, or Rs with **bright colours** (on all or part of the feather): yellow, green, blue, reddish, etc.	p. 51
2. Ps, Ss, or Rs **white to pale grey**	p. 55
3. Ps, Ss, or Rs **uniform** dark colour (no contrasting edge): black, dark grey, dark brown (iridescence possible)	p. 59
4. Ps, Ss, or Rs with paler 'wedges' on a darker background	p. 62
5. Ps, Ss, or Rs with several bars	p. 64
6. Ps, Ss, or Rs bicoloured with contrasting black and white	p. 67
7. Ps, Ss, or Rs bicoloured with contrasting brown and white	p. 70
8. Ps, Ss, or Rs with well-defined white marks	p. 72
9. Ps, Ss, or Rs with irregular marks	p. 73
10. Rs barred or with obvious markings	p. 76
11. Ps, Ss, or Rs grey with black tip or black with grey tip	p. 78

List of colour criteria

The table below offers a series of characters based on the general colour of the feather or obvious patterns. The diagrams are valid for any feather (primaries, secondaries, and tail feathers); the details are given in each paragraph. These list species by general colour category, after which it is necessary to refer to the individual descriptions to confirm or refute any identification.

Important: for small feathers (TL < 10 cm) and those without specific shape criteria, also consult the keys to the passerine 'large' feathers (pp. 119 and subsequent pages).

Species groups correspond to the different categories described in Chapter 8: passerines, near-passerines, coastal and wetland waterfowl, other groups.

1. Ps, Ss and Rs with a bright colour (on all or part of a feather): yellow, green, blue, reddish, etc.

Few European species have bright colours on their 'large' feathers. Their presence is therefore a useful feature.

On the other hand, do not discount the possibility of escaped exotic or feral species: parakeets and parrots in particular, which often have mutant colour forms.

The following table shows species with striking colours (more or less showy and bright). Xs indicate what type of feather may be concerned (at least one feather per bird).

Sizes are then specified, in orange if the type of feather is not of this colour type.

Finally, eventual details are given, with the visible colour (the feather is not necessarily entirely of this colour). (f = female, m = male, y = young)

Species	P	S	R	Pmax	Pmin	Smax	Smin	Rmax	Rmin	Colour present
Khaki green										
Golden Oriole		X		13.5	8.3	9	6.7 (6)	10	8.2	S khaki (f)
Grey-headed Woodpecker	X	X	X	13	9	14	8	12	7	Ps, Ss & Rs khaki-green
Green Woodpecker	X	X	X	15	10	12	8	12	8	Ps, Ss & Rs khaki-green
Metallic and/or dark green										
Mandarin Duck	X	X		21	11	13	11	14	10	Ps & Ss green-blue
Northern Pintail		X		22 (25)	11	12	9	14 f/22 m	8	Ss brown-green
Eurasian Wigeon		X		22	11	13	9	12 (16)	9	Ss green
Northern Shoveler		X		19 (22)	10	11	7	13	8	Ss green
Black Stork	X	X	X	50 (54)	30	33	26	26 (32)	20	Ps, Ss & Rs green sheen
Domestic Chicken (m)	X	X	X	> 19	< 13	> 19	< 12	> 20	< 12	Ps, Ss & Rs possible dark green sheen
Common Starling	X	X		11,3	7.5	7.8	6.4	7.3	6.3	Ps & Ss green (blue) sheen
Spotless Starling	X	X		11	7.5	7.6	6.4	7.2	6.7	Ps & Ss green (blue) sheen
Common Goldeneye		X		22	10	11	8	13	5	Si dark green
European Bee-eater	X	X	X	13	7.5	8	6	14	9	Ps, Ss & Rs green
Glossy Ibis	X	X	X	24	15	17	14	13	10	Ps, Ss & Rs green to purple sheen
Egyptian Goose		X		45	24	30	27	16	10	Ss green or purple
Common Magpie		X	X	18 (19)	11	15	11	27	11	Ss & Rs green sheen
Eurasian Teal		X		16 (18)	8	9	6	9 (10)	6	Ss green

Species	P	S	R	Pmax	Pmin	Smax	Smin	Rmax	Rmin	Colour present
Common Shelduck		X		26	12	16	10	14	10 (8.5)	Ss greenish-bronze
Metallic and/or dark blue										
Wood Duck	X	X		n.d.	n.d.	n.d.	n.d.	n.d.	n.d.	Ps & Ss blue
Mallard		X		22	13	15	11	12	8	Ss metallic blue
Domestic duck		X		> 25	12	> 20	10	20	8	Ss blue
Lady Amherst's Pheasant		X		23	17	18	16	f 60/m 105	11	Si (m) bluish-black
Indian Peafowl		X		41	32	37	28	56	34	Ss blue iridescence
Common Magpie	X	X	X	18 (19)	11	15	11	27	11	Ps, Ss & Rs blue iridescence
Purple Swamphen	X	X	X	20 (25)	16	16.5 (20)	14.5	9.5	5.5	Ps, Ss & Rs blue iridescence
Passerines with blue colours	X	X	X	< 10		< 10		< 10		Ps, Ss & Rs pale blue or dark blue iridescence: see passerines
Pale blue or green-blue										
Eurasian Jay	X	X		18 (20)	12 (11)	16 (17)	9	17 (18)	12	Ps & Ss blue barred with black
European Bee-eater	X	X	X	13	7.5	8	6	14	9	Ps, Ss & Rs blue-green
Common Kingfisher	X	X	X	7.5	5	6,5	4	4.5 (6)	3.5	Ps, Ss & Rs blue
Rose-ringed Parakeet	X	X	X	18	10	10	5.5	23	8	Ps, Ss & Rs blue
Budgerigar	X	X	X	9	4	5	3.5	11	4	Ps, Ss & Rs blue
Azure-winged Magpie	X	X	X	16	8	14	8	25	9	Ps, Ss & Rs pale blue
European Roller	X	X	X	19	11	12	9	17	12	Ps, Ss & Rs pale blue
Pale green to green-yellow										
Rose-ringed Parakeet	X	X	X	18	10	10	5,5	23	8	Ps, Ss & Rs yellow-green
Budgerigar	X	X	X	9	4	5	3,5	11	4	Ps, Ss & Rs yellow-green
Yellow to golden										
Grey Wagtail			X	7.7 (8.2)	5.0	5.2	4.7	11.7	9.5	Rs yellow
Finches with yellow	X	X	X	< 10		< 10		< 10		Ps, Ss & Rs yellow: see passerines
Pin-tailed Sandgrouse		X		17	8,5	8,5	7.5	18	6.5	Si yellow
Bohemian Waxwing	X		X	10.8	6,7	6.7	5.5	7.4	6.2	Ps & Rs yellow at tip
Golden Oriole			X	13.5	8.3	9	6.7 (6)	10	8.2	Rs yellow
Little Bustard		X		18	12.5	13.5	11.5	12	10	Si golden
Rose-ringed Parakeet	X	X	X	18	10	10	5.5	23	8	Ps, Ss & Rs yellow
Budgerigar	X	X	X	9	4	5	3.5	11	4	Ps, Ss & Rs yellow
Red, rufous or reddish-brown										
Diurnal raptors / Osprey		X	X	75	12	49	9	56	10	Ss & Rs reddish
Larks	X	X	X	11.5	5	8	5	8	5	Ps, Ss & Rs reddish
Little Bittern		X		12	9	9	8	6	5	Si reddish
Eurasian Bittern	X	X	X	30	20	22	17	12	8	Ps, Ss & Rs reddish
Wood Duck		X		n.d.	n.d.	n.d.	n.d.	n.d.	n.d.	Si reddish
Woodcock and snipes	X	X	X	15.5	6	12	6	10.5	5	Ps, Ss & Rs with reddish wedges or markings

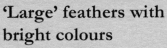

Species	P	S	R	Pmax	Pmin	Smax	Smin	Rmax	Rmin	Colour present
Mandarin Duck		X		21	11	13	11	14	10	Si rufous
Great Spotted Cuckoo	X			20 (22)	10	11 (13)	7	23	10	Ps rufous (y)
Common Cuckoo	X	X	X	19 (22)	10	12	8	19	10	Ps, Ss & Rs rufous
Common Pheasant			X	23	15	19	15	60	13	Rs rufous to pink
Lady Amherst's Pheasant			X	23	17	18	16	f 60/m 105	11	Rs (f) with golden-red
Golden Pheasant			X	21	14	18	14	70	12	Rs golden-red
Reeves's Pheasant		X	X	27	17	18	14	140	13	Ss & Rs golden-red
Falcons	X	X	X	37	9	23	9	30	12	Ps, Ss & Rs rufous
Eurasian Jay		X		18 (20)	12 (11)	16 (17)	9	17 (18)	12	Si dark red
Song Thrush	X	X		10.2 (11.5)	7.5	7.9 (8.5)	6.5	9.8 (10.3)	8	Ps & Ss pale rufous
European Bee-eater		X		13	7.5	8	6	14	9	Ss orange
Eurasian Hoopoe		X		14 (17)	9.5	13	9	12.5	9	Si rufous
Glossy Ibis	X	X	X	24	15	17	14	13	10	PSR with green to purple iridescence
Bohemian Waxwing		X		10.8	6.7	6.7	5.5	7.4	6.2	SS with red wax at shaft tip
Baillon's Crake	X			7.5	5	5.5	4	?	?	PS rufous
Spotted Crake			X	9.5	7	7.5	6	6	4	RS rufous border
Little Crake	X	X	X	9	6	6.5	5	6	4.5	Ps, Ss & Rs rufous border
Common Kingfisher	X	X		7.5	5	6.5	4	4.5 (6)	3.5	PS & Ss rufous edge inner vane
Siberian Jay	X	X	X	14	9	12	8	15	12	Ps, Ss & Rs rufous
Common Rock Thrush			X	8.9	7 ?	9.5 ?	6.5 ?	8 ?	5 ?	Rs rufous
Great Bustard			X	61	33	43	21	29	21	Rs orangish
Indian Peafowl	X			41	32	37	28	56	34	Ps rufous
Rock Partridge			X		12	n.d.	n.d.	n.d.	11	Rs?
Chukar			X	n.d.	n.d.	n.d.	n.d.	n.d.	n.d.	Rs?
Grey Partridge			X	14 (15)	9	11	9	9 (10)	7	Rs?
Red-legged Partridge			X	13 (15)	9	11 (12)	9 (7)	11	7	Rs?
Small passerines with rufous colours	X	X	X	< 10		< 10		< 10		Ps, Ss & Rs rufous: see passerines
Water Rail			X	10.5	7	8.5	7	7	4	Rs + Si rufous border
Corncrake	X	X	X	12 (14)	7.5	9	6.5	7.5	5	Ps & Ss rufous + Rs rufous border
Owls	X	X	X	43	6	30	5	35	5	Ps, Ss & Rs pale reddish-yellow (velvety surface)
European Roller		X		19	11	12	9	17	12	Si rufous
Common Shelduck		X		26	12	16	10	14	10 (8.5)	Si rufous
Wallcreeper		X		9.6	8.4	8.2	5.6	6.2	5.2	Ss pink and/or orange
Turtle Dove		X		15	7.5	10	8	14	10	Si rufous tip
Pink (often pale) or pink iridescence										
Pine Grosbeak	X	X		12	8.4	8.4	6	10.6	9.6	Ps & Ss pale pink edge
Rose-coloured Starling			X	n.d.	n.d.	n.d.	n.d.	n.d.	n.d.	Re pale pink

Species	P	S	R	Pmax	Pmin	Smax	Smin	Rmax	Rmin	Colour present
Common Pheasant			X	23	15	19	15	60	13	Rs reddish to pink
Greater Flamingo		X	X	33	20	22	16	16	11	Si + Rs ad pale pink
Wallcreeper	X	X		9.6	8.4	8.2	5.6	6.2	5.2	Ps pink Ss pink and/or orange
Purple (often iridescence)										
Red-breasted Merganser		X		19 (21)	10	13 (14)	8	10 (12)	7	Si purple-blue
Egyptian Goose		X		45	24	30	27	16	10	Ss green or purple
Common Magpie			X	18 (19)	11	15	11	27	11	Rs purple iridescence
European Roller	X	X	X	19	11	12	9	17	12	Ps, Ss & Rs purple-blue

2. Ps, Ss and Rs white or pale grey

Few species have white or very pale grey 'large' feathers. It is necessary to think in particular of domestic varieties, whose albino or partly white forms can lose feathers in the wild, usually near villages or farms: Columbidae, Anatidae, peafowl, turkey, etc. These feathers are sometimes confusing, but three-dimensional and silhouette criteria usually quickly distinguish them from other groups.

All 'large' feathers white:

Very few species are concerned. They are detailed in the following table. In domestic species, birds may have just a few white feathers or be completely white.

SPECIES WITH WHITE FLIGHT AND TAIL FEATHERS							
Species	Pmax	Pmin	Smax	Smin	Rmax	Rmin	Remarks
Domestic species: pure or spotted white							
Domestic geese	40 (45)	< 23	24	< 20	19	< 13	Ps with tegmen. Ps, Ss & Rs: see wild birds for shape
Domestic ducks	> 25	12	> 20	10	20	8	Ps with tegmen. Ps, Ss & Rs: see wild birds for shape
Domestic chickens	> 19	< 13	> 19	< 12	> 20	< 12	Ps and Rs curved; Ps & Ss rigid
Indian Peafowl, white form	41	32	37	28	56	34	Size and shape often diagnostic; fine lines along top of rachis
Domestic turkey, white form	> 35	n.d.	n.d.	n.d.	> 40	n.d.	See size, shape and colour
Domestic doves and pigeons	22	7	15	8	19	10	See wild birds for shape of rachis thick at the base.
Wild species							
Swans (three species)	47	22	36	21	24	13	Tegmen present on underside of Ps & Rs quite curved
Squacco Heron	n.d.	n.d.	n.d.	n.d.	12	8	outer Ps notched quite high; Rs quite short and curved
Cattle Egret	22	14	16 (17)	12	11	8	outer Ps notched quite high; Rs quite short and curved

Species	Pmax	Pmin	Smax	Smin	Rmax	Rmin	Remarks
Little Egret	24	17	19	14	13,5	9	outer Ps notched quite high; Rs quite short and curved
Great Egret	35 (38)	25 (22)	28 (30)	21	20	16	outer Ps notched quite high; Rs quite short and curved
Eurasian Spoonbill	31	21	22	17	15	12	outer Ps well notched; Rs quite squat
Snowy Owl: (male)	35	23	24	20	34	23	Ps, Ss & Rs and body feathers have velvety surface

The primaries and secondaries are white or pale grey:
Few species have white or very pale grey flight feathers. Look especially at terns, some gulls, a few waders, etc.

For T < 13 cm and typical wader shape: see wader key (p. 255) and particularly Pied Avocet (Ps: T < 13 cm, Ss: T < 10·5 cm), Eurasian Oystercatcher (Ss: T < 12 cm) as well as small waders (Ss: T < 8 cm).

Species	Type concerned	Pmax	Pmin	Smax	Smin
Diurnal birds of prey: see group	Si			49	9
Northern Gannet	S			19	14
Common Goldeneye	S			11	8
Grebes: see group	(P) S	11 (14)	7,5	11	5.5
Rock Ptarmigan	PS	16 (20)	11	13 (15)	10
Willow Ptarmigan	PS	18	11	12	9
Gulls: see group	PS	50	8,5	24	7,5
Waders: see group	PS	13	10	12	4
Velvet Scoter	S			13	9
Little Bustard	S			13.5	11.5
Passerines: see group	PS	< 10	< 10	< 10	< 10
White Pelican	S			38	30
Dalmatian Pelican	S			38	30
Terns: see group	PS	32	5,5	15	4,5

The rectrices are white or pale grey:
White rectrices (Rs) are more common. The tail is sometimes entirely white, but often it is just the outer feathers that are white, in which case, they sometimes have dark markings, depending on the individual. White tails often serve a signal function for maintaining contact between individuals in a group, or as a warning signal when they are spread out, for example when taking flight.

Species	Rmax	Rmin	The potentially affected Rs	Special cases and clues
Anatidae				
Greylag Goose	19	13	outer	Squat, quite curved
Common Shelduck	14	10 (8.5)	outer	Squat, quite curved

1

2

3

4

5

6

7

8

9

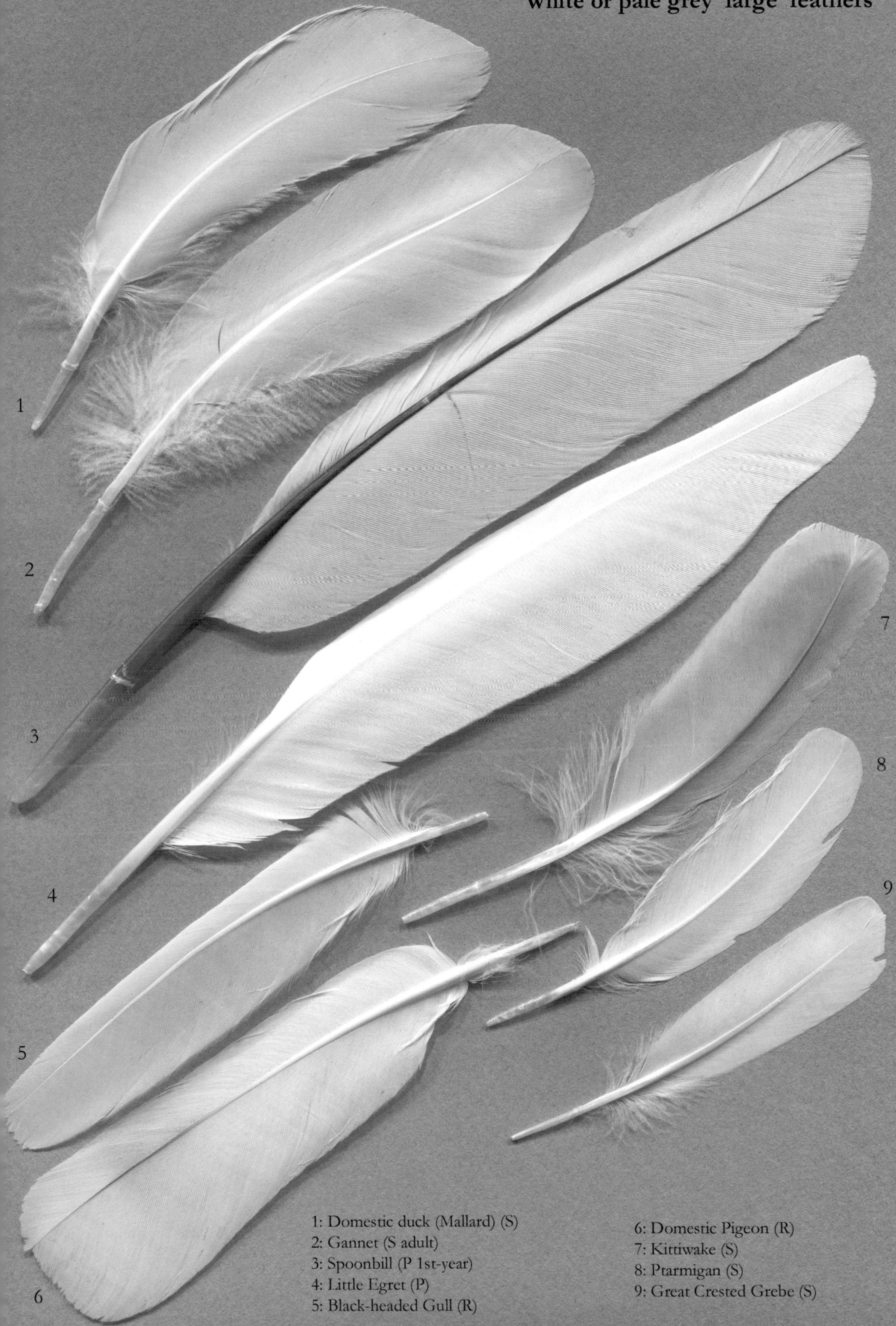

1: Domestic duck (Mallard) (S)
2: Gannet (S adult)
3: Spoonbill (P 1st-year)
4: Little Egret (P)
5: Black-headed Gull (R)

6: Domestic Pigeon (R)
7: Kittiwake (S)
8: Ptarmigan (S)
9: Great Crested Grebe (S)

Life size

Species	Rmax	Rmin	The potentially affected Rs	Special cases and clues
Mallard	12	8	outer (male)	squat, quite curved
Long-tailed Duck	14 f / 28 m	8	outer	squat, quite curved
Other aquatic species				
Northern Fulmar	15	10	all	rectangular with rounded end
Northern Gannet	29	12 (11)	all	pearly, straight, elongated spearhead shape, calamus not upturned
White Pelican	22	15	all	pearly, short, quite wide, outers curved
Dalmatian Pelican	28	20	all	pearly, short, quite wide, outers curved
White Stork	27	18	all	matt white, rectangular and quite short
Sacred Ibis	n.d.	n.d.	all	quite short and curved
Greater Flamingo	16	11	all (young)	short, outers curved, pale pink in adult
Gulls: see group	23	8	all	rectangular, quite supple; those of immatures have a terminal bar or dark markings
Terns: see group	20	4	all	shape ranges from oval to very long and narrow, often with silvery or dark grey hue
Waders: see group	14	4	all	often a rectangular shape
Various species				
Willow Ptarmigan	14	11	central	rectangular; place of discovery is significant
Rock Ptarmigan	14 (15)	11	central	rectangular; place of discovery is significant
Black-winged Kite	15	10	all	often pale grey
Egyptian Vulture	35 (37)	19	all	quite rigid
White-tailed Eagle	40	24	all	dark markings at base
Barn Owl	14	11	outer	short, velvety surface
Wagtails: see family	12	7	outer	elongated, fragile appearance
Shrikes: see family	12.5	8	rarely all, often outer	elongated, fragile appearance
Other passerines	< 10	< 10	all	small, fragile appearance

White-tailed passerines: due to their small size and flexibility, the rectrices of white-tailed passerines can generally be distinguished from those of non-passerines. The table on pp. 123-125 gives a list of passerines that can have white in their tails. White rarely covers the whole rectrix but usually occurs on the outer vane.

3. Ps, Ss and Rs uniform dark colour (no contrasting edge): black, dark grey, dark brown (iridescence possible)

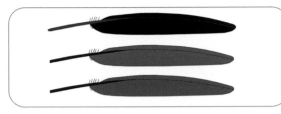

No markings or contrasting border. The edge and base of the inner vane may be paler, but without sharp contrast. Some birds have all their 'large' feathers of a similar colour (Common Raven, Black Woodpecker, cormorants, etc.), but in most only some of their dark feathers are uniform.

For dark 'large' feathers (as for white ones), it is important to consider shape criteria in order to arrive at a smaller choice of species. For example, we can distinguish between waders, auks, divers, pigeons, swifts or hirundines from the shape of some of their feathers. Black rectrices may also have a characteristic silhouette: scoters, woodpeckers, auks, Common Magpie, etc.

The following table summarises the main groups and species that have uniformly dark individual 'large' feathers. Group descriptions can be referred to for more detail.

Species	P	Pmax	Pmin	S	Smax	Smin	R	Rmax	Rmin
UNIFORMLY DARK 'LARGE' FEATHERS (BLACK, BROWN OR GREY)									
Passerines									
Wagtails							bl	12	7
Spotted Nutcracker	n	17 (18)	11	bl	14	11	bl	14	11
Black corvids	n	47	13	bl	28	11	bl	30	9
Eurasian Jay				i: bl	16 (17)	9	(bl)	17 (18)	12
Mistle Thrush	o: br	13.4 (14)	9.6 (8.5)	i: br	9.5 (10)	7.6			
Fieldfare	(br)	12.4 (14)	9.2 (8.5)	(br)	9.3 (10.5)	7.5	bl	12.3 (14)	10.5 (10)
Redwing				i: b	7.7	6.3	br	9.1 (11)	7.9 (7)
Hirundines	br bl	12	4.5	br bl	6	3	br bl	13	4
Bohemian Waxwing	o	10.8	6·7						
Ring Ouzel	br bl	12.8	9.1				br bl	12.5	8
Blackbird	br bl	11.4 (13)	8.7	br bl	9.6	7.5	br bl	11.8 (13)	10.2
Blue Rock Thrush	x	10.1 (11)	8	x	8	6.5	x	8.7 (9.5)	8.2 (7.5)
Rock Thrush	(x)	8.9	7?	(x)	9.5?	6.5?			
Common Magpie				bl	15	11	n	27	11
Shrikes							bl or br	12·5	6·5
Other passerines: see group	x	< 10	< 10	x	< 10	< 10	x	< 10	< 10
Near passerines									
Great Spotted Cuckoo	x	20 (22)	10	x	11 (13)	7	x	23	10
Pin-tailed Sandgrouse	g?	17	8.5						
Black-bellied Sandgrouse	b	16.5	10						
Alpine Swift	x	22	5	x	6.5	4	x	10	5
Common Swift	x	15	4.5	x	5	4	x	9.5	4.5
Pallid Swift	x	15	4.5	x	5.5	4	x	9	5
Ring-necked Parakeet	o: g bl	18	10						
Budgerigar	o: g bl	9	4						
Black Woodpecker	n	21 (23)	13.5	n	17 (21)	12	n	21	11.5

Key: For feather types: x = the concerned feather; bl = black; br = brown; c = central; o = outer; g = grey; i = internal (inner); m = median. For feather size: f = female; m = male.

Species	P	Pmax	Pmin	S	Smax	Smin	R	Rmax	Rmin
Black-and-white woodpeckers							n	12	4
Rock Dove	g	20	12						
Feral Pigeon	br g bl	20.5	11	br g bl	13	8	br g bl	14	12
Stock Dove	g	19.5	11						
Wood Pigeon	g	21.5	13.5	g	14.5	9.5			
European Roller	o	19	11						
Turtle Dove	g br	15	7.5	g br	10	8	c: g br	14	10
Laughing Dove	g br	15?	7.5?	g br	10?	8?	c: g br	14?	10?
Collared Dove	g br	15	9	g br	10	9	c: g br	16	12
Some aquatic birds of shore and wetlands									
Auks	x	16	5·5	x	10	4·5	x	10	4
Brent Goose	bl	32 (44)	14	bl	18 (24)	12	bl	18	9
Canada Goose	br bl	48	23	br bl	29	20	br bl	20	16
Barnacle Goose	g br	39	18	g br	22	18	bl	18	11
Black-crowned Night Heron	g	26	19	g	20	14	g	13	10
Little Bittern	br	12	9	br	9	8	bl	6	5
Dabbling ducks	x	25	8	x	20	7	x	16 (22)	6
Diving ducks	x	23	8	x	15	7	x	14 (28)	5
White Stork	x	49	28	x	31	26			
Black Stork	x	50 (54)	30	x	33	26	x	26 (32)	20
European Shag	br bl	21	17	br bl	18	13	br bl	17	12
Pygmy Cormorant	br bl	16	12	br bl	14	11	br bl	15?	9?
Greater Flamingo	bl	33	20	bl	22	16			
Northern Gannet	br bl	37	18	br bl	19	14	br bl	29	12 (11)
Red-knobbed Coot	g bl	20	14?	g bl	16?	12?	g bl	10?	6?
Eurasian Coot	g	16.5	10.5	g	12.5	9	n	8	5
Northern Fulmar	g	26	11.5	g	12.5	9			
Common Moorhen	br bl	15	10	br bl	11	8.5	br bl	9	6
Great Cormorant	br bl	27 (31)	19	br bl	20	16	br bl	20 (23)	14
Little Grebe	br	8 (8.5)	6						
Great Crested Grebe	br (imm)	14 (18)	10						
Red-necked Grebe	br	14 (18)	9						
Common Crane	g bl	45	26	g bl	32	26	g	24	21
Marsh terns	x	21	8	x	10	8	x	11	7
Grey Heron	bl	37	25	g	28	20	g	20	15
Purple Heron	br bl	34	24	br bl	26	19	g bl	15 (20)	12
Glossy Ibis	br bl	24	15	br bl	17	14	br bl	13	10
Skuas							br	33	12
Gulls	x	50	8.5	x	24	7.5	x	23	8
Waders	x	24	4.5	x	13.5	4	x	14	4

Key: For feather types: x = the concerned feather; bl = black; br = brown; c = central; o = outer; g = grey; i = internal (inner); m = median. For feather size: f = female; m = male.

1

2

3

4

5

6

7

8

10

9

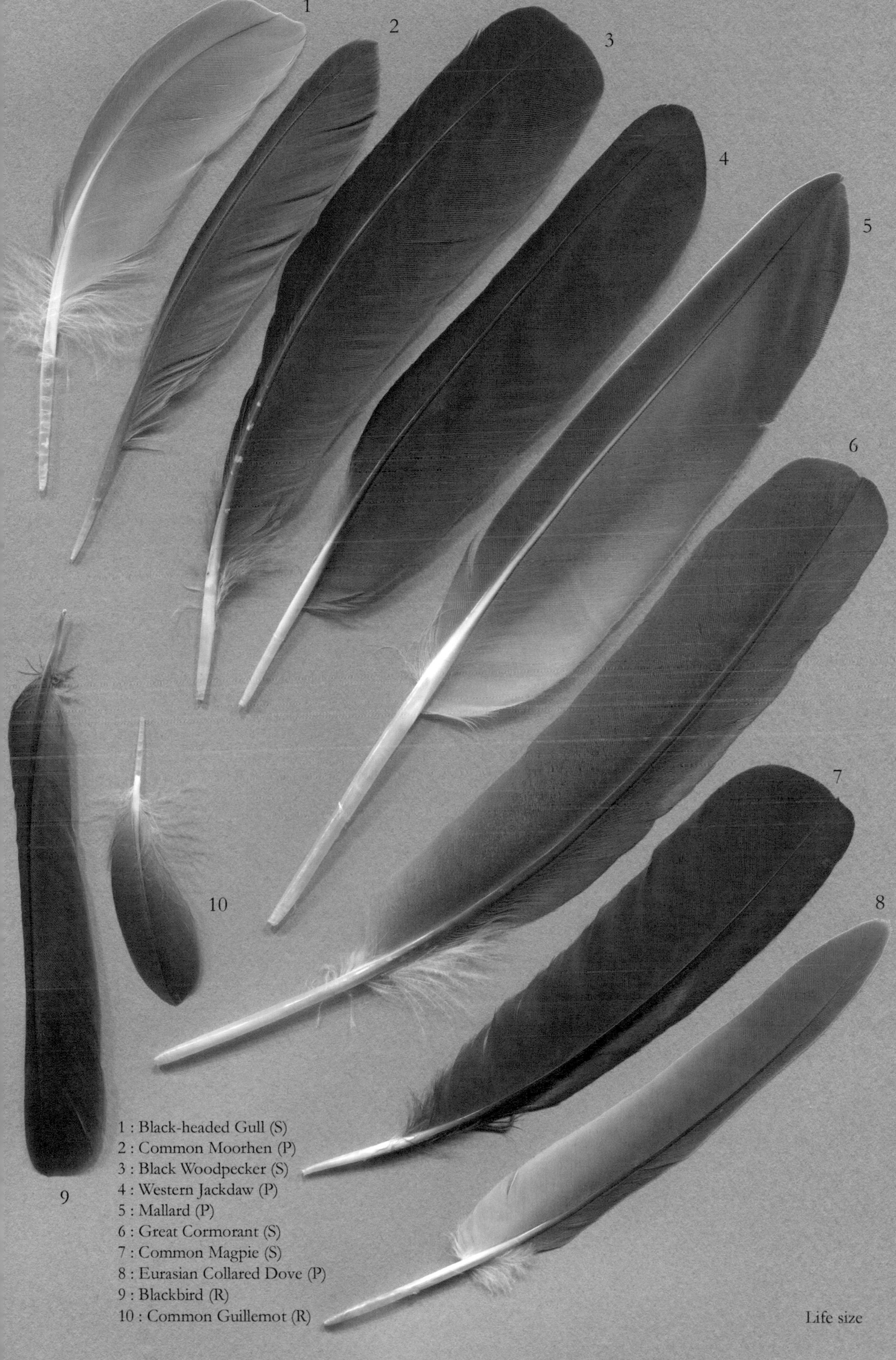

1 : Black-headed Gull (S)

2 : Common Moorhen (P)

3 : Black Woodpecker (S)

4 : Western Jackdaw (P)

5 : Mallard (P)

6 : Great Cormorant (S)

7 : Common Magpie (S)

8 : Eurasian Collared Dove (P)

9 : Blackbird (R)

10 : Common Guillemot (R)

Life size

Species	P	Pmax	Pmin	S	Smax	Smin	R	Rmax	Rmin
Baillon's Crake	br	7.5	5	br	5.5	4			
Spotted Crake	br	9.5	7	br	7.5	6	br	6	4
Little Crake	br	9	6	br	6.5	5			
European Storm Petrel	x	10.5	5	x	4	4	x	6.5	5.5
Grey geese, *Anser* spp.	g br	40 (45)	20	g br	24	15			
Egyptian Goose	bl	45	24	bl	30	27	bl	16	10
White Pelican	g	50 (60)	33	g	38	30	g	22	15
Dalmatian Pelican	g	50	32	g	38	30	g	28	20
Divers	bl or br bl	30	11	bl or br bl	16 (21)	8	br bl or br	9	6
Shearwaters	x	28	9	x	13	7	x	15	8
Water Rail	br	10.5	7	br	8.5	7	br	7	4
Terns	o: g, bl	32	5.5						
Purple Swamphen	bl	20 (25)	16	bl	16.5 (20)	14.5	bl	9.5	5.5
Other groups									
Diurnal birds of prey	x	75	12	x	49	9	x	56	10
Common Quail	o	8.5	5.5						
Domestic chicken (m)	x	> 19	< 13	x	> 19	< 12	x	> 20	< 12
Falcons	x	37	9	x	23	9	x	30	12
Hazel Grouse	m: br	14	9						
Capercaillie	m	f 27 / m 33 (37)	f 16 / m 21	m	f 21 / m 24	f 15 / m 19	m	f 24 / m 36	f 18 / m 21
Rock Ptarmigan							x	14 (15)	11
Red Grouse	x	18 (19)	11	x	12	10	x	13	11
Willow Ptarmigan							x	14	11
Indian Peafowl				x	37	28			
Red-legged Partridge	br	15	9	b	12	7	b	12	7
Black Grouse	x	20 (25)	12				m	f 15 / m 23	11

Key: For feather types: x = the concerned feather; bl = black; br = brown; c = central; o = outer; g = grey; i = inner; m = median. For feather size: f = female; m = male.

4. Ps, Ss and Rs with pale 'wedges' on darker background

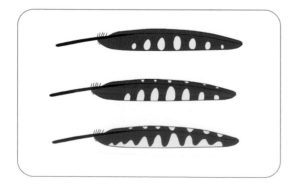

'Wedge' or notch is the term used here to describe a triangular or rounded spot that touches the vane's outer edge. These wedges are sometimes separated from the edge and thus become oval spots. Conversely, they sometimes join together on the vane's edge and thus form a sort of comb-shaped pattern. Sometimes these species are also found with barred feathers. Indeed, when the wedges of the two vanes meet, they form a separation between two spots, and the wedges progress to being bars.

1 : Common Cuckoo (P of young)
2 : Common Kestrel (P of female)
3 : Eurasian Curlew (P)
4 : Common Pheasant (P of male)

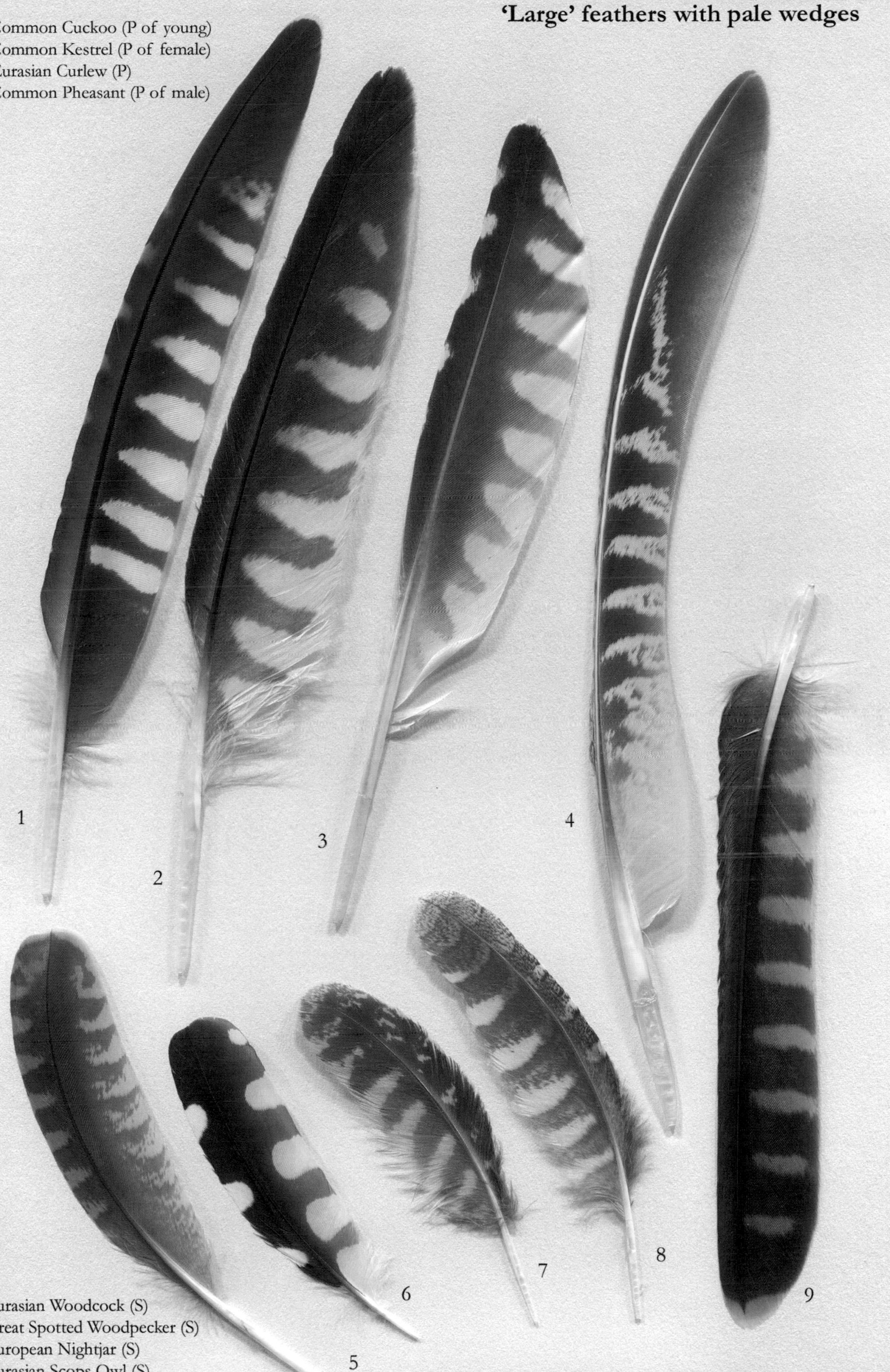

1

2

3

4

5 Eurasian Woodcock (S)
6 Great Spotted Woodpecker (S)
7 European Nightjar (S)
8 Eurasian Scops Owl (S)
9 Eurasian Hobby (R of immature)

6

7

8

9

5

Life size

SPECIES WITH 'LARGE' FEATHERS WITH PALE WEDGES ON A DARKER BACKGROUND									
Species	P	Pmax	Pmin	S	Smax	Smin	R	Rmax	Rmin
Diurnal birds of prey	x	75	12	x	49	9	x	56	10
Eurasian Woodcock	x	15.5	10.5	x	12	10	x	10.5	7.5
Common Quail	x	8.5	5.5						
Northern Pintail							f	14	8
Spotted Redshank	x	13.5	7	x	8	6.5			
Common Cuckoo	x	19 (22)	10	x	12	8	x	19	10
Eurasian Curlew	x	24	13	x	13.5	11			
Whimbrel	x	19.5	8.5	x	10.5	7			
Red-necked Nightjar	x	18 (21)	8	x	12	8	x	18 (20)	15
European Nightjar	x	18	9	x	11	8	x	16	12
Common Pheasant	x	23	15						
Reeves's Pheasant	x	27	17						
Falcons	x	37	9	x	23	9	x	30	12
Pin-tailed Sandgrouse							x	18	6,5
Hazel Grouse	x	14	9						
Eurasian Hoopoe				x	13	9			
Waders	x	24	4.5	x	13.5	4	x	14	4
Grey-headed Woodpecker	x	13	9	x	14	8			
Great Spotted Woodpecker	x	13 (15)	8	x	10	7.5			
Lesser Spotted Woodpecker	x	9.5	6	x	7	5			
Middle Spotted Woodpecker	x	11.5	8.5	x	9	6.5			
Syrian Woodpecker	x	11.5	9	x	10	8			
Three-toed Woodpecker	x	12	7	x	8.5	7			
Green Woodpecker	x	15	10	x	12	8	x	12	8
Owls	x	43	6	x	30	5	x	35	5
Eurasian Wryneck	x	8.5	5.5	x	7.5	5			

5. Ps, Ss and Rs with several bars

At least two bars are visible on one or both vanes.

In most cases, the bars are regular in width and spacing. Sometimes the thickness of the bars or the intervals between them can be irregular. The last bar is sometimes much wider than the others, and is then called the terminal bar, or subterminal bar if it is not quite at the tip of the feather. The bars are more or less contrasting: brown or black on a white or buff background, rufous on a brown background, black on a rufous background, etc. Generally, the background pales towards the base of the inner vane (Common Buzzard, for example), but it can remain uniform (Eurasian Wren, etc.).

The background or the bars themselves are not necessarily of a uniform tint; contrasting areas can be marked in different ways (striped, speckled, etc.). This phenomenon is common in species that need good camouflage when they are at rest (owls, nightjars, etc.).

Finally, to clarify the wording; 'brown bars on a rufous background' generally indicates that the thinnest bands are brown, and the background colour is rufous, but the opposite is referred to as 'rufous bars on a brown background' if brown is the predominant colour. The proportions between dark and light may vary depending on the position of the

1 : Tawny Owl (P)
2 : Eurasian Sparrowhawk (S of female)
3 : Eurasian Hoopoe (S)
4 : Eurasian Wryneck (P)
5 : Common Kestrel (R of young male)
6 : Common Pheasant (R of rufous female)

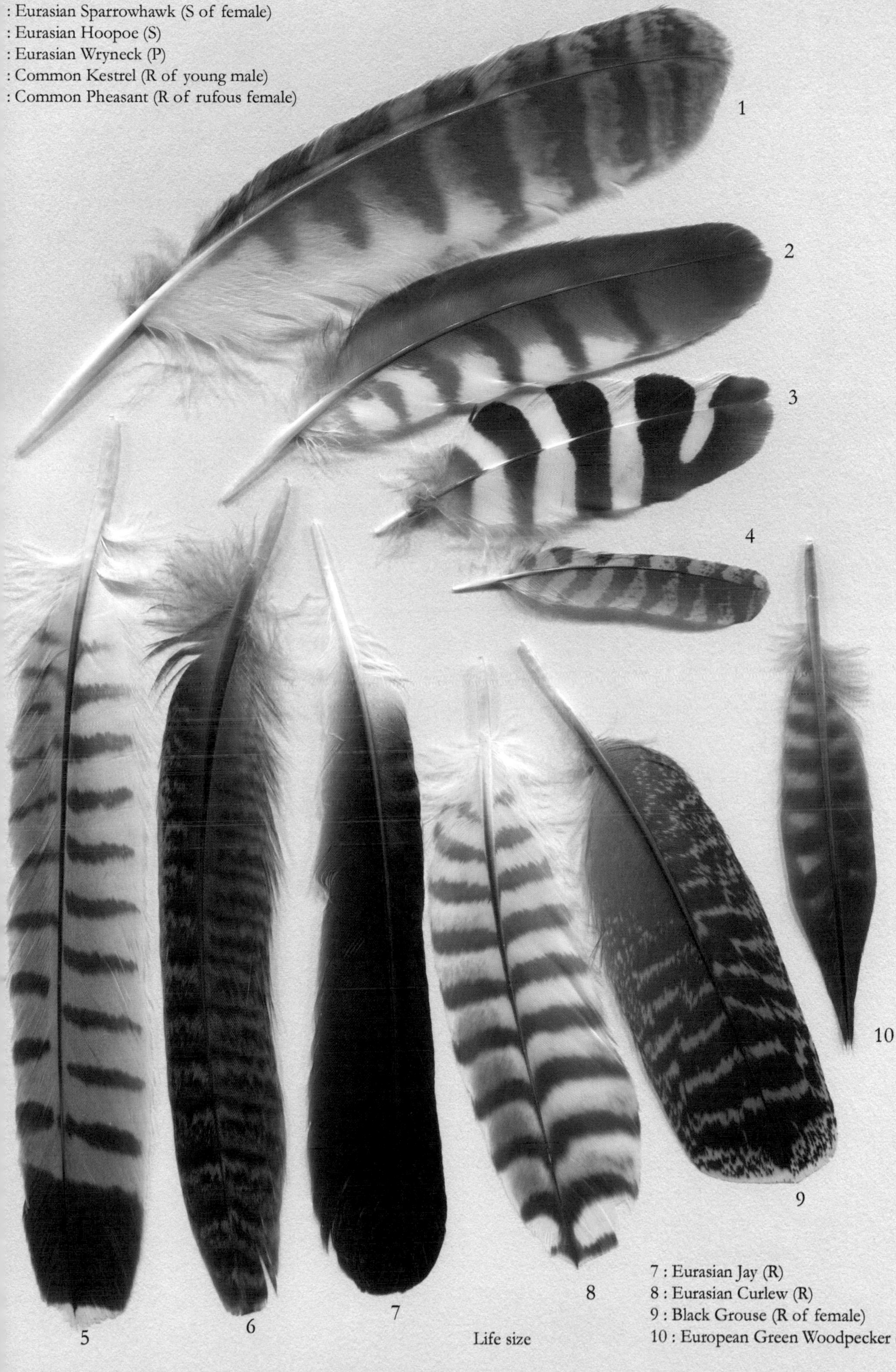

1

2

3

4

5

6

7

8

9

10

Life size

7 : Eurasian Jay (R)
8 : Eurasian Curlew (R)
9 : Black Grouse (R of female)
10 : European Green Woodpecker (R)

feather (between the outside and inside of the tail, for example), but also between individual feathers. It is sometimes a discriminating criterion between two similar looking species. Here are some additional criteria to combine with that of the bars to find a species group:

If the upper surface is velvety (covered in small 'hairs'), prioritise owls p. 362 . If the criterion is not obvious and is visible mainly at the base of the outer web, see also diurnal raptors (Accipitriformes) p. 346 or nightjars p. 218.

If a primary clearly shows wedges near the base (generally much clearer on the inner vane); look at diurnal raptors first p. 346 . Likewise, if the rectrices are narrow and elongated look at falcons, or on the contrary wide and quite solid look at gliding diurnal raptors.

If the primaries or secondaries are rigid and concave, look at the Galliformes (p. 324) or bustards (p. 338), for example.

If the rectrices are barred, the observer should also consult the section concerning barred or obviously marked rectrices (page 76).

Species	P	Pmax	Pmin	S	Smax	Smin	R	Rmax	Rmin
Diurnal birds of prey	x	75	12	x	49	9	x	56	10
Eurasian Bittern	x	30	20				x	12	8
Common Quail	x	8.5	5.5	x	6.5	5	x	4.5	3
Mallard							(f)	12	8
Northern Pintail							(f)	14	8
Domestic chicken	x	> 19	< 13	x	> 19	< 12	x	> 20	< 12
Common Cuckoo				x	12	8	x	19	10
Red-necked Nightjar	x	18 (21)	8	x	12	8	x	18 (20)	15
European Nightjar	x	18	9	x	11	8	x	16	12
Common Pheasant	x	23	15	x	19	15	x	60	13
Lady Amherst's Pheasant	x	23	17	x	18	16	x	f 60 / m 105	11
Golden Pheasant	x	21	14	x	18	14	x	70	12
Reeves's Pheasant	x	27	17	x	18	14	x	140	13
Falcons				x	23	9	x	30	12
Pin-tailed Sandgrouse				i	8.5	7.5	x	18	6.5
Black-bellied Sandgrouse							x		
Eurasian Jay	x	18 (20)	12 (11)	x	16 (17)	9	x	17 (18)	12
Hazel Grouse	x	14	9	x	12	9	x	14	12
Black Grouse	f	27	16				f	24	18
Eurasian Hoopoe				x	13	9			
Rock Ptarmigan							c	14 (15)	11
Willow Ptarmigan							c	14	11
Waders	x	24	4,5	x	13.5	4	x	14	4
Great Bustard							x	29	21
Little Bustard	(x)	18	12.5	x	13.5	11.5	x	12	10
Indian Peafowl				i	37	28			
Grey Partridge	x	14 (15)	9	x	11	9	x	9 (10)	7
Great Spotted Woodpecker							x	10 (11.5)	6.5
Lesser Spotted Woodpecker				i	7	5	x	7	4

Species	P	Pmax	Pmin	S	Smax	Smin	R	Rmax	Rmin
Middle Spotted Woodpecker							x	9,5	6
Syrian Woodpecker	x	11.5	9	x	10	8			
Three-toed Woodpecker							x	9	6
Green Woodpecker				i	12	8	x	12	8
Owls	x	43	6	x	30	5	x	35	5
Garganey							(f)	9	7
Black Grouse							f	15	11
Eurasian Wryneck	x	8.5	5.5	x	7.5	5	x	8	6

6. Ps, Ss and Rs bicoloured, with contrasting black and white

This contrast is present in the form of:
– white wedges on a black background: see type 4. 'Ps, Ss, or Rs with paler 'wedges' on a darker background' p. 62;
– black bars on a white background or white bars on a black background: see type 5. 'Ps, Ss and Rs with several bars' p. 64 ;
– several black and white patches (usually longitudinal): often Ps, see waders (T < 20 cm), see gulls (T > 11 cm);
– a 'simple' pattern; species are shown in the table on the next page.

The following patterns can be distinguished in this table:
• a smaller or larger well-defined white base, a vane partly white from the base: code 'wb';

• a terminal bar: the tip of the feather is lighter or darker, the bar may be subterminal (not completely at the end of the feather): code 'tbb' if the bar is black and code 'twb' for a white bar;

• a rounded or drop-shaped mark, terminal or lateral, on one or both vanes, more or less extensive: code 'wdm' for a white mark and 'bdm' for a black mark;

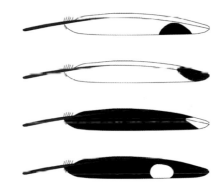

• an elongated mark, either diagonal (code 'dm') or longitudinal (thin pale or dark edges to the sides of the vanes are not included here): code 'lm';

• the code 'x' indicates the possible presence of several different patterns or patterns other than those above (more complex or compounded).

SPECIES WITH CONTRASTING BLACK AND WHITE 'LARGE' FEATHERS
Showing feathers and patterns (codes indicated above)

Species	P	Pmax	Pmin	S	Smax	Smin	R	Rmax	Rmin
Passerines									
Wagtails							x including dm	12	7
Spotted Nutcracker	wdm	17 (18)	11				twb	14	11
Bohemian Waxwing				wb	6.7	5.5			
Hirundines							wdm / dm	13	4
Common Magpie	x	18 (19)	11	x including wdm	15	11			
Azure-winged Magpie	o : wb	16	8						
Shrikes	wb	11	6	x	8	5	x	12.5	7
Wallcreeper	wdm (often 2)	9.6	8.4	i : dm	8.2	5.6	twb	6.2	5.2
Other passerines	x	< 10	< 10	x	< 10	< 10	x	< 10	< 10
Near passerines									
Common Cuckoo	(grey/white base)	19 (22)	10		12	8	x	19	10
Pin-tailed Sandgrouse	i : wb	17	8.5						
Black-bellied Sandgrouse	grey/white base?	16.5	10						
Eurasian Hoopoe	x	14 (17)	9.5	x	13	9	x	12.5	9
Budgerigar	o : wdm	9	4				x	11	4
Black-and-white woodpeckers	wedges	14	6	wedges	11	5	x	12	4
Turtle Dove							twb	14	10
Laughing Dove							twb	14 ?	10 ?
Collared Dove							twb / o : lm	16	12
Aquatic birds of coasts and wetlands									
Auks	twb / wb	16	5.5	twb / wb	10	4.5	twb	10	4
Barnacle Goose							(wb)	18	11
Wood Duck	x	n.d.	n.d.	twb	n.d.	n.d.			
Mandarin Duck	x	21	11	twb	13	11			
Northern Shoveler							x	13	8
Diving ducks				x	14	7			
White Stork	lm	49	28	lm	31	26			
Black Stork	tbb	38	26	x	27	23	wb	14	12
Greater Flamingo				im	22	16			
Northern Gannet				x	19	14	x	29	12 (11)
Long-tailed Duck							male	28	8
Sacred Ibis	tbb	30	24	tbb	28	20			
Gulls	o : x	50	8.5	bdm (im)	24	7.5	bdm (im)	23	8
Waders	x	24	4.5	x	13,5	4	x	14	4
Red-crested Pochard	tbb	19	11						
European Storm Petrel	wb	10.5	5	wb	5	4	wb	6.5	5.5
White Pelican				lm	38	30			
Dalmatian Pelican	lm	50	32	lm	38	30			
Black-throated Diver	lm	22 (23)	11						

'Large' feathers with contrasting black and white

1: Black-headed Gull (P)
2: Eurasian Oystercatcher (P)
3: Common Magpie (P)
4: Eurasian Hoopoe (P)
5: Northern Gannet (S of subadult)
6: Ruddy Turnstone (S)
7: Great Spotted Woodpecker (R)
8: Common Chaffinch (R)
9: European Turtle Dove (R)
10: Common Cuckoo (R)

Life size

Species	P	Pmax	Pmin	S	Smax	Smin	R	Rmax	Rmin
Great Northern Diver				lm	16 (21)	11			
Shearwaters	wb	28	9	wb	13	7	wb	15	8
Eurasian Teal				(o)	9	6			
Eurasian Spoonbill	y	31	21	y	22	17			
Terns	o : lm	32	5.5				lm	20	4
Common Shelduck	wb	26	12	wb	16	10	tbb	14	10 (8.5)
Other groups									
Diurnal birds of prey	o: wb or x	75	12				x	56	10
Domestic chicken (m)	x	> 19	< 13	x	> 19	< 12	x	> 20	< 12
Lady Amherst's Pheasant							x	105	11
Capercaillie	lm(male)	33 (37)	21						
Snowy Owl	bdm	35	23	bdm	24	20	bdm	34	23
Rock Ptarmigan							twb	14 (15)	11
Willow Ptarmigan							twb	14	11
Great Bustard	tbb	61	33	tbb	43	21	x	29	21
Little Bustard	tbb	18	12.5	x	13.5	11.5			
Indian Peafowl				lm	37	28			
Helmeted Guineafowl	spotted	28	20	spotted	25	13	spotted	20	14
Black Grouse	tbb	20 (25)	12	tbb	16	12			

7. Ps, Ss and Rs bicoloured with contrasting brown and white

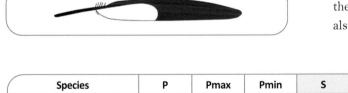

The same types of patterns can be found as those in black and white (see above). Beware, as worn black and white feathers may appear brown and white. If there are any doubts about the original colour, look also at black and white pattern data also.

Species	P	Pmax	Pmin	S	Smax	Smin	R	Rmax	Rmin
Diurnal birds of prey	x (o often)	75	12	x	49	9	x	56	10
Gadwall	x	21	10	x	12	8			
Northern Pintail				o	12	9			
Eurasian Wigeon	x	22	11						
Domestic chicken (m)	x	> 19	< 13	x	> 19	< 12	x	> 20	< 12
Greater Flamingo				im	22	16	im	16	11
Northern Gannet	poorly defined	37	18	x	19	14	x	29	12 (11)
Grebes	x	14 (18)	6	x	11	5			
Mistle Thrush	x	13.5 (14)	10 (8.5)	x	9.5 (10)	7.5	x	13 (13.5)	9 (8)
Long-tailed Duck							female	14	8
Skuas	x	32 (34)	12	x	20	11	x	33	12
Gulls	x	50	8.5	x	24	7.5	x	23	8
Waders	x	24	4.5	x	13.5	4	x	14	4
Grey geese *Anser* spp.							x	19	< 12

'Large' feathers with contrasting brown and white

1: Common Buzzard (P)
2: Great Skua (S)
3: Northern Gannet (S of immature)
4: large gull sp. (S of immature)
5: Gadwall (P)
6: Great Crested Grebe (P)
7: Mistle Thrush (P)

1

2

3

4

5

6

7

Life size

Species	P	Pmax	Pmin	S	Smax	Smin	R	Rmax	Rmin
Great Bustard	x	61	33						
Dalmatian Pelican				x	38	30			
Small passerines	x	< 10	< 10	x	< 10	< 10	x	< 10	< 10
Red-backed Shrike	x	8.5 (9.5)	5.5						
Feral Pigeon	x	20,5	11	x	13	8	x	14	12
Great Northern Diver				x	16 (21)	11			
Garganey				o	10	7			
Eurasian Teal				o	9	6			
Common Shelduck				x	16	10	j	14	10 (8.5)
Black Grouse	x	20 (25)	12	x	16	12			

8. Ps, Ss and Rs with well-defined white marks

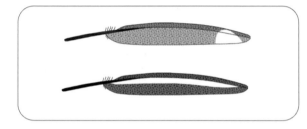

The base of the inner vane is often paler than the rest of the feather; this subtlety is not taken into account here. If several white spots are regularly spaced and form a series of ovals, see paragraph:

Type 4. 'Ps, S, and Rs with pale 'wedges' on dark background', p. 62.

If the white spot is on a uniform black background: see type 6. 'Ps, S, and Rs bicoloured, contrasting black and white', p. 67 .

If the white spot is on a uniform brown background: see type 7. 'Ps, S, and Rs bicoloured, contrasting brown and white', p. 70.

Species	P	Pmax	Pmin	S	Smax	Smin	R	Rmax	Rmin
Diurnal birds of prey	x	75	12						
Wood Duck	ov	n.d.	n.d.						
Gadwall	i (grey)	21	10						
Mandarin Duck	ov	21	11						
Spotted Nutcracker	x	17 (18)	11						
Domestic chicken (m)	x	> 19	< 13	x	> 19	< 12	x	> 20	< 12
Red-necked Nightjar	o	18 (21)	8				male	18 (20)	15
European Nightjar	o	18	9				male	16	12
Lady Amherst's Pheasant	lm	23	17						
Eurasian Coot				tip	12.5	9			
Greater Scaup	i (grey)	17	9						
Pin-tailed Sandgrouse							tip	18	6.5
Black-bellied Sandgrouse	i base	16.5	10						
Eurasian Jay				x	16 (17)	9			
Mistle Thrush							x	12.8 (13.3)	8.8 (8)
Bohemian Waxwing				tip	6.7	5.5			
Red Grouse	o : lm	18 (19)	11						

Species	P	Pmax	Pmin	S	Smax	Smin	R	Rmax	Rmin
Gulls	x	50	8.5						
Waders	x	24	4.5	x	13.5	4	x	14	4
Golden Oriole				tip	9	6.7 (6)			
Budgerigar	x	9	4	x	5	3.5	x	11	4
Small passerines	x	< 10	< 10	x	< 10	< 10	x	< 10	< 10
Rock Dove							o : lm	14	11
Feral Pigeon							o : lm	14	12
Cory's Shearwater	base iv	26	11.5	base iv	12.5	10	base iv	15	12
Eurasian Teal	i	16 (18)	8						
Wallcreeper	x	9.6	8.4	x	8.2	5.6	x	6.2	5.2

Key: For feather types: x = the concerned feather; bl = black; br = brown; c = central; o = outer; g = grey; i = internal (inner); lm = longitudinal mark; m = median. For feather size: f = female; m = male.

9. Ps, Ss and Rs with irregular marks

These markings usually serve to break-up a bird's outline and camouflage it in its environment. They are therefore more common in species or individuals subject to high predation (females that incubate on the ground, young birds, etc.). Irregular patterns may appear on 'transitional' feathers between two categories, which are visible at rest, for example on inner secondaries (colour may be intermediate between median secondaries and scapulars). They are also common on the rectrices of female ducks.

Some young birds also have camouflaging plumage, for example, on the rectrices of young gulls. They are very common on the outer vanes of the visible feathers of Galliformes, nightjars and owls.

If the pattern appears to be regular, see also type 5. 'Ps, Ss and Rs with several bars' p. 64 and type 4. 'Ps, Ss and Rs with paler 'wedges' on a darker background' p. 62.

If the marks are white on a uniform black background, see also type 6. 'Ps, Ss and Rs bicoloured with contrasting black and white' p. 67.

If the spots are white on a solid brown background, see also type 7. 'Ps, Ss and Rs bicoloured with contrasting brown and white' p. 70 .

In other cases, the following table lists species with irregular marks on their 'large' feathers.

SPECIES WITH FEATHERS MARKED WITH IRREGULAR SPOTS									
Species	P	Pmax	Pmin	S	Smax	Smin	R	Rmax	Rmin
Passerines and near passerines									
Common Cuckoo	i	19 (22)	10				x	19	10
Red-necked Nightjar	x	18 (21)	8	i	12	8		18 (20)	15
European Nightjar	(x)	18	9	i	11	8		16	12
Pin-tailed Sandgrouse				x ?	8.5	7.5		18	6.5
Mistle Thrush							o	13 (13.5)	9 (8)
Eurasian Hoopoe				(x) i	13	9		12.5	9
Small passerines	x	< 10	< 10	x	< 10	< 10	x	< 10	< 10

Species	P	Pmax	Pmin	S	Smax	Smin	R	Rmax	Rmin
Black-and-white woodpeckers							x	12	4
Eurasian Wryneck				i	7.5	5		8	6
Aquatic birds of coasts and wetlands									
Eurasian Bittern	x	30	20	x	22	17	x	12	8
Gadwall				x	12	8		10	8
Mallard							x	12	8
Northern Pintail							x	14 f / 22 m	8
Northern Shoveler							x	13	8
Common Eider							x	14	8
Little Grebe				x	7.5	5.5		-	-
Gulls							imm	23	8
Waders	x	24	4.5	x	13.5	4	x	14	4
Spotted Crake				i	7.5	6	x	6	4
Garganey							x	9	7
Eurasian Teal							x	9 (10)	6
Terns				y	15	4.5	y	20	4
Galliformes									
Common Quail	x	8.5	5.5	x	6.5	5	x	4.5	3
Domestic chicken (m)	x	> 19	< 13	x	> 19	< 12	x	> 20	< 12
Common Pheasant	x	23	15	x	19	15	x	60	13
Golden Pheasant	x	21	14	x	18	14	x	70	12
Hazel Grouse	x	14	9	x	12	9	x	14	12
Capercaillie	x	f 27 / m 33 (37)	f 16 / m 21	x	f 21 / m 24	f 15 / m 19	x	f 24 / m 36	f 18 / m 21
Rock Ptarmigan	(x)	16 (20)	11	i	13 (15)	10		14 (15)	11
Red Grouse				x	12	10	x	13	11
Willow Ptarmigan	(x)	18	11				x	14	11
Indian Peafowl	x	41	32	x	37	28	x	56	34
Grey Partridge	x	14 (15)	9	x	11	9	x	9 (10)	7
Red-legged Partridge	x	13 (15)	9	x	11 (12)	9 (7)		11	7
Black Grouse	x	20 (25)	12	x	16	12	x	f 15 / m 23	11
Other groups									
Diurnal birds of prey	x	75	12	x	49	9	x	56	10
Golden Eagle (young)	y	55 (68)	34	y	39 (42)	26	y	37 (41)	33
Falcons	x	37	9	x	23	9	x	30	12
Great Bustard							x	29	21
Little Bustard				x	13.5	11.5	x	12	10
Owls	x	43	6	x	30	5	x	35	5

'Large' feathers with irregular markings

large gull sp. (R of immature)
large gull sp. (S of immature)
Common Pheasant (S of male)
Lesser Spotted Woodpecker (R)

5: Jack Snipe (R)
6: Mallard (R of female)

'Large' feathers
with white marks

Common Sandpiper (P)
Budgerigar (P)
Hawfinch (R)
): Barn Swallow (R of young)
: Yellowhammer (R)
2: Mistle Thrush (R)
: Eurasian Coot (S)

Life size

10. Rs barred or with obvious markings

The rectrices can be strongly marked, and within the same species there are sometimes several different patterns depending on rectrix, or even on the same rectrix. This paragraph gives a non-exhaustive list of species with clearly marked rectrices.

• If TL > 14 cm, look especially at the diurnal raptors, p. 338 and p. 346.

• If the upper surface is velvety, and very soft to the touch, look especially at owls, p. 362 .

• If one or three wide bands in shades of grey or brown, and a thick rachis at its base, it may well be a pigeon or dove, p. 219.

• Not confined to the coast, immature large gulls occur in many places (see p. 227): their feathers are less robust than those of diurnal raptors, the dark marks are well-defined on a white background (sometimes greyish), and never tinged rufous. A wider terminal bar is often present and the markings are irregular in shape, see p. 230.

• For Galliformes and Otididae with banded tails: grouse, ptarmigans, pheasants, bustards, see p. 324 and p. 338.

• For waders with banded tails: see also p. 246.

Other species have marks on their rectrices and for these consult the previous criteria tables.

SOME SPECIES WITH WELL-MARKED RECTRICES								
SIMPLIFIED PATTERNS			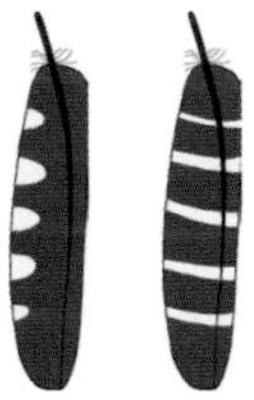					
Species	Rmax	Rmin	pale mark/ wedge/bar on dark background	regular black/dark bars on pale background	irregular black/dark bars on pale background	wide dark terminal (t) or subterminal (s) band	tip paler than the rest	Irregular marks or pattern (no real bars)
Waders: shape rectangular to short								
Bar-tailed Godwit	10	7		x				x
Eurasian Woodcock	10.5	7.5	x				x	
Common Snipe	7 (9)	5						x
Jack Snipe	6.5	5						x
Spotted Redshank	8.5	7		x				
Green Sandpiper	7.5	6		x				
Common Redshank	8	6		x				
Ruff	8.5	5.5		x			x	o
Eurasian Curlew	14	11.5		x				
Whimbrel	12	10		x				
Stone Curlew	14	9			x	t		x
Grey Plover	9.5	7.5		incomplete				
European Golden Plover	9.5	7	x	incomplete				

1

2

3

4

5

6

7

9   8   Life size

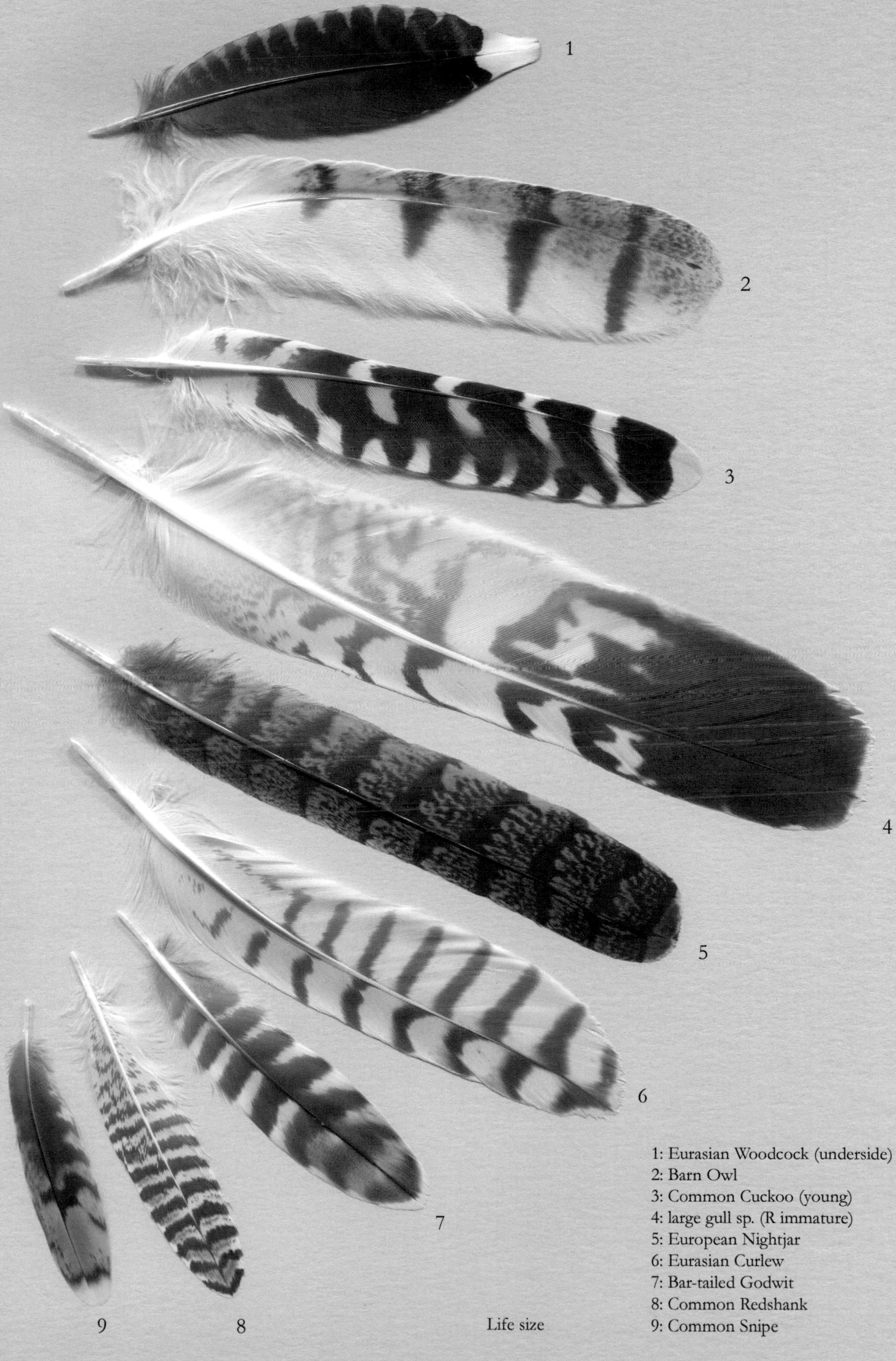

1: Eurasian Woodcock (underside)
2: Barn Owl
3: Common Cuckoo (young)
4: large gull sp. (R immature)
5: European Nightjar
6: Eurasian Curlew
7: Bar-tailed Godwit
8: Common Redshank
9: Common Snipe

Species	Rmax	Rmin	pale mark/ wedge/bar on dark background	regular black/dark bars on pale background	irregular black/dark bars on pale background	wide dark terminal (t) or subterminal (s) band	tip paler than the rest	Irregular marks or pattern (no real bars)
Shape narrow or elongated								
Common Cuckoo	19	10	x				x	x
Red-necked Nightjar	18 (20)	15		x			x	x
European Nightjar	16	12		x			x	x
Common Pheasant	60	13		x				x
Other pheasants	140	11		x				x
Indian Peafowl	56	34						x
Shape short or rectagular								
Eurasian Bittern	12	8						x
Common Quail	4.5	3						x
Pin-tailed Sandgrouse (special shape)	18	6,5		x				
Capercaillie (f)	f 24 / m 36	f 18 / m 21		x				c
Rock Ptarmigan	14 (15)	11		central summer				
Red Grouse	13	11		central summer				
Great Bustard	29	21			x	s	x	x
Little Bustard	12	10			x		x	
Grey Partridge	9 (10)	7						x
Black Grouse (f)	f 15 / m 23	11			x		x	
Eurasian Wryneck	8	6		x				x
Eurasian Wren	3.5	2.5		x				

11. Ps, Ss and Rs grey with black tip or black with grey tip

This patterning is particularly seen in gulls and pigeons; other species are also shown in the following table.

Species	black tip, grey base	grey tip, black base	P	Pmax	Pmin	S	Smax	Smin	R	Rmax	Rmin
Diurnal birds of prey	x		x	75	12	x	49	9			
Falcons	x								x	30	12
Common Pochard	x		x	16	11	x	11	9			
Greater Scaup	x		inner	17	9						
Tufted Duck	x		inner	16	9						
Pin-tailed Sandgrouse	x		x	17	8.5	x	8.5	7.5			
Gulls	x		x	50	8.5	x	24	7.5			

Species	black tip, grey base	grey tip, black base	P	Pmax	Pmin	S	Smax	Smin	R	Rmax	Rmin
Red-crested Pochard	x		x	19	11	x	13	11			
Small passerines	x	x							x	< 10	< 10
Rock Dove	x		x	20	12	x	13	8	x	14	11
Feral Pigeon	x		x	20.5	11	x	13	8	x	14	12
Stock Dove	x					x	12	9	x	14	10
Wood Pigeon	x					x	14.5	9.5	x	19	15.5
European Roller	x								x	17	12
Terns	x		x	32	5.5	x	15	4.5			
Wallcreeper		x							x	6.5	5
Laughing Dove		x							x	14 ?	10 ?
Collared Dove		x							x	16	12

CRITERIA OF THE 3-DIMENSIONAL STRUCTURE

These criteria are generally common to related species. They do not allow for species identification but are very effective in finding a species group, either from the same family or subject to the same biological constraints (heavy, silent flight, etc.).
It will be necessary to check whether the found feather has:
– a velvety upper surface, soft to the touch;
– a shiny patch on the underside if a primary;
– a marked natural curvature when viewed in profile (especially primaries);
– a well-developed and structured aftershaft (or hyporachis).

• UPPER SURFACE 'VELVETY'

In general, the upper surface of a feather appears smooth to the touch, or only very slightly rough.
In some cases, they are extremely soft to the touch, resembling cat or rabbit fur. On closer inspection, the surface is not smooth but covered with tiny 'hairs'. These structures play a role in reducing noise in flight.
If this criterion is very obvious, visible over the entire feather surface (and on all feathers), the species belongs to the order Strigiformes, and is one of the owls. The feather colours are usually quite characteristic, with alternating dark and light marks that facilitate the camouflage of these birds during the day.

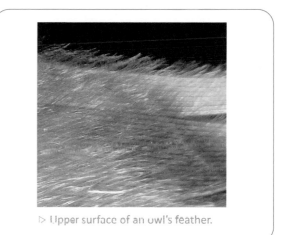

▷ Upper surface of an owl's feather.

Refer to descriptions on pp. 362 and 363.
There is a very wide range of sizes: TP = 6 - 43 cm, TS = 5 - 30 cm, TR = 5 - 35 cm.
If this criterion is not clearly defined and visible, especially on the inner vane, with fairly delicate markings, usually regular, it might be from a nightjar, see p. 218 .
If this criterion is not at all distinct, look especially at fresh feathers and at the base of the inner vane; it may belong to a diurnal bird of prey, particularly one of the Accipitridae (see descriptions p. 346).

• PRIMARY WITH SHINY AREA ON THE UNDERSIDE

In species with a high wing-loading (heavy birds, small wings), flight feathers can be reinforced to support a bird during flight. In addition to the concave curvature that compensates for the pressure exerted on the feather, the barbs close to the rachis are thickened and this is detectable to the touch and common to many species. In certain cases, this thickening is more accentuated, its presence being revealed by a bright area around the rachis, on the underside of the feather.

This reinforced structure is termed the 'tegmen'; in Europe it is mainly observed in two groups, the Anatidae and the Galliformes. The tegmen is only present on primaries. In these two groups, the boundary between the areas of thickened and non-thickened barbs is abrupt and well-defined. See Anatidae, p. 290 and the Galliformes, p. 324. In other groups, Laridae, for example, and more generally in large species, there is a bright area, but it is less clearly defined and contrasts less.

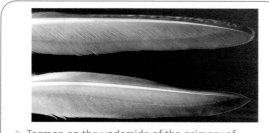

▷ Tegmen on the underside of the primary of a galliform (top) and a duck (bottom)

▷ Shiny area around the rachis on the underside of a primary of a gull.

• PRIMARY (OR SECONDARY) CONCAVE WHEN VIEWED IN PROFILE

This criterion is associated with a high wing-loading or species that fly infrequently, so often rather terrestrial or aquatic.

Primaries can be very concave; their curvature is sometimes so developed and the rachis so rigid that it is impossible to flatten them by hand. The secondaries of these species may have an equally marked curvature, but this is not always the case. A few groups with this character are listed below, and can be further identified using colour and silhouette criteria:

– **Galliforms**: the outer primaries are emarginated and notched very low, giving them a narrowed outline except at the base. The secondaries are often also distinctly curved in profile. See p. 324;

– **Bustards**: the size, colouring and area where found of the two European species is usually sufficient to identify them (see p. 338). TP > 30 cm for Great Bustard;

– **Rallids**: especially the larger species: Common Moorhen and coots have a blackish to greyish-brown tint. Primaries neither notched nor emarginated. See p. 321;

– **Grebes**: the outer primaries are highly notched, high up the feather, and have a brown to brown-grey tint, eventually with some white near the base of the inner vane. The secondaries often have more white, or are totally white. TP = 6 - 14 (18) cm and TS = 5.5 - 11 cm. See p. 288;

– **Auks**: the primaries are brown-black, neither notched nor emarginated, sometimes with white at the tip or the base. TP = 5.5 - 16 cm. See p. 243;

– **Diving ducks**: the primaries have a tegmen on the underside. See p. 290.

• THE AFTERSHAFT (OR HYPORACHIS)

The aftershaft is a duplication of the rachis, a supposedly archaic character. It can be seen on many body feathers, but it is generally unstructured and appears as a small tuft of long barbs.

It is sometimes also present on the 'large' feathers or coverts of certain groups, but always in the form of a rather loose tuft.

In the Galliformes, it is highly developed with an obvious rachis and parallel barbs. It often has a rectangular and elongated shape, sometimes widened at the tip. It is present on nearly all body feathers, but is often also present on 'large' feathers, especially secondaries. As it is fragile, it quickly disappears after a feather is moulted. For this group, see p. 326.

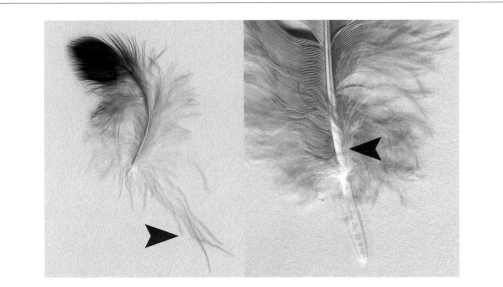

▷ The fluffy aftershaft of a Carrion Crow's body feather (left) and under a Grey Heron's secondary covert (right).

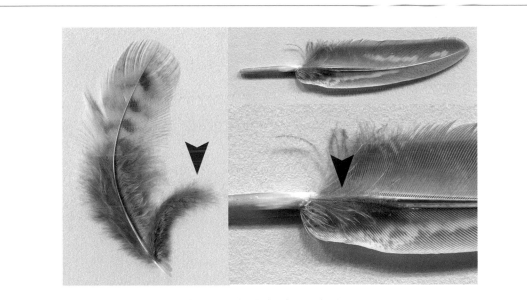

▷ The developed aftershaft of a Common Pheasant's body feather and primary covert

CRITERIA OF SHAPE AND SILHOUETTE

• COMPARING SILHOUETTES

The silhouette of the primaries in particular betrays the type of flight most frequently used by a bird. The relative sizes of primaries and secondaries are indicative of wing shape (see graphs for each group), but there are also other clues to be had from isolated feathers: proportion of the calamus, position of the notch, shape of the tip, etc. It is therefore necessary to take into account any unusual details about 'large' feathers, because not only can it guide the observer to a group of species, but it can also avoid gross errors, eliminating confusion between species easily distinguished by this means.

Note: as a very general rule, the primaries of young birds (juveniles and early immature

▷ **Comparison of the shape of tail feathers in passerines depending on age.**
R5 of a Blackcap (left) and R5 of a Song Thrush (right), (different scale). Adults (top); 1st-year birds, with narrower tail feathers and a more pointed tip (bottom).

stages), are slightly narrower and more pointed than those of adults. Likewise, the young bird's rectrices often have a wedge-shaped end with a small, pointed tip. These details may help to clarify a bird's age in some cases.

• THE PRIMARIES

Various criteria can be found on primaries, including size, proportion of the calamus, and general silhouette or shape of the tip. The table below gives some details that may well help in identification.

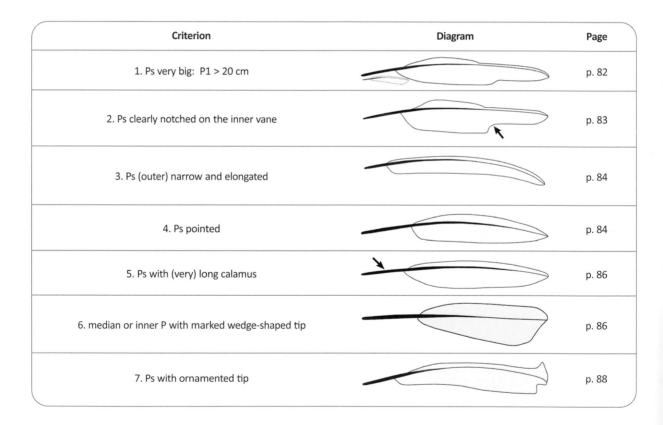

Criterion	Diagram	Page
1. Ps very big: P1 > 20 cm		p. 82
2. Ps clearly notched on the inner vane		p. 83
3. Ps (outer) narrow and elongated		p. 84
4. Ps pointed		p. 84
5. Ps with (very) long calamus		p. 86
6. median or inner P with marked wedge-shaped tip		p. 86
7. Ps with ornamented tip		p. 88

1. Ps very big: P1 > 20 cm

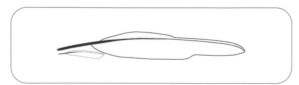

With a P1 of at least 20 cm, we are definitely dealing with a large species. But it is sometimes difficult to know exactly where the feather comes from on the wing. Smaller species may have a median primary of more than 20 cm. Thus, the groups listed below are given as an indication, in order to guide the observer, but it will obviously be necessary to look at the range in size of other suspected species. In this category, the largest primary of a bird often exceeds 40 cm, and even reaches 75 cm in the Lammergeier.

For very large primaries, silhouette is important, especially the position of notches and emarginations.

Outer Ps clearly emarginated/notched a long way down (Em/Not can extend as far as 35% from the base):

Numerous diurnal birds of prey: vultures, eagles, kites, buzzards, etc., with a uniform dark coloration in some or with clear markings in others.

Birds that walk most of the time: Great Bustard, Capercaillie, or even Indian Peafowl.

Other large birds: pelicans, storks, Common Crane.

Just one passerine: Common Raven, with all 'large' feathers black.

Outer Ps clearly emarginated/notched high up (Em/Not > 75%)

Large aquatic birds such as Great Egret, Grey Heron, Purple Heron, Sacred Ibis, Eurasian Spoonbill, Greater Flamingo, etc.

Anatidae such as swans and geese, etc. Verify that there is a tegmen on the underside.

Some large owls such as: Snowy, Eurasian Eagle, Ural and Great Grey; check that the upper surface is 'velvety'.

None of the Ps are emarginated or notched:

Great Bittern (barred or spotted), large gulls (Great Black-backed Gull, Glaucous Gull, etc.).

2. Ps clearly notched on the inner vane

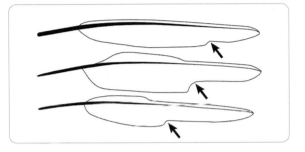

Emargination on a primary's outer vane is common. Obvious notching on the outer web is less common and may be a useful criterion for identification. Depending on species, it is more or less visible on the outer primaries and is located more or less close to the base. A feather described as having a low notch has a low Not value (Not < 45%), see measuring method, p. 18.

Species	Pmax	Pmin	affected primaries	placing on feather	other criteria
SPECIES WITH CLEARLY NOTCHED OUTER PRIMARIES					
Aquatic birds of coasts and wetlands					
Swans, geese	48	15	P10 and P9, generally P8	high: > 65%	tegmen
Ducks	26	12	P10, often P9	high: > 75%	tegmen
Herons, egrets	38	16	P10 - P8, more or less obvious depending on species (P7 sometimes slightly)	high: > 75%	white, grey or black colour
Storks	50 (54)	30	P10 - P6	low: > 40%	black to black-brown, possibly a pale base
Cormorants	27 (31)	12	P10 - P8	very high: > 60%	quite uniform brown
Greater Flamingo	33	20	P10 - P9, very marked	high: > 70%	black
Northern Gannet	37	18	P10 obviously; P9 and P8 slightly	high: > 75%	bronze-brown colour, paler towards inner base
Grebes	14 (18)	6	P11 - P10, P9 more or less markedly	high: > 75%	Curved in profile, grey to grey-brown with white
Common Crane	45	26	P10 - P7	very low: > 35%	black, pale at base
Glossy Ibis	24	15	P10 - P9	high	iridescent black-brown

SPECIES WITH CLEARLY NOTCHED OUTER PRIMARIES					
Species	Pmax	Pmin	affected primaries	placing on feather	other criteria
Sacred Ibis	30	24	P10 - P9 (P8 a little)	high	white with black tip
Pelicans	50 (60)	35	P10 - P6	very low: > 35%	black, grey towards the base
Eurasian Spoonbill	31	21	P10 - P8 (P7 a little)	high	white, sometimes with black tip
Other groups					
Diurnal birds of prey	75	12	P10 - P8 at least, often as far as P6, sometimes P5	low: > (35) 45%	variable colour, often barred, sometimes uniformly dark
Common Quail	8,5	5,5	P10 (P9), not very marked	high	curved in profile, note colour
Black corvids	47	13	P10 -P7 markedly, P6 and P5 less clear	quite low: > 50 % (45 % on P10)	black, slight blue, green, purple iridescence
Other wild Galliformes	33 (37)	9	P10- P6 or more often P5. The base of the vane has a 'domed' silhouette	very low: > 30 %	the Ps appear widened at their base as they also have a low emargination.
Falcons	37	9	P10, P9 more or less pronounced, P8 sometimes	high : > 75 %	often dark-coloured with regular marks on the inner vane.
Great Bustard	61	33	P10 - P6	quite low: > 45 %	bicoloured, blackish-brown and white
Little Bustard	18	12,5	P10 - P7	quite low:	black and white
Owls	43	6	P10 - P9 at least, sometimes up to P6	quite high : > 60 %	upper surface 'velvety' soft
European Roller	19	11	P10 - P9, P8 slightly	high : > 70 %	black-brown, small pale base possible

3. Ps (outer) narrow and elongated

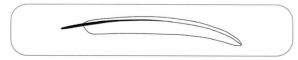

This shape is found in species with pointed wings; very marked on the outer primaries, and less so towards the middle of the wing. The primaries look stretched; they are more or less arched along their length and pointed at the tip.

For dark primaries that are black, dark brown or dark grey, the most obvious choice is hirundines (TP = 4.5-12 cm), swifts (TP = 7-22 cm), storm petrels (TP = 5-14 cm) and less obviously shearwaters (TP = 10-29 cm) or sandgrouse (outer Ps, TP = 12-17 cm).

Terns (TP = 5.5-32 cm) generally have a lot of white, or a white area along the length, and at least in adults a silvery hue.

The European Bee-eater (TP = 7.5-13 cm) has green-blue primaries with a dark grey edge on the inner vane.

Finally, the Galliformes have primaries that are emarginated and notched quite low on both sides, which gives them a narrow look; they are brown to black with pale barring, the presence of a tegmen on the underside helps with identification (TP = 9-33/37 cm).

4. Ps pointed

The primaries can have a pointed tip, sometimes very obviously so, sometimes not so much depending on the group (species, age, wear, etc.). The criterion is more obvious on outer primaries; the inner ones have a more wedge-shaped tip.

There are a few groups of species in which this shape is commonly observed; their wings are also pointed:
– diving ducks (TP = 10-23 cm) such as: Common Eider, Velvet Scoter and Common Goldeneye, brown to black colour with tegmen on the underside;
– storm petrels (TP = 5-14 cm), black;
– auks (TP = 5.5-16 cm), black, sometimes with white near the base;

Markedly notched primaries

1: Red-legged Partridge
2: Eurasian Sparrowhawk (young male)
3: Western Jackdaw
4: Great Cormorant
5: Common Kestrel (female)
6: Common Pochard
7: Great Crested Grebe

Life size

– terns (TP = 5.5-32 cm), but less frequently, with a grey or black colour with some white along the length.

Other aquatic species have slightly pointed outer primaries, quite arched. On one or two outer primaries the tip can be pointed, particularly in the Great Bittern, Little Bittern, and the Rallidae, but the primaries remain otherwise quite wide. See also species with narrow and elongated outer primary, (p. 84).

5. Ps with (very) long calamus

A long calamus provides for better leverage for the feather and improves its attachment. It can therefore support a greater load, but it is not always directly related to the wing-loading.

The calamus is particularly long on all the 'large' feathers of divers, and on those of some species of Anatidae. Numerous other wetland species have a well-developed calamus.

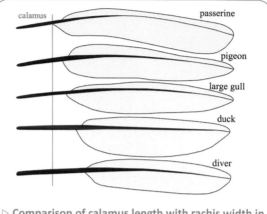

▷ **Comparison of calamus length with rachis width in a few species (primaries)**

SPECIES WITH A LONG CALAMUS				
Species	**Pmax**	**Pmin**	**Long calamus (> 20%)**	**Very long calamus (> 30%)**
Auks	16	5.5	15-30 %	
Swans, geese, shelducks	48	12	25-35 %	25-35 %
Ducks	25	8	20-35 %	20-35 %
Cormorants	27 (31)	12	20-25 %	
Northern Gannet	37	18	20-25 (30) %	
Northern Fulmar	26	11.5	20-25 %	
Grebes	14 (18)	6	15-25 %	
Common Crane	45	26	x	
Skuas	32 (34)	12	x	
Gulls	50	8.5	20-30 %	
Waders	24	4.5	20-25 (30) %	
Great Bustard	61	33	x	
Little Bustard	18	12.5	x	
Pelicans	50 (60)	32	20-25 %	
Divers	> 30	11		30-35 %
Shearwaters	28	9	x	
Rails, crakes and gallinules	20 (25)	5	≤ 25 %	
Terns	32	5.5	15-20 %	

6. Median or inner P with well-defined wedge-shaped tip

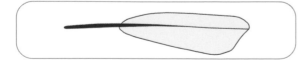

This shape is associated with wings with a pointed hand, as in Charadriiformes, Anatidae, Columbidae, Falconidae, etc.

The table (p. 88) provides a non-exhaustive list of these species.

2

3

4

5

6

7

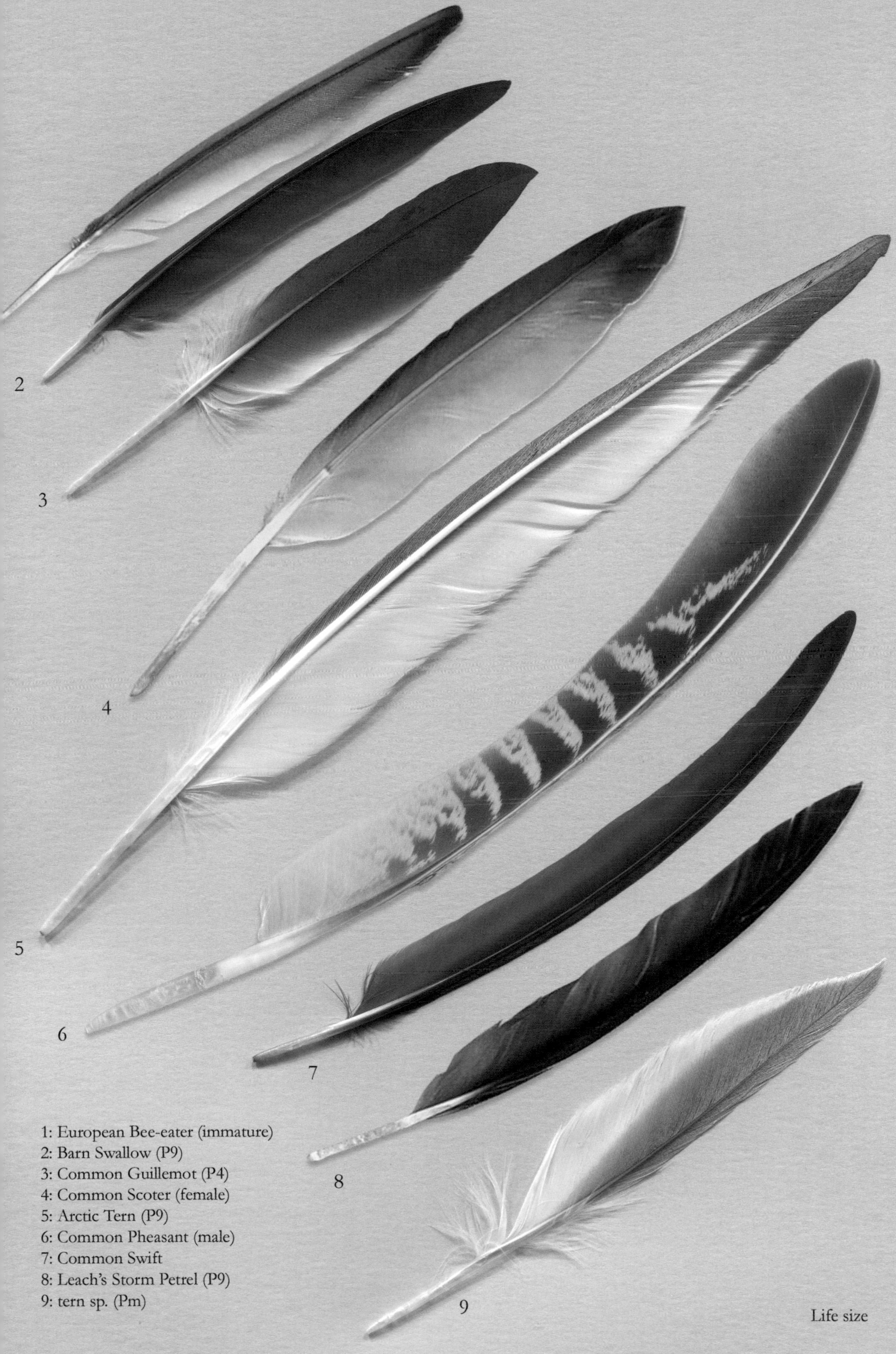

1: European Bee-eater (immature)
2: Barn Swallow (P9)
3: Common Guillemot (P4)
4: Common Scoter (female)
5: Arctic Tern (P9)
6: Common Pheasant (male)
7: Common Swift
8: Leach's Storm Petrel (P9)
9: tern sp. (Pm)

8

Life size

9

SPECIES WITH MEDIAN OR INNER PRIMARY WITH MARKED WEDGE-SHAPED TIP		
Species	**Pmax**	**Pmin**
Auks	16	5.5
Ducks, geese and swans	48	8
Pigeons and doves	21.5	7.5
Falcons	37	9
Northern Gannet	37	18
Sandgrouse	17	8.5
Skuas	32 (34)	12
Gulls	50	8.5
Waders	24	4.5
Swifts	22	4.5
Pelicans	50 (60)	32
Shearwaters and storm petrels	29	5
Terns	32	5.5

7. Ps with ornamented tip

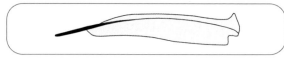

Refer to the criteria for secondaries.

• THE SECONDARIES

Some secondaries have a particular shape. For example, some are only slightly curved (Columbidae, Rallidae, median secondaries of different groups, etc.). However, some show a particular feature, especially: very large size (> 20 cm), a long calamus with a prominent lateral curvature, or a tip with a special shape.

List of criteria	Diagram	Page
1. S very large (T > 20 cm)		p. 88
2. S (curved) with very long calamus		p. 90
3. S with special tip: pointed, bevelled, split, ornamental		p. 90

1. Ss very large (T > 20 cm)

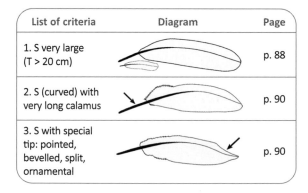

For secondaries measuring at least 20 cm, large species can be targeted. The size of the secondaries along the forearm varies much less than those of the primaries on the hand. However, there is still variability between individuals that can cause problems. Only a selection of species with very large secondaries are shown in the following table, but this already helps guide the reader to most of the most likely species. The general shape and colour may need to be studied more precisely to refine the identification.

SELECTION OF SPECIES WITH VERY LARGE Ss		
Species	**Smax**	**Smin**
Diurnal and nocturnal birds of prey		
Diurnal birds of prey	49	16
Saker Falcon	21	15
Eurasian Eagle Owl	30	22
Snowy Owl	24	20
Herons and other large aquatic birds		
Black-crowned Night Heron	20	14
Eurasian Bittern	22	17
White Stork	31	26
Black Stork	33	26
Greater Flamingo	22	16
Great Egret	28 (30)	21
Common Crane	32	26
Grey Heron	28	20
Purple Heron	26	19
Sacred Ibis	28	20
Eurasian Spoonbill	22	17
Swans and geese	36	18
Great Cormorant	20	16
White Pelican	38	30
Dalmatian Pelican	38	30
Gulls and allies		
Great Skua	20	16
Iceland Gull	> 19	< 17
Herring Gull	20	13
Glaucous Gull	24	17
Great Black-backed Gull	24	18
Other species		
Common Raven (and other black corvids)	16 (28)	15
Great Bustard	43	21
Indian Peafowl	37	28

Primaries with a long calamus and/or bevelled tip

1: Common Blackbird
(Pi of normal shape)
2: Red-throated Diver
3: Northern Fulmar
4: Black-headed Gull
5: Eurasian Collared Dove
6: Great Crested Grebe

7: Gadwall
8: Red-crested Pochard
9: Whimbrel
10: Common Swift (P3)
11: Northern Gannet (P3)

Pi of normal proportions

Life size

2. Ss (curved) with very long calamus

Many groups of wetland birds, as well as some terrestrial species, have secondaries with a long calamus. Sometimes there is also a marked lateral curvature. Divers have a particularly long calamus. Generally, the calamus is also long on the primaries of these species.

SPECIES WITH Ss WITH A LONG CALAMUS			
Species	Smax	Smin	Long calamus (> 20%)
Auks	10	4.5	20-30 %
Swans and geese	48	12	25-30 %
Ducks	16	6	20-30 %
Cormorants	20	11	20-25 %
Northern Gannet	19	14	20-30 %
Sandgrouse	> 8.5	< 7.5	curved
Grebes	11	5.5	15-25 %
Skuas	20	11	20-25 (30) % and curved
Gulls	24	7.5	20-25 (30) % and curved
Waders	13.5	4	20-25 (30) % and curved
Great Bustard	43	21	curved
Little Bustard	13.5	11.5	curved
Pelicans	38	30	25-30 %
Rose-ringed Parakeet	10	5.5	x
Budgerigar	5	3.5	x
Divers	> 16	8	25-35 %
Procellariiformes	12.5	4	20-25 % and curved
Terns	15	4.5	15-25 % and curved

3. Ss with particular tip shape: pointed, bevelled, split or ornamental

S with pointed tip: one species of wader has secondaries that have markedly pointed tips, the Jack Snipe. (TS = 6-7.5 cm). The secondaries are brown, with a broad, blurred white tip, and have the characteristic wader shape (long calamus, marked lateral curvature).

Ss with a bevelled tip: instead of being rounded or vaguely square, the tip is bevelled in the Rallidae. (TS = 4-16.5/20 cm). In addition, the secondaries are very much curved inwards, which gives them a rather characteristic silhouette (see p. 321).

Ss with a split tip: the tip of each vane is longer than the rachis making the feather tip look split at the tip, with two humps. Few species have secondaries of this shape in an obvious way. They include the European Bee-eater which has rufous secondaries with a black tip and are tinted green (TS = 6-8 cm), the hirundines with black-brown secondaries (TS = 3-6 cm) and the larks. In this last group, the overall hue is brown, or with rufous on the border of the inner vane.

This type of tip is sometimes apparent in waders and gulls, etc., or other species, but is never so marked.

Ss with ornamental tip: these secondaries have a modified tip designed to 'decorate' a bird.

This feature can be seen in just a few species, and it is usually the tail feathers that are modified and used during the male's courtship display. A well-known but exotic example is the Mandarin Duck; the male's secondaries are elongated to form a 'sail' and tinged with red (about 10-15 cm in size). These tertials point upwards when the male is at rest. The male Wood Duck also has modified inner secondaries, but not so exuberant. The King Eider has this same form of 'sail' but less marked and without a particular coloration.

One passerine in the region covered here also has secondaries with ornamented tips, the Hawfinch; the primaries and especially secondaries have a flared and wavy tip, tinged with a dark blue sheen. This feature is present in both sexes but is more developed in the male.

• THE RECTRICES

The rectrices are more homogeneous than the remiges in their shape but criteria of size, stiffness or shape can be useful. Particularly noteworthy are whether they have a general elongated or curved shape, a pointed end and the length of the calamus…

Secondaries with a long calamus

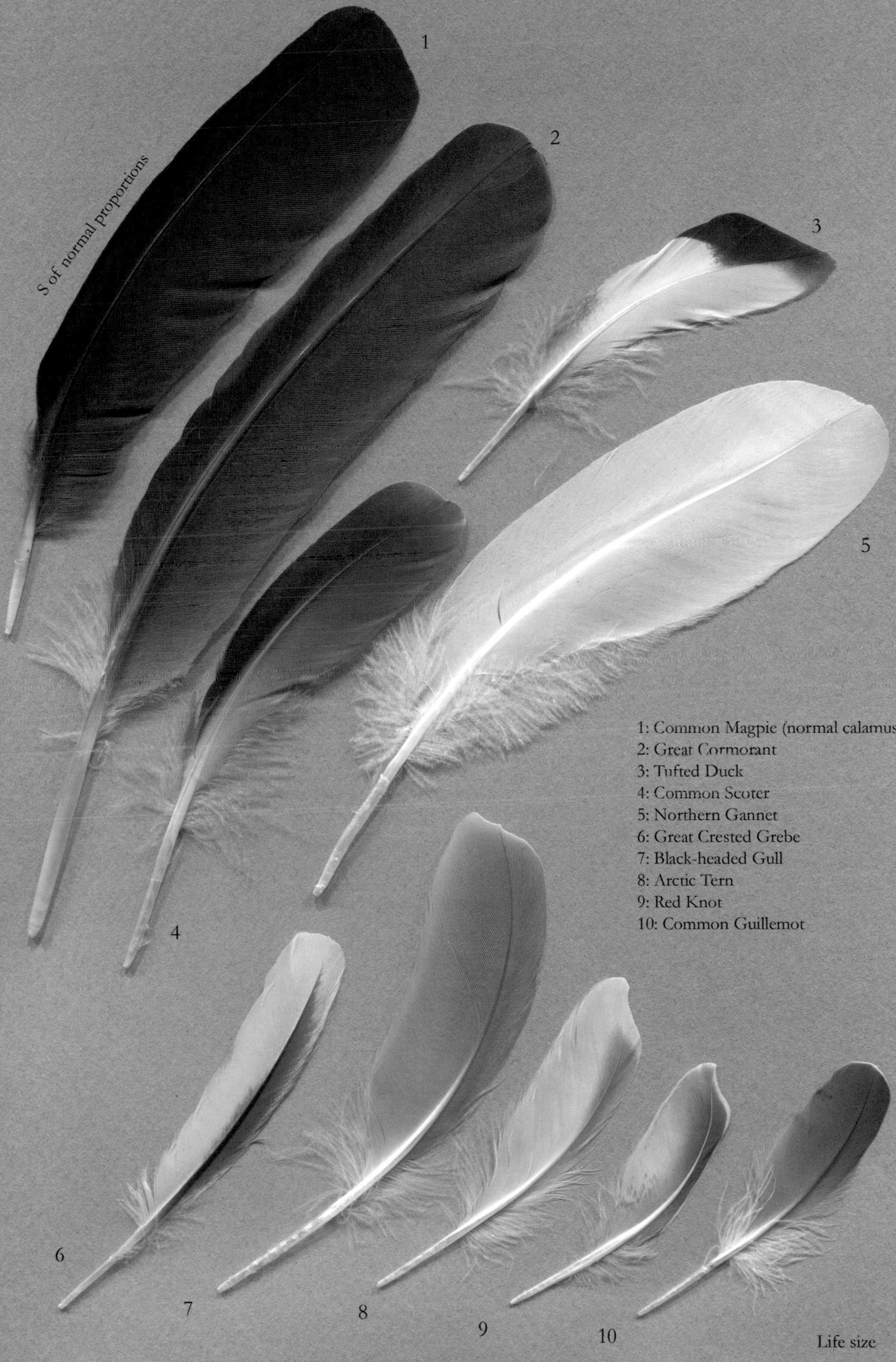
S of normal proportions
1
2
3
4
5
6
7
8
9
10

1: Common Magpie (normal calamus)
2: Great Cormorant
3: Tufted Duck
4: Common Scoter
5: Northern Gannet
6: Great Crested Grebe
7: Black-headed Gull
8: Arctic Tern
9: Red Knot
10: Common Guillemot

Life size

Hawfinch

P

S

undulated tip

Flared tip

Apparently split tip

European Bee-eater (*immature*)

Jack Snipe

Pointed tip

Common Skylark

Common Moorhen

Bevelled tip

broken base

Sand Martin

Life size

List of criteria	Diagram	Page
1. R very large (T > 20 cm)		p. 92
2. R narrow and elongated (tapering shape)		p. 93
3. R rigid and/or pointed and/or straight		p. 94
4. R spearhead-shaped and straight		p. 96
5. R curved, pointed or rounded		p. 96
6. R with long calamus (C > 20%)		p. 98

1. Rs very large (T > 20 cm)

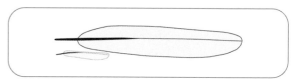

Large rectrices are of course found in large species such as raptors, swans, large grouse, storks, etc. But one must also think about medium-sized species with a pointed or graduated tail, whose central rectrices are elongated and sometimes quite big (magpies, Great Spotted Cuckoo, parakeets, etc.). Pheasants and various domestic Galliformes (including peafowl and turkeys) also have very large rectrices. The following table shows most species with a rectrix longer than 20 cm (sometimes only central or outer feathers) and indicates what other criteria are important for their identification.

SPECIES WITH LARGE RECTRICES										
Species	Rmax	Rmin	other criteria		Species	Rmax	Rmin	other criteria		
			shape	colour				shape	colour	
Passerines and near passerines					Common Crane	24	21	x	x	
Black corvids	30	9		x	Long-tailed duck	28	8	x		
Great Spotted Cuckoo	23	10	x	x	Skuas	33	12	x	x	
Rose-ringed Parakeet	23	8	x	x	Gulls (large)	23	15	x	x	
					Pelicans	28	15	x	x	
Black Woodpecker	21	11.5	x	x	Terns	20	4	x	x	
Common Magpie	27	11	x	x						
Azure-winged Magpie	25	9	x	x	**other groups**					
Aquatic birds of coasts and wetlands					Diurnal raptors	56	10		x	
Herons (large)	20	12	x	x	Pheasants	140	11	x	x	
Canada Goose	20	16	x	x	Falcons	30	12		x	
Domestic duck	20	8	x		Domestic galliforms	56	12	x	x	
Northern Pintail	22	8	x							
White Stork	27	18		x	Western Capercaillie	36	f 18 / m 21	x	x	
Black Stork	32	20		x						
Swans	24	13	x	x	Great Bustard	29	21		x	
Northern Gannet	29	12 (11)	x	x	Owls	35	5	surface		
Great Cormorant	23	14	x	x	Black Grouse	23	11	x	x	

2. Rs narrow and elongated (tapering shape)

Narrow and elongated rectrices are associated with a long, rectangular or graduated tail, and serve to provide balance (magpies, wagtails, etc.); they are in this case quite symmetrical in shape with parallel sides. They are also seen in small passerines (Long-tailed Tit, Bearded Reedling, etc.)

This type of rectrix is also found in the centre of pointed tails (pheasants, parakeets, etc.) or on the outer edge of forked tails or those with streamers. In these cases, they may be narrower towards the tip forming a tapered shape (tip narrower than the base).

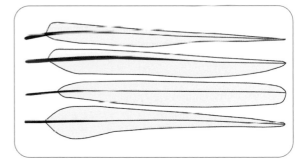

The main species that have these forms of rectrices are given in the following table.

SPECIES WITH NARROW AND ELONGATED RECTRICES				
Species	Rmax	Rmin	R affected	Colour of affected R
Wagtails	12	7	x	white, black or black and white (yellow at the tip)
Northern Pintail	14 f / 22 m	8	c	brown to black, sometimes with a pale edge
Great Spotted Cuckoo	23	10	x	black-grey (olive), with or without a white tip
Pheasants	140	11	x	often barred
Pin-tailed Sandgrouse	18	6.5	c	Brown, barred yellow-buff
Pratincoles	12	5	x	black and white
European Bee-eater	14	9	c	grey tinted green-blue
Long-tailed Duck	28	8	c male	black, white edge

SPECIES WITH NARROW AND ELONGATED RECTRICES				
Hirundines	13	4	x	black-brown, with a pale mark at the end
Swifts	10	4	o	black-brown
Other passerines	< 10	< 10	x	variable
Rose-ringed Parakeet	23	8	x, especially c	green-yellow to blue
Budgerigar	11	4	x, especially c	green-yellow to blue, with or without dark markings
Common Magpie	27	11	x	black with: green, blue, purple, etc. iridescence
Azure-winged Magpie	25	9	x	bluish-grey
Shrikes	12.5	6.5	x	brown, black or black and white
Terns	20	4	o	grey to white

3. Rs rigid and/or pointed and/or straight

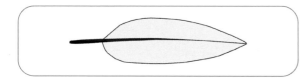

The central rectrices are straight in the majority of birds. The vanes are of the same width or nearly so, their sides are symmetrical, and the tip is rounded or square. Rigidity is the norm for rectrices, with a fairly short and upturned calamus.

In some species, they are excessively stiff, have a very pointed tip or are very straight, even the outer rectrices. They are listed in the following table.

SPECIES WITH RIGID AND/OR POINTED AND/OR STRAIGHT RECTRICES					
Species	Rmax	Rmin	R pointed	R rigid	R straight (even outer ones)
Auks	10	4	x		
Northern Pintail	14 f / 22 m	8	x		
Eurasian Wigeon	12 (16)	9	x		
Diving ducks	14	5	x		
Cormorants	20 (23)	9		x	x
Pheasants	140	11	x		
Northern Gannet	29	12 (11)	x	x	x
Pin-tailed Sandgrouse	18	6.5	x		
Treecreepers	7	4.5	x	x	
Long-tailed Duck	14 f / 28 m	8	x		
Woodpeckers	21	4	x	x	x
Terns	20	4	o		
Eurasian Wryneck	8	6			x

9

8

7

6

5

4

3

2

1

1: Budgerigar (R2, wild type)
2: Common Swift (R6)
3: Barn Swallow (R6)
4: Common Magpie (R1)
5: Common Pheasant (R1 of female)
6: Northern Pintail (R1 of male)
7: Common Tern (R6, very worn tip)
8: White Wagtail (R5)
9: Long-tailed Tit (R3)

Life size

In several groups, all the rectrices are straight and particularly rigid with a straight and fairly long calamus. These species use their tails for moving, climbing on a support, or for diving.

If the tip is rounded, often very worn, and the feather is a dark brown colour, it may well be from a cormorant.

If there is a more or less clearly defined tip and it is spearhead-shaped, it may well be that of a Gannet (white and/or brown colour).

If there is a narrowed and sharp or forked tip, it is of a woodpecker. Treecreepers also have this shape of rectrice, which are quite narrow and reddish-brown.

Diving waterfowl often have a fairly rigid tail with sharp and spearhead-shaped rectrices: pochards, Scaup, scoters and goldeneyes. Two dabbling ducks also have ornamental, elongated central rectrices, the other rectrices rather pointed: Eurasian Wigeon and Northern Pintail.

Auks have brown-black rectrices, sometimes with white spots; they are short.

Pheasants have several pairs of very straight and quite elongated central rectrices and the outer rectrices of terns can be quite narrowed and pointed (see also type 2. 'rectrices narrow and elongated', p. 93).

4. spearhead-shaped and straight

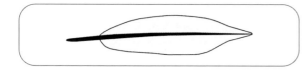

See criteria type 3. 'Rs rigid and/or pointed and/or straight' previously, especially diving ducks, woodpeckers, treecreepers and Gannet.

5. R curved, pointed or rounded

The rectrices are increasingly curved at the base the more they are located towards the outside of the tail.

Generally, this curvature is visible towards the base of the vanes, and the rachis then becomes quite straight. Sometimes this curvature is visible over a greater length, especially for short, rounded tails, as in most Anatidae or Rallidae.

These rectrices can be confused with secondaries, and it is necessary to check the appearance of the calamus (curved upwards and quite short on a rectrix). The outer vane is clearly narrow on the outer rectrices; this can also be a criterion because secondaries have vanes of similar width. Many species have curved rectrices, so the size, colour and shape of their tips can be good criteria for a more refined identification.

The main groups concerned are the owls (velvety surface), ducks, herons and other long-legged species (Common Crane, Eurasian Spoonbill, storks, etc.). Other species are listed in the following table.

A FEW SPECIES WITH CURVED RECTRICES				
Species	Rmax	Rmin	pointed	rounded
Auks	10	4	x	x
Ducks	24 (28)	5	x or o only	x
Herons	20	5		x
Eurasian Woodcock	10.5	7.5		x
Common Quail	4.5	3		x
White Stork	27	18		x
Black Stork	26 (32)	20		x
Pheasants	140	11	x	
Greater Flamingo	16	11		x
Red-knobbed Coot	10 ?	6 ?		x
Eurasian Coot	8	5		x

1: Tufted Duck (R2)
2: Common Scoter (R5)
3: Razorbill (R2)
4: Black Woodpecker (R5)
5: Northern Gannet (immature)
6: Common Pheasant (extremity of R1 of male)
7: Great Cormorant
8: Tern sp.
9: European Green Woodpecker
10: Short-toed Treecreeper (R2)

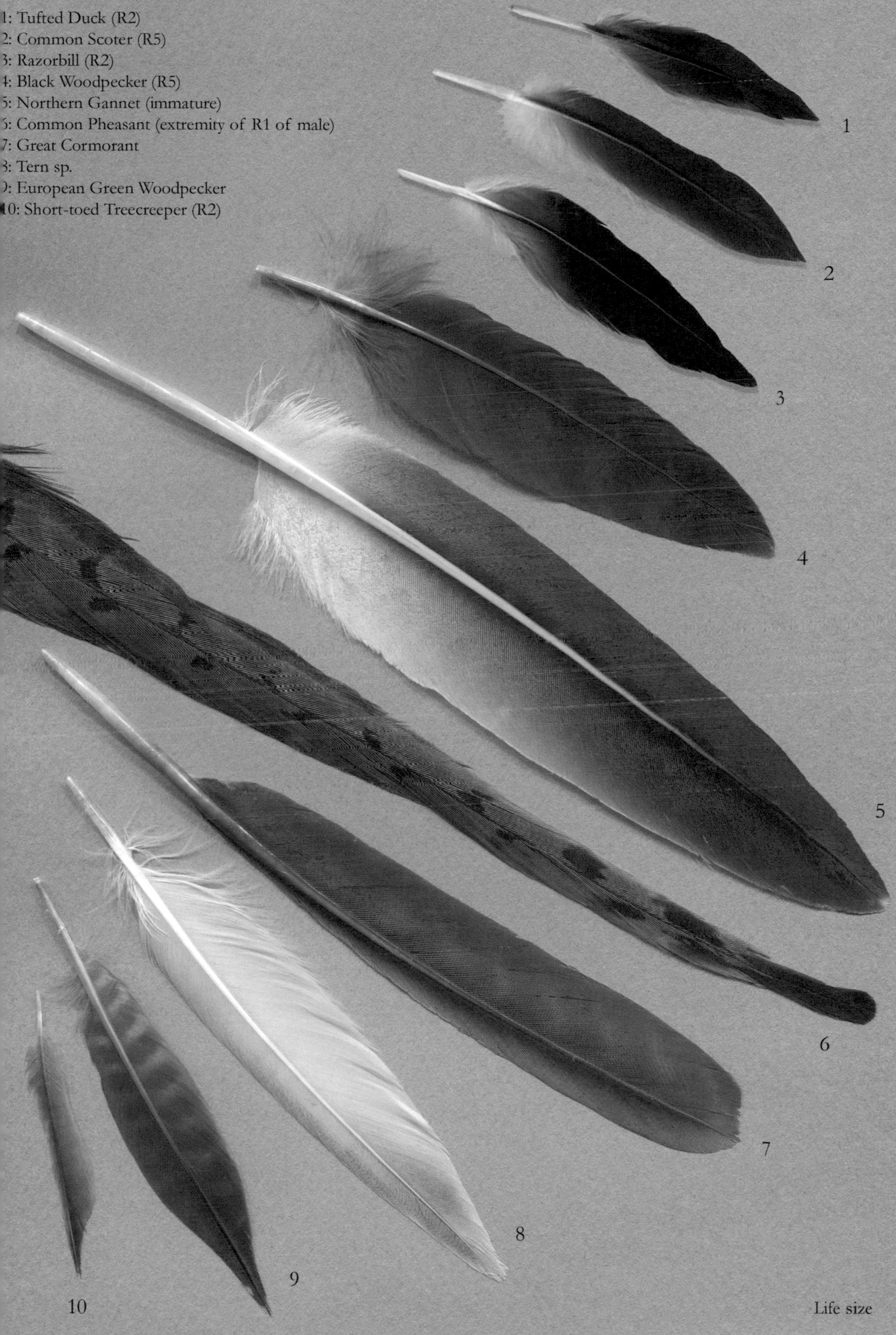

1

2

3

4

5

6

7

8

9

10

Life size

SOME SPECIES WITH CURVED RECTRICES				
Species	Rmax	Rmin	pointed	rounded
Capercaillie	f24 / m36	f18 / m21		x
Common Crane	24	21		x
Glossy Ibis	13	10		?
Sacred Ibis	n.d.	n.d.		?
Waders (others)	14	4	(x)	x
Grey Partridge	9 (10)	7		x
Corncrake	7.5	5	x	
Owls	35	5		x
Eurasian Spoonbill	15	12		x
Black Grouse	f 15 / m 23	11		x

Anatidae: the rectrices are generally short and spearhead-shaped. Some males have elongated central rectrices (Northern Pintail, Long-tailed Duck, etc.) and some are curved upwards (Mallard). Those of diving ducks are generally narrower and more pointed, and often more uniform. They may be confused with those of auks or divers. See Anatidae p. 290.

Ardeidae: the rectrices are rounded at their tip, relatively uniform (often white or grey) except for the Great Bittern (blackish-brown and reddish-buff).

Rallidae: the rectrices can be rounded, rectangular or pointed, but are always short, the outer ones being very curved towards the centre of the tail.

6. Rs with long calamus (C > 20%)

Few species have this character. They are species that use their tails in diving or climbing, or some aquatic birds.

SPECIES WITH RECTRICES WITH A LONG CALAMUS		
Species	Rmax	Rmin
Auks	10	4
Ducks (some)	24 (28)	5
Cormorants	20 (23)	9
Northern Gannet	29	12 (11)
Pelicans	28	15
Woodpeckers	21	4
Divers	> 9	5

1: Mute Swan
2: Mallard (female)
3: Black Grouse (male)
4: Eurasian Coot
5: Common Guillemot

6: Dunlin (R1)
7: Eurasian Scops Owl
8: Common Pheasant (male)
9: Little Egret
10: Black-tailed Godwit

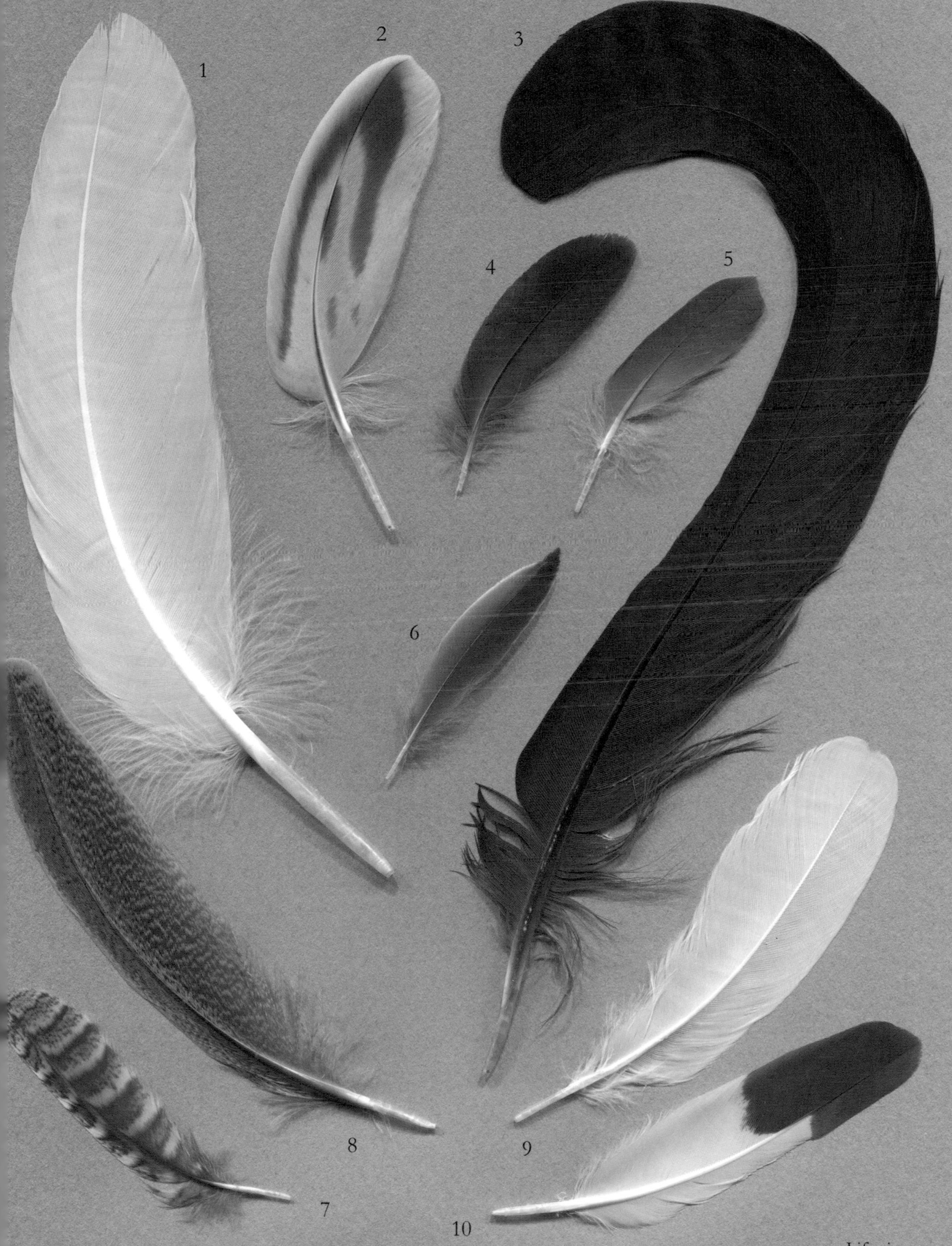

1

2

3

4

5

6

8

7

9

10

Life size

6. Some examples of identifiable body feathers

Body feathers pose a particular problem for the naturalist. Even more than 'large' feathers, they show a significant diversity within a species, not only because of their location on a bird, which affects their shape and colour, but also because of their very varied functions, which result in an almost infinite variation in colour, depending on a species' needs: camouflage, indication of sex or age, display, etc.

Certain body feathers have easily identifiable patterns, but this is quite rare; usually it is their shape or a small detail that will allow recognition to a group of species. As with 'large' feathers, *identifying the original location of a body feather as much as possible will make identification easier* by focusing attention on a certain part of a bird. For example, if the feather is identified as a white uppertail-covert, it is necessary to look for a species showing a white tail-base in flight (but not necessarily a white rump), such as the Eurasian Jay, most wheatears and some waders.

Ventral white body feathers with black spots are more likely to be thrushes, and black with white tips those of starlings, etc.

Size: it may be useful for separating species in extreme cases, but it is amazing to note the differences in size between the body feathers of the same individual bird. Generally speaking, for an individual, the smaller the area

▷ **The 'large' and principal body feathers of a Blackbird (young male)**

The following table lists the general criteria for the shape and outline of most of the normal-looking body feathers; some criteria may vary depending on the adaptations of the groups concerned (see examples below).

Placement or name	General shape	Specific details
forehead, throat	short, quite wide	proportionately very small
ear-coverts, lores	quite short, more or less wide	small, barbs quite well spaced (some need to let sound pass to the ears)
crown, crest	quite elongated	very elongated in species with a crest
nape, neck	'classic' feather shape; sometimes a little elongated but often rounded	certain ornamental ones well-developed
mantle (top of back)	more or less oval; more or less rounded tip, increasingly long towards the back	certain ornamental ones well-developed
scapulars (base of wings)	larger, lanceolate feathers with little down at base, sometimes more oval; rachis often quite rigid	sometimes without a special shape (passerines, other small species); resemble those of mantle and back; sometimes of contrasting colour forming bands at the wing-base
back	many linked barbs (for waterproofing); more or less oval or elongated	generally, the same colour as mantle; mainly quite dark in most species
rump (bottom of back)	often with oval tip, sometimes wide	colour sometimes contrasts with that of back
Uppertail-coverts (base of tail, above)	normally quite elongated with a slightly upturned calamus	quite characteristic shape; slightly less marked than the undertail-coverts
breast	quite wide, square or rounded tip	often quite curved in profile (they cover a domed surface)
belly	generally, quite supple, with fewer linked barbs than those of the back, but largely coloured at tip	normally clearly paler than those of upperside
flank	very large and fluffy; generally barbs not linked	sometimes same colour as axillaries and underwing-coverts
vent	similar to those of the belly, but shape sometimes wider or more rounded	colour sometimes contrasts with that of belly
undertail-coverts (base of tail, below)	shape quite similar to uppertail-coverts but more concave, raised borders, upturned calamus (as do the rectrices they press on); tip often a little triangular	characteristic spoon-shape when viewed in profile (see key p. 28)
primary coverts and greater (secondary) coverts	few linked barbs; PC rigid with long calamus, bent at tip; GC oval with a short calamus, see identification key	shape similar to that of the 'large' feathers (see key p.29 and 30)
thigh (tibia), legs	normally very downy as those on the flanks; often very short and wide; sometimes well developed (e.g. birds of prey)	sometimes same shape as flank feathers; often uniform in colour
median and lesser (secondary) coverts	quite rounded, with many linked barbs; those on outer edge are curved	little down at base, but can be confused with those of body
underwing-coverts	very 'flat' in profile; more or less elongated depending on species and their position	characteristic flattened shape and outline
axillaries	as underwing-coverts, but often more almond-shaped	flattened shape; sometimes coloured like flanks

of the body covered the smaller the feathers. Those of the head (ear-coverts, throat, forehead) are very short, those of the crown longer (clearly if a bird has a crest), those of the breast even longer and especially wider. Those of the upperside (from the mantle to the uppertail-coverts) are usually quite large, with many linked barbs (this reduces the plumage's permeability). Those of the underside (from belly to vent) are much more flexible, with fewer linked barbs, the largest often those of the flanks. The latter are surprisingly long for body feathers; those of the Great Tit, for example, can exceed 4 cm, while most of the 'large' feathers of this species measure between 5 and 7 cm. In contrast, the head or throat feathers are only between 0.1 and 0.3 cm long in this same passerine.

Therefore, size alone is not a useful criterion, but it does give indications of the size of a bird if the position of the body feather has been placed fairly precisely.

Structural details: exist only in certain groups and lead directly to these species. Several are listed below. They concern the general shape, the proportion of linked barbs, the presence of a hypocalamus or a well-developed rachis, etc.

OWLS

The 'velvety' surface of the body feathers, combined with their colour and pattern, almost always makes it possible to identify a feather to this group. Generally speaking, compared with other groups, the downy part is quite developed, and the feathers feel very soft. They are also a little squatter in shape, as are the remiges and rectrices. See structural criteria (p. 79) and species' descriptions (p. 362).

It should be noted that owls often have feathers covering the whole of the feet as far as the claws, which helps to reduce the sound made by the talons before catching the prey. These feathers on the tarsus and feet are very elongated and look a lot like hairs, with few barbs.

Other body feathers are also very different, such as those of the face. They form a parabola that concentrates sound and light to optimise prey location. They therefore have a unique shape, with a widened tip, rather stiff barbs and a rather long calamus. They are rigid and implanted in concentric circles; their surface is not downy. Those on the edge of the facial disc have double lateral curvature; those close to the ears have almost no barbules, and their barbs are wide apart (the bottom three on the illustration opposite).

life size

▷ Facial disc feathers of a
Long-eared Owl

However, these feathers are rarely discovered in isolation, mainly because of their small size.

PARTRIDGES, PHEASANTS AND GROUSE

Most of their body feathers have a structured aftershaft (hyporachis) with a developed shaft and parallel barbs. See also structural criteria, aftershaft (p. 80). The colours of the dorsal body feathers in some species are very different between females (with camouflage colours) and males, which display and may have feathers that are patterned or with contrasting colours. The body feathers often appear quite 'solid' compared to those of other species of the same size. See species descriptions (p. 324).

Regarding domestic species whose feathers might be found in the wild, one of the most frequent is the Indian Peafowl; most of its body feathers (at least in the male) are of a different size and/or colour.

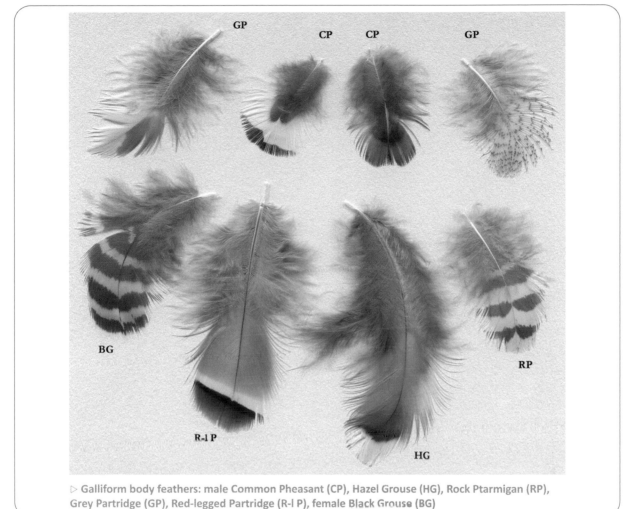

▷ Galliform body feathers: male Common Pheasant (CP), Hazel Grouse (HG), Rock Ptarmigan (RP), Grey Partridge (GP), Red-legged Partridge (R-l P), female Black Grouse (BG)

PIGEONS AND DOVES

As with their 'large' feathers, the rachis of the body feathers of this group is well developed and appears wide at the base. Even the down feathers of this group have a rigid rachis. On the larger body feathers, the 'down' at the base is very dense and 'velvety'. The rachis is thick at its base but narrows sharply, usually in the area of the linked barbs. Generally, they are of uniform colour, but in domestic pigeons there is a wide variety of markings and colours. In the European Turtle Dove, some of the body feathers (scapulars and

▷ Columbidae body feathers: European Turtle Dove (the top 3) and Common Wood Pigeon.
A: axillary, N: neck, Ba: back, L: lesser covert, S: scapular, W: white mark on the neck, U: uppertail-covert, Be: belly.

wing-coverts) have contrasting colours, with a black centre and rufous edges. In the five common species, the neck feathers have a recognisable shape and a contrasting pattern or metallic sheen. The uppertail- and undertail-coverts of these species are for the large part made up of linked barbs and can be confused with inner secondaries or scapulars in some cases.

AQUATIC BIRDS

In swimming species, the body feathers, especially those on the underside, are fairly curved in profile with down feathers that are dense and soft at their base. This form of feather traps more air in the plumage and increases its thermal insulation properties, useful for birds that spend much of their time in cold water. Thus, unlike most 'terrestrial' birds that do not usually swim, their ventral body feathers are not essentially downy but generally have linked barbs. However, there is abundant down under these body feathers (with free barbs).

• ANATIDAE

Their body feathers are often wider towards the tip, especially those of the belly. Their coloration is very variable, uniform or spotted, sometimes with a metallic sheen (dark green of male ducks, for example). For uniform white feathers, swans are an obvious choice, but be careful because many other species have white bellies.

Two patterns in particular point to a smaller number of species.

Geese: on the belly and often also on the back, the feathers are very frequently a uniform grey to grey-brown, with a pale border, sometimes underlined with a darker (black or dark brown) colour. Secondary coverts may also have this pattern. Their pale border partly disappears with abrasion.

Many species of duck with a part of their body grey: males (in particular) have a pattern of alternating fine black and white vermiculations (or stripes) on the relevant body feathers. This pattern is sometimes also visible on the secondary coverts, usually on the lesser coverts, sometimes on the median ones.

From a distance, this type of pattern appears grey but is very distinctive close to. A few species have a distinctive coloration, but most are very similar. Photographs can be examined in an attempt to distinguish the species.

The relative proportions of white and black give them an appearance of lighter or darker grey.

▷ Goose body feathers: Canada Goose, back, worn, (left); Brent Goose, belly (centre); Barnacle Goose, secondary covert (right).

▷ Northern Pintail, male: scapulars

Duck scapulars sometimes have a specific pattern (much as the inner secondaries). Scapulars are very large in this group and when a bird is at rest, they partly cover the wing and are clearly visible. Their shape is quite characteristic (often lanceolate or leaf-shaped) and in many species they are narrow and elongated (teal and other dabbling ducks in particular). Those of males are easier to identify with the frequent presence of edges or borders that contrast with the rest of the feather.

• ARDEIDAE

The most recognisable body feathers in this family are those which play a role in display. However, once a good number of them have been examined, others can generally be recognised, without their structure being easily described. If the feathers were recently moulted and have not yet become too wet, they may appear dusty, as if powdered. This powder is the result of the disintegration of special down feathers; it prevents the mucus of fish or other captured prey from dirtying the plumage. Other than the Great and Little Bitterns, the members of this family have ornamental feathers, which grow during the summer moult or a little later. Some species have special, very elongated feathers on the nape, two on each bird (sometimes four). They are black in Grey and Purple Herons, and white in Black-crowned Night Heron and Little Egret. Elongated nape feathers are more numerous in the Cattle Egret (yellow to orange) and the Squacco Heron (buff to yellow with a black border). The scapulars and body feathers of the mantle and nape are also

▷ Body feathers of Little Egret (2 white) and Grey Heron (6 grey, black or bicoloured).

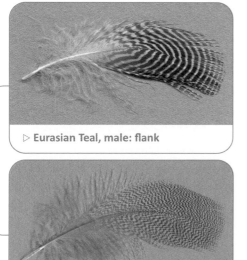

▷ Eurasian Teal, male: flank

▷ Mallard, male: flank

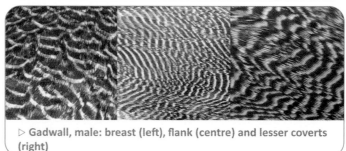

▷ Gadwall, male: breast (left), flank (centre) and lesser coverts (right)

▷ Northern Pintail, male: flank (left), Common Pochard, male: flank (centre) and lesser coverts (right).

modified in certain cases. In general, their structure makes them easily recognisable, being more or less elongated with unlinked barbs depending on species and their position. Finally, some neck feathers are identifiable in Grey and Purple Herons, with distinctive longitudinal black markings. Beware, however, of possible confusion with the feathers of diurnal birds of prey.

BRIGHT COLOURS

Much commoner on body feathers than 'large' feathers, and frequently intended for communication between individuals of the same species. They are most often found on the underside of a bird, while the upperside is usually dark or streaked (camouflage against predators). Bird identification guides, preferably those with photographs, can be consulted to try to find the species concerned. Care should be taken, however, when comparing, because in isolated feathers the colours sometimes seem rather dull compared to the colours on the living bird. This is because the body feathers of a bird are numerous and superimposed on one another, and the down at their base is often dark. Their colour is more intense due to the number and contrast on a dark background (which can be partly reproduced by placing them on a black sheet of paper or other support).

Due to the larger number of brightly coloured species, it is impossible to detail them all here. However, guides with photographs and species descriptions can be consulted. It may also be useful to consult the key to brightly coloured 'large' feathers (p. 50), especially for wing- and uppertail-coverts.

For feathers in shades of yellow, orange, red or pink priority should be given to passerines (tits, wagtails, European Robin, redstarts, finches, etc.). For colours between orange and dark brown, consider also waders, certain diurnal birds of prey or the Galliformes. Ducks, and particularly the dabbling ducks, also show a great diversity of colour, including metallic greens, reds, sky blue, pink, etc.

LIGHT/DARK CONTRASTS

The multitude of different markings and colours makes differentiation difficult. However, before looking at photographs or using guides illustrating feathers, describing the colours and markings accurately can be useful: narrow bars, wide bars, central mark, several longitudinal marks, a spot, a border or an edge, etc. This objective research may then help to eliminate species whose patterns differ greatly from the studied feather. The following groups of species have contrasting patterns on their body feathers.

• DIURNAL BIRDS OF PREY

The body feathers of the underside are frequently white or reddish, marked with dark brown, sometimes black. The markings are often regular in successive bars or triangles, or dark 'drops' in the centre of the feathers. The underwing-coverts have similar markings. Distinguishing between species is difficult, but species can be deduced from some patterns by comparison with photographs.

• WETLAND BIRDS

Various waders have ventral markings for part of the year. Also, young gulls have a camouflaged plumage for one or more years. The females of many duck species have body feathers of camouflage colour, often reddish-brown or dark brown.

• PASSERINES

Black with white markings: the Common Starling has black body feathers with a white tip; they show green or purple iridescence. The white is a spot or V-shaped mark at the tip of the feather, which is lost to abrasion during the winter. In the Spotless Starling, the white tips are very small. In the Spotted Nutcracker, the white mark is rounded or drop-shaped. In the Ring Ouzel, the ventral feathers are black edged with white, with the centre often pale and without any iridescence.

White with black markings: in thrushes the ventral feathers are white or reddish with a black, dipped-in–ink, drop-shaped, arrowhead-shaped, or blurred mark at the tip. However, other passerines have marks of this type: Barred Warbler, Sedge Warbler, pipits, etc.

7. Exercise in the identification of some feathers of common species

This chapter presents groups of moulted feathers found in four different habitats and enables the reader to train in the recognition of the specific origin of frequently found feathers. Plates of more than fifty feathers are then presented, followed by a brief description of the feathers and an aid to their identification.

The answers are given on p. 373.

A BEGINNER'S EXERCISE

These few illustrations allow readers, if they wish, to start using the identification keys. Look carefully at the feathers and then try to follow one of the keys.

It is a good idea to start by identifying the feather's placement. It can first be measured (for ease of use, the feathers are presented here at life size), and then the salient criteria of shape or colour can be ascertained.

Start with one of the keys (colour or silhouette key). It is important to follow the different stages of the process, so that one can go back if the conclusions of the search do not seem realistic. It is also possible to note the species or group of species that are indicated by one key, then use another key and see if the two different analyses match.

Finally, the species description plates should be consulted to make sure the identification is relevant. And, if the reader wishes to verify his findings, the answers are given on p. 373.

• IN TOWNS AND VILLAGES

Feathers can sometimes be found in the street under perching sites or nests (pigeons, swallows, sparrows, etc.) or at roosting sites, especially during the moulting period (small passerines such as wagtails, sparrows, starlings, etc.).

The plate shows fifteen feathers found in a small town in central France in late summer.

• IN PARKS AND IN THE COUNTRYSIDE

Urban parks are often wooded and often allow the collection of feathers from many common forest bird species. Outside towns, species of other habitats offer increasing interest; those of agricultural environments, hedgerows, wasteland, etc. Searching along edge habitats can be very productive.

The plate shows fifteen feathers collected in an area of hedgerows in northern France in the summer.

• ALONG LAKESIDES

These feathers can also be found on the banks of streams, in freshwater marshes, and even sometimes at the edges of urban waterways. Resting areas where birds preen are particularly rich in body feathers; banks open to the wind will provide more 'large' feathers.

The plate shows nine feathers collected on the shore of a lake in the French Alps in summer.

• ON THE SEASHORE

Coastal shorelines with large tidal ranges are more conducive to regular searching as the waves deposit a large amount of debris at each hide tide, including feathers of course. Walking the tide line and looking among the seaweed and other debris can produce many feathers, sometimes during a very short search.

The plate shows eleven feathers discovered along a beach on the Atlantic coast of France in winter.

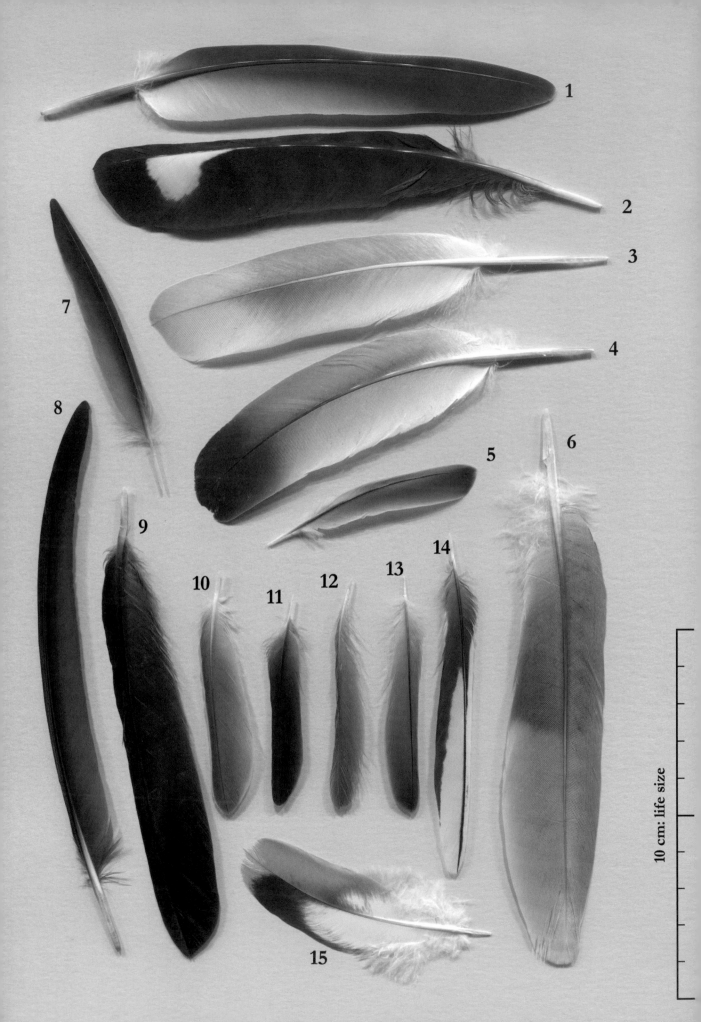

Feathers found in towns and villages

10 cm: life size

21 22 23 24 25 26 27 28 29 30 31 32 33 34 35

10 cm: life size

Feathers found in the countryside

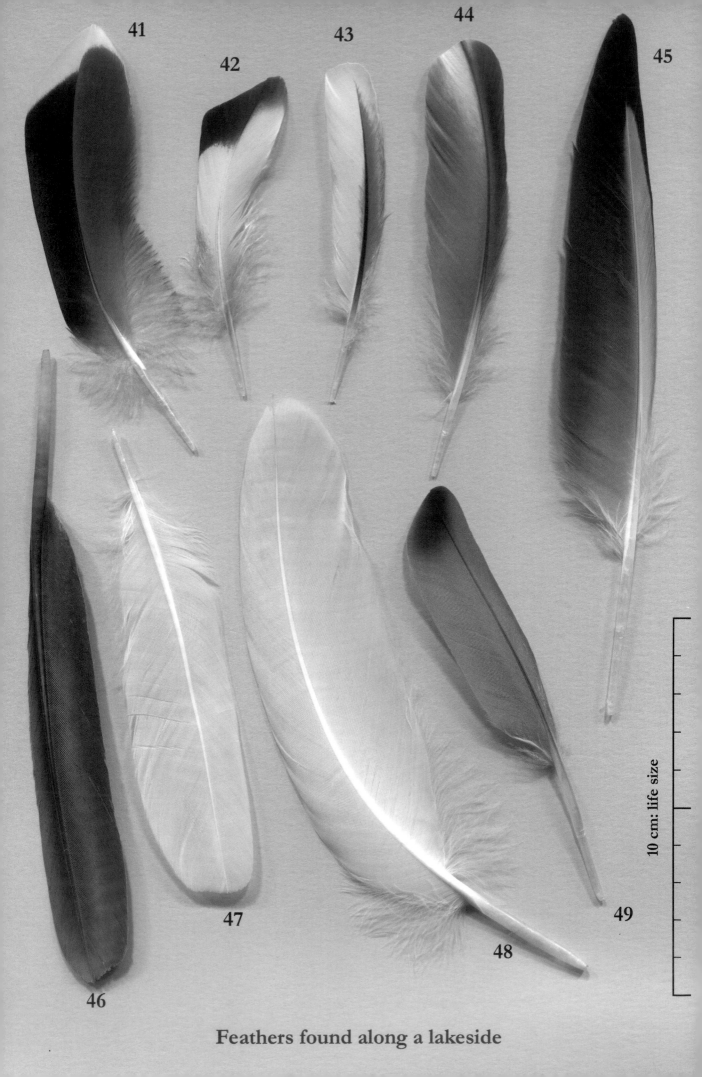

41

42

43

44

45

46

47

48

49

10 cm: life size

Feathers found along a lakeside

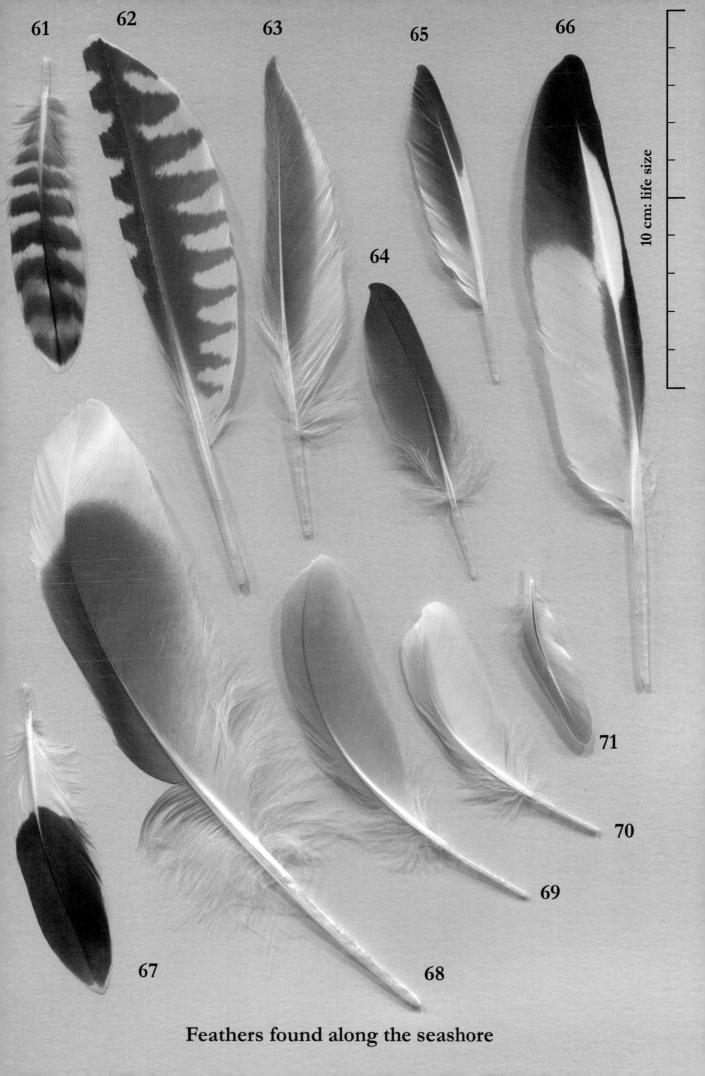

61 62 63 65 66

64

10 cm: life size

71

70

69

67 68

Feathers found along the seashore

AID TO IDENTIFICATION

Here are clues regarding the location, size and criteria observed for the feathers presented in the preceding plates. The answers are given on p. 373.

No.	In towns and villages: example of a small town in the Loiret (central France) in late summer
	Criteria
1	Right-hand primary. Arched outline, rachis thick at base, quite long calamus (nearly 20%). T = 13.9 cm.
2	Left-hand secondary. Contrasting black and white (one white mark), dark blue iridescence on outer vane (hardly visible here). T = 13.8 cm.
3	Left-hand primary. Quite long calamus, rachis thick at base, mixture of reddish-brown and white without an obvious pattern. T = 12.5 cm.
4	Left-hand secondary. Thick rachis at the base, grey with ill-defined black tip. T = 11.4 cm.
5	Right-hand primary. Brown, outer vane pale at the base (probably a trace of a wing-bar). T = 6 cm.
6	Tail feather. Rachis thick at its base, dense down, more contrasting on underside (not visible here) with a black base. T = 15 cm.
7	Left-hand primary. Dark brown, probably an outer primary, quite narrow shape. T = 8.6 cm.
8	Right-hand primary. Black, very elongated, and not very curved. T = 14.8 cm.
9	Left-hand tail feather. Quite external, black, with dark blue and green iridescence on outer vane (not visible here). T = 12.8 cm.
10	Left-hand tail feather. Quite squat (short tail?), brown with buff edge on outer vane. T = 6.6 cm.
11	Left-hand tail feather. Black-brown, narrowed tip (forked tail?). T = 5.7 cm.
12	Left-hand tail feather. Rufous! T = 6.2 cm.
13	Right-hand tail feather. Brown-grey, no particular criteria. T = 6.4 cm.
14	Right-hand tail feather. Elongated, black and white with long white markings. T = 9.2 cm.
15	Secondary covert (left). Rachis thick at base, dense 'down'. T = 7 cm. Be careful, this isn't a P, S or R; this feather doesn't feature in the keys, but it has a diagnostic shape and colour.

No.	In parks and in the countryside: example of hedgerows, Île-de-France (northern France) in summer
	Criteria
21	Right-hand primary. Brown, wide reddish border on the inner vane. T = 5.5 cm.
22	Right-hand primary. White wedges on black background. T = 10.2 cm.
23	Right-hand primary. Contrasting brown/white. T = 10.6 cm.
24	Right-hand secondary. Black with large white area and also some blue at extremities of the white (not obvious here). T = 12.9 cm.
25	Right-hand secondary. Buff with regular brown barring, quite rigid (not visible here). T = 12.2 cm.
26	Left-hand tail feather. Spearhead-shaped, regular dark barring, very rigid (not visible here). T = 9.8 cm.
27	Right-hand tail feather. Black, a little elongated. T = 10.9 cm.
28	Right-hand secondary. Black with some white at the base of the vanes. White on the outer vane indicative of a wing-bar. T = 5.7 cm.
29	Right-hand primary. Quite small, black with white border on both vanes, tip paler (slightly visible here). T = 4.9 cm.
30	Left-hand tail feather. Black with a large white mark. T = 5.5 cm.
31	Right-hand tail feather. Much yellow at the base, otherwise black-brown. T = 5.9 cm.
32	Body feather. Velvety surface (not visible here) and diagnostic coloration. T = 5.6 cm.
33	Left-hand secondary. Brown, buff to white border, and note especially the tip split in two bumps. T = 6.7 cm.
34	Left-hand secondary. Pale wedges on a brown background, rufous marks on the outer vane. T = 10.3 cm.
35	Left-hand secondary. Velvety surface (not visible here), regular brown barring with paler background towards edge of inner vane. T = 14.7 cm.

No.	On a lakeside: example of a lake in the Savoie (French Alps) in summer
	Criteria
41	Right-hand secondary. Black-brown, white tip, dark blue iridescence on outer vane (not visible here), long calamus. T = 11.6 cm.
42	Right-hand secondary. White with black tip, long calamus. T = 8.5 cm.
43	Left-hand secondary. White and brown, quite rigid for its size (not visible here). T = 9 cm.
44	Left-hand secondary. Grey with blurred white mark at tip. T = 11.6 cm.
45	Left-hand primary. Grey with black tip, long calamus. T = 18.5 cm.
46	Left-hand tail feather. Dark brown, very rigid, calamus straight but not upturned (not visible here), elongated and quite long. T = 16.7 cm.
47	Right-hand tail feather. White, rectangular, quite supple. T = 12.6 cm.
48	Right-hand tail feather. White, curved, rounded tip, calamus quite long. T = 17 cm.
49	Left-hand primary. Grey with black tip, with a tegmen (not visible here). T = 11.8 cm.

No.	On the seashore: example of a beach on the Ile de Ré (Atlantic coast of France) in winter
	Criteria
61	Central tail feather (left). Regular brown barring, oval shape. T = 8.4 cm.
62	Right-hand primary. Pale wedges on brown background, bevelled tip, long calamus. T = 15 cm.
63	Right-hand primary. White and grey with silvery appearance (not visible here). Bevelled tip, quite pointed and narrow, long calamus. T = 12.8 cm.
64	Left-hand primary in appearance (actually an S1). Black-brown, long calamus. T = 8.2 cm.
65	Left-hand primary. Black and white, pale rachis, bevelled tip, long calamus. T = 8.7 cm.
66	Left-hand primary. Several black and white patches, long calamus. T = 16.8 cm.
67	Central tail feather (left). Black and white, small pale tip, oval shape. T = 8.5 cm.
68	Right-hand secondary. Dark grey, big white tip, long calamus, quite curved. T = 18.3 cm.
69	Right-hand secondary. Uniform pale grey, curved, long calamus. T = 10.5 cm.
70	Right-hand secondary. White with a small pale grey zone, curved, long calamus. T = 7.5 cm.
71	Left-hand tail feather. Brown-grey, bicoloured rachis. T = 5.1 cm.

8. Species descriptions

This chapter presents descriptive texts and illustrations of 'large' feathers, and sometimes of body feathers, for different groups of European birds.

EXPLANATION OF THE GROUPING OF SPECIES IN THE DESCRIPTIONS

• 'GROUPS' OF SPECIES

The term 'group' brings together species related at different levels, such as order, family or genus. But a group can also include species at a different level (only part of a family, etc.); their phylogenetic alliance therefore varies. For example:

The term 'black corvids' groups the genera *Corvus* and *Pyrrhocorax*.

The term 'waders' groups diverse families of Charadriiformes, such as Charadriidae (ringed plovers, plovers, etc.), the Scolopacidae ('shanks, sandpipers, etc.), and Recurvirostridae (stilts, avocets, etc.), but not all the species in this order. The English terms sandpiper and 'shank both include species of the genus *Tringa*, but both also include other species and the separation of these two groups is not based in biological reality. There are many other examples.

Here, descriptions concern groups (whether they have a systematic reality or not) and apply in theory to all species and subgroups contained in each group.

Despite the constant evolution of the classification of birds, the species in the groups presented here share a certain lifestyle and therefore the shape and/or colour of their feathers share many criteria.

The order of presentation adopted here does not follow the established classification, as seen in identification guides, but allows the grouping of species whose feathers, and/or whose lifestyles, are most similar. The reader will find in the keys and identification tables that follow much information allowing species to be distinguished from one another.

• THE ORIGIN OF DESCRIPTIONS AND SAMPLES

In general, the descriptions have been based on the examination of several individuals of each species. In some cases, small sample sizes (three individuals) are indicated by the abbreviation 'SS'. In such cases caution needs to be exercised concerning the data, which reflect only part of the possible range of variation in the species concerned. Descriptions and measurements were established by consulting reference works, websites with illustrations, and public or private collections (see References, p. 400).

The feathers illustrated come from various sources:

– moulted: feathers found singly or in groups at preening or roosting sites;

– plucked bird: feathers found in groups from birds killed by natural (mammals, raptors, etc.) or domestic predators (cats, etc.). More often than not they belong to a single individual;

– bird remains, whole corpses or wings: birds found dead as a result of collisions (with vehicles, power lines, windows, etc.), birds killed and lost during hunting (including both protected and unprotected species!), birds intentionally or accidental trapped (in fishing nets, hollow poles, cages of domestic animals, etc.) or birds that have died of natural causes (rarely of age, more often by disease, malnutrition, etc.). Wings were also taken from birds found dead; no birds were mutilated or mistreated in producing the illustrations.

• SCALES USED IN THE PLATES

Due to the very large variation in size between species, feathers had to be represented at different scales. The number of scales used has been kept to a minimum to facilitate comparisons, while keeping the illustrations large enough to be legible. Five main scales were chosen, with the aim of simplifying the calculations needed to establish the actual size of a feather.

SCALES USED IN THE PLATES					
Real size of the 10 cm measure on the page:	10 cm	6.7 cm	5 cm	3.3 cm	2.5 cm
The relationship between the plate and reality is:	1/1	2/3	1/2	1/3	1/4
To calculate the actual size of the feather depicted on the plate, multiply its measurement on the plate by:	1	1.5	2	3	4

CONTENTS OF THE DESCRIPTIVE PAGES

Note: for obvious reasons, not all information is equally detailed for each group.

Group name: order, family...
Number of 'large' feathers (the term 'large' feathers is used to indicate primaries, secondaries and tail feathers) (example): NP = 10 (+1) / NS − 10-12 (14) / NR = 14
Indicates for the group concerned (and sometimes for species where they differ) the usual number of primaries NP, secondaries NS and tail feathers NR. The numbers in brackets indicate extremes sometimes encountered. When the outermost primary is very short compared to the next one, this is indicated (+1) after the number of normal-sized primaries, and similarly for very short tail feathers. For secondaries, the first number indicates the usual number, the second indicates a common variation in the number, and a number in brackets indicates the extremes encountered. In the case of a very short outermost primary, its size is not taken into account in the size ranges shown in the descriptions. Its size (if measurements are available) is shown in the summary table at the end of this guide; see Biometric Data Table (p. 374).
Distinctive criteria of the group: size, colour, structural criteria, etc.
Short description of the commonest wing and tail shapes in the group.
Possible confusion: similar species (eventual precision on the 'large' feathers concerned).
'Large' feather colour: main colours or characteristics for these feathers in the group.
Primaries: shape of the primaries, colour, etc.
Secondaries: shape of the secondaries, colour, etc.
Rectrices: shape of the rectrices, colour, etc.
Body feathers: description if they are remarkable or identifiable.
Note: Further information on moult, where feathers of this group can be found, etc.
One or more tables summarise the distinctive criteria of the species in question (size, colour, etc.). They provide an additional aid to identification and summarise the data presented elsewhere in more detail for each species.

In particular, the sizes of the 'large' feathers are shown, in cm:

– Pmax and Pmin: length of the biggest and the smallest primary (except reduced outer primary);

– Smax and Smin: length of the biggest and the smallest secondary (except inner ones if very different);

– Rmax and Rmin: length of the biggest and the smallest tail feather.

Species 1 English name (*Scientific name*): description of the 'large' feathers, and of some body feathers if useful.

Species 2 English name (*Scientific name*): description of the 'large' feathers, and of some body feathers, etc.

Graphs (primaries and secondaries at least, tail feathers when useful): show the relative sizes of the 'large' feathers for the group (a curve for the mean of the whole group or several curves where there are strong differences within the group). They can be used to calculate all the sizes of the 'large' feathers based on a found feather or the measurements indicated in the text. For how to use them see pp. 21 to 24.

For tails, when the rectrices of the same individual show a variation of less than 10%, the shape is called 'square' (short tail) or 'rectangular' (long tail). In such cases there is no graph of relative sizes, as all rectrices measure between 90 and 100% of the longest feather.

Photograph of the wing: depending on the material available and its value for identification, an entire wing or one reconstructed from a set of feathers from a plucked bird, may be illustrated. It allows the reader to see the shape and colours of the feathers on the wing. Due to the smaller number of tail feathers and for reasons of space, tails are not shown.

GROUPS DESCRIBED

• PASSERINES

Corvidae: crows, magpies, jays, etc. (p. 130)

Turdidae: thrushes, rock thrushes (p. 136)

Sturnidae: starlings (p. 141)

Other medium-sized passerines: Golden Oriole, Waxwing, Wallcreeper (p. 142)

Alaudidae: larks (p. 144)

White-throated Dipper, Eurasian Wren (p. 147)

Certhiidae: treecreepers (p. 148)

Laniidae: shrikes (p. 148)

Muscicapidae (1): chats, wheatears (p. 153)

Motacillidae: wagtails and pipits (p. 154)

Muscicapidae (2): flycatchers (p. 158)

Fringillidae: finches (p. 160)

Regulidae: 'crests (p. 170)

Paridae: tits (p. 172)

Other small passerines: Bearded Reedling, Long-tailed and Penduline Tits, nuthatches (p. 174)

Muscicapidae (3): Robin, redstarts, nightingales, etc. (p. 178)

Emberizidae: buntings (p. 180)

Passeridae: sparrows, Snowfinch (p. 185)

Prunellidae: accentors (p. 186)

Sylviidae: warblers (p. 188)

Hirundinidae: swallows, martins (p. 200)

• NEAR PASSERINES

Apodidae: swifts (p. 203)

Picidae: woodpeckers and Wryneck (p. 204)

Upupidae, Alcedinidae, Coraciidae, Meropidae: Hoopoe, Kingfisher, Roller, bee-eaters (p. 210)

Psittacidae: Rose-ringed Parakeet, Budgerigar (p. 212)

Cuculidae: cuckoos (p. 214)

Caprimulgidae: nightjars (p. 218)

Columbidae: pigeons and doves (p. 219)

Pteroclidae: sandgrouse (p. 224)

• AQUATIC BIRDS OF COAST AND WETLANDS

Charadriiformes: general criteria (p. 227)

Laridae: gulls (p. 227)

Sternidae: terns (p. 236)

Stercorariidae: skuas (p. 240)

Alcidae: auks (p. 243)

Charadriidae: curlews, godwits, sandpipers, plovers, 'shanks, etc. (p. 246)

Procellariidae, Hydrobatidae: shearwaters, petrels, storm petrels (p. 282)

Sulidae: Gannet (p. 284)

Phalacrocoracidae: cormorants and Shag (p. 285)

Gaviidae: divers (p. 286)

Podicipedidae: grebes (p. 288)

Anatidae: swans, geese, ducks. (p. 290)

Ardeidae: herons, egrets, etc. (p. 311)

Pelecanidae: pelicans (p. 316)

Other large, long-legged wetland birds: storks, ibises, Spoonbill, Flamingo, Crane (p. 318)

Rallidae: rails, crakes, gallinules (p. 321)

• TERRESTRIAL BIRDS OF DRY HABITATS

Otididae: bustards (p. 324)

Phasianidae: partridges, pheasants, grouse, etc. (p. 326)

• BIRDS OF PREY

Falconidae: falcons (p. 339)

Accipitridae, Pandionidae: buzzards, eagles, kites, harriers, vultures, Osprey, etc. (p. 346)

Tytonidae, Strigidae: owls (p. 362)

PASSERINES

The following pages provide a summary of the different criteria for identifying certain species or for restricting searches to a smaller number of species. Species that closely resemble passerines should also be examined if in doubt are indicated before the keys.

As with previous tables, the observer should cross-check the information. In passerines, the distinction between primaries, secondaries and tail feathers is quite easy. The tables therefore treat the remiges and the rectrices separately. There are many tables presented in Chapter 5 'Key to flight and tail feather determination' (p. 46) that will guide identification using obvious colour criteria. They can also be used according to the size of the feather being examined. They will allow the identification of certain 'large' feathers, especially when they exhibit a bright colour, distinct bars, a white mark, etc.

The tables presented below therefore give general information on the colour of the 'large' feathers of a wide variety of passerines, without necessarily allowing identification to species level every time. Indeed, many passerines have very similar feather sizes, and this criterion may be of little use on its own. In addition, several groups are very similar and do not show an obvious colour feature.

General remarks

Number of 'large' feathers: NP = 9 (+1); NS = 9; NR = 12.

Known exceptions: Cetti's Warbler: NR = 10; Corvidae: NS = 10-11; Eurasian Golden Oriole: NS = 9-10 (11). Some other groups sometimes also have a small S10.

Be careful when using the notes on the presence of notches in the general biometrics table (p. 374) or in the text. In passerines, they are sometimes very difficult to distinguish; worn feathers can be especially misleading.

The passerines contain a family of 'giants', the Corvidae, whose 'large' feathers can be confused with those of species of other orders. Their size and shape, as well as their colour, clearly distinguish them from other passerines.

Corvidae are not included in the following keys to the passerines but in the general keys in Chapter 5 (p. 46).

Wing shape and relative sizes of wing feathers

For the small passerines, there are two main wing shapes, rounded hand (type A) or a rather triangular hand (type B). In some cases, a variant of type A can be distinguished by its slightly longer outer primaries [type A(b)]. When the relative sizes of the feathers are well represented by one of these graphs, the wing shape is mentioned in the descriptive texts. Otherwise, they are illustrated in a specific graph.

Note. the proportion of P10 is highly variable within a species, and even more so within a group; the value given in the graphs is therefore very approximate.

The graphs opposite indicate the ranges of variation for the different standardised shapes [A, A(b) and B].

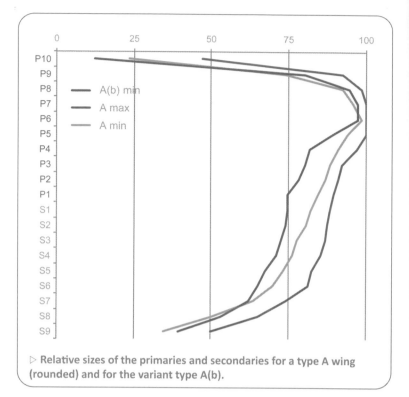

▷ **Relative sizes of the primaries and secondaries for a type A wing (rounded) and for the variant type A(b).**

Note that for secondaries of identical size, there is a decrease in proportion relative to Pmax as the size of the outer primaries increase; the wing then appears narrower. Directly comparing the wings of different species, it is rather the shape of the hand that changes; it is more elongated for a type B.

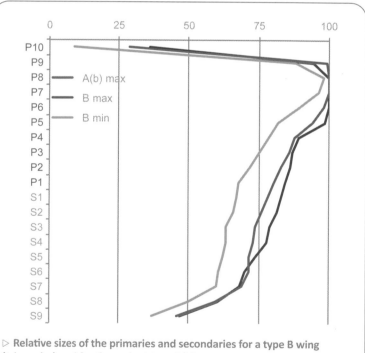

▷ **Relative sizes of the primaries and secondaries for a type B wing (triangular) and for the variant type A(b).**

Obvious criteria

Certain groups are easy to identify because of the size, colour or the shape of their 'large' feathers.

Size

The Corvidae aside (large size), size T is indicated to the nearest millimetre. However, these species are generally small and size is not always a very useful criterion. Nevertheless, it is still easy to establish several size categories (at least for primaries and secondaries);
– large size: for P > 8, for S > 8, for R > 9: see the Corvidae;
– medium size: see Turdidae, Golden Oriole, etc.;
– small-medium size: see starlings, larks, shrikes, Waxwing, crossbills, Hawfinch, etc.;
– small size: majority of species of European passerines;
– very small size: for P < 5.5, for S < 4.5, for R < 5: see 'crests, Wren, Zitting Cisticola, other small warblers, etc.

Shape of tail feathers

The general shape of the tail feathers is quite elongated; the tip of the outer ones sometimes curves outwards.
Elongated tail feathers: some with an obviously more elongated shape are found in various groups: for example, wagtails and shrikes (T = 7-12 cm), Long-tailed Tit and Bearded Reedling (T = 4-10 cm), and Dartford and Marmora's Warblers (T = 5-7 cm). Swallows may also have elongated outer tail feathers but they are very much narrower towards the tip.
Short tail feathers: in other cases, the tail feathers may sometimes be short, as in Common Starling or Eurasian Nuthatch; they then appear wider than normally shaped feathers.
Spearhead-shaped tail feathers: the treecreepers have a rigid tail, their rectrices (rufous) are spearhead-shaped and elongated, with a long tip. They cannot be confused!

Shape of primaries and secondaries

The flight feathers of swallows are easily recognised; the outer primaries are elongated and the others are pointed, and bevel-shaped. Their secondaries have a split tip that forms two bumps at the feather extremity. This criterion is also seen in the Alaudidae, and less clearly in some of the Fringillidae and Emberizidae.

The Hawfinch has black and white remiges, some of which have a widened tip, making them easily recognisable.

Feather colour

The Wren has very barred 'large' feathers. Treecreepers have a large, pale wing-bar crossing their remiges. Others have bright colours in their plumage: Golden Oriole, some finches, Waxwing, Wallcreeper, etc. (see in the following text, as well as in the table on pp. 51-55). Bright colours are sometimes restricted to the feather edges and may disappear with wear, but if they are still visible, they can be very useful in identification.

• KEY TO THE IDENTIFICATION OF PASSERINE PRIMARIES, SECONDARIES AND TAIL FEATHERS

Several colour criteria for species that have feather shapes resembling those of passerines
These criteria concern species whose feathers cannot be distinguished from those of passerines by form, at least when they are found in isolation.
– **Contrasting black and white:** in the form of white wedges or bars: see woodpeckers, Hoopoe.
– **Bright colours:** see table (pp. 51-55). Medium size: see European Roller, European Bee-eater, etc. Small size: pale-blue on at least the outer vane, reddish edge to the inner vane of Ps and Ss: see Common Kingfisher.
– **Brown or rufous tints:** Ps may be slightly hooked, Ss with bevelled tip, Rs short and curved: see Rallidae (p. 321).
– **Buff bars or wedges on a brown background:** see Wryneck, Common Quail.

– **Uniform black-brown**, Ps and Rs with an elongated shape, Ss very squat: see swifts.

– **Brown, black or white** (uniform or mixed): see also the Ps of small waders.

It is possible that other species may be confused with passerines. If in doubt, the 'large' feather identification keys may also be of use (p. 46).

Summary table of the colours and markings of passerine 'large' feathers.

Reminder: the criterion in a box it is not necessarily present on all the 'large' feathers of the species. On the other hand, for a 'large' feather with a certain criterion, species with similar boxes must be considered (if the sizes match). The criteria are treated independently, so the same 'large' feather can display one or more of them.

As the criteria have sometimes been found in a small number of individuals, it is possible that some individuals of species not listed in the tables also exhibit these criteria.

List of criteria:

1. Primary or secondary with some white .... p. 120
2. Tail feather with some white .... p. 122
3. Remige or rectrix with some bright colour: yellow, red, orange, pink, blue .... p. 125
4. Remige or rectrix with a different contrasting mark .... p. 127
5. Remige or rectrix with contrasting margin on at least one of the vanes .... p. 128

1. Primary or secondary with some white

Two species of passerine have wing feathers with a lot of white: Snow Bunting (Ps and Ss with more or less white, some Ss totally white) and the appropriately named White-winged Snowfinch (Ss with much white, but at least the base is black).

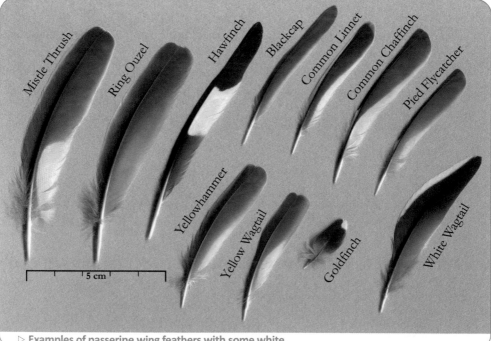

▷ **Examples of passerine wing feathers with some white.**

Other white marks come in the form of edges or borders on one or both of the vanes, a small white spot on the outer vane, a white base or a white tip. With regard to these last two criteria, the extent of white is very variable, from very reduced to almost the whole of the feather. The reader should therefore look at the descriptions or illustrations to find out more.

PASSERINES WITH WHITE IN THEIR PRIMARIES OR SECONDARIES
Part 1: species with a white margin or other white mark

Species	Pmax	Pmin	Smax	Smin	Outer vane: white margin	Inner vane: white margin or border or other mark	White mark on outer vane	Only a white spot at tip (size variable)	White base (size variable)
Alpine Accentor	10	7	7.2	5.4				x	
Calandra Lark	11.2	7.8 (7 juv)	7.8	6.3	x			x	
Horned Lark	9.3	6.3 (5.6 juv)	6.5	5.7 (4.8 juv)	P	S indistinct		x	
Two-barred Crossbill	8.3	5.2	5.6	4.6		(x)		x	
Grey Wagtail	7.7 (8.2)	5.0	5.2	4.7					x
White/Pied Wagtail	8.1	5.4	5.8	5.2	x		x		x
Yellow Wagtail	7.5	5.2	6.0	4.9			x		vi
Snow Bunting	9.8 (10.7)	5.8	6.7 (7.3)	5		x			x
European Goldfinch	7	4.9	5.1	4.4		x		x	
White-throated Dipper	7.6 (9.5)	6.0	6.2	4.2		x		x	
Collared Flycatcher	7.4	5.3	5.4	4.1	Si ov white				x
Semicollared Flycatcher	7.4	5.3	5.4	4.3	SI ov white				x
Pied Flycatcher	7.2	5.3	5.2	3.7			x		
Mistle Thrush	13.4 (14)	9.6 (8.5)	9.5 (10)	7.6					oval
Hawfinch	9.2 (10.5)	6.3	7 (7.3)	5.4		spot			
House Martin	11	4.5	5.5	4	tip		tip		
Bohemian Waxwing	10.8	6.7	6.7	5.5			tip		
Twite	6.9	4.8	5.1	4.1	x	x		(x)	
Common Linnet	7	4.7	5	4	x	x		(x)	
Long-tailed Tit	5.8 (6.7)	4	4.8	3.6	x		x		
Azure Tit	6.3	5.2	5.8	4	x	x		x	
Blue Tit	6.1	4.9	5.1	3.9		x		x	
Siberian Tit	6.3	5.3	5.5	4	x		x		
Sombre Tit	6.7	5.7	6	4.6	x		x		
Coal Tit	6.1	4.7	5.1	3.5		x		x	
Blue Rock Thrush	11	8	8.2	6.5				(pale)	
Common Rock Thrush	10.7	7.3	7.5	5.8				pale	
White-winged Snowfinch	10.9	6.2	6.8	5.3				x	
Bearded Reedling	5.8	4	4.8	3.9 (3.2 juv)	x	x	indistinct		Si
Shrikes	10.4 (11)	5.9	7.8 (9)	4.5	x			x	x
Common Chaffinch	8.1	5.6	6.3	4.6	x	x	x		
Brambling	8.1 (9.4)	5.7	5.8 (6.5)	4.7	x	x and mark (cream)	x		
Firecrest	5 (5.6)	3.9	4.3	2.8		margin		x	
Goldcrest	5	4	4.2	3		margin		x	
Nuthatches	7.8	5.3	6.5	4.3			x		x
Whinchat	6.5 (7.3)	5.2	5.3	3.7				x	Si
Common Stonechat	5.7 (6)	4.6	5	3.6				x	Si
Wallcreeper	9.6	7.2	8.2	5.2		2 oval marks	Si	x	

					Outer vane: white margin	Inner vane: white margin or border or other mark
Species	**Pmax**	**Pmin**	**Smax**	**Smin**		

PASSERINES WITH WHITE IN THEIR PRIMARIES OR SECONDARIES
Part 2: species with only a white margin or border on one or both vanes

Species	Pmax	Pmin	Smax	Smin	Outer vane: white margin	Inner vane: white margin or border or other mark
Short-toed Lark	7.6	5	5.7	5.3		blurred
Woodlark	7.2	6	6.5	5.8	P	very blurred
Crossbills	9.6	5.3	6.8	4.8		(x)
Eurasian Bullfinch	8.4	5.8	6.4	4.8	x	x
Buntings	8.3	4.9	6.7	4.4	x	x
Zitting Cisticola	4.8	3.6	4.5	3.2		x
Pine Grosbeak	10 (12)	7	8.2	6	x	
Sylvia warblers	7.5 (8.6)	4.2	6.2	3.5	x	x
Fieldfare	12.4 (14)	9.2 (8.5)	9.3 (10.5)	7.5	outer P	
Redwing	10.5 (12)	7.4	7.7	6.3	outer P	
Sand Martin	11	5	5.5	4	x	x
Hippolais warblers	7.7	4.4	5.7	3.4	x	x
Golden Oriole	13.5	8.3	9	6.7 (6)	P grey	
Ring Ouzel	12.8	9.1	9.5	7.2	blurred	
Tits (others)	7	4.6 (4)	6.1	3.5	x	x
Phylloscopus warblers	6.5	4.1	4.9	3		x
Penduline Tit	5	4.1	4.3	3.2	x	x
Black Redstart	7.8	6	6.3	4.4	x	x
Arctic Redpoll	7 (7.6)	4.8	5.2	3.8	x	x
Lesser/Common Redpoll	6.6 (7.5)	4.7	4.8	3.7	x	x
Isabelline Wheatear	8.4	6	6.6	5		x
European Greenfinch	7.7	5.3	5.8	4.6		x

2. Tail feathers with some white

The white markings on a passerine's tail feathers are very varied and quite common, especially on the outer feathers. They serve as a visual signal (especially when taking flight) but are also used in territorial defence or other communication functions.

The following have been distinguished in the next table: white margins and borders (one or both vanes), white tip or base, a white oval or elongated mark, small white non-basal mark, all-white tail feathers or white with a dark base.

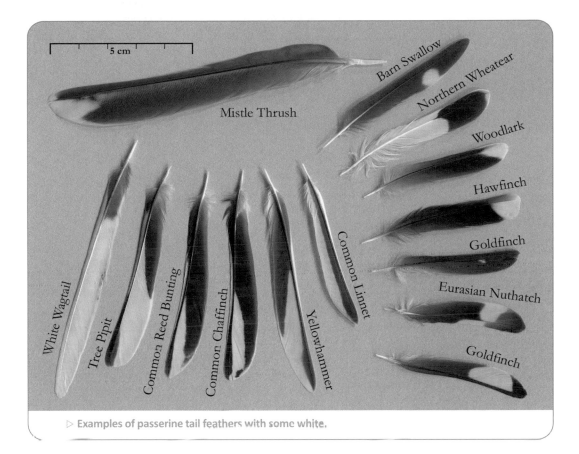

▷ Examples of passerine tail feathers with some white.

PASSERINES WITH SOME WHITE IN THEIR TAIL FEATHERS
Part 1: species that may have a white margin or other white mark

Species	Rmax	Rmin	outer vane: white margin	Inner vane: white margin or border	White tip	White base	Small non-basal white mark	white oval or long mark	R white with dark base	R entirely white
Alpine Accentor	8	6.7			x					
Rufous Bush Robin	8	6.3			x					
Calandra Lark	7.3	5.4	x		x			outer		o
Short-toed Lark	6.4	5.6	tip	tip				outer		
Eurasian Skylark	7.9	6	x					outer		
Woodlark	5.9	5.1			x			outer blurred		
Lesser Short-toed Lark	7.2	5.1 (4.1)	x					outer		
Wagtails	11.7	7.1	x	x					x	x
Eurasian Bullfinch	7.5	6.3						(x)		

Species	Rmax	Rmin	outer vane: white margin	Inner vane: white margin or border	White tip	White base	Small non-basal white mark	white oval or long mark	R white with dark base	R entirely white
Snow Bunting	7.9 (8.3)	5.9	x	x	x	x	x	x		(x)
Lapland Bunting	7.1	6.1 (5)	x	x				x		
Buntings (others)	8.7	5.3	x					x		
European Goldfinch	5.7	4.7			x		x	x		
Zitting Cisticola	5	2.9			x					
Sylvia warblers	8.2	5	x	x	x		x	x	x	
Semicollared Flycatcher	6	5.5 (5.2)	x				x			
Red-breasted Flycatcher	6.2	5					x			
Mistle Thrush	12.8 (13.3)	8.8 (8)			x			outer blurred		
Hawfinch	6.5	5.3			x					
Crag Martin	7	5					x			
Barn Swallow	12 (13)	4.5					x	x		
Olive-tree Warbler	7.6	6.2	x	x	x					
Long-tailed Tit	9.6	4	x					x		
Azure Tit	6.9	6.1			x			x		
Great Tit	7.2	5.3			x			x		
Rock Sparrow	6.1	4.8					x			
White-winged Snowfinch	8.3	6.8	x		x		x		x	(x)
Bearded Reedling	9.6	4.3 (4 j)						(x)		
Shrikes	12.2	6.9	x		x	x		black		x
Common Chaffinch	7.8	6.1	x					x		
Brambling	7.5	5.5					x	x		
Pipits	8	5.6			x			x		
Dupont's Lark	7.3	5						outer		
Corsican Nuthatch	4.3	3.9					x			
Eurasian Nuthatch	5.5	4.5					x			
Whinchat	5.3	4.6				x				
Wallcreeper	6.4	5.2			x					
Wheatears	8 (8.2)	5.5			(x)	x				x

PASSERINES WITH SOME WHITE IN THEIR TAIL FEATHERS
Part 2: species with only a white margin or border on one or both vanes

Species	Rmax	Rmin	Outer vane: white margin	Inner vane: white margin or border
Horned Lark	7.9	6.5 (5.2 juv)	tip	
Crossbills	8	5.7		(x)
Pine Grosbeak	10.6	8.6	x	
Collared Flycatcher	5.9	5	x	
Pied Flycatcher	6.1	5.2	x	
Fieldfare	12.3 (14)	10.5 (10)	(tip)	(tip)
House Martin	7	4	x	tip
Sand Martin	6	4	x	x
Hippolais warblers (others)	6.5 (7.6)	4.5	x	x
Twite	7.1	4.6	x	x
Common Linnet	6.2	4.6	x	x
Moustached Warbler	5.4	4.3	x	
Blue Tit	5.8	5	x	
Willow Tit	6.4	4.8	x	
Siberian Tit	7.3	6.2	x	
Sombre Tit	7.1	6.3	x	
Marsh Tit	6.1	5.2	outer	
Aquatic Warbler	5.4	3.8	x	x
Sedge Warbler	5.6	4.4	x	x
Eurasian Penduline Tit	5.4	4.2		x
Common Rosefinch	6.4	5.5		x
Trumpeter Finch	5.9	4.6	x	x
Arctic Redpoll	6.8 (7.5)	5	x	x
Lesser/Common Redpoll	6.5	4.8	x	x

3. Remiges or rectrices with some bright colour: yellow, red, orange, pink, blue

A key in Chapter 5 (p. 51) includes the main species with a bright colour on their 'large' feathers.

Here are details of passerines that show this feature, which can be very obvious (e.g. red tail of the redstarts) or more discreet (e.g. yellow edging of the primaries of buntings, etc.).

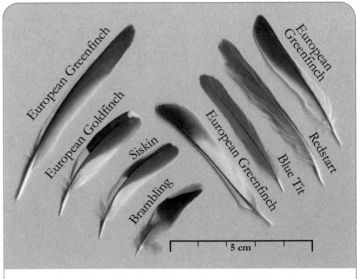

▷ Examples of 'large' feathers of passerines with bright colours

'LARGE' FEATHERS OF PASSERINES WITH SOME PINK

Species	P S	R	Pmax	Pmin	Smax	Smin	Rmax	Rmin
Trumpeter Finch	x	x	7.6	4.9	5.2	4.4	5.9	4.6
Common Rosefinch	x	x	7.3	5.5	5.6	4.8	6.4	5.5
Pine Grosbeak	x	x	10 (12)	7	8.2	6	10.6	8.6
Rose-coloured Starling		x					8.1	6.6
Wallcreeper	x		9.6	7.2	8.2	5.2		

'LARGE' FEATHERS OF PASSERINES WITH SOME BLUE OR GREEN

Species	P S	R	Pmax	Pmin	Smax	Smin	Rmax	Rmin
Eurasian Bullfinch	blue irid.		8.4	5.8	6.4	4.8		
Rose-coloured Starling	green-blue irid. ve		11.1	7.5	7.7	5.8		
Common Starling	green irid. ov		11.3	7.5	7.8	6.4		
Spotless Starling	green irid. ov		11	7.4	8.2	6.1		
Hawfinch	blue irid.		9.2 (10.5)	6.3	7 (7.3)	5.4		
Barn Swallow	blue irid.		12	4.5	6	4		
Golden Oriole	Si: greenish		13.5	8.3	9	6.7 (6)		
Azure Tit	blue	blue	6.3	5.2	5.8	4	6.9	6.1
Blue Tit	blue	blue	6.1	4.9	5.1	3.9	5.8	5
Red-flanked Bluetail	blue	blue	7.1	5.4	5.5	4	6.4	5.6
European Greenfinch	green-grey		7.7	5.3	5.8	4.6		

'LARGE' FEATHERS OF PASSERINES WITH SOME YELLOW

Species	P S	R	Pmax	Pmin	Smax	Smin	Rmax	Rmin
Crossbills	margin (♀)		9.6	5.3	6.8	4.6		
Grey Wagtail		x					11.7	9.5
European Goldfinch	x		7	4.9	5.1	4.4		
Golden Oriole		x					10	8.2
Common Chaffinch	x		8.1	5.6	6.3	4.6		
Tree Pipit	(x)	(x)	7.8	5.4	6.6	5.4	7.4	5.9
Meadow Pipit	(x)	(x)	7.9	5.6	6.5	5.1	7	5.6
Phylloscopus warblers	x	x	6.5	3.6	5	3	5.6	3.9
Firecrest	outer margin	outer margin	5 (5.6)	3.9	4.3	2.8	4.7	3.9
Goldcrest	outer margin	outer margin	5	4	4.2	3	4.4	3.6
European Serin	x		6.6 (7.2)	4.4	5.2	3.7		
Eurasian Siskin	x	x	6.6 (7.1)	4.4	4.8	3.7	5.5	3.9
Corsican Finch	x		n.d.	n.d.	n.d.	n.d.		
Citril Finch	x		6.6	5.1	5	4.1		
European Greenfinch	x	x	7.7	5.3	5.8	4.6	6.4	4.7

Species	Yellow on P S	Reddish on P S	Yellow on R	Reddish-orange on R	Pmax	Pmin	Smax	Smin	Rmax	Rmin

'LARGE' FEATHERS OF PASSERINES WITH SOME YELLOW AND/OR REDDISH-ORANGE

Species	Yellow on P S	Reddish on P S	Yellow on R	Reddish-orange on R	Pmax	Pmin	Smax	Smin	Rmax	Rmin
Buntings	margin	margin	margin		8.3	4.9	6.7	4.4	8.7	5.3
Eurasian Treecreeper	x	x		x	6 (6.5)	4.1	5.1	3.7	6.7	4.7
Short-toed Treecreeper	x	x		x	6	4.5	5	3.8	6.7	4.1
Bohemian Waxwing	tip ov	S red wax	x		10.8	6.7	6.7	5.5	7.4	6.2
Brambling	x	ov			8.1 (9.4)	5.7	5.8 (6.5)	4.7		

'LARGE' FEATHERS OF PASSERINES WITH SOME ORANGE OR RUFOUS

Species	P S	R	Pmax	Pmin	Smax	Smin	Rmax	Rmin
Rufous Bush Robin	pale	x	7.3 (8.2)	5.9	6.2 (7)	5.1	8	6.3
Lesser Short-toed Lark	iv		8.3	5.2	6.1	4.7		
Cetti's Warbler	tinted	tinted	5.6	4.4	5.4	3.5	7	4.9
Thekla's lark	iv	o	9	6.3	7.4 (7.7)	6.7	6.8	5.8
Crested Lark	iv	(o)	9.5	6.4	7.8	5.8	7.5	5.9
Bluethroat		x					6.3	4.7
Song Thrush	base iv	tinted	10.2 (11.5)	7.5	7.9 (8.5)	6.5	9.8 (10.3)	8
Common Rock Thrush		x					7.2	6.2
Bearded Reedling	x	x	5.8	4	4.8	3.9 (3.2 j)	9.6	4.3 (4 j)
Common Nightingale		x					7.4	6.2 (5·5)
Thrush Nightingale		x					7·5	6·2
European Robin	x		6.6	4.9	5.7	4.2		
Common Redstart		x					6.5	4.8
Black Redstart		x					6.8	5.8
Great Reed Warbler	tinted		8.3 (9.2)	6.1	7	4.8		
Acrocephalus warblers (others)	tinted		6 (6.3)	4.2	5	3.4		
Wallcreeper	x		9.6	7.2	8.2	5.2		

4. Remiges or rectrices with other contrasting marks

Can be well-defined, dark or pale (but not white), etc. This list is not exhaustive.

'LARGE' FEATHERS WITH DIVERSE CONTRASTING MARKS

Species	P S	R	Pmax	Pmin	Smax	Smin	Rmax	Rmin
Alpine Accentor		buff tip					8	6.7
Zitting Cisticola		subterminal black mark					5	2.9
Common Starling	P dark tip		11.3	7.5	7.8	6.4		
Dartford Warbler		pale marks at tip					6.9	5.2
Collared Flycatcher	Si black and white		7.4	5.3	5.4	4.1		

Species	P S	R	Pmax	Pmin	Smax	Smin	Rmax	Rmin
Semicollared Flycatcher	Si black and white	irregular marks	7.4	5.3	5.4	4.3	6	5.5 (5.2)
Pied Flycatcher		irregular marks					6.1	5.2
Treecreepers	single bar, pale tip		6 (6.5)	4.1	5.1	3.7		
Locustella warblers		pale tip					6.8	4.1
Brambling	cream at base of outer vane		8.1 (9.4)	5.7	5.8 (6.5)	4.7		
Eurasian Nuthatch		tricoloured					5.5	4.5
Wallcreeper	pink at base of outer vane	grey tip	9.6	7.2	8.2	5.2	6.4	5.2
Eurasian Wren	bars	bars	4.2	3.2	4	2.9	3.7	2.7

5. Remiges or rectrices with contrasting margin on at least one of the vanes

Many species have a contrasting margin or border on one or both vanes (if it is white: see criteria 1 and 2, pp. 120-125). Some are indicated in the colour criteria in the previous paragraphs.

For reasons of space, the following table does not give the sizes of those 'large' feathers concerned but only the presence and colour of these marks. Be careful, however, as depending on the wear of the feather and the width of the coloured band, this criterion is not necessarily visible on the feather (especially on outer vanes). The colour may also vary slightly from the one given here.

'LARGE' FEATHERS WITH A MARGIN OR A BORDER (NOT WHITE)				
Species	P S outer vane	P S inner vane	R outer vane	R inner vane
Alpine Accentor	buff	pale buff	buff	
Dunnock	fawn	buff	fawn	
Rufous Bush Robin	rufous	buff		
Short-toed Lark	pale		R1 rufous	R1 rufous
Eurasian Skylark	buff to grey	buff	R1 reddish-brown	R1 reddish-brown
Horned lark	S rufous		R1 rufous	
Woodlark	buff to grey			
Lesser Short-toed Lark	buff	buff		
Crossbills (4 species)	buff			
Grey Wagtail	pale			
Yellow Wagtail	cream-buff		yellowish	
Cetti's Warbler		fawn		
Black-headed Bunting	buff		yellow-buff	
Corn Bunting	buff		grey to buff	
Other buntings (10 species)	variable		variable	
Zitting Cisticola	reddish-buff		buff	fawn
Crested & Thekla's Larks	buff	buff	R5 reddish-buff	
Pine Grosbeak	buff		buff	
Rose-coloured Starling	buff			

Species	P S outer vane	P S inner vane	R outer vane	R inner vane
Common Starling	buff		buff	buff
Spectacled Warbler	rufous	buff		
Blackcap	brown-grey	white and buff	brown-grey	
Lesser Whitethroat	buff		buff	
Rüppell's Warbler	buff-grey	buff		
Garden Warbler	brown	buff to white	brown	
Barred Warbler	grey			
Common Whitethroat	reddish-buff	buff	buff	
Sardinian Warbler	grey to rufous	buff		
Subalpine Warbler	buff-grey			
Dartford Warbler	buff-grey	buff	grey	grey
Marmora's Warbler	grey	buff		
Spotted Flycatcher	buff	buff	buff	buff
Red-breasted Flycatcher	pale	pale		
Bluethroat		grey		
Treecreepers (2 species)	tip buff			
Mistle Thrush			buff	buff
Red-rumped Swallow		buff		
Hippolais warblers (others)	buff		cream	cream
Western Olivaceous Warbler	buff-grey		cream	cream
Linnet and Twite	fawn		fawn	fawn
River Warbler	pale brown	buff		
Savi's Warbler	buff	buff		
Grasshopper Warbler	reddish-brown	buff	fawn	fawn
Moustached Warbler	fawn	buff	fawn	fawn
Great Tit	grey			
Coal Tit	green-brown		green-brown	
Marsh Tit	brown			
Tree Sparrow	buff to rufous	buff	buff to rufous	
Rock Sparrow	cream to yellowish	buff (pale)	yellowish	
Sparrows (3 species)	rufous to buff		buff to rufous	
Bearded Reedling	buff to rufous	cream		
Aquatic Warbler	buff to yellowish	buff		
Sedge Warbler	buff	buff	buff	buff
Woodchat Shrike			buff (young)	buff (young)
Red-backed Shrike	S reddish-buff		buff	
Common Chaffinch	yellow		buff	
Brambling	yellow		yellow	
Red-throated Pipit	buff	reddish	buff	buff
Tree Pipit	yellowish	grey	yellowish	yellowish

Species	P S outer vane	P S inner vane	R outer vane	R inner vane
Meadow Pipit	yellowish	buff-grey	yellowish	yellowish
Rock Pipit	buff-grey	brown-grey	buff	
Tawny Pipit	reddish-buff	pale	reddish-buff	reddish-buff
Water pipit	buff-grey	brown-grey	buff	
Phylloscopus warblers (9 species)	greenish-yellow	buff		greenish-yellow
Penduline Tit	pinkish-buff	grey	buff	
Red-flanked Bluetail		grey-buff		
Common Rosefinch	buff		buff	
Trumpeter Finch	buff	buff	buff	buff
Nightingales (2 species)		buff		
European Robin		buff		
Common Redstart	pale	buff		
Black Redstart		grey		
Great Reed Warbler	buff	fawn		
Acrocephalus warblers (other: 5 species)	fawn-buff	cream	fawn-buff	fawn-buff
European Serin	yellow		pale yellow	buff
Dupont's Lark	buff	buff		
Eurasian Nuthatch	grey			
Lesser/Common Redpoll	fawn		fawn	fawn
Stonechat and Whinchat	buff	pale		
Isabelline Wheatear Northern Wheatear	cream fawn	pale		
Citril finches (2 species)	yellow		pale yellow	buff

The following pages describe species by family or by similarity. The order followed is not systematic order.

• CORVIDAE: CROWS, MAGPIES, JAYS

NP = 9 (+1) / NS = 10-11 / NR = 12. Rook, Raven and crows have 11 Ss.

Distinctive criteria of group: shape, size, colour.

This family has rounded or rectangular wings, well-fingered in some species. The tail is square to wedge-shaped, graduated in magpies.

The size of the 'large' feathers varies from that of Blackbird to that of Common Buzzard.

Possible confusion: for the primaries of black species, especially with diurnal birds of prey.

'Large' feather colour: certain species have the whole of these feathers black; others have contrasting or bright colours.

Primaries: passerine-type shape (but large), outer primaries clearly emarginated/notched in the black species. Beware, in round-winged species, the smallest primary is often P9 (excluding the reduced P10).

Secondaries: quite wide. In the medium-sized species there are nine developed secondaries, S10 being generally very small and supple (2-3 cm). In the larger species, S10 is well developed and there is a small S11.

Tail feathers: quite long, rectangular; central ones longer in the Common Raven, very long in the magpies.

Corvidae, other than black species

Species	Pmax	Pmin	Smax	Smin	P S colour	Rmax	Rmin	R colour
SUMMARY OF 'LARGE' FEATHER CRITERIA OF NON-BLACK CORVIDS								
Common Magpie	18 (19)	11	15	11	black; much white on inner vane of Ps, little or none on Ss; green-blue irid.	27	11	black with blue, purple and bronze irid. on outer vane (both vanes on R1)
Azure-winged Magpie	16	8	14	8	black-grey Ps; white inner base, outer vane white and pale-blue; Ss grey with blue outer vane	25	9	pale blue with white tip or margin
Eurasian Jay	18 (20)	12 (11)	16 (17)	9	black-grey; outer vane partly white (except inner Ss), marked with blue/black bars at the base (especially Ss); inner Ss maroon with black tips	17 (18)	12	black; base lightly marked with blue-grey
Spotted Nutcracker	17 (18)	11	14	10	black-grey; white mark on inner vane of certain Ps	14	11	black with white tip
Siberian Jay	14	9	12	8	brown, some with rufous base	15	11.5	central ones brown; others rufous, often with brown tip

COMMON MAGPIE *Pica pica*: Ps black with wide white patch on inner vane (confusion possible with outer Ps of Black-headed Gull - but note different size), inner Ps almost all black. Black secondaries (sometimes with a small white spot on outer Ss) with broad area of green or blue iridescence on inner vane. Inner Ss with iridescence on both vanes. Rs become very long towards the centre (pointed, wedge-shaped tail). *Possible confusion*: with the Ss and Rs of the other Corvidae, but the iridescence is always much more obvious in Common Magpie.

AZURE-WINGED MAGPIE *Cyanopica cyanus cooki*: Ps dark grey, outer ones with white at base of inner vane. The outer vane is azure blue at the base (until the emargination for P4-P9) and then white, bicoloured then all blue for the inner Ps (P1 and P2). Ss ash-grey with azure-blue outer vane. Rs become very long towards the centre, tail rounded but elongated to wedge-shaped depending on the individual (curve of relative sizes similar to that of Common Magpie). Rs azure-blue; central Rs usually blue, others may have a little white at the tip. In the Asian races of the Azure-winged Magpie *Cyanopica cyanus*, certain central Rs have a large white tip.

Possible confusion: with Eurasian Jay for the outer Ps, but the blue areas are barred black in the Jay.

EURASIAN JAY *Garrulus glandarius*: Ps dark grey with white inner vane or vane margin, usually blue with black bars at the base. Sometimes only white or pale grey on outer vane, sometimes barred blue on the whole outer vane (inner Ps); rarely with a white tip. Ss dark grey with marks on the outer vane. S1 to S5 with black tip, then white patch and black and blue barred base. S6 generally without white but barred at the base. S7 to S9 are grey to black, more or less marked with rufous. S9 is often dark rufous with oblique black tip. S8 often has less obvious rufous but with the same pattern. S7 is generally black-grey,

▷ **Wing of Eurasian Jay.**
Young: note the irregularity of the bars at the ends of all the primaries, greater coverts and alula feathers, and the abundance of blue barring on the inner Ps or the Ss.

rarely marked with rufous or blue (base of outer vane). Rs black-grey to black with pale blurred bars (grey to blue) towards the base, at least on the outer vane, and at least in the form of a few pale wedges. Some individuals have Rs barred to the tip. Rectangular to rounded tail (outer Rs measure 80–90% of longest Rs).

Alula, PCs and SCs coloured with black and blue bars, only on their visible part (no bars on base of inner vane). The black bars are thinner and more widely spaced in the young. Pigment abnormalities with white areas in usually striped parts sometimes occur.

Possible confusion: with the Ps of pigeons, but those of the Jay are much less robust and there are emarginations and notches on the outer Ps; the Ps of Azure-winged Magpie (a little smaller than those of the Jay); the Ss of other Corvidae if these don't have marks (often S7). Rs normally distinguishable by the markings at their base.

SPOTTED NUTCRACKER *Nucifraga caryocatactes* : Ps and Ss black, more grey-brown towards the base, often with a very small white tip, with dark blue iridescence on outer vane (sometimes difficult to see). At least two inner Ps (most often P4 and P5, sometimes P2 and P3) have a white oval on the edge of the inner vane, towards the basal third. Rs black with a white tip, discreet on central Rs becoming bigger towards outer Rs.

Primary coverts black, with a small white tip, the white descending a little on the rachis. Body feathers quite recognisable, black-brown, generally with a white drop-shaped spot (like a thrush's ventral feather in negative).

SIBERIAN JAY *Perisoreus infaustus* : Ps and Ss brown with reddish markings. The markings are found mainly at the base of the outer and inner vanes, going up towards the tip on the sides of the vanes, the brown forming a point descending along the rachis. The inner Ss (S9 to S6 or S5) have little red, but red is still visible at the base of the rachis. Rs rufous, with a grey-brown corner at the tip of the outer vane. R2 has a diagonal brown-grey tip. R1 is almost without rufous, except at its base. The tail is rectangular, but outer Rs shorter (R6 measures about 80-85% of R1).

The PCs and greater coverts are rufous, with a brown corner on the inner vane.

Possible confusion: with the Ps and Ss of Song Thrush, but those of Siberian Jay are bigger and the red colour darker (yellow-rufous in Song Thrush). Rs: see the key to bright colour criteria, Chapter 5.

Black Corvidae

The smallest is the Western Jackdaw, the largest the Common Raven. The 'large' feathers are black, the body feathers black or grey. There is almost always more or less obvious iridescence coloured blue, violet, green or purple. Feathers are sometimes discoloured by the sun and then appear brown. Some have pigmentation defects and have lighter, even white patches.

The tail is rectangular (feather size variation less than 10%), except in the Common Raven (wedge-shaped). The tip of the Rs can be clearly conical or bevelled in young birds..

SUMMARY OF CRITERIA OF THE 'LARGE' FEATHERS OF BLACK CORVIDE								
Species	Pmax	Pmin	Smax	Smin	Iridescence	Rmax	Rmin	Iridescence
Common Raven	47	26	26 (28)	16	Ps, blue-green; Ss, purple-red	30	20	green/purple-red
Hooded Crow	32	18	21	13	Ps, blue-green; inner P and S, purple-red	20 (22)	16	green-blue/ purple-red
Carrion Crow	31	17	21	12	P, purple-green; Ss, green (iv) purple-red (ov)	21	15	green/purple-blue
Rook	29 (35)	18	21	15	Ps, purple-green; Ss, green (iv) purple-red (ov)	20 (23)	16	red/purple-green
Red-billed Chough (SS)	27 (31)	17	18	14	no data	17	16	no data
Alpine Chough (SS)	24	15	16	12	no data	19	16	no data
Western Jackdaw	20 (28)	13	16	11	Ps, green-blue; Ss, purple-red	17	9	green-blue

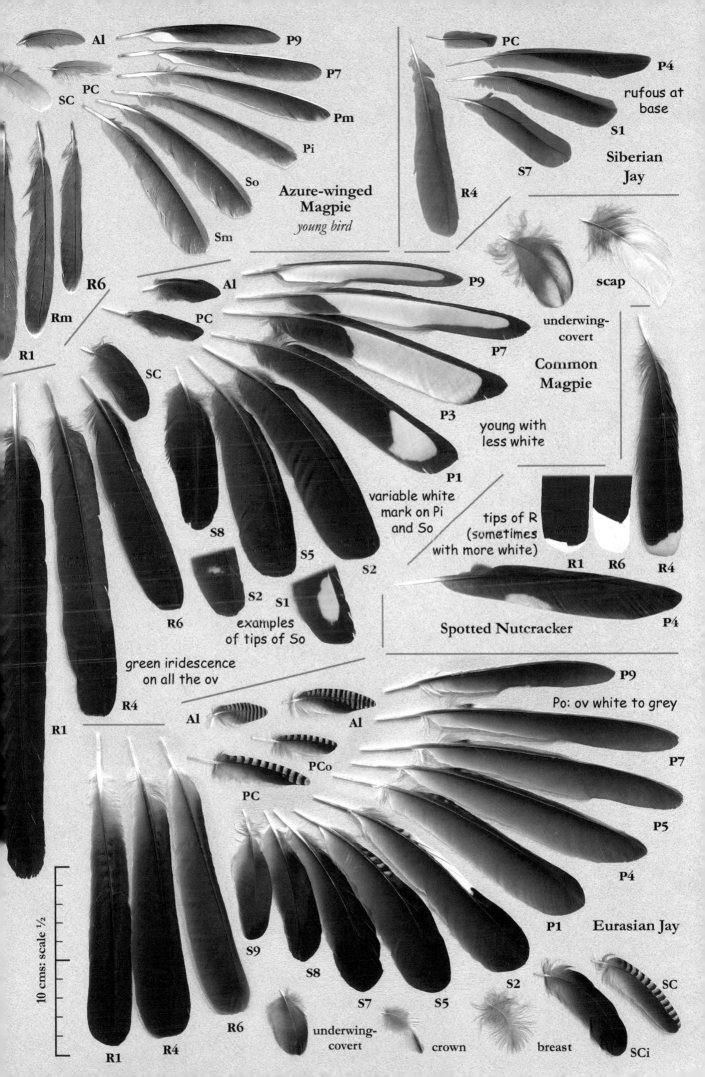

Al

SC

PC

P9

P7

Pm

Pi

So

Sm

**Azure-winged
Magpie**
young bird

R6

Rm

R1

Al

PC

SC

P9

P7

P3

P1

R4

PC

P4

rufous at
base

S1

S7

**Siberian
Jay**

scap

underwing-
covert

**Common
Magpie**

young with
less white

variable white
mark on Pi
and So

tips of R
(sometimes
with more white)

R1   R6   R4

Spotted Nutcracker

P4

S8

S5

S2

R6

S2

S1

examples
of tips of So

R4

R1

green iridescence
on all the ov

Al

Al

PCo

PC

P9

Po: ov white to grey

P7

P5

P4

P1   **Eurasian Jay**

S9

S8

S7

S5

S2

SC

SCi

R6

R4

R1

underwing-
covert

crown

breast

10 cms: scale ½

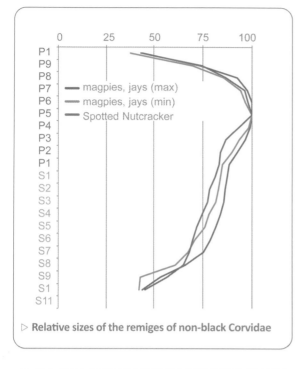

▷ **Relative sizes of the remiges of non-black Corvidae**

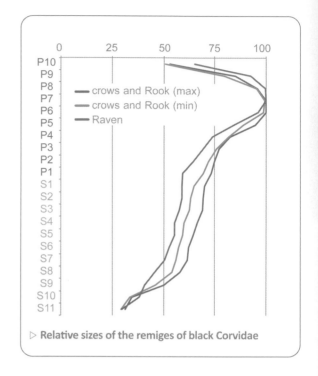

▷ **Relative sizes of the remiges of black Corvidae**

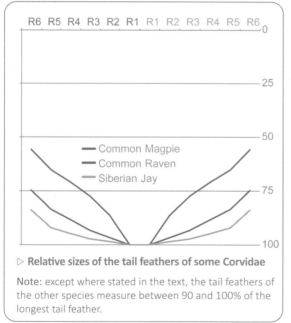

▷ **Relative sizes of the tail feathers of some Corvidae**

Note: except where stated in the text, the tail feathers of the other species measure between 90 and 100% of the longest tail feather.

▷ **Grey tint at the base of a Hooded Crow's secondary**

COMMON RAVEN *Corvus corax*: the size of its flightfeathers distinguishes it from all the other black Corvidae considered here. The hand is slightly more pointed and the forearm narrower than in other black species. The tail is clearly wedge-shaped, with outer Rs measuring about 75% of the central ones.

Possible confusion : with diurnal birds of prey (although they are not so black), storks (but their iridescence is different, sometimes white); see colour criteria 'Ps, Ss and Rs uniformly dark' (p. 59) and 'species with obviously notched outer Ps' (p. 84).

HOODED CROW *Corvus cornix* and **CARRION CROW** *Corvus corone*: the 'large' feathers are very similar, but the base of the vane is often tinged with grey in the Hooded Crow (at least on outer Ps and Ss, and on inner Rs). The absence of grey does not eliminate Hooded Crow, some individuals having all-black flight feathers.

ROOK *Corvus frugilegus*: compared to crows, the hand and forearm are very slightly narrower.

RED-BILLED CHOUGH *Pyrrhocorax pyrrhocorax* is only slightly bigger than **ALPINE CHOUGH** *Pyrrhocorax graculus*; both are slightly larger than the Jackdaw, and significantly smaller than the Rook and crows. These two species live in particular habitats (mountains, rocky coasts) and are rarely found elsewhere. The environment where feathers are found can therefore aid in their identification.

WESTERN JACKDAW *Corvus monedula*: its 'large' feathers measure about 65-75% of those of crows. It is the smallest of the black Corvidae; Ps and Ss similar in size to the Common Magpie.

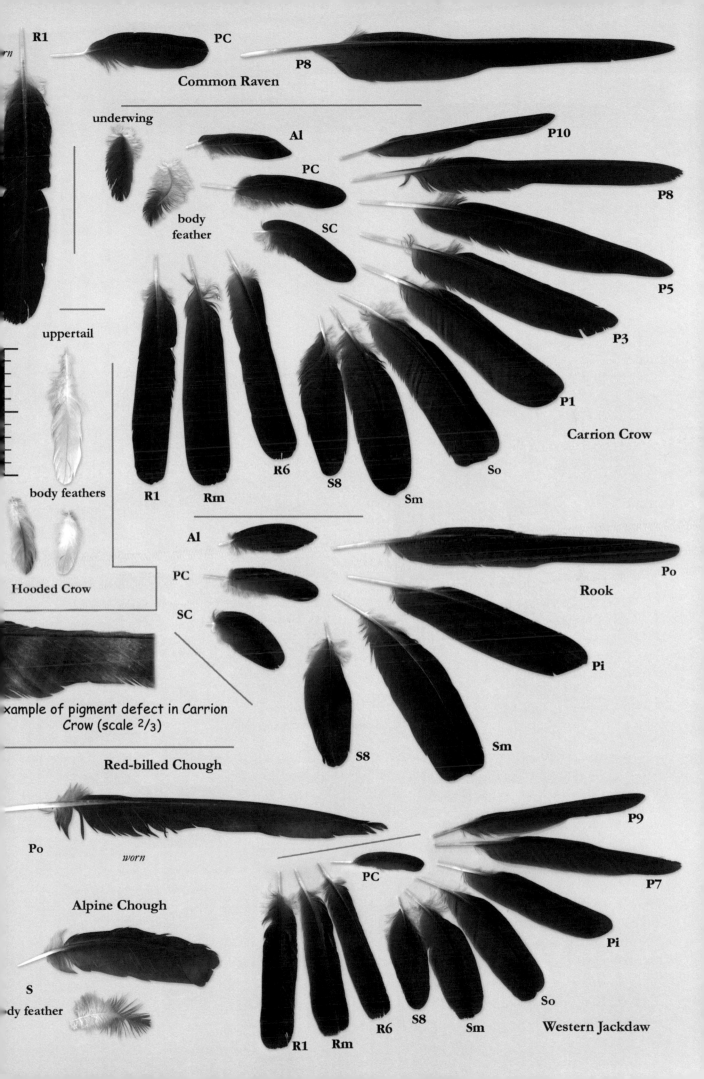

R1

PC

P8

Common Raven

underwing

Al

PC

P10

body
feather

PC

SC

P8

uppertail

P5

P3

body feathers

P1

R1

Rm

R6

S8

Sm

So

Carrion Crow

Hooded Crow

Al

PC

SC

Po

Rook

xample of pigment defect in Carrion
Crow (scale 2/3)

Pi

S8

Sm

Red-billed Chough

Po

worn

P9

PC

P7

Alpine Chough

Pi

S

R1

Rm

R6

S8

Sm

So

dy feather

Western Jackdaw

• TURDIDAE: THRUSHES, ROCK THRUSHES

General criteria of group: size, colour.

This group has rounded or slightly pointed wings; the tail is rectangular, more rounded in Mistle Thrush. The 'large' feathers have the typical passerine shape.

Possible confusion: with Common Starling (Ps and Ss), Golden Oriole (Ps and Ss), and some shrikes (Ps and Ss) but smaller in size.

'Large' feather colour: generally grey or brown to black, frequently with a different margin or a basal mark. Rufous markings in the Common Rock Thrush (R) and Song Thrush (P, S). Female colour slightly darker or browner compared to males in rock thrushes and Blackbird; sexes the same in the thrushes.

Primaries: outer Ps slightly emarginated and notched.

Secondaries: Ss with square or rounded tip.

Tail feathers: in most of the Turdidae, Rs quite elongated; Rs of young in fresh plumage with a small, pale tip (see pp. 81 and 82), growth-bars sometimes quite visible. In rock thrushes Rs shorter and more rounded.

SUMMARY OF CRITERIA OF THE 'LARGE' FEATHERS OF THRUSHES, INCLUDING ROCK THRUSHES

Species	Pmax	Pmin	Smax	Smin	P S Colour	Rmax	Rmin	R Colour
Common Rock Thrush (SS)	10.7	7.3	7.5	5.8	brown-grey; pale margin	7.2	6.2	rufous; central ones partly brown
Blue Rock Thrush	11	8	8.2	6.5	black with slate-blue margin on outer vane, a little white at tip	9.3	8.2 (7.5)	black with blue margin on outer vane
Blackbird	11.4 (13)	8.7	9.6	7.5	black or brown; possible slight pale margin on outer vane	11.8 (13)	10.2	black or brown
Ring Ouzel	12.8	9.1	9.5	7.2	black-grey; pale grey margin on outer vane	12.5	8	black-grey; outer ones with pale grey margin on outer vane
Redwing	10.5 (12)	7.4	7.7	6.3	brown-grey; pale fine margin on outer vane	9.1 (11)	7.9 (7)	brown; outer ones can have a buff corner
Song Thrush	10.2 (11.5)	7.5	7.9 (8.5)	6.5	brown; reddish base of inner vane	9.8 (10.3)	8	reddish-brown
Mistle Thrush	13.4 (14)	9.6 (8.5)	9.5 (10)	7.6	brown-grey; white zone on inner vane; pale margin	12.8 (13.3)	8.8 (8)	brown-grey; white margin and tip on inner vane
Fieldfare	12.4 (14)	9.2 (8.5)	9.3 (10.5)	7.5	black-brown; white to reddish margin on outer vane	12.3 (14)	10.5 (10)	grey-brown; base tinted grey

COMMON ROCK THRUSH *Monticola saxatilis* (SS): brown-grey remiges; off-white margin on the outer vane and tip, larger in young and females. Rs rufous, the central ones with distal part brown-grey (often with reddish tip). Sometimes some brown on outer vane.

BLUE ROCK THRUSH *Monticola solitarius* (SS): 'large' feathers black-grey, paler at the base. Quite dark in males, lighter in females. 'Large' feathers sometimes with slate-blue margin on the outer vane, especially on the Rs of the male. Remiges of young and females with an off-white tip.

BLACKBIRD *Turdus merula*: Ps, Ss and Rs are a brown to reddish-brown colour in females and young, charcoal black in the male. No sharp contrast, but the outer Ps can be edged with grey or light brown on the outer vane. Young males often have black Rs in their first year. Tail feathers large for the bird's size. The Blackbird has the most rounded wings in the group (it is the most sedentary and the most attached to densely vegetated habitats).

RING OUZEL *Turdus torquatus*: Ps and Ss are brown-grey, with grey edges on the outer vane, wider on the Ss. Inner vane paler, greyish along approximately its basal two thirds. Rs black-grey; outer Rs have a thin grey margin on the outer vane.

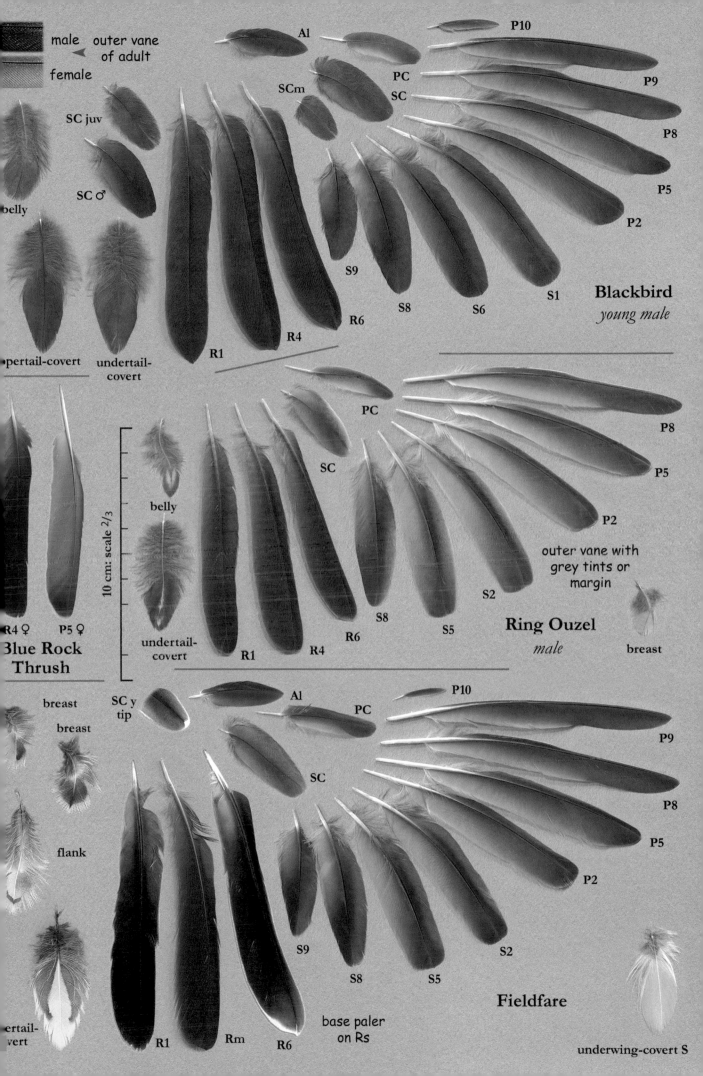

male outer vane
of adult
female

SC juv

belly

SC ♂

overtail-covert

undertail-
covert

Al

SCm

PC

SC

P10

P9

P8

P5

P2

S9

R6

S8

S6

S1

R1

R4

Blackbird
young male

R4 ♀   P5 ♀

**Blue Rock
Thrush**

10 cm: scale ⅔

belly

undertail-
covert

R1

R4

PC

SC

R6

S8

S5

S2

P8

P5

P2

outer vane with
grey tints or
margin

Ring Ouzel
male

breast

breast

breast

flank

overtail-
covert

SC y
tip

Al

PC

SC

R1

Rm

R6

base paler
on Rs

P10

P9

P8

P5

P2

S9

S8

S5

S2

Fieldfare

underwing-covert S

REDWING *Turdus iliacus*: Ps brown-grey; outer Ps with fine, buff margin on outer vane. Inner vane paler, buff-grey, on about the basal two thirds. Si with pale tip. Rs brown; greyish iridescence on undertail. R6 of young sometimes with pale corner on inner vane.

SONG THRUSH *Turdus philomelos*: Ps and Ss brown with rufous tint on inner vane. Outer Ps with pale (buff to rufous) on outer margin. Rs warm brown, often red-hued, reddish-yellow iridescence underneath.

MISTLE THRUSH *Turdus viscivorus*: Ps and Ss brown-grey, buff-grey edge on outer vane, sometimes a pale tip on inner Ss. The inner vanes of Ps and Ss have a clearly defined white zone, elongated on the Ps, more rounded in the Ss. The white/brown limit is blurred on the inner Ss. Rs brown-grey, buff margin on outer vane, white margin and tip to inner vane. Sometimes also a small white tip to outer vane. R6 with more white towards tip and large, more or less white zone along the rachis. The R6 may be largely pale (especially young birds), but in general the margin and base of the inner vane remain dark (adults and young).

FIELDFARE *Turdus pilaris*: Ps and Ss black-brown, inner vane paler at base. Outer Ps with thin white margin, becoming buff to rufous, wider and more blurred towards inner Ps and Ss. Inner vane partly greyish, becoming white towards the base. Generally a very thin pale margin towards the tip of the remiges. Very small white base to outer vane (except sometimes outer Ps). Rs black, tinged with grey at the base, sometimes reddish-brown on the outer vane.

Most useful criteria for differentiating the six species of Turdus.

– *3 size ranges* can be distinguished for remiges, and especially for the Ps: Mistle Thrush and Fieldfare > Blackbird, Ring Ouzel, Blue and Common Rock Thrushes > Song Thrush and Redwing.

– *Colour*: for remiges, the presence of a pale border or margin on the outer vane especially distinguishes Ring Ouzel and Fieldfare. An often pale tip can differentiate between the two rock thrushes. The colour of the base of the inner vane is a good feature (not always obvious on the outer Ps): white, reddish, greyer or little different from the rest of the feather.

– *Differentiating the rectrices of Song Thrush and Redwing*: their size is almost the same, usually slightly smaller in the Redwing. The upperside is vaguely tinged reddish in Song Thrush but this isn't always easy to see. On the other hand, the underside is clearly tinged with reddish-yellow in Song Thrush and grey-brown in Redwing.

– *Body feathers*: those from the breast to the belly are generally identifiable in the thrushes (see illustrations). Beware, however, of possible confusion with those of other passerines with striped underparts, such as pipits.

SUMMARY OF THE PRINCIPAL CRITERIA FOR DIFFERENTIATING THE SIX SPECIES OF *TURDUS*					
Comparison of remiges	Fieldfare (F)	Mistle Thrush (MT)	Song Thrush (ST)	Redwing (R)	Blackbird (B)
Ring Ouzel (RO)	T: F > RO, colour: ov brown on the Ss of F	T: MT > RO	colour	colour	colour: uniform in B; pale margin in RO
Blackbird (B)	T: F > B, colour: contrast ov/ base iv(S) in F	T: MT > B, colour: base iv white in MT	colour: ST redder, iv rufous at the base	colour	
Redwing (R)	T: F > R	T: MT > R, colour: base iv white in MT	colour: base of iv rufous in ST		
Song Thrush (ST)	T: F > ST	T: MT > ST, colour: base iv white in MT / rufous in ST			
Mistle Thrush (MT)	white base to iv in MT				

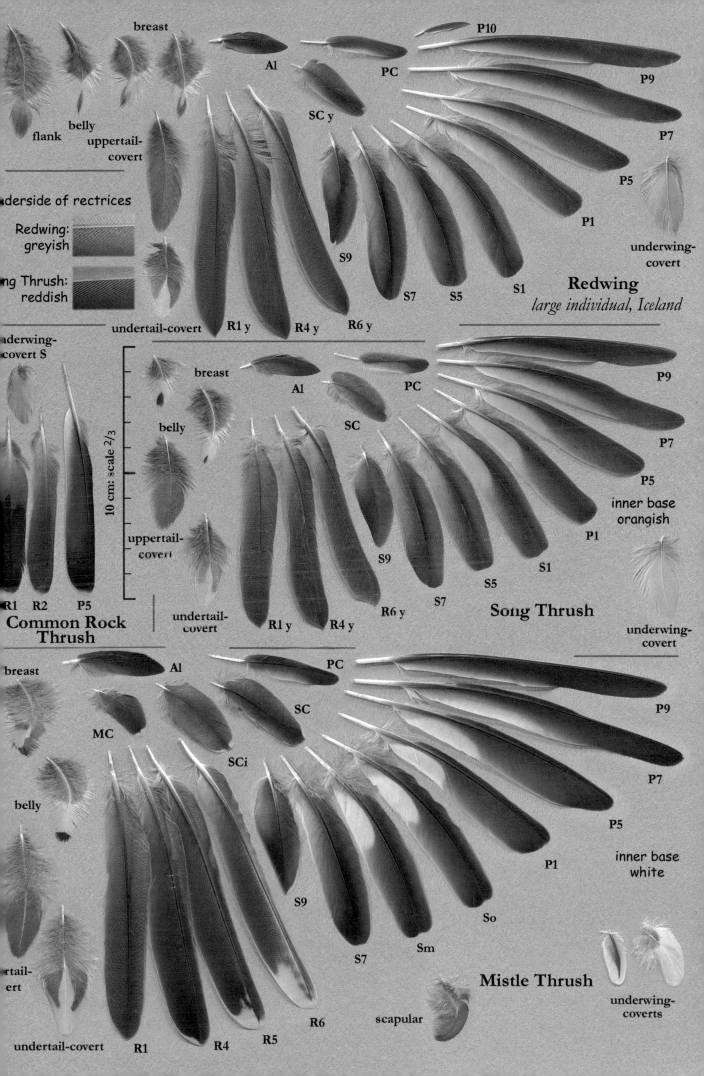

breast

flank

belly

uppertail-covert

underside of rectrices

Redwing: greyish

ng Thrush: reddish

underwing-covert S

undertail-covert

R1 y   R4 y   R6 y

Al

SC y

S9

S7   S5   S1

PC

P10

P9

P7

P5

P1

underwing-covert

Redwing
large individual, Iceland

R1   R2   P5

Common Rock Thrush

10 cm: scale 2/3

breast

belly

uppertail-covert

undertail-covert

R1 y   R4 y

Al

SC

PC

S9

S7

R6 y

S5

S1

P1

P9

P7

P5

inner base orangish

underwing-covert

Song Thrush

breast

belly

rtail-ert

undertail-covert

MC

Al

SCi

R1   R4   R5

R6

S9

S7

Sm

So

scapular

PC

SC

P9

P7

P5

P1

inner base white

Mistle Thrush

underwing-coverts

Fieldfare

Mistle Thrush

Ring Ouzel

Blackbird

Song Thrush

Redwing

▷ Thrush wings (Mistle Thrush reconstructed from a plucked bird)

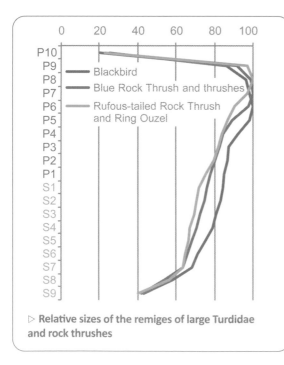

Blackbird
Blue Rock Thrush and thrushes
Rufous-tailed Rock Thrush and Ring Ouzel

▷ Relative sizes of the remiges of large Turdidae and rock thrushes

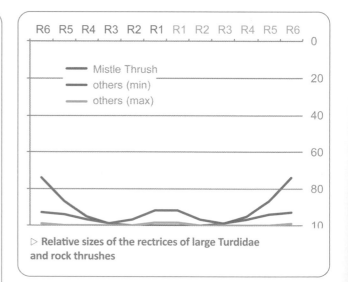

Mistle Thrush
others (min)
others (max)

▷ Relative sizes of the rectrices of large Turdidae and rock thrushes

• STURNIDAE: STARLINGS

Distinctive criteria of the group: colour. This family has quite pointed wings and a short tail.

Possible confusion: with Blackbird and thrushes. Ps and Ss of comparable size to Blackbird, Rs much shorter (see p. 136).

'Large' feather colour: general shade brown to black; possibly margins or marks, no colour difference on inner vane.

Primaries: outer Ps slightly emarginated and notched.

Secondaries: Ss with square, rounded or undulated tip.

Tail feathers: Rs quite wide; those of young with narrower tip. Square tail, very slightly forked.

COMMON STARLING *Sturnus vulgaris:* Ps and Ss brown, darker towards the tip, with a buff to rufous margin on the outer vane. Ps with a paler mark towards the tip, as if 'worn'. Ss often have a blue-green iridescence on the outer vane; the more inner ones with a large, paler area with a 'worn' appearance. Rs wide and short, brown, often with a pale margin on the outer vane, and also a 'worn' appearance. Young· Ps, Ss and Rs light brown with buff or reddish margin on outer vane.

Possible confusion: with thrushes or Blackbird for outer Ps.

SPOTLESS STARLING *Sturnus unicolor* (SS): Ps, Ss and Rs similar to Common Starling but darker, black-brown, without buff margin, with dark green iridescence and a pale 'worn' area that's quite brown. Some individuals show almost the same coloration, but with less contrast in the details indicated for Common Starling. The 'large' feathers of the young are very similar to those of Common Starling; wings a little wider than that species.

ROSE-COLOURED STARLING *Pastor roseus* (FE): no data on individual 'large' feathers of adults. The 'large' feathers of the young are very similar to those of Common Starling, but margins are whitish. Ps, Ss and Rs of adult are blackish-brown; Ps browner and Ss blacker, with blue or dark green iridescence. Outer Rs sometimes with pale pink mark towards the tip.

MEASUREMENTS OF THE 'LARGE' FEATHERS OF STARLINGS						
Species	Pmax	Pmin	Smax	Smin	Rmax	Rmin
Common Starling	11.3	7.5	7.8	6.4	7.3	6.3
Spotless Starling (SS)	11.6	7.4	8.2	6.1	7.9	5.8
Rose-coloured Starling	11.1	7.5	7.7	5.8	8.1	6.6

▷ **Common Starling wing.**

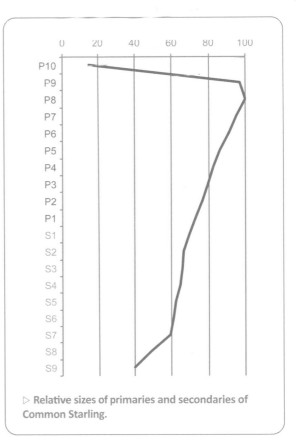

▷ **Relative sizes of primaries and secondaries of Common Starling.**

• OTHER MEDIUM-SIZED PASSERINES: GOLDEN ORIOLE, WAXWING, WALLCREEPER

Eurasian Golden Oriole *Oriolus oriolus*, Oriolidae: has quite long, pointed wings and a rectangular tail.

Distinctive criteria: colour.

Possible confusion: with the thrushes (Ps and Ss, but markings less well-defined in the oriole).

TP = 8.3-13.5. TS = (6) 6.7-9. TR = 8.2-10.

'Large' feather colour: dark and may go unnoticed, but the tail feathers are remarkable for their size and for the yellow on them. Ps and Ss black-grey with a pale margin on the outer vane (white and more or less well-defined) and at the tip (quite blurred). Base and margin of inner vane grey (often paler in female), sometimes vaguely tinged with yellowish (male). Ss of males with small white spot at tip (or thin white margin). Ss of females with a margin that becomes wider and yellowish to greenish towards the inner Ss, eventually extending to the entire vane. Nine or 10 Ss in all. Rs of male black with bright yellow tip, often with base tinted yellow. Yellow at the tip (central Rs) becomes more extensive and covers a third (sometimes half) of the feather towards the outer Rs. Rs of female grey tinged with yellow, with a yellow mark towards the tip. Yellow often extends further on the inner vane (both sexes). The yellow spot on the outer Rs is rounded in young becoming angular from the second year onwards. Some primary coverts are black with yellow tips.

Note: Golden Oriole undergoes a partial moult before the autumn migration to Africa.

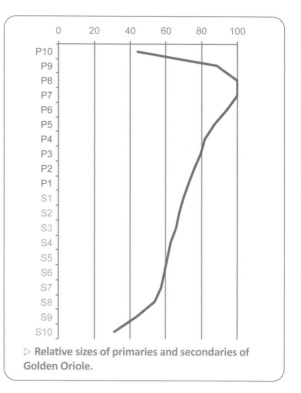

▷ **Relative sizes of primaries and secondaries of Golden Oriole.**

Bohemian Waxwing *Bombycilla garrulus*, Bombycillidae: has quite pointed wings and a rather short rectangular tail. The shape of the 'large' feathers is reminiscent of that of starlings.

Distinctive criteria: colour.

Possible confusion: colour is usually diagnostic (except P9). The yellow spot on the oriole's Rs is a different shape.

TP = 6.7-10. TS = 5.5-6.7. TR = 6.2-7.4.

'Large' feather colour: 'large' feathers black-brown. Inner vane of the Ps and Ss largely grey, less so on

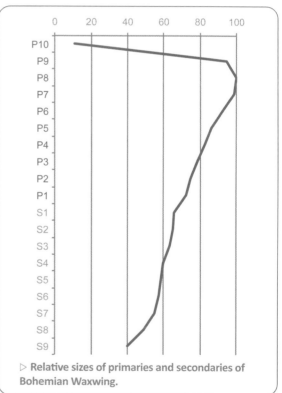

▷ **Relative sizes of primaries and secondaries of Bohemian Waxwing.**

the Rs. P1 – P6 with obvious yellow or white edge at the tip of the outer vane; rounded at the base. On P7 and P8 the white mark is smaller, and absent on P9. In adults the pale mark passes the inner vane at its tip (white), forming a characteristic 'L'. The Ss have an oval white spot at the end of the outer vane (except S8 and S9). The rachis of the Ss is tipped with a red wax spot (sometimes absent on S7 – S9 and in the young). The inner Ss are reddish-grey. The Rs black-grey with yellow tip.

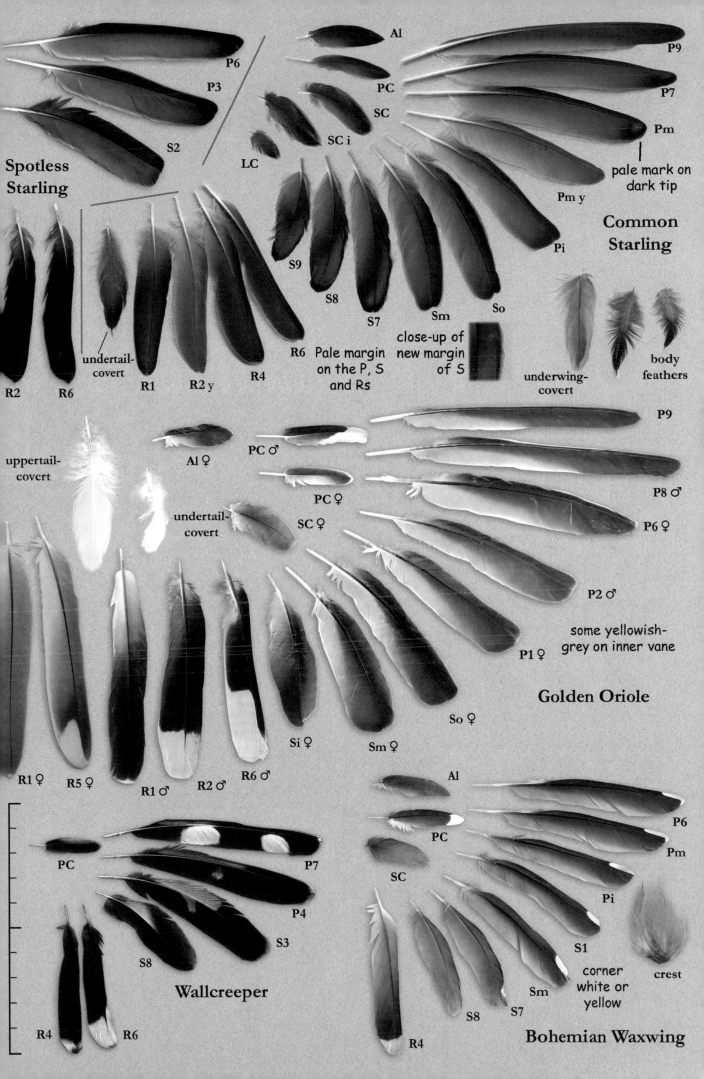

Spotless Starling

P6
P3
S2

Al
PC
SC
SC i
LC

P9
P7
Pm
pale mark on
dark tip
Pm y
Pi

Common Starling

R2   R6

undertail-
covert

R1   R2 y   R4

S9
S8
S7
R6   Pale margin
on the P, S
and Rs

Sm   So

close-up of
new margin
of S

underwing-
covert

body
feathers

uppertail-
covert

Al ♀

PC ♂
PC ♀
SC ♀

P9

P8 ♂

P6 ♀

undertail-
covert

P2 ♂

P1 ♀

some yellowish-
grey on inner vane

Golden Oriole

R1 ♀   R5 ♀

R1 ♂   R2 ♂   R6 ♂

Si ♀

Sm ♀

So ♀

Al

P6

PC

Pm

SC

Pi

PC

P7

P4

S3

S8

S1

Sm

S8   S7

corner
white or
yellow

crest

R4   R6

Wallcreeper

R4

Bohemian Waxwing

WALLCREEPER *Tichodroma muraria*, Tichodromidae: has rounded wings and a short, square tail.

Distinctive criteria: colour.

Possible confusion: for Rs, with the nuthatches (a little smaller) and Hawfinch (more white at tip, without grey); Ps and Ss characteristic.

TP = 7.2-9.6. TS = 5.2-8.2. TR = 5.2-6.4.

'Large' feather colour: Ps and Ss black with a thin white tip and especially white, pink and/or orange marks. Generally there are two rounded white marks on the inner vane of P6-P9; often only one on P4 and P5. Outer vane tinged with pink on basal half to 2/3 on P1-P6 (P8) and on all Ss, increasingly pale towards inner Ss. Inner and median Ss (often S1-S5) with an orange mark on the inner vane, touching the rachis. Rs black with white or grey tip: grey on R1 and becoming whiter towards the outer ones; larger on R6, but still with a little grey at the tip.

Alula, primary and greater coverts black, secondary coverts with some pink, at least on the outer vane.

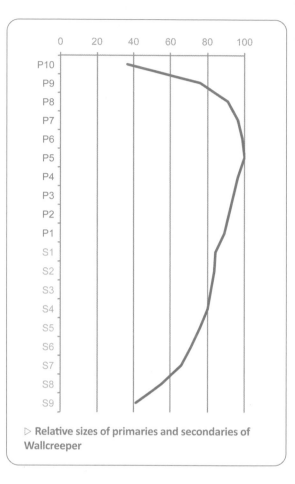

▷ Relative sizes of primaries and secondaries of Wallcreeper

• ALAUDIDAE: LARKS

Distinctive criteria of group: size, colour.

This family has quite triangular or rounded wings depending on the species; tail rectangular or square or very slightly forked (central Rs a little shorter).

Possible confusion: numerous for Ps and Rs; Ss however have a characteristic shape. For example, Skylark / young Common Starling.

Colour of 'large' feathers: variable; grey-brown, often with pale margin; inner vane tinted rufous in Thekla's and Crested Larks. Often pale marks on outer tail feathers.

Shape: the Ss and often the inner Ps [P1-P3 (P4)] have a tip as if split, forming two bumps. The vanes are longer than the rachis. This shape is also found in the hirundines, but their is colour different (black-brown in hirundines). S7 is often longer than the other Ss, a character hardly discernible in the young. There are nine or 10 Ss (S10 is quite small).

MEASUREMENTS OF THE 'LARGE' FEATHERS OF LARKS						
Species	Pmax	Pmin	Smax	Smin	Rmax	Rmin
Calandra Lark (SS)	11.2	7.8 (7 juv)	7.8	7.1	7.3	5.8
Greater Short-toed Lark	7.6	5	5.7	5.3	6.4	5.6
Eurasian Skylark	9.9	6.5 (5.4 juv)	6.9	5	7.9	6
Horned Lark (SS)	9.3	6.3 (5.6 juv)	6.5	5.7 (4.8 juv)	7.9	6.5 (5.2 juv)
Woodlark	7.2	6	6.5	5.8	5.9	5.1
Lesser Short-toed Lark (SS)	8.2	5.4	5.7	5.3	7	5.3
Thekla's Lark	9	6.3	7.4 (7.7)	6.7	6.8	5.8
Crested Lark	9.5	6.4	7.8	5.8	7.5	5.9
Dupont's Lark (SS)	8.8	6.4	7	5.4	7.3	5
Note: some measurements of young birds are also given.						

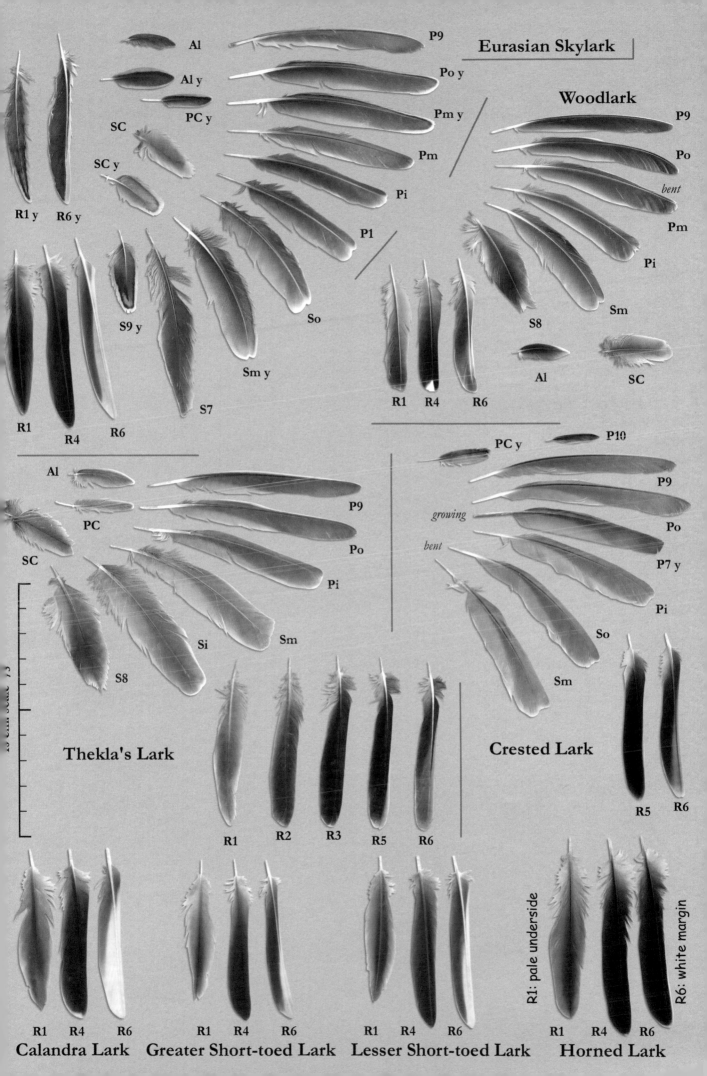

Eurasian Skylark

Al

Al y

PC y

SC

SC y

R1 y   R6 y

P9

Po y

Pm y

Pm

Pi

P1

So

Sm y

S7

S9 y

R1   R4   R6

Woodlark

P9

Po

bent

Pm

Pi

Sm

S8

Al   SC

R1   R4   R6

S8

R1   R4   R6

Thekla's Lark

Al

PC

SC

P9

Po

Pi

Si

Sm

Crested Lark

PC y   P10

P9

Po

P7 y

Pi

So

Sm

R5   R6

growing

bent

R1   R2   R3   R5   R6

Calandra Lark
R1   R4   R6

Greater Short-toed Lark
R1   R4   R6

Lesser Short-toed Lark
R1   R4   R6

R1: pale underside

Horned Lark
R1   R4   R6

R6: white margin

CALANDRA LARK *Melanocorypha calandra* (SS): Ps and Ss dark brown-grey, browner towards the inner Ss. White margin on outer vane and at tip. Pale tip to inner Ps (quite ill-defined); thin or absent on inner Ss. R1 brown, darker towards the centre, the others black-grey. R6 with a lot of white descending from the tip; black only at the base of the outer vane and a little along the inner vane (sometimes nearly totally white). R4 and R5 with small white tip; sometimes also R3 and R2. Ps, Ss and Rs of young are browner, with less white.

GREATER SHORT-TOED LARK *Calandrella brachydactyla*: Ps and Ss brown. Pale margin to outer vane and at the tip, wider on Ss. Pale to white border on inner vane. Inner Ss black-brown with broad reddish-brown border on the outer vane and at the tip. R1 black with wide reddish border. R6 with a lot of white, the others without white or with a white border towards tip (distal third).

EURASIAN SKYLARK *Alauda arvensis*: Ps and Ss brown with buff margin on the outer vane and with ill-defined pale border towards tip, more extensive on the Ss. Grey margin on outer Ps. Inner vane of Ss with a more or less clear buff margin. Rs bordered or lined with pale margin or border towards tip. R1 brown with reddish-brown border; R2–R6 black-brown to black. R6 with a lot of white diagonally (dark base of outer vane and side of the inner vane, white at the tip). R5 with white margin on the outer vane, R4 and R3 with white or black margins. Ps and Ss of young with large buff margin around the entire feather, the pale colour underlined with dark. Rs as in the adult or with buff margin; the white of the outer Rs can be buff. R1 sometimes reddish with dark markings.

HORNED LARK *Eremophila alpestris* (SS): Ps and Ss brown, Ps darker towards the tip. light grey edge on the outer vane of Ps, not always clearly visible but wide on P9. Pale edge at tip of Ps and Ss. Ss with reddish-brown margin on outer vane and pale edge on inner vane. R1 black in the centre with large brown to reddish-brown border. Other Rs black, R6 with white margin at the end of the outer vane and at the tip. R5 (sometimes R4) with white border towards tip of outer vane. The Ps, Ss and Rs of young are browner.

WOODLARK *Lullula arborea*: Ps and Ss brown. White to cream margin on the outer vane of the outer and median

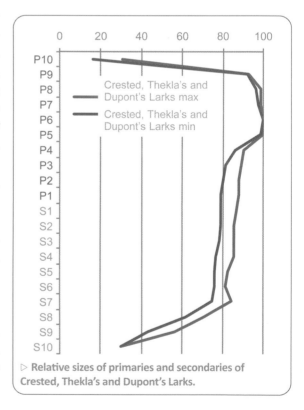

▷ **Relative sizes of primaries and secondaries of larks.**
Note: in Greater Short-toed Lark, S7 is very long and can measure 90% of Pmax.

▷ **Relative sizes of primaries and secondaries of Crested, Thekla's and Dupont's Larks.**

Ps, more buff to grey-brown on the inner Ps and the Ss. Edge of the inner vane paler, buff to white-grey (quite indistinct). Rs brown-grey to black with pale markings. R1 and R2 brown, darker in the centre; tip paler. R3–R5 grey at the base becoming black towards the tip, with a triangular white spot, largest on R5.

R6 brown-grey with a broad, pale, ill-defined distal spot. In the young, Ps and Ss are more brown, but R6 more contrasting in black and white.

LESSER SHORT-TOED LARK *Calandrella rufescens* (SS): Ps and Ss resemble those of a small Crested Lark. Ps and Ss brown with grey or buff margin on outer vane. Ps and Ss with a tawny margin on the inner vane; wide on the Ss, narrower and not reaching the tip on the Ps. R1 and R2 brown, darker in the centre; R1 largely bordered with reddish-brown. R3–R6 brown-black or black. R6 is largely white (outer vane and centre and tip of inner vane), the others towards R3 have less and less white, only on the outer vane and that near the tip; may be absent on R3.

CRESTED LARK *Galerida cristata* and **THEKLA'S LARK** *Galerida theklae*: Ps and Ss brown, with a reddish area on the inner vane, going from the edge towards the tip and joining the rachis before the base. Fine buff margin on the outer vane and at the tip. Often a darker, thin band before the light margin on the inner Ss. R1 and R2 brown (R2 darker), vaguely more rufous around the edges. R3–R6 black with base of inner vane paler, grey to reddish. R6 with a large pale area, buff to rufous, from the tip and larger on the outer vane. R5 also with pale mark on outer vane; only a small pale mark on R4 and sometimes at the end of other Rs (especially in young birds). R1, R2 and R6 may show vague dark barring. Crested and Thekla's Larks are very difficult to differentiate: Thekla's Lark is on average a little smaller but there is much overlap. On the underside of the Ps and Ss, Crested is slightly more reddish in shade and Thekla's more greyish at the base of the inner vane, at least on worn feathers. Rs very similar; those of Thekla's Lark often more contrasting, a little redder on outer Rs.

DUPONT'S LARK *Chersophilus duponti* (SS): Ps and Ss brown, with a pale margin. Broad fawn-coloured margin on inner vane. The inner Ss are more rufous. Rs with same pattern as Crested and Thekla's Larks, except outer Rs that are black and white (without any rufous). The R1 is a little shorter (about 85% of Rmax).

• TWO ATYPICAL SPECIES:
WHITE-THROATED DIPPER & EURASIAN WREN

WHITE-THROATED DIPPER *Cinclus cinclus*, Cinclidae:
wings rounded, rectangular tail.
Distinctive criteria: size, colour.
Possible confusion: few comparable species, possibly the crossbills.
TP = 6-7.6 (9.5). TS = 4.2-6.2. TR = 5-6.

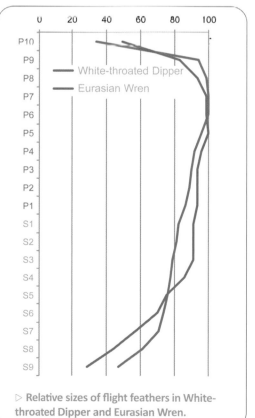

▷ Relative sizes of flight feathers in White-throated Dipper and Eurasian Wren.

▷ **Wing of Eurasian Wren.**

Colour of 'large' feathers: Ps and Ss chocolate brown; pale margins (often white) on the inner vanes, wider towards the tip; outer vane slightly slate-grey. Rs brownish slate-grey.

EURASIAN WREN *Troglodytes troglodytes*, Troglodytidae:
very much smaller, with very rounded wings and square tail.
Distinctive criteria: size, colour.
Possible confusion: none.
TP = 3.2-4.2. TS = 2.9-4. TR = 2.7-3.7.
Colour of 'large' feathers: Ps, Ss and Rs brown to rufous, with regular dark barring. Ps and Ss have outer vane reddish with black barring, more buff on outer Ps. The inner Ss and Rs have barring on both vanes.

• CERTHIIDAE: TREECREEPERS

Distinctive criteria of group: colour of primaries and secondaries; shape and colour of tail feathers.
Two species of this family occur in Europe, **EURASIAN TREECREEPER** *Certhia familiaris* and **SHORT-TOED TREECREEPER** *Certhia brachydactyla*.

They have short, rounded wings (type A) and especially a rigid tail that serves as a support when moving on tree trunks. The two species are almost impossible to differentiate from a feather found in isolation. On the other hand, their 'large' feathers are easily distinguishable from those of other passerines.
Possible confusion: shrikes (black and white) for the Rs, but they are rufous and more slender.
Remiges: nearly all brown with pale tip and a short buff edge on the tip of the outer vane, crossed by a wide buff to yellowish band. Outer Ps brown. Inner Ss brown marked with vague yellow to rufous spots. Often a small, pale tip (except outer Ps).
TP = 4.1-6 (6.5). TS = 3.7-5.1.

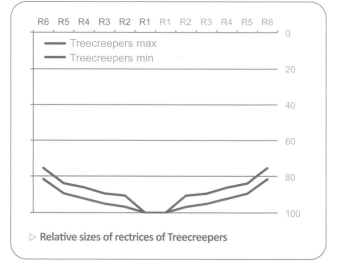

▷ **Relative sizes of rectrices of Treecreepers**

Rectrices: Rs buff to rufous, pointed, increasingly tapered at the tip towards the tail's centre. The tail acts as a support when moving on trunks, so the rectrices are rigid. Their shape resembles those of the woodpeckers, but they are more slender. The tip is pointed but breaks quite easily and can form a forked end.
TR = 4.1-6.7.

• LANIIDAE: SHRIKES

Distinctive criteria of group: colour.
This family has fairly broad wings; rather triangular in migratory species, much more rounded in more sedentary ones, such as the Great Grey Shrike. The tail is long, rectangular and the outer feathers very slightly shorter.
Possible confusion: the Rs are quite long, like those of wagtails, for example. Rs with markings sometimes similar to those of finches, buntings, larks, etc.
Colour of 'large' feathers: quite brown, except in the Red-backed Shrike; the white wing bar is seen on the Ps as a wide white base with black tip. The Ss are usually uniform, sometimes with pale markings. The tail feathers are also black and white, with varying patterns, but generally black for the central feathers, white for the outer ones, and bicoloured for the others.

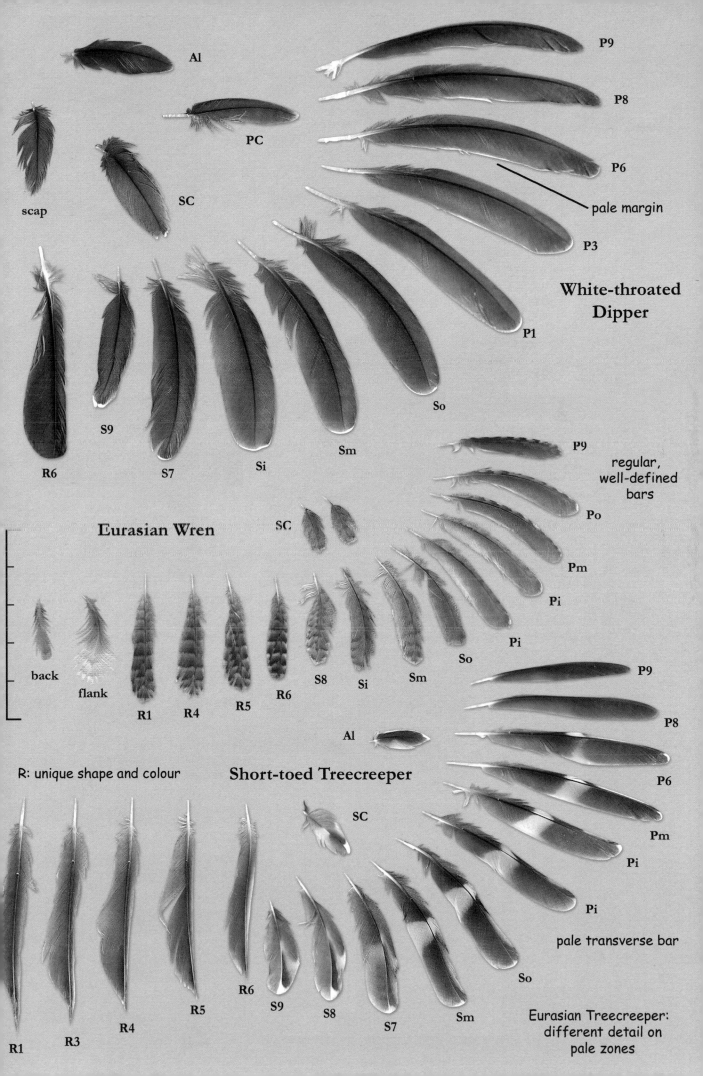

Al

PC

scap

SC

P9

P8

P6

pale margin

P3

P1

White-throated Dipper

So

R6   S9   S7   Si   Sm

SC

Eurasian Wren

P9

regular,
well-defined
bars

Po

Pm

Pi

Pi

back   flank   R1   R4   R5   R6   S8   Si   Sm   So

R: unique shape and colour

Al

Short-toed Treecreeper

P9

P8

P6

Pm

Pi

Pi

SC

pale transverse bar

So

R1   R3   R4   R5   R6   S9   S8   S7   Sm

Eurasian Treecreeper:
different detail on
pale zones

MEASUREMENTS OF THE 'LARGE' FEATHERS OF SHRIKES							
Species	Pmax	Pmin	Smax	Smin	Rmax	Rmin	Wing shape
Red-backed Shrike	8.1 (9.3)	5.9	6.5	5	8.9 (9.3)	6.9	A(b)
Lesser Grey Shrike (SS)	10.3	6.9	7.4	5.6	10	7.5	B
Woodchat Shrike	8.5	6.3	6.5	4.5	8.8	7 (6.5)	A(b)
Great Grey Shrike	10.4 (11)	7.7	7.8 (9)	6.2	12.2	9.5	A
Southern Grey Shrike	10	7	8.1	6.1	11	8.5	A
Masked Shrike	8	6.2	6.5	5.3	9.6	7.4	A(b)

The primaries are often black-brown with a white base. The white extends along the side of the inner vane, at least on the outer Ps. The area of white increases towards the inner Ps.

COMPARISON OF PRIMARIES		
Species	White base	Paler tip
Red-backed Shrike	small, especially in the male	thin, wider on outer vane
Lesser Grey Shrike (SS)	1/3 (outer) to 1/2 (inner)	on median and inner Ps
Woodchat Shrike	1/3 (outer) to 1/2 (inner)	especially on inner Ps
Great Grey Shrike	1/4 (outer) to 1/2 (inners)	on inner Ps
Southern Grey Shrike	1/3 (outer) to 1/2 (inner)	on inner Ps or absent
Masked Shrike	1/3 (outer) to 2/3 (inner)	thin, on inner Ps

The secondaries are often black-brown with a small pale tip, sometimes with a pale base or paler margin to the inner vane.

COMPARISON OF SECONDARIES			
Species	White base	Paler tip	Remarks
Red-backed Shrike	no	buff	buff to rufous margin on outer vane
Lesser Grey Shrike (SS)	sometimes (outer)	small	
Woodchat Shrike	base of inner vane paler, sometimes white (especially on inner Ss)	small	more contrasting in male; Ss sometimes brown with pale margin at tip
Great Grey Shrike	sometimes basal half of outer and median Ss; often paler margin on inner vane	well-defined, wider on inner Ss (rarely ill-defined)	varying contrast depending on age and sex
Southern Grey Shrike	basal border of inner vane (or simply paler)	small	
Masked Shrike	border of inner vane (basal half) except for inner Ss	small	white border at tip and on distal half of outer vane (thin on outer Ss)

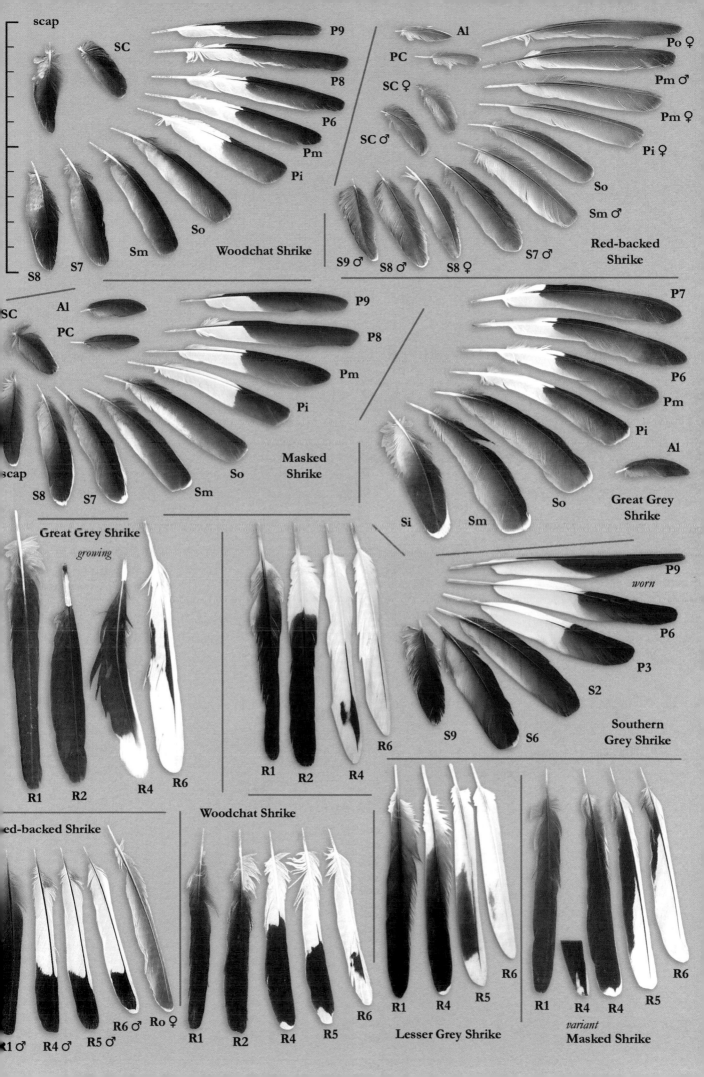

scap

SC

P9
P8
P6
Pm
Pi

So

Sm

S7

S8

Woodchat Shrike

Al

PC

SC ♀

SC ♂

S9 ♂   S8 ♂   S8 ♀   S7 ♂

Po ♀
Pm ♂
Pm ♀
Pi ♀

So

Sm ♂

Red-backed
Shrike

SC

Al

PC

scap

S8   S7

P9
P8
Pm
Pi

So

Sm

Masked
Shrike

P7

P6
Pm
Pi

Al

So

Si   Sm

Great Grey
Shrike

Great Grey Shrike

growing

R1   R2   R4   R6

R1   R2   R4   R6

Woodchat Shrike

worn

P9

P6

P3

S2

S9   S6

Southern
Grey
Shrike

Red-backed Shrike

R1 ♂   R4 ♂   R5 ♂   R6 ♂   Ro ♀

R1   R2   R4   R5   R6

Lesser Grey Shrike

R1   R4   R5   R6

R1   R4   R4   R5   R6

variant
Masked Shrike

The tail feathers are black-brown to black, increasingly white towards the outer feathers. The white increases at the base and then at the tip.

COMPARISON OF RECTRICES						
Species	**R1**	**R2**	**R3**	**R4**	**R5**	**R6**
Red-backed Shrike (male)	black, sometimes pale tip		basal half (or more) white; rarely a little white or all black; rachis black; tip white		between 1/3 and 1/4 of distal part black; large white tip; rachis black. Rarely a little black on R6	
Lesser Grey Shrike (SS)	black	base white	small white tip and white base	large tip and base white (rarely a little black)	iv partly black, sometimes only a small mark or entirely white	white or small black mark on iv
Woodchat Shrike	black	small white base	well-defined white base; small white tip	basal 1/3 white, white tip	basal 1/2 white, large white tip	white; black marking on iv (on distal 1/3)
Great Grey Shrike	(very) small white tip; sometimes small white base	small white tip and base	base and large tip white	often at least 1/4 of distal part white (and white base)	large black median mark, especially on iv	small black mark on the iv or on rachis; rarely totally white
Southern Grey Shrike	black	black	tip white	large white tip	distal 1/4 white	distal 1/3 or 1/2 white
Masked Shrike (SS)	black	black	small white tip and white base	small white tip and white base; white margin on ov (at least distal 1/3)	diagonal white mark, more or less large, extending from the tip; rachis black; ov white	base of iv black diagonally; rachis black

Note: ov = outer vane, iv = inner vane

RED-BACKED SHRIKE *Lanius collurio*: Ps, Ss and Rs quite brown, especially in females and young. Rs of female and young brown to reddish-brown, sometimes with vague darker barring. Pale margin at tip in young, also on outer vane of R6 (sometimes R5). In males, unlike other shrikes, R3 with much black, others mostly white. Black mark of varied shape and size on R4–R6.

WOODCHAT SHRIKE *Lanius senator*: young with brown Ps and Ss, base and margins paler (buff to cream). Rs brown to buff (outer), pale tip increasing in size towards outside. R6 can be quite pale.

GREAT GREY SHRIKE *Lanius excubitor*: Rs, white present on the central feathers, increasing regularly from the base and tip as far as outer Rs. When there is less black, it occurs along the rachis. On R6, the black may be absent (rarely) or as irregular markings along the rachis.

IBERIAN GREY SHRIKE *Lanius meridionalis*: Rs with little white compared to the other shrikes. The black occurs up the rachis into the white area.

MASKED SHRIKE *Lanius nubicus*: on average more white on Ps and less white on Rs compared to other shrikes.

▷ **Lesser Grey Shrike wing.**

• MUSCICAPIDAE (1): CHATS AND WHEATEARS

Distinctive criteria of group: colour.

These families have quite triangular to rounded wings. The tail is rectangular.

Possible confusion: with numerous species for the Ps and Ss; for Rs see the key of passerine tail feathers.

Colour of 'large' feathers: the remiges are often brown, sometimes with contrasting margin or base. The rectrices are often white with a black end.

MEASUREMENTS OF THE 'LARGE' FEATHERS OF CHATS AND WHEATEARS							
Species	Pmax	Pmin	Smax	Smin	Rmax	Rmin	Wing shape
Whinchat	6.5 (7.3)	5.2	5.3	3.7	5.3	4.6	A(b)
Common Stonechat	5.7 (6)	4.6	5	3.6	5.1	4.6	A
Isabelline Wheatear (SS)	8.4	6	6.6	5	6.9	5.5	B
Northern Wheatear	8.8	6.5 (6)	6.4	5	7	5.5	B
Black-eared Wheatear (SS)	8.2	6.1	6.4	4.7	7.2	5.7	A(b)
Pied Wheatear (SS)	8.1	5.9	6.4	4.7	7	6	A(b)
Black Wheatear (SS)	8.3	6.7	7.1	5.8	8 (8.2)	7.4	A

WHINCHAT *Saxicola rubetra*: Ps and Ss brown, pale margin at tip and rufous on the outer vane, broader on the Ss. Grey margin to outer Ps. Pale (white to grey or buff) border on inner vane. Pale base to outer vane (white or buff) on P5-P2 (P1), extensive on S9-S6 (S5), whiter and wider in female. Rs white at the base and dark brown towards tip, pale borders at tip and pale margins to vanes. R1 dark or with a small white base to the inner vane. R2 with about half its base white (on inner vane). The following Rs dark on distal half or third. Rachis with black markings on underside. In the young, the white of Rs

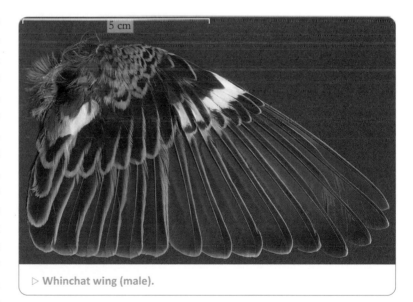

▷ **Whinchat wing (male).**

is more cream-coloured, the dark is a little more extensive, and the dark/pale limit is less well-defined.

Primary coverts are white with black tips, sometimes black at the base or on the outer vane. Inner greater coverts with a white base, black tip and buff margins.

COMMON STONECHAT *Saxicola rubicola*: Ps and Ss very similar to those of Whinchat, but on outer vane of Ps (female) the pale zone is not quite at the base. Rs dark brown to black, with more or less well-defined pale margins.

ISABELLINE WHEATEAR *Oenanthe isabellina* (SS): Ps and Ss brown, tips darker on the Ps. Cream margin on outer vane and at the tip, white border on inner vane. Rs with white base and black distal part, and pale margin towards tip. About half (R2) or third (R6) of distal part black, R1 with more black (about two thirds to three quarters, rarely more).

Northern Wheatear *Oenanthe oenanthe*: Ps and Ss brown, darker at the tip; tawny to rufous margin on the outer vane and tip, more extensive on the Ss. Inner vane with pale margin. R1 black with white base (less than a third). R2 with about half distal part black, reduced to a third on R3–R6. Tawny margin on outer vane and tip. The black often has a rounded edge and continues a little along the rachis.

Pied Wheatear *Oenanthe pleschanka*: Ps and Ss brown (female) to black (male). Lighter margin on outer vane and tip in female. Rs with white tips. The amount of black lessens from R1 towards median Rs then increases towards R6. R1with a small white base. R6 often with a black distal third, more extensive on the outer vane (sometimes less). R2–R5 with a variable extent of black and patterning; sometimes all white, with a large black tip or a mark on the tip of one or both vanes.

Black-eared Wheatear *Oenanthe hispanica*: Ps and Ss dark brown (female) to black (male), with pale margin at tip and on outer vane (quickly worn away). Inner vane with pale margin in female. R1 black with (very) small white base; the other Rs white with black tip. High variability in the proportion of black, individuals showing extreme amount of dark or pale are generally males.

R6 with extensive black tip, descending on the outer vane; other Rs with less black, sometimes white or with a small black mark at the tip of the outer vane. R2 can also have a black edge at the tip of the inner vane. The black may continue along the rachis.

Black Wheatear *Oenanthe leucura* (SS): Ps and Ss dark brown (female) to black (male); brown margin on outer vane and tip, bordered with brown to grey on inner vane. Rs white with black end. R1 white on a third to a half, the others with decreasing black, about a quarter and less than a fifth to R6. Sometimes small white tip (young?).

• MOTACILLIDAE: PIPITS AND WAGTAILS

Distinctive criteria of group: especially length of tail feathers, shape of tertials, colour (wagtails). This family spends a lot of time walking, but frequently makes short flights when feeding. The flight is undulating. They have triangular wings, with a more or less rounded hand and a fairly wide forearm. The tail is long to very long, especially in wagtails, since it serves as a counterbalance when the bird moves on the ground.

Possible confusion: with other brown passerines (larks, etc.) or with black and white 'large' feathers (shrikes, tits, etc.)

Colour of 'large' feathers: brown to grey, often with pale or white margins. Outer Rs often totally or partly white.

Secondaries: the S7 is often longer and more pointed than the other Ss.

Rectrices: all of similar size, the central feathers slightly shorter in the pipits (R1 > 90% of Rmax).

MEASUREMENTS OF THE 'LARGE' FEATHERS OF PIPITS AND WAGTAILS						
Species	Pmax	Pmin	Smax	Smin	Rmax	Rmin
Tawny Pipit	8	6.5	6.7	6	7.8	6.7
Tree Pipit	7.8	5.4	6.6	5.4	7.4	5.9
Meadow Pipit	7.9	5.6	6.5	5.1	7	5.6
Red-throated Pipit	7.7	5.3	5.7	5.2	7.3	5.8
Water Pipit	8.9	6.8	7.1	5.8	8.0	6.4
Rock Pipit	8.1	6.1	6.7	5.6	7.7	6
Yellow Wagtail	7.5	5.2	6.0	4.9	8.3	7.1
Grey Wagtail	7.7 (8.2)	5.0	5.2	4.7	11.7	9.5
Pied/White Wagtail	8.1	5.4	5.8	5.2	10	8.4

Al

PC

SC × 3

Si: white base

P9

P8

bent

P6

Pi

So

Sm

Si

S9

Common Stonechat

R1

Rm

R5

P10

P7

P5

P3

S1

S4

S8

Whinchat

R: white base

R1

R2

R4 ♀

R6

R: white base

bases broken

PC

P9

P8

P6

Black-eared Wheatear

Si

Sm

Pi

R1

Rm

R5

P9

P6

Al

SC

PC

Pm

R: white base

Pi

uppertail-covert

R1

Rm

S9

R6

Si

Sm

So

Northern Wheatear

TAWNY PIPIT *Anthus campestris*: Ps and Ss brown with buff margin on outer vane and at tip, wider on the Ss. Margins wider and more rufous towards the inner Ss. Inner vane with pale margin (ill-defined). R1 dark brown bordered with buff to rufous. R2–R4 brown to black with pale margin on outer vane. R5 and R6 black-brown with a long diagonal white mark (sometimes cream) descending from the tip; and continuing lower on the outer vane.

TREE PIPIT *Anthus trivialis*: Ps and Ss brown with buff to yellow margin on outer vane and tip; broader on the Ss and on both vanes on the inner Ss. Inner vane bordered with buff-grey (blurred). R1 dark brown with buff to yellowish border. R2–R4 brown to black with yellowish margin on the outer vane. R4 sometimes with a small white spot on the inner vane. R5 as R4 but with a white oval mark towards tip of the inner vane, sometimes very small. R6 black-brown with a long white (sometimes cream) diagonal mark descending from the tip; lower on the outer vane, becoming a pale mark sometimes tinged with grey.

MEADOW PIPIT *Anthus pratensis*: Ps and Ss as in Tree Pipit, but border of inner vane grey. Rs as in Tree Pipit, but R6 often with a little black along the tip of the outer vane.

RED-THROATED PIPIT *Anthus cervinus* (SS): Ps and Ss brown with buff margin on outer vane and tip, wider on the Ss. Inner vane bordered with rufous. Rs as in Meadow Pipit but with buff, not yellow margin.

WATER PIPIT *Anthus spinoletta*: Ps and Ss brown. The Ps have a grey margin and the Ss a buff one on the outer vane (inner Ss with broader border - from grey to buff). The Ps have a darker tip with a pale margin. The inner vanes of Ps and Ss have a vague brown-grey border. Rs as in the Tree Pipit with buff margins but the mark on R5 is more pointed and triangular-shaped (often very reduced). That on R6 is generally less extensive than in Tree Pipit, sometimes tinted with grey.

ROCK PIPIT *Anthus petrosus* (SS): 'large' feathers as in Water Pipit, but the pale markings of the Rs are generally tinted with grey and less extensive.

YELLOW WAGTAIL *Motacilla flava*: Ps and Ss brown with buff to cream margin on the outer vane. The border of the inner vane paler on P9-P7 (P6) becoming white at the base on P6-P1 and the Ss. The white does not touch the rachis, even on the Ss. Inner Ss with wider pale border on outer vane. Rs long. R1–R4 dark brown to black, yellowish margin on the outer vane. R5 and R6 with a lot of white (less in young); diagonal mark descending from the tip, some black on the border of the inner vane, sometimes black along the rachis, R6 generally with more white.

GREY WAGTAIL *Motacilla cinerea*: Ps and Ss black-brown, sometimes with grey margin on outer vane. Border of inner vane paler on P9-P7 (P6), rarely as far as P2. On (P6) P5 and towards inner Ps and Ss, the inner vane is largely white towards the base, with a small ill-defined grey edge. Towards the middle and inner Ss, the white extends onto the outer vane and covers about half the base of the Ss. Inner Ss with a diagonal white zone mainly on inner vane and a white border on the distal half of the outer vane (sometimes yellowish). Rs very long. R1–R3 black with a grey base and white to yellowish margin on the outer vane (especially at the base). R4–R6 with increasing white, a small grey base and often some yellow on the outer vane (base). R4 and R5 usually with a lot of black on the outer vane (sometimes a little on the edge of the inner vane). R6 can be almost all white. Sometimes more irregular black, especially on R4 (especially young birds).

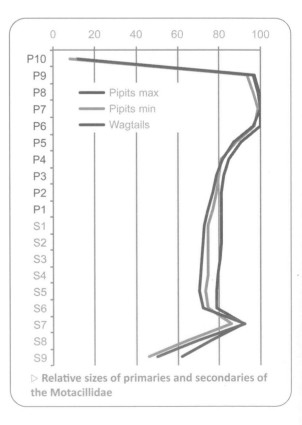

▷ Relative sizes of primaries and secondaries of the Motacillidae

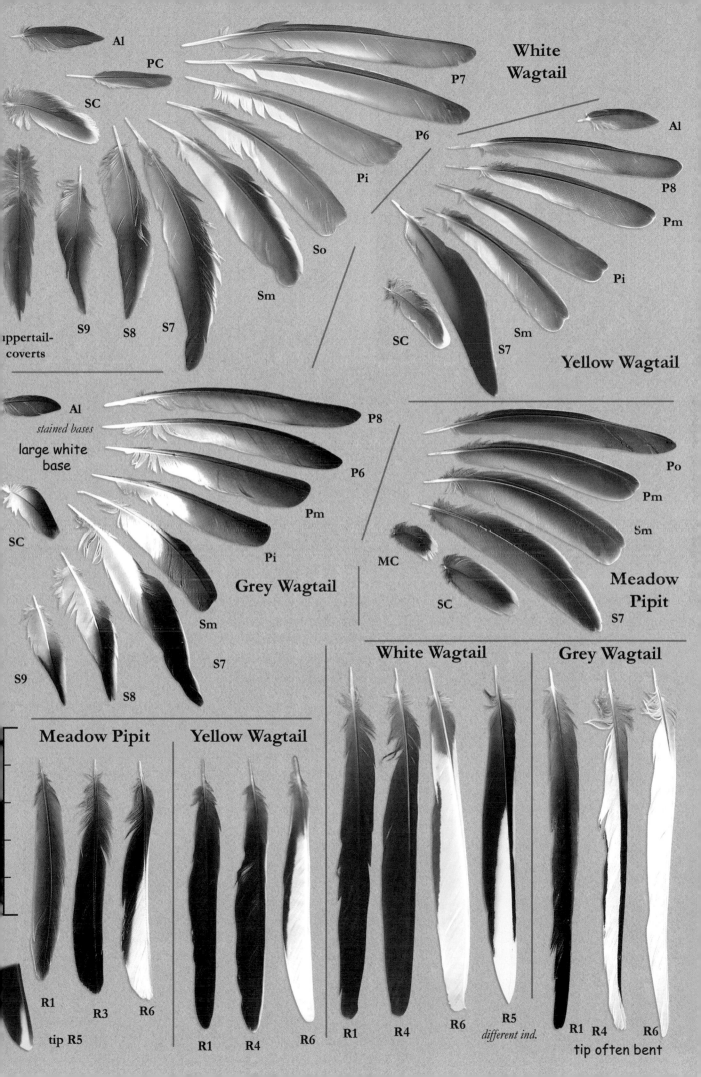

White Wagtail

Al

PC

SC

P7

P6

Pi

So

Sm

uppertail-
coverts

S9   S8   S7

Al

P8

Pm

Pi

SC

Sm

S7

Yellow Wagtail

Al

stained bases

large white
base

SC

P8

P6

Pm

Pi

Sm

S7

Grey Wagtail

S9

S8

Po

Pm

Sm

MC

SC

S7

Meadow Pipit

Meadow Pipit   **Yellow Wagtail**

White Wagtail   **Grey Wagtail**

R1

R3

R6

tip R5

R1

R4

R6

R1

R4

R6

R5

different ind.

R1   R4   R6

tip often bent

WHITE/PIED WAGTAIL *Motacilla alba alba / M. a. yarrellii*: Ps and Ss black-brown, grey to white margin, broader on the inner Ss. Border of the inner vane paler on P9-P8 (P7), then white at the base; the limit is quite indistinct on the Ps towards the edge of the change in colour. The white does not touch the rachis even on the Ss. The shade of colour is darker, especially on the inner Ss, in the *yarrellii* (Pied Wagtail) subspecies. Rs long. R1–R4 black, possibly with white margin on outer vane. R5 and R6 with lots of white, but usually remaining black at the base and on the border of the inner vane (basal half or more).

• MUSCICAPIDAE (2): FLYCATCHERS

Distinctive criteria of group: colour.

This family has triangular wings (type B); the tail is square.

Possible confusion: other passerines, including chats and wheatears, the Fringillidae, etc.

Colour of the 'large' feathers: the remiges are black or brown, with or without white at the base, possibly with a pale border. The rectrices are black, brown or black-and-white.

MEASUREMENTS OF THE 'LARGE' FEATHERS OF FLYCATCHERS						
Species	Pmax	Pmin	Smax	Smin	Rmax	Rmin
Spotted Flycatcher	7.8	5.6	5.5	4.5	6.8	5.7
Red-breasted Flycatcher (SS)	6.3	4.9	5	3.7	6.2	5
Semicollared Flycatcher	7.4	5.3	5.4	4.3	6	5.5 (5.2)
Collared Flycatcher	7.4	5.3	5.4	4.1	5.9	5
Pied Flycatcher	7.2	5.3	5.2	3.7	6.1	5.2

SPOTTED FLYCATCHER *Muscicapa striata*: Ps and Ss brown, with buff border on inner vane, and buff margin on outer vane and the tip; more extensive on the inner Ss. Rs brown-grey with buff margin. Margin and tip sometimes grey or white.

RED-BREASTED FLYCATCHER *Ficedula parva*: Ps and Ss brown, bordered with buff-grey on the outer vane and at tip, pale border on the inner vane. R1 and R2 black; R3 with the middle part of the outer vane white. The white generally extends onto the inner vane. The white zone is more or less extensive depending on the individual, and can extend down along the rachis. R4–R6 have a black tip, white median part and a small grey base, increasingly reduced towards R6. The black at the tip may descend along the edge of the outer vane on R6 and along the rachis on R5 and R4.

SEMICOLLARED FLYCATCHER *Ficedula semitorquata*: Ps and Ss dark brown to black, with a white basal half and a small grey base. White base a little smaller on P9. On the Ps, the white extends a little further along the inner than the outer vane at the rachis. Rachis black. Inner Ss with much more white on the outer vane (sometimes in irregular marks). Quite small pale base in young. R1–R3 usually black; white appears on R4 and increases to R6. Generally some white on the middle part of the outer vane of R4, becoming more extensive to R6 but still with black tip. Markings often irregular with large variations in pattern and proportions.

COLLARED FLYCATCHER *Ficedula albicollis*: Ps and Ss as in Semicollared Flycatcher, with a little less white on the inner Ss. Rs dark brown (female) to black (male). R6–R5 (R4) may have some white on the median or basal part of the outer vane.

PIED FLYCATCHER *Ficedula hypoleuca*: Ps and Ss dark brown to black, with white margin on the inner vane. From P6 inwards onto the other Ps and Ss, there is an almost basal small white to cream mark on the outer vane. On the inner Ss, the regular or irregular pale mark can be quite extensive on the outer vane (base and edge, sometimes whole vane), and possibly also the base of the inner vane as well. R1–R3 black. Some white or cream (sometimes grey) appears on R4 (sometimes R3) in the middle of the outer vane, then becomes increasingly extensive towards R6, covering almost all the vane (except the tip), and sometimes extending onto the inner vane.

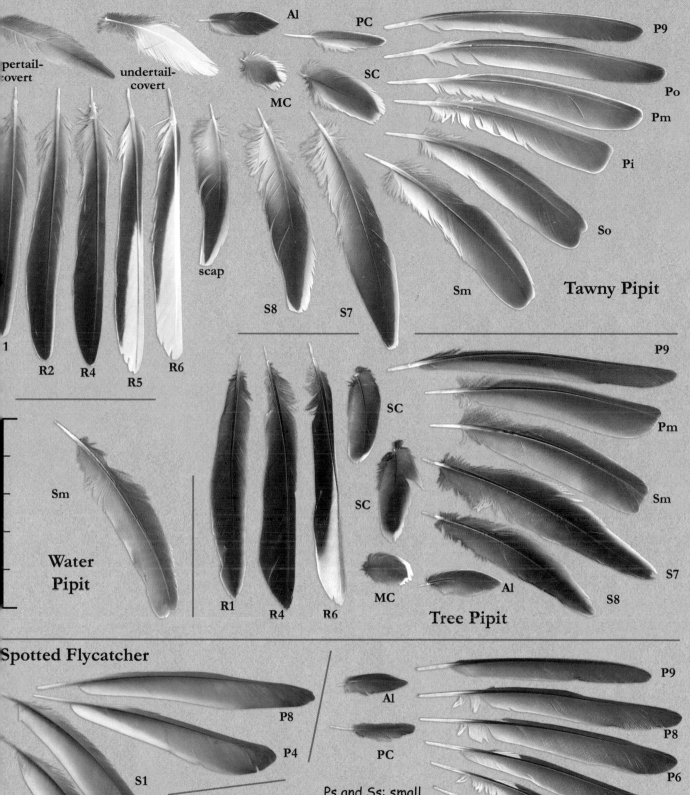

pertail-covert

undertail-covert

Al

PC

SC

MC

scap

S8   S7

P9

Po

Pm

Pi

So

Sm

Tawny Pipit

1

R2   R4   R5   R6

Sm

Water Pipit

R1   R4   R6

SC

SC

MC

Al

Tree Pipit

P9

Pm

Sm

S7

S8

Spotted Flycatcher

P8

P4

S1

s broken

S5

R1   R6   R2   R3   R4   R5

R3 often black

Al

PC

Ps and Ss: small white base ov

SC

R6   S9   S8   S7

P9

P8

P6

Pm

Pi

So

Sm

Pied Flycatcher

• FRINGILLIDAE: FINCHES

Distinctive criteria of group: shape, size, colour.

This family has short but pointed wings, with a triangular hand; some species, however, have more rounded wings (Bullfinch, Serin, etc.). Their flight is often undulating (flapping of the wings alternating with short glides). The tail is very frequently forked.

Many species have characteristic markings on their 'large' feathers, often with obvious yellow or white colours. Their feathers can be found in a variety of habitats, often in proximity to man, in gardens and parks.

Possible confusion: with the other passerines. Within this group, some species are very similar to each other (e.g. Linnet and Twite, and the crossbills).

Colour of the 'large' feathers: very variable (see following descriptions).

Primaries and secondaries: frequently have a mark corresponding to the wing bar. The tips of the outer and median Ss are slightly split (two-bump shape) in several species (not consistently indicated in the descriptions).

Rectrices: regularly have contrasting colours, such as yellow or white patches alternating with black. The central feathers are generally a little shorter than the others.

Body feathers: have bright and varied colours in many species: yellow, pink, red, etc. They can be important when finding feathers, so do not neglect them when collecting.

Here species are grouped according to the general colour criteria visible on the 'large' feathers:
– with some white: chaffinches, Linnet, Hawfinch;
– without bright colours: redpolls, crossbills, Bullfinch;
with some pink: rosefinches and Pine Grosbeak;
– with some well-defined yellow: Greenfinch, Goldfinch, Siskin;
– with ill-defined yellow patches: Serin, citril finches.

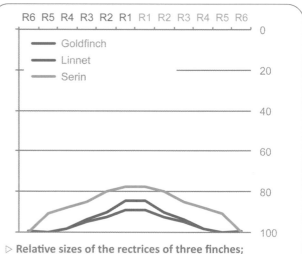

▷ **Relative sizes of the rectrices of three finches; European Goldfinch, Common Linnet and European Serin.**

MEASUREMENTS OF THE 'LARGE' FEATHERS OF FINCHES								
Species	Pmax	Pmin	Smax	Smin	Rmax	Rmin	Wing shape	Tail shape (Rmin/Rmax)*
Two-barred Crossbill	8.3	5.2	5.6	4.6	7.2	5.7	B	forked (85-90 %)
Scottish Crossbill (SS)	8.4	5.7	5.5	5	6.8	5.8	B	forked (85-90 %)
Common Crossbill	8.5	5.3	5.7	4.8	7.3	5.7	B	forked (85-90 %)
Parrot Crossbill (SS)	9.6	5.9	6.8	5.1	8	6.1	B	forked (85-90 %)
Eurasian Bullfinch	8.4	5.8	6.4	4.8	7.5	6.3	A(b)	rectangular (> 95 %)
European Goldfinch	7	4.9	5.1	4.4	5.7	4.7	B	slightly forked (88-92 %)
Pine Grosbeak	10 (12)	7	8.2	6	10.6	8.6	B	slightly forked (89-93 %)
Hawfinch	9.2 (10.5)	6.3	7 (7.3)	5.4	6.5	5.3	B	(very) slightly forked (86-92%)
Twite	6.9	4.8	5.1	4.1	7.1	4.6	B	forked (79-87 %)

Species	Pmax	Pmin	Smax	Smin	Rmax	Rmin	Wing shape	Tail shape (Rmin/Rmax)*
Common Linnet	7	4.7	5	4	6.2	4.6	B	forked (84-88 %)
Common Chaffinch	8.1	5.6	6.3	4.6	7.8	6.1	A(b)	slightly forked (90-95 %)
Brambling	8.1 (9.4)	5.7	5.8 (6.5)	4.7	7.5	5.5	A(b)	forked (83-87 %)
Common Rosefinch (SS)	7.3	5.5	5.6	4.8	6.4	5.5	B	forked (85-90 %)
Trumpeter Finch (SS)	7.6	4.9	5.2	4.4	5.9	4.6	B	forked (85-90 %)
European Serin	6.6 (7.2)	4.4	5.2	3.7	5.8	4.3	A(b)	highly forked (75-85 %)
Arctic Redpoll	7 (7.6)	4.8	5.2	3.8	6.8 (7.5)	5	B	highly forked (80-86 %)
Common Redpoll	6.6 (7.5)	4.7	4.8	3.7	6.5	4.8	B	highly forked (80-84 %)
Eurasian Siskin	6.6 (7.1)	4.4	4.8	3.7	5.5	3.9	B	highly forked (80-85 %)
Corsican Finch (SS)	n.d.	n.d.	n.d.	n.d.	5.7	4.7	A(b)	–
Citril Finch (SS)	6.6	5.1	5	4.1	6.1	4.4	A(b)	highly forked (80-85 %)
European Greenfinch	7.7	5.3	5.8	4.6	6.4	4.7	B	forked (84-88 %)

*: variation measured from a sample

Finches with some white

The white on the remiges occurs as a basal zone on the outer vane (Chaffinch and Brambling), as a line (Linnet and Twite) or as a large zone on the inner vane (Hawfinch).

COMMON CHAFFINCH *Fringilla coelebs*: Ps and Ss dark grey to black, with a white border on the inner vane and straw-yellow border on the outer vane. P9-P7 with white border (ill-defined) on the notched part of the inner vane, with pale yellow margin on the emarginated part of the outer vane and white margin towards the tip. P6-P1 with white at base of outer vane, the rest with pale yellow margin which disappears in the middle part of the feather towards the inner Ps. White border on the inner vane, widening towards the base and touching the rachis lower than the white on the outer vane; a little pale grey at the base of both vanes.

Ss with white base, going up the inner vane and becoming narrower; a little grey at the base near the rachis. Outer web bordered with pale yellow in the distal half (or less). Inner Ss grey at the base becoming darker towards the tip, bordered with yellowish-buff on the distal part of the outer vane and the tip. R1 dark greenish-grey with buff margin; sometimes with white margin on the inner vane. R2–R4 black, with or without a white or yellowish margin on outer vane. R5

5 cm

▷ **Chaffinch wing (male)**

and R6 black with white margin on the outer vane, and a white mark at the tip. R5 with a triangular white mark on the inner vane, sometimes descending a little along the edge of the vane. On R6 there is a diagonal white mark from the tip of the inner vane to the rachis, passing onto the middle part of the outer vane. This white mark is sometimes smaller, with irregular edges or inclusions of black marks. Greater coverts black with white or pale yellow tips.

Possible confusion: for the Rs, with buntings in particular; Ps and Ss, with Pied Flycatcher (no border on the outer vane).

BRAMBLING *Fringilla montifringilla* : Ps and Ss dark grey to black, bordered with white on the inner vane and with a pale margin on the outer vane. P9-P7 bordered with white (ill-defined) on the notched part of the inner vane, with white or pale yellow margin on the outer vane. Other Ps with a white or cream area towards the base of the outer vane, but with grey at the base of the vane, with a complete white or pale yellow margin. Inner vane bordered with white widening towards the base but not touching the dark grey at the base along the rachis. Ss with a small white or cream diagonal mark on the basal third of the rachis; base of the vane is dark grey with the distal half bordered with white, cream or orange (inner and median Ss). Inner Ss grey at the base and black towards the tip, with orange border on the distal half of the outer vane. R1 ash grey, with cream to pale yellow margin on inner vane and border on outer vane. R1 is sometimes noticeably darker with light yellow border at the end and on the distal half (or less) of the outer web. R2–R4, and sometimes R5, are black (dark grey at the base) with pale yellow margin to distal part of the outer vane; margin descends increasingly lower towards R5. R5 may also have a lighter grey mark towards the tip of the inner vane (generally quite ill-defined). R6 has more or less contrasting colours, black with a paler diagonal band. The most obvious mark is a thin white stripe starting from the margin of the inner vane on its distal third and passing the rachis as it descends onto the outer vane by 'spreading' over a large part of its centre. In contrast, in some cases there is only a wide, diagonal, light grey band on the inner vane. Intermediate patterns also occur (see diagram).

Greater coverts black with yellow-buff or orange tips.

Possible confusion: for the Ps, some individuals have a lot of white at the base, reminiscent of Chaffinch. For the Ps and Ss, Pied Flycatcher, although that species does not have a pale margin on the outer vane of those feathers.

5 cm

▷ **Brambling wing (male)**

▷ **Examples of the pattern of R6 of Brambling.**

COMMON LINNET *Linaria cannabina* : Ps and Ss dark brown to black, inner vane with white border, tip with white to pale brown border (small on outer Ps). Ps with white margin on outer vane (not at tip); margin sometimes tinted with buff towards tip. Ss with fawn border, white at the base (S1 without fawn). Inner Ss without white, grey at the base and bordered with fawn at the tip. The tip of the Ss is slightly forked. Rs black with large white

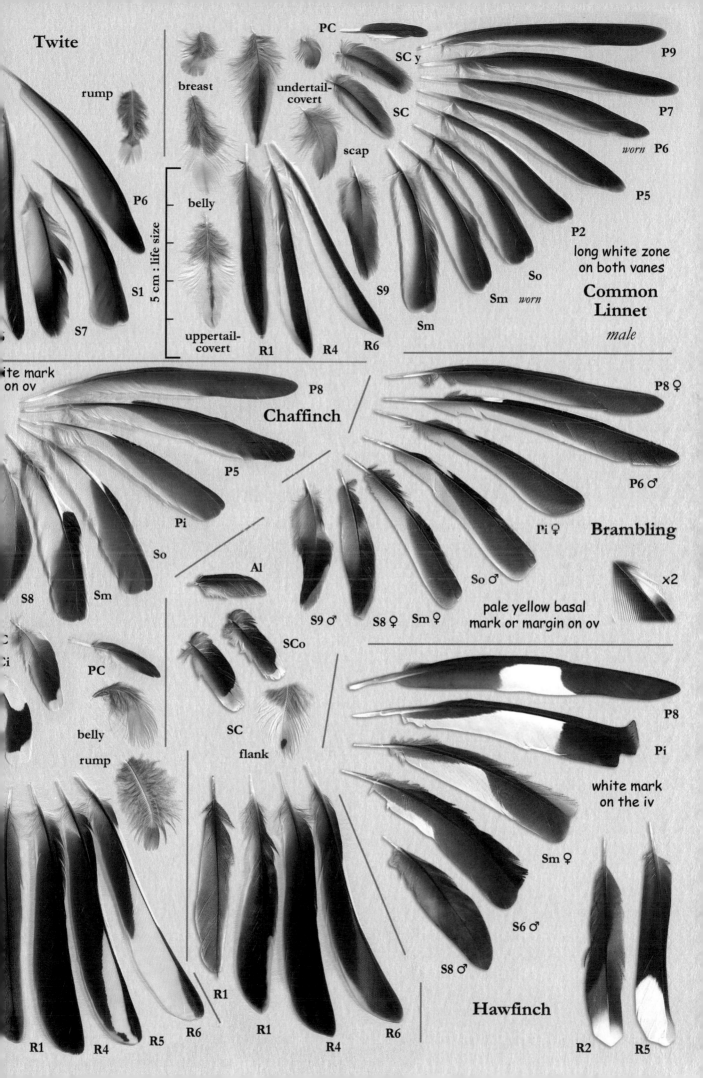

Twite

rump

breast

undertail-covert

scap

PC

SC y

SC

P9

P7

worn P6

P5

P2

long white zone
on both vanes

Common
Linnet

male

So

Sm *worn*

Sm

belly

5 cm : life size

uppertail-covert

R1

R4

R6

P6

S1

S7

S9

ite mark
on ov

P8

Chaffinch

P5

Pi

So

Sm

S8

Al

S9 ♂

S8 ♀

Sm ♀

P8 ♀

P6 ♂

Pi ♀

Brambling

So ♂

×2

pale yellow basal
mark or margin on ov

PC

SCo

SC

flank

belly

rump

R1

R6

R1

R4

R6

P8

Pi

white mark
on the iv

Sm ♀

S6 ♂

S8 ♂

Hawfinch

R2

R5

R1

R4

R5

border on the inner vane. Outer vane with white margin or all white. At the tip the white can have a fawn tint becoming better defined from R6 towards R2. R1 is black with a thin buff to tawny margin all round. An indication of the sex is given by the width of the white on the outer vane of P2-P4: in the male, the white reaches the rachis or reaches within less than 0.5 mm of it; in the female, very narrow white, at 1 mm or more from the rachis (rarely 0.5 mm).

▷ **Common Linnet wing (female)**

TWITE *Linaria flavirostris*: the 'large' feathers are similar to those of Common Linnet, but overall with less white. Ps and Ss dark brown to black, with a white border on the inner vane (grey on outer Ps) and with a pale margin at the tip. Ps with white margin on the outer vane, broader on the inner Ps of the male. Ss with fawn border on the distal part of the outer vane (sometimes with white on S1). Inner Ss grey towards the base, dark grey towards tip, with fawn border. Rs dark grey with white margin; broader on the inner vanes of R6–R5 (R4). Increasing fawn in the white from R4 towards R1. R1 with fawn margin. R2 almost without white.

Compared to Common Linnet, R3–R6 have less white on the inner vane. The margin is very narrow on the R3 of Twite and the centre is dark grey-brown (black in Common Linnet). On R6, white covers less than half the width of the inner vane (more than half in Linnet). Some subspecies not present in the area covered by this guide may have more white on the 'large' feathers (the remiges may then resemble those of the Common Chaffinch, for example).

HAWFINCH *Coccothraustes coccothraustes*: Ps and Ss impossible to confuse; black with a broad white mark on the inner vane, rather trapeze-shaped on the Ps and triangular on the Ss. The spot may have a more irregular shape and margin in young birds. Sometimes little or no white on S1. S5 sometimes with little white. S6-S9 generally without white, with increasing rufous-grey on the outer vane and tip. The shape of the tip of the remiges is very different (not visible in young): P5-P1 and S1-S5 have an extended and wavy tip on the Ps. In the male, the tips of (P8) P7-P2 (P1) have a green-blue to purple iridescence and S1-S5 (S6) more of a purple iridescence. In the female, the iridescence is often weaker; the outer vane of the Ss and P9-P5 (distal part) have a pale grey border. Rs black with broad white tip on inner vane. From R2 to R5, the white decreases (from about one third to one fifth of feather length), the outer vane is increasingly tinged with red towards

▷ **Hawfinch wing (male)**

the tip and white at the tip. R2 has a white tip, the rest of the inner vane black and a large part of the outer vane reddish (at least the distal half, often more). R2 sometimes with black reduced to a subterminal mark on the inner vane and replaced with grey elsewhere. R1 is dark grey at the base; the grey brightens and turns reddish towards the tip, which does or does not have white. R1 can also have a wide dark grey diagonal base, the rest reddish with a white tip. In the female, the tips of R2 and R3 are sometimes grey and not white, R1 is often less rufous.

Some of the male's primary coverts are black with white in an arc on the inner vane. The greater coverts are black at the base and on the inner vane; the outer ones are all white on the outer vane, on the others rufous increasingly takes the place of white and black going inwards.

Finches without bright colours

EURASIAN BULLFINCH *Pyrrhula pyrrhula*: Ps, Ss and Rs black, more or less obviously with dark blue iridescence on the outer vane (more visible on inner Rs and inner Ss). Ps and Ss finely bordered with white on the inner vane (sometimes quite ill-defined). The emarginated Ps can have a white or grey margin from the emargination to the tip. Inner Ss black with grey base. S9 may have some peony-pink on the outer vane in the male and pinkish buff in the female. R6 has rarely a little white along the rachis on the inner vane, as a thick line or a narrow oval (especially in north and east Europe). Birds in northern Europe are the largest (up to 15-20% larger than those in the south). Greater coverts black with a large grey tip or buff in the young. Uppertail-coverts black with blue sheen. Pink-red ventral body feathers in the male.

Possible confusion: with various passerines, particularly crossbills, but in general in the Bullfinch the outer vane is very dark (inner vane is paler and contrasts).

5 cm

▷ **Bullfinch wing (male)**

COMMON REDPOLL *Acanthis flammea* and **LESSER REDPOLL** *Acanthis cabaret*: Ps and Ss dark brown; cream to buff edge towards the tip with white edge to outer vane (mainly in *flammea*) or buff (outer Ps) to fawn (mainly in *cabaret*). Inner vane bordered with paler brown on the Ps and with a white arc on the Ss. The margin of the outer vane becomes wider towards the inner Ss and fades out in the median part (sometimes absent). The inner Ss have this border, are grey at the base and darker towards the tip. Rs are dark brown with margin on both vanes varying from white to buff.

Greater coverts dark grey with white or reddish tips (wide on the outer vane).

Possible confusion: with House Sparrow for the outer Ps.

ARCTIC REDPOLL *Acanthis hornemanni* (SS): pattern of 'large' feathers much as Common Redpoll, but the margins and borders are usually white (without tawny or reddish tint). Rs with wide white border, especially on the inner vane.

The crossbills *Loxia* spp.: four species with very similar 'large' feathers. Ps, Ss and Rs dark grey-brown, outer vane with pink-buff margin in the male, buff in the young, and greenish yellow in the female (the shades are difficult to appreciate on isolated worn feathers). Margins towards the tip of Ps and Ss white or cream. Inner vane of Ps, Ss and Rs sometimes with white border.

COMMON CROSSBILL *Loxia curvirostra*, **SCOTTISH CROSSBILL** *Loxia scotica* and **PARROT CROSSBILL** *Loxia pytyopsittacus*: no salient criteria for the 'large' feathers. The margins of the wing-coverts and the colour of the body feathers can give indications of the age and sex of a plucked bird.

TWO-BARRED CROSSBILL *Loxia leucoptera*: the remarks for the other three species apply. Ps and Ss (especially S3-S6) often have a wider white margin at the tip than in other species. S7-S9 have a white tip (in a wide or double arc). Greater coverts black with a white tip.

Finches with some pink

TRUMPETER FINCH *Bucanetes githagineus* (SS): localised to southern Spain. Ps and Ss brown, inner vane bordered with buff or light grey, sometimes with a slight pinkish shade. Outer vane and tip with buff, cream or pale pink (especially male) margin. Rs grey with darker tip, tinged with pale pink, and with a white cream or pinkish buff margin. The middle part of the outer vane is more extensively tinged with pale to candy-pink in the male (except on R1).

Greater coverts brown; outer vane tinted with pink (especially male).

COMMON ROSEFINCH *Carpodacus erythrinus* (SS): Ps, Ss and Rs dark brown; margin of outer vane and tip buff, yellowish-buff (female) or pinkish-brown (male). S7-S9 have outer vane with margin varying from light grey to yellowish-buff. Inner vane of Rs with more or less well-defined, white or cream margin. Some males have pinkish patches at the tip of the Rs.

Greater coverts brown; may have pink-tinged outer vane in the male (light brown if not).

PINE GROSBEAK *Pinicola enucleator* (SS): Ps, Ss and Rs dark brown, paler towards the base and inner vane margin. Ps with buff, dull pink, candy pink or white margin on the outer vane, the margin becoming white towards and at tip. Outer vane of Ss with white to cream border on the distal half, wider on the inner Ss, and often pale pink on the outer Ss in the male. Rs with white, yellowish or buff margin on the outer vane, brown-pink in the middle part of feather in some males.

Greater coverts black, with a white arc (sometimes pink) on the border of the outer vane as far as the tip of the two vanes.

Finches with well-defined yellow

EUROPEAN GOLDFINCH *Carduelis carduelis*: Ps and Ss black with white arc-shaped borders at the base on the inner vane, except on P9 (ill-defined with grey margin) and inner Ss. Outer vane bright yellow towards the base, on half to two thirds of the feather. Rounded or drop-shaped white tip, very small or absent on P9-P6, often tinged with buff on (S5) S6-S9. P9 without any yellow and with an ill-defined greyish inner border. S8 and S9 grey on the basal half, without yellow or only with a hint of yellow at the base of the outer vane. Rs black with a white mark.

R6 and R5 black with a large oval mark on the distal third of the inner vane (R5 rarely with white on the margin of the vane at the same level as the mark). R4 sometimes with a smaller rounded or more irregular mark, if not all black. R3–R1 black, usually with a sharp white tip, sometimes tinged with buff or grey.

5 cm

▷ **Reconstructed Goldfinch wing.**

rump ♂

Al

PC

SC

P9

P8

P6

P4

ov: pale margin

P1

undertail-covert

uppertail-covert

So

Sm

S8

S7

R1

R3

R4

R6

Common Crossbill

P8 ♀

MC

Two-barred Crossbill

SC

S8

S7

white at the tip of the Ps, Ss, greater and median coverts

Parrot Crossbill
young male

P8

SC

S7

R1

R5

quite big

belly

P7

ov: pale margin

P2

S4

R1

R5

Common/Lesser Redpoll

belly

♂

Al

PC

SC

P9

P8

P6

Pm

Bullfinch

Common Rosefinch

breast ♂

So

Sm

S9 ♂

S7

R1

Rm

R5

ov: pale margin

P6

P3

S3

R5

From R1–R3 the white tip decreases; it can still be present on R4, sometimes on R5 and R6 but very small. Sometimes a black mark at the end of the white tip.

Greater coverts a mix of black and bright yellow. On outer ones the black is on the outer vane, and at the base on the median coverts and also on the inner vane on inner coverts.

Confusion possible with Barn Swallow for Rs (central and neighbouring Rs).

EURASIAN SISKIN *Spinus spinus*: 'large' feathers a mix of black and yellow, with fairly bright yellow in the male but much paler (and a little less extensive) in the female and young. Black Ps and Ss with very pale yellow margin in an arc shape on inner vane. Basal part of outer vane yellow on P6 (about a quarter) to S6 (about a half). Greenish-yellow margin on all Ps; only a border on the distal third or quarter on the Ss. On the Ss, the middle part of the outer vane is thus totally black.

Inner Ss black towards the tip, yellowish-grey towards the base with little or no yellow at the base of the outer vane; a wide yellow margin at the tip of the outer vane. Rs yellow with black tip (with a more or less sharp and regular separation); rachis black; outer vane with yellow border. R1 greenish-grey, darker at the tip, tinged with yellow or yellow-green on both vanes. The black has an undulating edge in adult males and is quite reduced. In the female and young, the black is quite extensive (very extensive in the young female) and descends in a drop-shape towards the rachis. On the outer vane of R6, the black continues along at least two thirds of the feather towards the base; it can also extend a little on R5. The Rs resemble those of the Greenfinch but are in general smaller.

EUROPEAN GREENFINCH *Chloris chloris*: in the male the contrast in colours is quite marked; they are not so sharp in the female or the young. Ps and Ss dark grey to black,

▷ Siskin wing (female)

▷ Siskin wing (male)

▷ Reconstructed Greenfinch wing (young male)

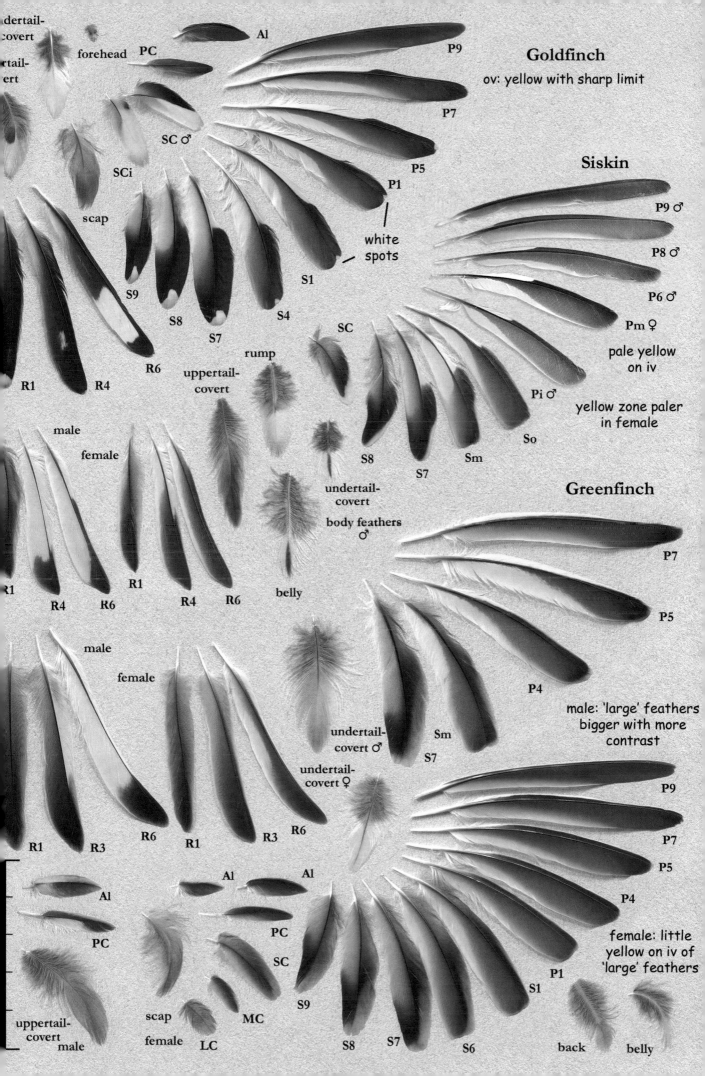

undertail-covert

forehead PC

Al

P9

Goldfinch

ov: yellow with sharp limit

SC ♂

SCi

scap

Siskin

P9 ♂

P8 ♂

P6 ♂

Pm ♀

pale yellow
on iv

yellow zone paler
in female

P7

P5

P1

white
spots

S1

S9

S8

S7

S4

R1

R4

R6

rump

uppertail-covert

SC

S8

S7

Sm

Pi ♂

So

male

female

undertail-covert

body feathers
♂

Greenfinch

R1

R4

R6

R1

R4

R6

belly

P7

P5

P4

male: 'large' feathers
bigger with more
contrast

male

female

undertail-covert ♂

undertail-covert ♀

Sm

S7

P9

P7

P5

P4

R1

R3

R6

R1

R3

R6

Al

Al

Al

PC

PC

SC

S9

MC

P1

S1

female: little
yellow on iv of
'large' feathers

PC

scap
female

LC

S8

S7

S6

uppertail-covert
male

back

belly

the inner vane with white arc-shaped border, rather ill-defined on P9-P7 (P6); tip with buff-grey margin. Outer vane of Ps has bright yellow, covering (almost) the whole feather in adult male, P1 with little yellow. In the young male, yellow touches the rachis at least towards the base (except sometimes on P1-P2). The yellow is restricted to a margin on the outer vane in the female, and does not touch (or only slightly) the rachis, near the base. In both sexes, the outer vane of the Ss has a greenish-yellow border in its basal part and with grey to greenish-grey on its distal part; the amount of grey increases towards the inner Ss. S7-S9 dark grey at the tip of the inner vane and near the rachis on the outer vane; base of inner vanes grey; greenish-grey tip. The outer vane with grey-green margin towards tip and green-yellow towards the base. Ss with a split tip.

Rs a mix of black and yellow. R1 and R2 dark greenish-grey, darker at the tip; the outer vane bordered with greenish-yellow margin on basal half and greyish on the distal half. R1 with identical vanes. The yellow margin is brighter in the male. R3–R6 of the male are yellow with black distal part, about half on R3 and quarter on R6; rachis black. In the female and young, the yellow is reduced to the basal part of the outer vane with a pale yellow margin on the inner vane, quite wide and extensive. Depending on the individual, their sex and age, the proportions of yellow and dark vary markedly.

In the adult male, alula feathers and primary coverts are grey with yellow on the outer vane.

Finches with ill-defined yellow patches

EUROPEAN SERIN *Serinus serinus*: Ps and Ss brown with ill-defined paler brown area on inner vane. Ps, S1 and S2 have a straw-yellow margin on the outer vane. S3-S9 with the outer vane a little greyer at the base, with light brown margin and straw-yellow (or buff) on the distal part.

On the middle part, the yellow is only visible at the end of the margin or absent.

The inner Ss have a large grey base and dark tip with a pale to yellow border towards the tip of the outer vane. The Rs are brown with a straw-yellow margin on the outer vane and buff on inner vane.

5 cm

▷ **Serin wing (young)**

CITRIL FINCH *Serinus citrinella* (SS): Ps and Ss brown. Ps and S1 with pale yellow or white margin on the outer vane. Outer vane of S2-S9 yellowish-grey at the base, with border varying from pale yellow (female) to greenish-yellow (male) on their distal part. In the middle part, the yellow is only visible as a fine margin or is absent. The inner Ss have a large grey base (sometimes yellowish) with a dark tip, a yellow margin on the outer vane and at the tip. The brown Rs have a white, pale yellow or greenish-yellow margin on the outer vane, and a cream to white margin on the inner vane. Greater coverts are dark grey with yellow to pale yellow tip on the outer vane.

CORSICAN FINCH *Serinus corsicana* (SS): like the Citril Finch but the margin on the outer vane of the Ps and Rs is often white.

• REGULIDAE: 'CRESTS

Distinctive criteria of group: size, colour.
This family has rounded wings (type A) and a square tail; very small.
Possible confusion: for the Rs, with some of the Sylviidae, particularly the *Phylloscopus* warblers (similar colour and size).

'Large' feather colour: browns, with yellow margins (see opposite).

MEASUREMENTS OF THE 'LARGE' FEATHERS OF REGULIDAE						
Species	**Pmax**	**Pmin**	**Smax**	**Smin**	**Rmax**	**Rmin**
Goldcrest	5	4	4.2	3	4.4	3.6
Firecrest	5 (5.6)	3.9	4.3	2.8	4.7	3.9

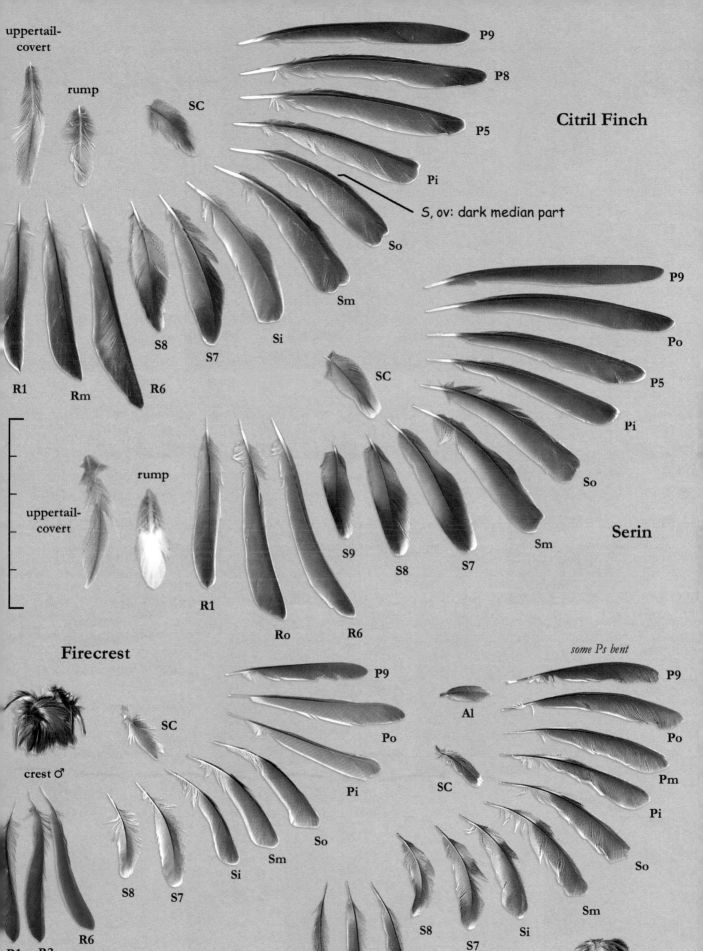

uppertail-
covert

rump

SC

P9

P8

P5

Pi

Citril Finch

S, ov: dark median part

So

Sm

Si

S8

S7

R1

Rm

R6

P9

Po

P5

Pi

So

SC

Sm

Serin

uppertail-
covert

rump

R1

S9

S8

S7

Ro

R6

Firecrest

SC

P9

Po

Pi

So

Sm

Al

SC

some Ps bent

P9

Po

Pm

Pi

So

Sm

crest ♂

Si

S8

S7

S8

S7

Si

R1 R2

R6

R1

Ro

R6

Goldcrest

crest ♂

black bar of ov of Ps and Ss:
- is sharper on Ps of Goldcrest
- is higher in the Goldcrest
(more than half, but often less
than 50% in Firecrest)

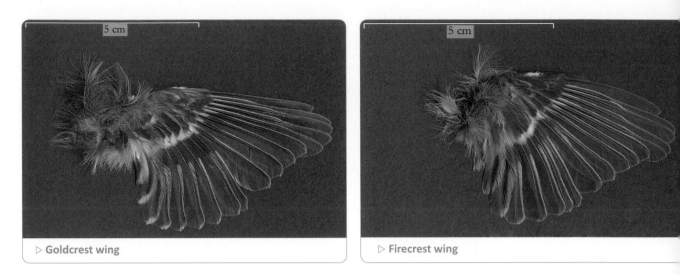

▷ Goldcrest wing

▷ Firecrest wing

GOLDCREST *Regulus regulus* and **FIRECREST** *Regulus ignicapilla* have very similar 'large' feathers so that generally individual feathers cannot be identified to species. Ps and Ss are black-brown with a white margin on the inner vane. The outer vane has a yellow margin, brighter towards the base. (P5) P4–P1 and the Ss have a black zone on the outer vane, almost basal; the vane is pale yellow between the black and the base. The inner Ss have no black zone.

On the median Ss there is a white tip that becomes larger towards the inner Ss. This mark is generally more extensive in Goldcrest. The Rs are brown with a yellow to pale green margin on the outer vane.

The underside of the Goldcrest's 'large' feathers are slightly more shiny and the shiny bases of the Rs and inner Ss stand out more than those of the Firecrest.

• PARIDAE: TITS

Distinctive criteria of group: colour.

This family has rounded wings (type A). The tail is rectangular to slightly rounded.

Possible confusion: with other small passerines.

Colour of 'large' feathers: generally, the remiges are grey to brown with a pale border on the inner vane, quite regular on the Ps, quite often towards the base of the Ss. The rectrices are most often uniform with a different coloured margin, the outer Rs sometimes paler. Certain species have blue in the feathers.

MEASUREMENTS OF THE 'LARGE' FEATHERS OF TITS						
Species	Pmax	Pmin	Smax	Smin	Rmax	Rmin
Azure Tit	6.3	5.2	5.8	4	6.9	6.1
Blue Tit	6.1	4.9	5.1	3.9	5.8	5
Willow Tit	6.1	4.6 (4)	5.3	3.5	6.4	4.8
Great Tit	7	4.9	6.1	4.1	7.2	5.3
Crested Tit	6.4	4.9 (4.2)	5.1	3.5	6	4.9
Siberian Tit (SS)	6.3	5.3	5.5	4	7.3	6.2
Sombre Tit (SS)	6.7	5.7	6	4.6	7.1	6.3
Coal Tit	6.1	4.7	5.1	3.5	5.2	4.6
Marsh Tit (SS)	6	4.8	4.9	3.5	6.1	5.2

AZURE TIT *Cyanistes cyanus*: readily distinguished from the other species. Ps and Ss dark grey with azure to cobalt-blue outer vane and inner vane with white border. Outer vane white: a border only on P9, white in the emarginated part for P8–P4 and on the distal part on other Ps. White tip on inner Ps and Ss, becoming larger towards inner Ss. On S7–S9, white on the inner vane and increasing on the outer vane from the tip. Rs obviously bluish dark grey, with white at the tip. White tip or small white spot on R1 which rapidly increases in area towards the outer Rs, consisting of more or less irregular marks. R6 white, except at base of inner vane.

Primary and greater coverts bluish-grey with broad white tip.

BLUE TIT *Cyanistes caeruleus*: Ps and Ss dark grey with white to grey border on the inner vane. Outer vane blue, brighter when viewed at an angle, sometimes with pale margin at tip. S6-S9 with small white tip. Rs dark grey, blue on outer web, and on both in R1. R6 with outer vane often grey, with light grey or white margin. Sometimes a small white tip to the Rs. Greater coverts bluish-grey with a white tip, yellowish in the young.

COAL TIT *Periparus ater*: Ps and Ss black-brown, with white border on inner vane. Possibly a pale margin at the tip from median Ps to Ss, small white tip on inner Ss. Rs black-brown. Outer vane of Ps, Ss and Rs with olive-brown to greenish margin. Size similar to Blue Tit.

GREAT TIT *Parus major*: Ps and Ss black-grey, with white border on inner vane. On the outer vane, outer Ps with pale grey or white margin on the emarginated part; the other Ps and Ss with silvery-grey margin, white at the tip. The margin is wider and white to yellowish on the inner Ss. All Rs have a black rachis. R1 ash-grey, the others black on the inner vane and ash-grey on outer vane; quite recognisable. Often one or two white marks at the tip of R5, sometimes also on R4. R6 is grey, paler and white or very grey; outer vane white and inner vane with a diagonal or triangular white mark. (Confusion possible with Rs of certain warblers.)

MARSH TIT *Poecile palustris*: Ps and Ss dark brown-grey, bordered with white to cream on inner vane. Tip and outer vane with margin greyish (outer Ps)

5 cm
▷ **Blue Tit wing**

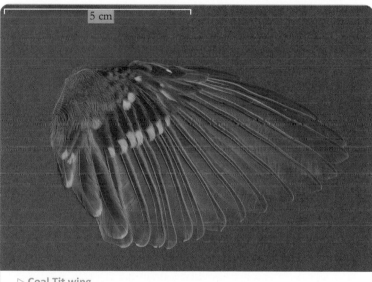
5 cm
▷ **Coal Tit wing**

5 cm
▷ **Great Tit wing (young)**

to light brown (Ss), broader on the inner Ss. Rs black-brown, outer vane quite brown; pale brown or pale grey margin on R6, sometimes on R5.

WILLOW TIT *Poecile montanus*: 'large' feathers as in Marsh Tit but the median and inner Ss have a more or less wide pale grey border (more obviously grey border also sometimes on the Ps). Rs with greyer outer vane, with wide pale grey border.

SOMBRE TIT *Parus lugubris* (SS): no data on isolated 'large' feathers. Resembles Willow Tit. Ps and Ss dark brown-grey, with white to cream border on inner vane. Outer vane with greyish border. Rs black-brown with pale grey margins.

SIBERIAN TIT *Parus cincta* (SS): no data for 'large' feathers. They resemble those of Willow Tit. Ps and Ss dark brown-grey, with creamy white border on the inner vane. Outer vane with greyish to white margin, wider on the Ss. Rs black-brown with pale grey margin.

CRESTED TIT *Lophophanes cristatus*: Ps and Ss dark brown with white to pale grey (ill-defined) margin on the inner vane (except inner Ss). Sometimes a small pale tip at the end of the outer vane. Outer vane browner; pale grey in the emarginated part of outer Ps. Rs dark brown-grey with quite brown outer vane.

• OTHER SMALL PASSERINES:
BEARDED REEDLING, LONG-TAILED AND PENDULINE TITS, NUTHATCHES

BEARDED REEDLING *Panurus biarmicus*, Panuridae. It has rounded wings (type A) and a long graduated tail.
Distinctive criteria: colour.
Possible confusion: with Redstart for the Rs.
TP = 4-5.8. TS = 3.9 (3.2 y)-4.8. TR = 4.3 (4 j)-9.6.
Colour of 'large' feathers: Ps and Ss black-brown with a wide white to cream border on the inner vane, increasing towards the inner Ss. P9-P6 (P5) with cream to buff margin and then wide buff border. Outer vane with margin that is white towards the base and buff towards the end on the outer Ps; buff on the other Ps; and rufous, increasing in width on the Ss towards the inner Ss. Base of outer vane with increasingly extensive white. S6-S8 with increasingly more contrast; black in the centre, the rest rufous, white towards the base and the centre. In general no black on S9. Central Rs more elongated. R1–R4 rufous with small area of ill-defined white at base; sometimes a grey tip. Ill-defined white or grey drop-shaped mark at the tip of R4 which is more extensive on the outer vane of R5. R5 often with ill-defined black on the basal third of the outer vane.

On R6 a mix of blurred areas of grey, white, rufous or black, the black more extensive than on R5. In the young, black can be present as for R3, and is more extensive and well-defined. Greater coverts black and white with pale rufous border.

LONG-TAILED TIT *Aegithalos caudatus*, Aegithalidae. It has rounded wings (type A) and a very long, graduated tail.
Distinctive criteria: shape of Rs, size.
Possible confusion: with wagtails but they have wider Rs, and other small species.
TP = 4-5.8 (6.7). TS = 3.6-4.8. TR = 4-9.6.
Colour of 'large' feathers: Ps and Ss black-brown with white border on inner vane; the white reaches the rachis towards the base. White margin on Ss, becoming increasingly wider towards the inner Ss (cream-white). Ss have white base on outer vane. Rs very

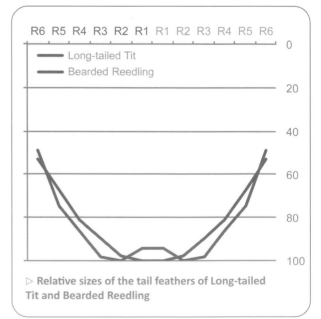

▷ Relative sizes of the tail feathers of Long-tailed Tit and Bearded Reedling

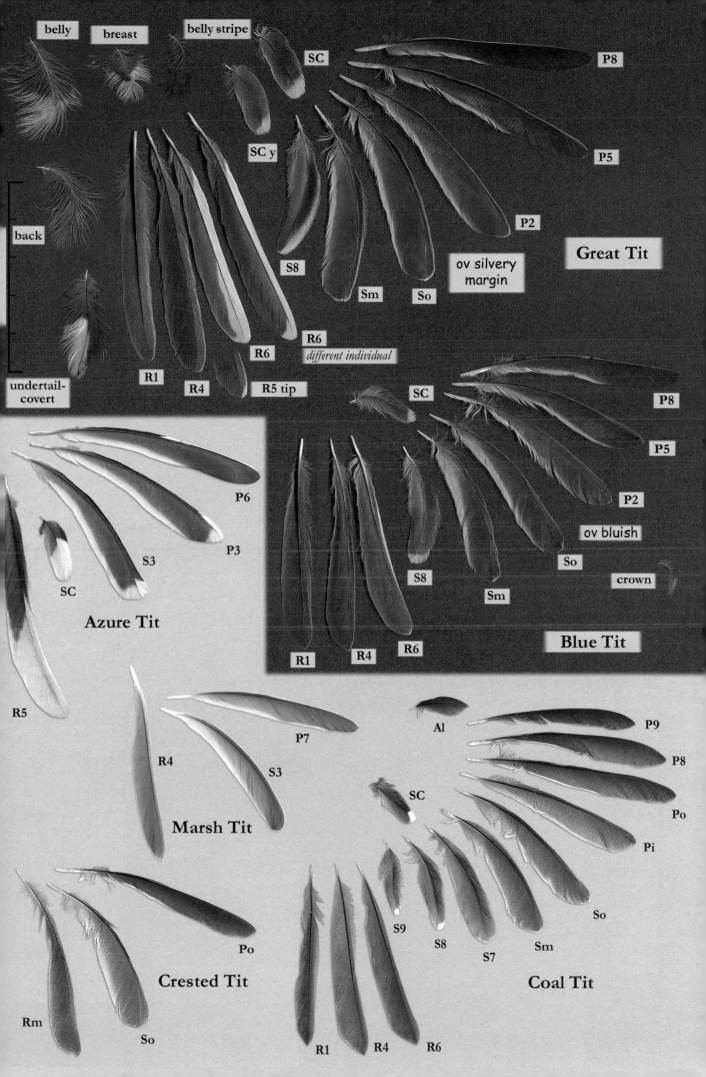

belly

breast

belly stripe

SC

SC y

back

undertail-covert

R1

R4

R5 tip

R6

R6
different individual

S8

Sm

So

ov silvery margin

P8

P5

P2

Great Tit

SC

P8

P5

P2

ov bluish

So

crown

S8

Sm

R1

R4

R6

Blue Tit

P6

P3

S3

SC

Azure Tit

R5

P7

R4

S3

Marsh Tit

Al

SC

P9

P8

Po

Pi

S9

S8

S7

Sm

So

Coal Tit

Po

Crested Tit

Rm

So

R1

R4

R6

elongated. R1–R3 black, with white or grey margin at base of the outer vane. R4–R6 with increasing amount of white on outer vane from the tip and a white triangular mark on the inner vane.

PENDULINE TIT *Remiz pendulinus*, Remizidae.
It has rounded wings (type A), a rectangular, slightly forked tail (R1 > 85% of Rmax).
Distinctive criteria: colour.
Possible confusion: with other small passerines, for example Linnet for the Rs (but Penduline Tit is smaller). TP = 4.1-5 (5.5). TS = 3.2-4.3. TR = 4.2-5.4 (SS).
'Large' feather colour: Ps and Ss black-grey, with wide pale grey to white border on inner vane, increasing in size towards the inner Ss. Outer vane and tip with white then pink-buff margin, broader on the inner Ss. Rs black-brown to black (outer ones more grey), bordered with white on the inner vane and with buff (or white) margin on the outer vane (on both vanes in R1).

• SITTIDAE: NUTHATCHES

These species have rounded wings (type A), and short, square tail.
Distinctive criteria: shape of the Rs, colour.
Possible confusion: with Wallcreeper and Hawfinch for the Rs, with other passerines for the Ps and Ss, for example Pied Flycatcher (for the Ps).

MEASUREMENTS OF THE 'LARGE' FEATHERS OF NUTHATCHES						
Species	Pmax	Pmin	Smax	Smin	Rmax	Rmin
Corsican Nuthatch (SS)	6.3	5.3	5.4	4.3	4.3	3.9
Rock Nuthatch (SS)	7.1	6.5	6.4	5.1	5.3	4.5
Eurasian Nuthatch	7.8	5.8	6.5	4.7	5.5	4.5

EURASIAN NUTHATCH *Sitta europaea*: Ps and Ss dark grey, with vague pale border on the inner vane, and possible pale margin towards tip. A little white at base of P9-P5, with white extending a little on the edge of the inner vane on P4-P1. An area of ill-defined pale grey at the level of the emargination on the outer web of the outer Ps. Ash-grey border to outer vane of the Ss that covers whole feather on the inner Ss. Rs quite short, with same width as far as the tip. R1 ash-grey; others black with ill-defined grey base and sharp grey tip (narrow on R2, widening to R6). Wide subterminal white band on inner vane of R4–R6 (though sometimes smaller on R3) and on outer vane of R6 (further towards the base).

CORSICAN NUTHATCH *Sitta whiteheadi* (SS): no data for isolated feathers. Very slightly smaller than Eurasian Nuthatch. Ps and Ss apparently very similar to those of Eurasian Nuthatch but Rs darker. R1 ash-grey, others black with an ill-defined grey base. Well-defined small grey tip on R4 becoming larger towards R6 (about one quarter to one third of the feather length), possibly with a little white in the centre of the grey.

ROCK NUTHATCH *Sitta neumayer* (SS): no data for isolated feathers. All Ps and Ss are grey; Ps quite dark with white border on inner vane (no data for outer vane), Ss more uniform. Rs uniform grey, with an ill-defined pale area at the tip of the inner vane of R5, and sharper, often crescent-shaped, on that of R6.

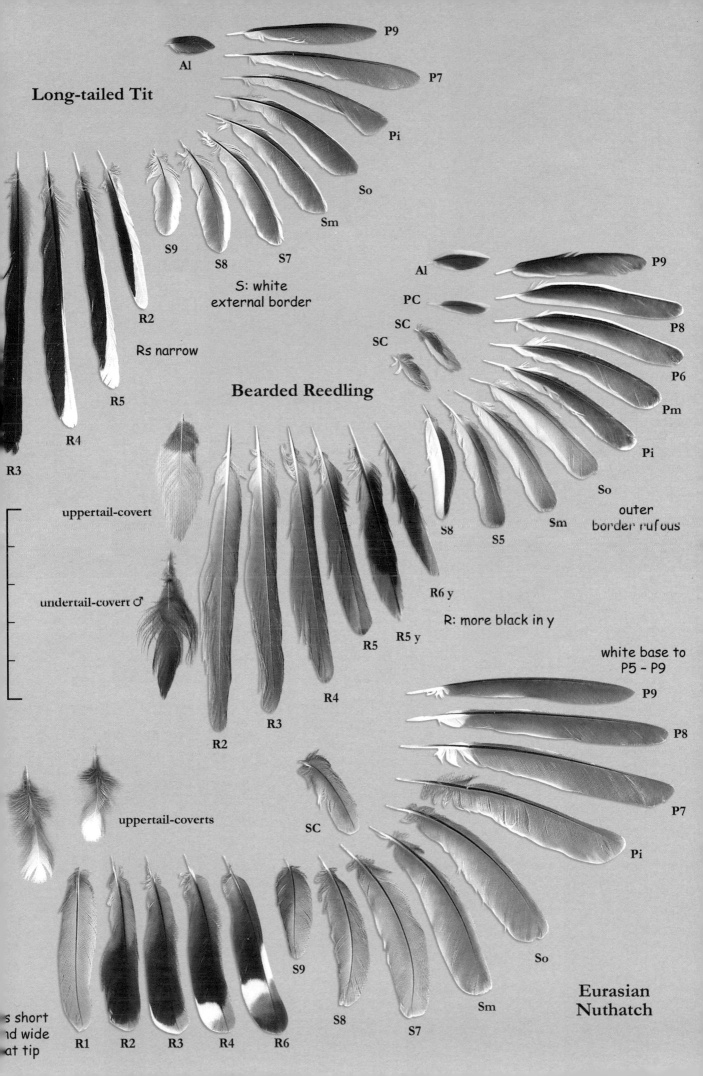

Long-tailed Tit

Al

P9

P7

Pi

So

Sm

S9

S8

S7

S: white
external border

R2

Rs narrow

R5

R4

R3

Bearded Reedling

Al

PC

SC

SC

P9

P8

P6

Pm

Pi

So

Sm

S8

S5

outer
border rufous

uppertail-covert

undertail-covert ♂

R6 y

R: more black in y

R5

R5 y

R4

R3

white base to
P5 – P9

P9

R2

P8

P7

Pi

uppertail-coverts

SC

So

s short
nd wide
at tip

R1

R2

R3

R4

R6

S9

S8

S7

Sm

Eurasian
Nuthatch

• MUSCICAPIDAE (3): ROBIN, REDSTARTS, NIGHTINGALES, ETC.

Distinctive criteria of group: colour.

This family has rounded wings, slightly narrower in the migratory species. The tail is rectangular in most species. *Possible confusion:* with species that have rufous Rs (Rock Thrush, etc), and with other passerines.

Colour of 'large' feathers: Ps and Ss often brown, sometimes with warmer brown markings. Rs often reddish.

MEASUREMENTS OF THE 'LARGE' FEATHERS OF OF ROBIN, REDSTARTS, NIGHTINGALES ETC.							
Species	Pmax	Pmin	Smax	Smin	Rmax	Rmin	Wing shape
Rufous Bush Robin (SS)	7.3 (8.2)	5.9	6.2 (7)	5.1	8	6.3	A
Bluethroat	6.6 (7.3)	5	5.4	4.2	6.3	4.7	A
Red-flanked Bluetail (SS)	7.1	5.4	5.5	4	6.4	5.6	A
Common Nightingale	7.5 (8.5)	5.5	5.9	4.3	7.4	6.2 (5.5)	A
Thrush Nightingale (SS)	7.5	5.9	6	4.6	7.5	6.2	B
European Robin	6.6	4.9	5.7	4.2	6.5	5.8	A
Common Redstart	7.3 (8.2)	5.4	5.6	4.2	6.5	4.8	A
Black Redstart	7.8	6	6.3	4.4	6.8	5.8	A

RUFOUS BUSH ROBIN *Cercotrichas galactotes:* Ps and Ss brown-grey, with tawny to reddish margin on the outer vane (wider towards the base) and bordered with buff to cream on the inner vane. Tip of inner Ps and Ss with pale border. R1 reddish-grey, more rufous towards the base or all rufous. The other Rs are rufous with a black terminal bar (R2–R3) or subterminal bar; black descending or not onto the outer vane of R6. Large white tip on R6, becoming smaller towards R4; reduced to a border or absent on R3 and R2. Rs of the young with little white and less black; the tail is also rounded.

Possible confusion: with redstarts and Rock Thrush for R1, with Crested and Thekla's Larks, etc. for the remiges.

EUROPEAN ROBIN *Erithacus rubecula:* Ps and Ss dark brown; buff border on inner vane, outer vane warm brown. Generally a very small orangish tip on the Ss and often on the inner Ps (sharper on the inner Ss). In young, the coloured tip is sharper and clearly visible on the inner Ss. P1-P5 and Ss often have a red or yellow-red margin on the distal half of the outer vane. Rs brown, outer vane warm brown.

COMMON NIGHTINGALE *Luscinia mega-rhynchos:* Ps and Ss dark brown, bordered with buff on the inner vane; outer vane warm brown vaguely tinged with red. Rarely a very small red tip. Rs dark rufous; R1 sometimes brown. R6 is very rarely bright rufous (in which case it resembles that of the redstarts). *Possible confusion:* for the Rs of paler individuals, with those of redstarts, for example.

THRUSH NIGHTINGALE *Luscinia luscinia* (SS): no data for isolated 'large' feathers. Similar to Common Nightingale but with the outer vane of the Ps and Ss less rufous, quite brown and the Rs darker reddish-brown.

▷ **European Robin wing**

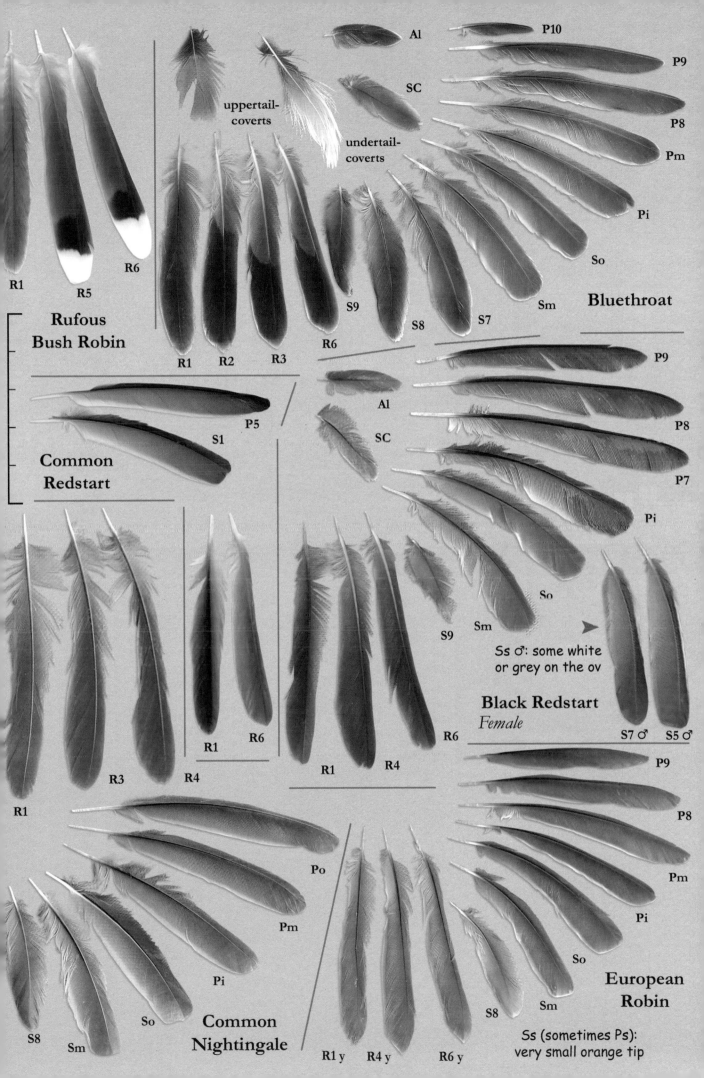

R1

R5

R6

**Rufous
Bush Robin**

uppertail-
coverts

undertail-
coverts

Al

SC

P10

P9

P8

Pm

Pi

So

Sm

Bluethroat

R1   R2   R3

S9

S8   S7

R6

**Common
Redstart**

P5

S1

Al

SC

P9

P8

P7

Pi

So

Sm

S9

Black Redstart
Female

Ss ♂: some white
or grey on the ov

R6

S7 ♂   S5 ♂

R1

R3   R4

R1   R6

R1   R4

P9

P8

Pm

Pi

So

Sm

**European
Robin**

S8

Po

Pm

Pi

So

Sm

**Common
Nightingale**

R1 y   R4 y   R6 y

S8

Ss (sometimes Ps):
very small orange tip

BLUETHROAT *Luscinia svecica*: Ps and Ss dark brown bordered with grey on the inner vane and a more or less clearly defined margin of buff on the outer vane. R1 black-brown or brown with the rachis rufous at its base; the other Rs rufous at the base with a black tip, a little lighter at their extremity. Black tip widens towards the R2 (distal half or more) and narrows towards the R6 (distal third).

RED-FLANKED BLUETAIL *Tarsiger cyanurus* (SS): no data for isolated 'large' feathers. Ps and Ss brown-grey, bordered with buff to grey on the inner vane; browner in the young and female. In males, outer vane bluish to ash-grey. In the young and the female the outer vane is brown. Rs black-grey in the male, otherwise dark brown. Outer vane (both vanes on R1) tinged with azure to cobalt-blue.

BLACK REDSTART *Phoenicurus ochruros*: Ps and Ss dark brown-grey, with white to grey on the inner vane; tip with pale, ill-defined border. Outer vane with brown margin in the young and female, pale grey in the male. Ss with white margin on the outer vane in the male, becoming more extensive towards the inner Ss and towards the base. R1 brown-grey with tawny margin. Other Rs bright red, sometimes darker or paler, often with dark markings towards the tip (one or two marks), more on the outer vane.

COMMON REDSTART *Phoenicurus phoenicurus*: Ps and Ss dark brown, with buff border on the inner vane; tip with ill-defined pale margin. Outer web with margin ranging from tawny to buff or light grey. Rs as in Black Redstart but normally paler, less often with dark markings at tip. R1 usually with wide rufous border on outer vane and rufous margin on the inner vane.

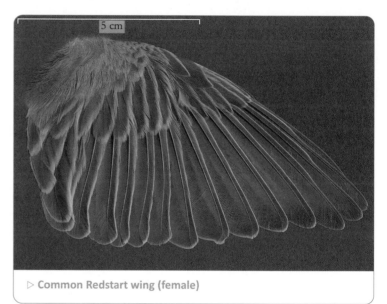

▷ **Common Redstart wing (female)**

• EMBERIZIDAE: BUNTINGS

Distinctive criteria of group: colour.

This family has rounded wings with quite a square hand, type A(b); more rectangular (type B) in Lapland and Snow Buntings.

Possible confusion: with other species with white-spotted Rs (Chaffinch, larks, etc.), and other passerines.

Colour of 'large' feathers: most species have dark brown remiges with a pale margin and more contrasting inner Ss. The Rs are often black; the outer pairs have a diagonal white area.

SNOW BUNTING *Plectrophenax nivalis*: Ps, Ss and Rs dark brown (young and female) to black (male) with some white. Depending on age and sex, the white is more or less extensive, with more white overall in males and adults. P9 -P4 (P3) dark with a white base; small in females, up to half or more of the base in males. The white often goes up and along the edge of the inner vane, but not always. P3-P1 with increasing amount of white and white at the tip. P1 and P2 may be entirely white in the *insulae* subspecies. Ss all white, or with some dark which is more extensive on the outer Ss and on the outer vanes of the Ss. Often also some dark on the inner vane. S7-S9 dark, possibly with white on the edge of the outer vane especially and a pale edge towards the tip. S7 with much white at the base in *insulae*. Rs of variable pattern with the three central pairs fairly black and the three outer pairs fairly white. R3 may have white (base of outer vane) and R4 can be almost black. The white is unusually distributed: in wide borders (R3), in the centre (R4) or in a broad, non-apical, diagonal mark. More often than not, R4–R6 are white with elongated black mark at the tip (especially on the outer vane) and with black towards the base on the edge of the inner vane. In *insulae*, R3 can be white at the base and on the outer vane and R2 may have white at the base of the outer vane.

Possible confusion: with Snowfinch.

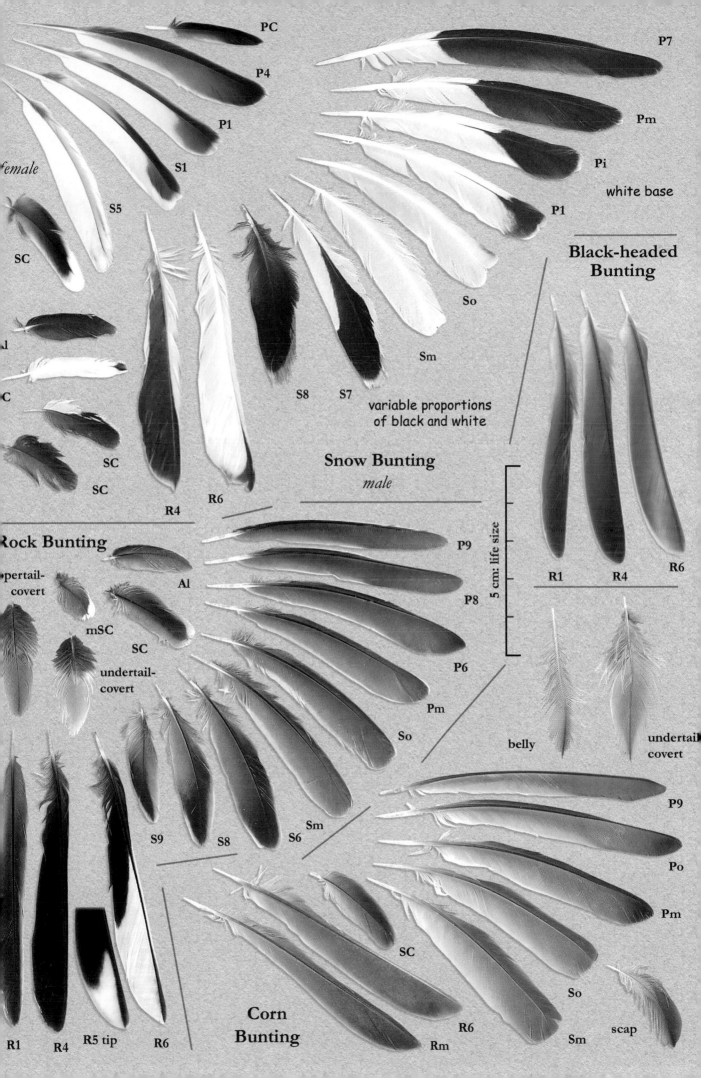

PC

P4

P1

S1

Female

S5

SC

SC

I

C

SC

SC

R4

R6

P7

Pm

Pi

P1

white base

So

Sm

S8   S7   variable proportions
of black and white

Snow Bunting
male

**Black-headed
Bunting**

R1   R4   R6

5 cm: life size

Rock Bunting

pertail-
covert

mSC

SC

undertail-
covert

Al

P9

P8

P6

Pm

So

belly

undertail
covert

S9   S8   S6   Sm

P9

Po

Pm

SC

R6

R1   R4   R5 tip   R6

**Corn
Bunting**

So

Sm   scap

Rm

MEASUREMENTS OF THE 'LARGE' FEATHERS OF BUNTINGS

Species	Pmax	Pmin	Smax	Smin	Rmax	Rmin
Yellow-breasted Bunting (SS)	6.5	5.4	5.4	5.2	6.6	5.8
Cretzschmar's Bunting (SS)	7.5	5.6	6	5.5	7.7	6.6
Snow Bunting	9.8 (10.7)	5.8	6.7 (7.3)	5	7.9 (8.3)	5.9
Reed Bunting	7	5.9	6	5.3	7.6	6
Rock Bunting	7.2	5.9	6.1	5.1	8.3	7.4
Yellowhammer	8	5.8	6.7	5.4	8.5	6.7
Lapland Bunting	8.3	5.4	6	4.8	7.1	6.1 (5)
Black-headed Bunting (SS)	7.5	5.9	6.2	5.3	8.2	6.7
Little Bunting (SS)	6.4	4.9	5.2	4.4	6.6	5.3
Ortolan Bunting	7.9	5.6	6.2	5.5	7.7	6.7
Corn Bunting	8.7	6.3	6.6 (7.3)	5.9	7.7 (8.4)	6.8
Rustic Bunting (SS)	6.7 (7.1)	5.1	5.5	4.5	6.8	5.6
Cirl Bunting	7.5	5.8	6.2	5	8.7	7.1

CORN BUNTING *Emberiza calandra*: Ps and Ss brown; buff margin on the outer vane and ill-defined greyish border on the inner vane. Pale margin at tip (especially inner Ps and Ss). Ss with more extensive pale border towards inner ones. S7-S9 darker, especially at tip. Rs brown with grey to buff margin. R6 with a paler zone (grey to buff) diagonally and towards the tip, especially on the inner vane. R5 can have pale drop-shaped mark at tip and on the outer vane. Rachis brown.

BLACK-HEADED BUNTING *Emberiza melanocephala*: Ps and Ss brown, white borders on inner vane. Buff margin on the outer vane of Ps, wider on the Ss. Inner Ss dark brown, fairly regularly bordered with buff. R1 brown with paler border. R2–R4 brown with buff margin. R5 as R4 or a little paler at the tip. R6 brown with a paler area diagonally from the tip.

The other species are very similar to each other. The following table summarises the colour criteria. Geographic locality and habitat are also important criteria for identifying buntings.

The Ps and Ss are dark brown. The inner Ss are generally grey at the base of the inner vane, darker towards the tip and on the outer vane, with a reddish or pale edge towards the tip and on the outer vane. This pale edge is narrower on the central part of the feather, which distinguishes them from those of sparrows or accentors, for example (so-called 'classic' pattern in table p. 184). R2–R4 are black, often with a pale margin. R1 is brown or

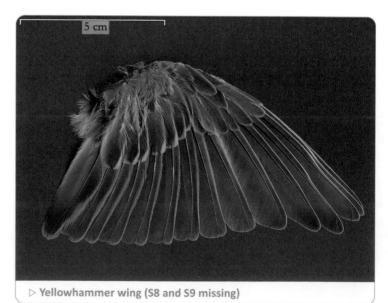

5 cm

▷ **Yellowhammer wing (S8 and S9 missing)**

Al

PC

P9

P7

P5

So

S9

S8   S7   S6

throat   **Yellowhammer**

P9

P7

P4

P1

SC

scap

Sm

MC   SC

S8   S7   **Reed Bunting**

uppertail-
covert

breast

uppertail-
covert

rtolan Bunting

back

undertail-
covert

R6
another
tip

R1

R4   R5   R6

R1

R4   R5

R6

undertail-
covert

R1   R3   R5   R6

Little Bunting   **Cirl Bunting**

rump

Al

SC

undertail-
covert

P9

P8

P6

Pi

So

R1

R4   R5   R6

R1   R4   R5   R6

scap

Si worn

S9   S8

Sm

undertail-
covert

black with a broad pale border and R6 and R5 are black with some white diagonally. On outer Rs, the extent of white varies quite widely between individuals of the same species, which can complicate the identification.

Species	Ps and Ss			Rs		
	Margin on outer vane	Border on inner vane	Inner Ss	R1	R2–R4	R5–R6
Yellow-breasted Bunting (SS)	Yellowish on Ps; reddish-yellow and wide on Ss	white	classic	brown with yellowish border	brown with yellowish margin	R5 with white along the rachis at the tip, sometimes a blurred, long mark towards the tip; R6 with a narrow diagonal white stripe from the tip of the inner vane
Cretzschmar's Bunting (SS)	White to cream on Ps; pinkish-rufous and wide on Ss	white to buff	classic	brown with rufous border	black-brown; pinkish-rufous margin, sometimes white at tip of R4	white hardly diagonal, more trapeze shaped
Reed Bunting	Buff-grey on Ps; rufous and wide on Ss.	well-defined white	classic	black in centre, rufous border, especially on ov	black, grey to buff margin or absent; pale base to iv	white diagonally
Rock Bunting	White to buff on Ps; rufous and wide on Ss	white to buff or grey, narrow	classic, but border thin and regular	black with rufous border	black; sometimes white tip on R4	white diagonally
Yellowhammer	Pale yellow on Ps; buff-yellow and wide on Ss	White to grey, not very well-defined	classic	black-brown with pale yellow or yellow-buff border	black with pale yellow margin on the ov	little white; ov with pale yellow margin
Lapland Bunting (SS)	White to buff on Ps; rufous-buff and wider on Ss	white	classic	dark brown with buff margin	brown to black with buff to white margin	R5 with small, white central mark and long mark at tip; R6 with long white mark, a little black on tip of ov
Little Bunting (SS)	Buff on Ps; buff to rufous on Ss	white	quite contrasting	brown, with paler margin	dark brown	white in very elongated mark, quite central on R5
Ortolan Bunting	Buff to pale yellow on Ps; more rufous and wider on Ss	white	classic	brown, with paler border	black, white margin at tip of ov; R4 sometimes with white mark at tip	white slightly diagonal, rather trapezoidal, white edge to ov
Rustic Bunting (SS)	Yellow-buff on Ps; rufous-buff and wide on Ss	white	quite contrasting	black with wide rufous-brown border	black with pale margin on the ov	white in very elongated mark, quite central on R5
Cirl Bunting	White to yellow on Ps; yellow-rufous on Ss, redder and wider on inner ones	white, well-defined	classic	dark brown with wide yellowish, rufous or greenish border	black with yellowish, greenish or reddish margin on ov; sometimes white tip to R4	white mark, ov with white margin (especially at tip)

PRINCIPAL COLOUR CRITERIA OF THE 'LARGE' FEATHERS OF BUNTINGS

• PASSERIDAE: SPARROWS, SNOWFINCH

Distinctive criteria of group: colour.

This family has rounded or triangular wings. The tail is rectangular.

Possible confusion: with other passerines, for example Spotted Flycatcher, buntings, accentors, etc.

Colour of 'large' feathers: brown in the sparrows, black and white in Snowfinch.

MEASUREMENTS OF THE 'LARGE' FEATHERS OF SPARROWS AND SNOWFINCH							
Species	Pmax	Pmin	Smax	Smin	Rmax	Rmin	Wing shape
House Sparrow	6.6	5.2	5.5	4.4	6.1	5.6	A(b)
Italian Sparrow (SS)	6.9	4.8	5.7	4	6.4	5	A(b)
Spanish Sparrow	6.8	4.8	5.3	4.9 (4 j)	6.2	5.4 (4.4 j)	A(b)
Tree Sparrow	6.4 (6.8)	4.7	4.9 (5.2)	4.2	5.8	5	A(b)
Rock Sparrow	8.4	5.6	5.9	4.9	6.1	4.8	B
White-winged Snowfinch	10.9	6.2	6.8	5.3	8.3	6.8	B

HOUSE SPARROW *Passer domesticus* and **ITALIAN SPARROW** *Passer italiae:* 'large' feathers indistinguishable for the two species. Ps and Ss brown with buff to rufous margins on the outer vane, broader on Ss. The basal quarter of the outer vane is paler and reddish, less clearly so on the Ss. Inner Ss darker towards the tip and with a wide reddish-brown border on the outer vane and at the tip. Rs brown with buff to rufous margin on the outer vane, but on both vanes for R1. Rs paler and often shorter in young.

Greater coverts darker at the tip, with large rufous border.

SPANISH SPARROW *Passer hispaniolensis:* as House Sparrow but margins and borders a little less rufous.

TREE SPARROW *Passer montanus.* Ps and Ss brown with obvious buff to rufous margin on outer vane (often also at the tip) and with buff border on the inner vane, broader towards the base. Inner Ss darker towards the tip and with wide reddish-brown border on the outer vane and at the tip, often with pale point, sometimes white. Rs brown, with buff to rufous margin on the outer vane, but on both vanes for R1.

Greater coverts like those of House Sparrow, but with a white tip.

Compared to House Sparrow, the Tree Sparrow has smaller 'large' feathers (beware of young House Sparrows, sometimes smaller than adults). Its feathers show sharper contrast of the pale base of the outer vane of Ps, and well-defined (c.f. ill-defined in the House Sparrow) pale border to the inner vane of the Ps and Ss (except inner ones), and paler Rs.

ROCK SPARROW *Petronia petronia:* Ps and Ss brown, with pale margin, inner vane with ill-defined pale border. On the Ps, outer vane with grey to buff margin, margin paler towards the tip. On the basal third, but not at the base, this margin is wider and yellowish and the outer vane is paler.

On the Ss, outer vane and tip bordered with greyish-buff to yellowish. Inner Ss darker towards the tip, the very tip paler. Rs brown, darker on the distal quarter. R1 and outer web of other Rs with buff to yellowish borders. R2–R6 have a wedge-shaped or rounded white mark at the tip of the inner vane (the white descends a little on the border of the vane). Outer web of R6 largely yellowish.

5 cm

▷ **House Sparrow wing (young male)**

WHITE-WINGED SNOWFINCH *Montifringilla nivalis*: Ps and Ss dark brown (young and female) to black (male) with a greater or lesser amount of white on the inner Ps and the Ss. Ps dark with lighter (brown to grey) margins on the outer vane and at the tip. P3 is dark, with a more or less broad white mark along the edge and on the distal third (rarely half) of the inner vane. P2 is usually white on the distal third or half of the inner vane (limit of oblique black). The outer web is sometimes white towards the tip. P1 is black only on the basal third or half, with a slanting limit to the black (higher towards the rachis). In the young: P3 is often dark with a little white on P2 and P1 (only on the inner vane). S1-S6 white with black at the base, decreasingly extensive towards S6 (about a third on S1 and a quarter to a fifth on S6). In the young, the base is dark as is the tip which has a drop-shaped or more extensive mark, with dark descending on the rachis (outer Ss). S7-S9 dark brown with brown to buff border at the tip and on the outer vane. S7 more or less broadly bordered with white on the base of the outer vane. R1 brown to black, with pale margin. R2–R6 white with a black base, decreasing in size towards R6. R2–R5 also have black at the tip, becoming narrower towards R5. R2 is quite variable, with black on the inner vane along the rachis. The black sometimes covers almost all the inner vane, leaving only a white rounded spot in the distal quarter. R6 sometimes with a little black on the inner vane. *Possible confusion:* with Snow Bunting, but in that species the base of the 'large' feathers is usually white.

• PRUNELLIDAE: ACCENTORS

Distinctive criteria of group: colour.

This family has rounded wings (type A). The tail is rectangular.

Possible confusion: with other passerines, for example the sparrows, Robin, etc.

Colour of the 'large' feathers: brown.

MEASUREMENTS OF THE 'LARGE' FEATHERS OF ACCENTORS						
Species	Pmax	Pmin	Smax	Smin	Rmax	Rmin
Dunnock	6.3	4.7	5.6	4.3	6.8	5.9
Alpine Accentor	10	7	7.2	5.4	8	6.7

DUNNOCK *Prunella modularis*: Ps and Ss brown, with narrow buff border on inner vane. Outer vane with fawn border on the non-emarginated part of the outer Ps, and with tawny border on the median and inner Ss. Other Ps and Ss vaguely more rufous on the outer vane than on the inner one, especially towards the base. S8-S9 with tawny border on both vanes, with dark brown centre, sometimes a paler tip.

Rs and Ss brown, with tawny margin on outer vane. In the young, Rs are paler brown with a less well-defined border.

ALPINE ACCENTOR *Prunella collaris* (SS): Ps and Ss brown, with ill-defined buff-

▷ **Dunnock (young) wing.**

grey border on inner vane. Outer vane with buff to cream margin on the non-emarginated part of the outer Ps. Ss with wide buff to rufous border on the middle part of the outer vane. Pale, cream, buff or rufous margin (outer Ps) or border (other Ps and Ss) with a larger pale mark on the outer vane. Outer vane of Ss black between the pale border and tip. Inner Ss dark brown or black with reddish-buff border, and pale white or cream tip. Rs dark brown, paler towards the border of the inner vane, with buff margin on the outer vane. R1 with buff or

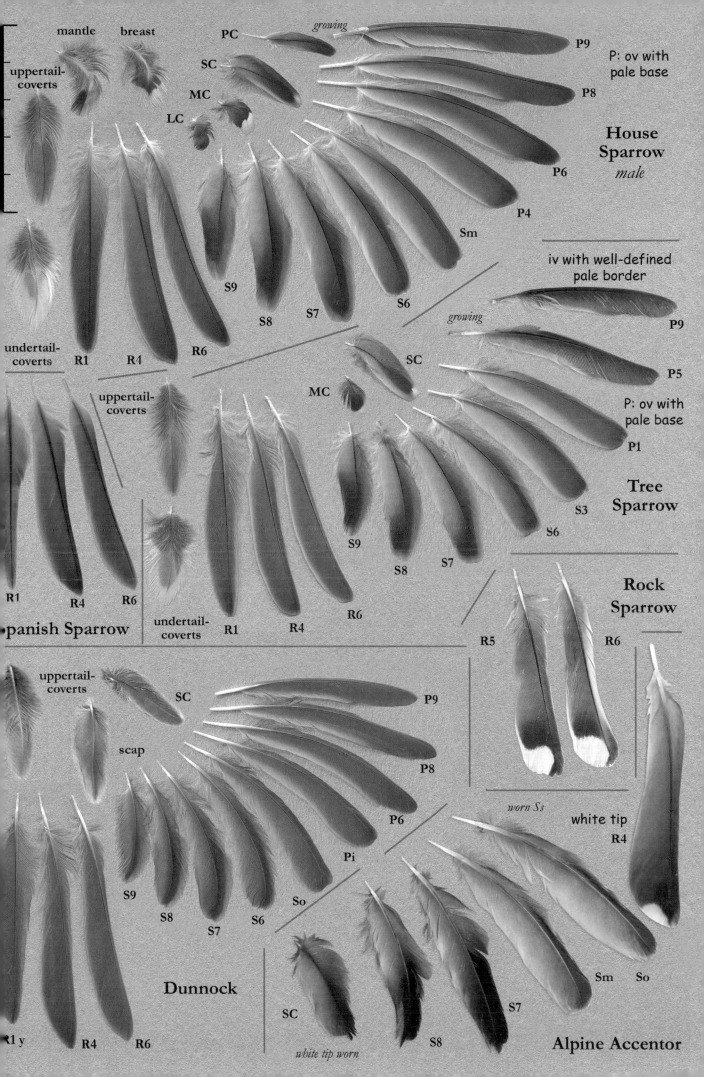

mantle   breast

uppertail-
coverts

PC

growing

SC

MC

LC

P9

P: ov with
pale base

P8

House
Sparrow
male

P6

P4

Sm

S9

S8   S7   S6

growing

iv with well-defined
pale border

P9

P5

undertail-
coverts   R1   R4   R6

SC

MC

P: ov with
pale base

P1

uppertail-
coverts

S3

S6

Tree
Sparrow

R1   R4   R6

S9

S8   S7

Rock
Sparrow

undertail-
coverts   R1   R4   R6

panish Sparrow

R5   R6

worn Ss

white tip
R4

uppertail-
coverts

SC

scap

P9

P8

P6

Pi

S9

S8   S7   S6

So

Sm   So

R1 y   R4   R6

Dunnock

SC

white tip worn

S8

S7

Alpine Accentor

tawny corner at the tip on both vanes. Other Rs with a buff, wedge-shaped tip to the outer vane and a rounded white or cream mark on the tip of the inner vane.

Greater coverts brown to black, with pale border towards their base and with a white tip.

• SYLVIIDAE: WARBLERS

Distinctive criteria of group: colour.

This family has rounded or triangular wings. The tail varies in shape, often rectangular but sometimes more or less clearly rounded.

Possible confusion: with other brown passerines, and for some Rs with other species that have white spots, etc.

Colour of 'large' feathers: primaries and secondaries often brown, sometimes with yellow or rufous margins. Rs similar, sometimes with pale or white marks towards the tip (*Sylvia* warblers, Zitting Cisticola, etc.)

From time to time, darker parallel bars occur on the tail feathers; these are growth bars, not pigmentation bars (only in certain individuals).

SYLVIA WARBLERS

Their wings are rounded, some species having a squarer hand. More often than not the tail is rectangular, sometimes more rounded: Spectacled (R6 shorter), Dartford, Marmora's, and Sardinian Warblers, etc. The margins of the outer vanes of the tail feathers quickly disappear if they are narrow. Isolated 'large' feathers of the different species are mostly difficult to differentiate, with rectrices causing fewer problems than remiges.

▷ Garden Warbler wing.

MEASUREMENTS OF THE 'LARGE' FEATHERS OF *SYLVIA* WARBLERS							
Species	Pmax	Pmin	Smax	Smin	Rmax	Rmin	Wing shape
Spectacled Warbler *(Sylvia conspicillata)* (SS)	5	4.2	4.5	3.5	5.8	5	A
Blackcap *(Sylvia atricapilla)*	6.5 (7.5)	5.3	5.3	4.5	6.8	6.5	A
Lesser Whitethroat *(Sylvia curruca)*	6.3 (6.6)	5.3	5.4	4.4	6.6	5.5	A
Rüppell's Warbler *(Sylvia rueppelli)*	6.2	5	5.3	3.9	6.8	5.8	A(b)
Garden Warbler *(Sylvia borin)*	6.6	4.9	5.1	3.7	6.2	5.7	A(b)
Barred Warbler *(Sylvia nisoria)*	7.5 (8.6)	5.8	6.2	4.2	8.2	7.1	A(b)
Common Whitethroat *(Sylvia communis)*	6.5	5.2	5.7	4.3	7.1	5.8	A(b)
Sardinian Warbler *(Sylvia melanocephala)*	5.3	4.3	4.8	3.8	6.6	5.4	A
Western Orphean Warbler *(Sylvia hortensis)*	6.8	5.7	5.7	4.7	7.5	6.4	A(b)
Subalpine Warbler *(Sylvia cantillans)*	5.6	4.2	4.6	3.5	6.4	5.1	A(b)
Dartford Warbler *(Sylvia undata)*	4.8	3.8	4.5	3.2	6.9	5.2	A
Marmora's Warbler *(Sylvia sarda)* (SS)	5.3	4.5	4.7	3.6	6.9	5.4	A

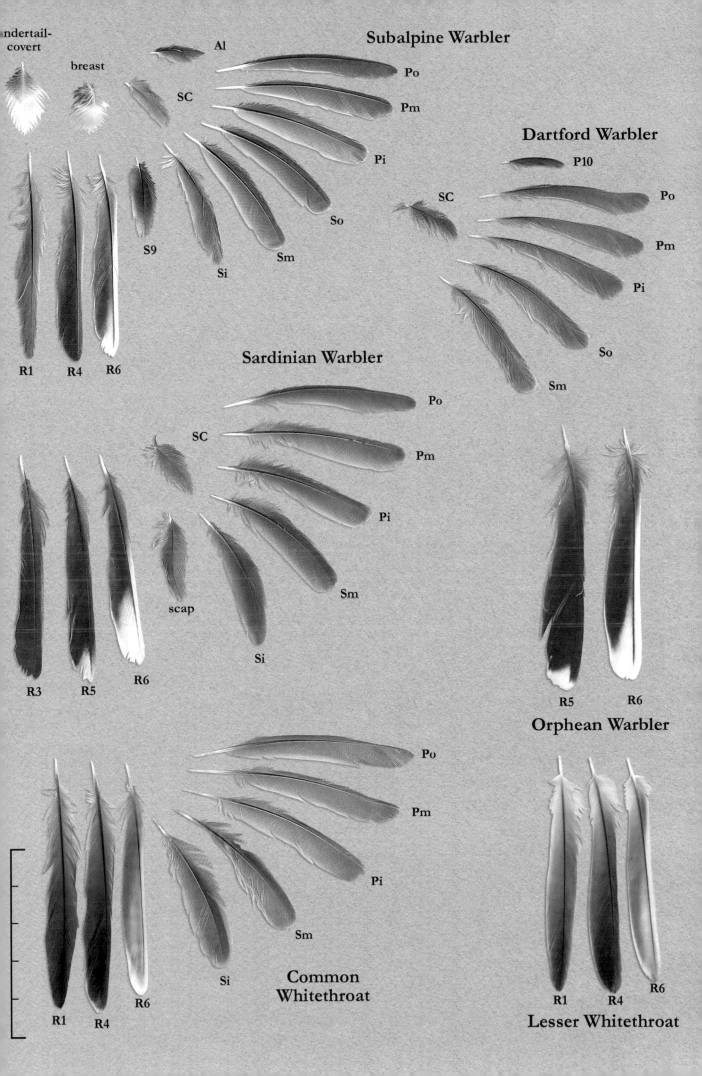

Subalpine Warbler

undertail-covert

breast

Al

SC

Po

Pm

Pi

So

Sm

Si

S9

R1

R4

R6

Dartford Warbler

P10

SC

Po

Pm

Pi

So

Sm

Sardinian Warbler

SC

Po

Pm

Pi

Sm

scap

Si

R3

R5

R6

Orphean Warbler

R5

R6

Po

Pm

Pi

Sm

Si

R1

R4

R6

Common Whitethroat

R1

R4

R6

Lesser Whitethroat

COLOUR CHARACTERISTICS OF THE 'LARGE' FEATHERS OF *SYLVIA* WARBLERS

Species	PS			R			
	Colour	Outer vane	Inner vane	Colour	R1	R median	R outer
Spectacled Warbler (SS)	Dark brown	rufous margin on the Ps; vane almost entirely rufous on Ss	buff border, large rufous border on inner Ss	dark grey to black	uniform	uniform; occasionally pale tip on R3 and R4	R5 with white tip or small white triangle at tip of iv; R6 black with diagonal white mark more extensive on iv; sometimes also black mark at tip along the rachis; R6 is shorter
Blackcap	Dark grey (m) to dark-brown (f, y)	greyish (m) or brown-grey (f, y); quite well-defined margin on Ps; ill-defined and less marked on Ss; buff margin at tip (y)	narrow white to buff border (f)	dark grey (m) to brown-grey (f, y)	thin brown-grey margin on ov		
Lesser Whitethroat	Dark brown-grey	white to cream margin becoming buff then rufous towards inner Ss; very thin at tip	white to cream border	dark grey to black	dark brown-grey with buff margin	R2–R5 with white to cream margin; tip with wider white border on iv (sometimes entirely black)	R6 white, with dark grey at base of vanes and on border of iv (on 2/3 at least); white or sometimes pale brown-grey with darker marks; dark rachis
Rüppell's Warbler (SS)	Dark brown or dark grey	grey or buff margin, paler and wider towards inner Ss (darker)	buff border?	black	uniform	R2–R3 often uniform; R4 sometimes with white drop-shaped mark at tip along the rachis	R5 with white tip or small white mark along the rachis on iv (sometimes oval); R6 with more white, diagonally or white with black at base continuing on border of iv
Garden Warbler	Dark brown	pale brown margin on outer Ps (paler at tip); pale brown vane on the other Ps	thin buff to white border	dark brown	thin buff margin on ov	pale brown margin widening to a border on the ov	
Barred Warbler (SS)	dark brown-grey	grey to brown-grey margin; also at tip	cream to white narrow border; also at tip	dark grey	uniform	R2 often uniform; if not a little white (or cream) at tip of iv, wider on R4, sometimes along border of iv	R5–R6 with white (or cream) at tip of iv, often triangular with longest side near the rachis (max. on the distal third); sometimes an irregularly shaped mark; white margin
Common Whitethroat	dark brown	buff to rufous margin on the Ps; rufous border on the Ss as far as tip	cream to buff border; rufous border on the inner Ss	dark brown-grey, buff margin	uniform	uniform, eventually with pale tip to R3–R5	R6 (like Orphean), dark grey with white diagonally, more extensive on iv; dark rachis; white often replaced by pale grey or buff in f.

Species	PS			R			
	Colour	Outer vane	Inner vane	Colour	R1	R median	R outer
Sardinian Warbler	Dark brown	grey margin (m) to rufous margin (f)	buff border	black with white margin	uniform	R2–R5 uniform or increasingly white at tip, descending on the border of iv; R4 and R5 with white at tip, especially on iv	R6 with a large white tip and white descending far along the ov
Western Orphean Warbler (SS)	Dark brown-grey	white to cream narrow margin	white to pale grey margin; pale margin at tip	black	dark grey to black	R2–R3 uniform; R4–R5 with a small pale mark at the tip of iv (sometimes passes onto inner Rs)?	R6 (sometimes dark grey) with white diagonally and more extensive on iv (sometimes entirely white); rachis black.
Subalpine Warbler	Dark brown-grey	grey (Ps) then buff (Ss) margin as far as tip.	white to cream border	dark grey to black	dark brown-grey	R2 uniform; R3–R5 can have a white tip	R6 with ov nearly all white and tip of the iv a white triangle or ill-defined rectangle (on distal 1/4 to 1/2); buff mark in y.
Dartford Warbler (SS)	dark brown to dark grey	buff margin; wider and greyer on the Ss of male, browner on Ss of female	cream to buff margin	dark grey, pale margin especially towards tip	uniform	R2–R5 uniform, sometimes with paler tip	R6 with pale mark on distal part of iv
Marmora's Warbler (SS)	Dark brown	margin of Ps and border of Ss grey in male, grey-brown in female	buff border?	dark grey with brown margin (f) to black with grey margin (m)	uniform	R2–R5 uniform or with a paler border towards tip	R6 with pale tip (buff, grey or white) and with some pale on ov (except dark base)

Note: f = female, m = male, ov = outer vane, iv = inner vane.

PHYLLOSCOPUS WARBLERS

Their wings are quite short and wide, more pointed in Wood and Arctic Warblers. The tail is rectangular to very slightly forked (R1 > 90% of Rmax).

Separating species within this group is particularly difficult, especially from isolated feathers. The site where the feather was found can discriminate several species, at least at certain times of the year. For plucked prey, the presence of yellow body feathers or greater coverts with pale tips can be important criteria, as well as the number of outer Ps that are emarginated (but be careful if any are missing!).

The 'large' feathers are brown; the outer vane is more greenish (olive), usually with a yellowish to greenish margin, sometimes well-defined. The Ps and Ss, and sometimes Rs, have a thin white to buff border on the inner vane. The two vanes are identical on R1 (brown with a margin). R6 usually has a grey or buff margin on the outer vane. In the Ps and Ss, the margin extends to the tip and becomes brighter, more markedly so in Bonelli's and Wood Warblers. According to *The Feather Atlas* (see References, 400), S7-S9 have different colours depending on the species: marked greenish hue in Greenish and Arctic Warblers, only a slight greenish hue in Willow Warbler, and Common and Iberian Chiffchaffs, and no greenish hue at all in species not considered in detail here. There is a pale border at the tip (and a dark grey background) in Western and Eastern Bonelli's, Yellow-browed and Wood Warblers.

MEASUREMENTS OF THE 'LARGE' FEATHERS OF *PHYLLOSCOPUS* WARBLERS								
Species	Pmax	Pmin	Smax	Smin	Rmax	Rmin	Wing shape	P emarginated
Yellow-browed Warbler (SS)	4.9	3.6	4.2	3	4.3	3.9	A	P5-P8
Arctic Warbler (SS)	5.7	4.6	4.6	3.7	4.8 (5.1)	4.4	B	P6-P8
Western Bonelli's Warbler	5.9	4.2	4.9	3.3	5.6	4.7	A	(P5 ill-defined) P6-P8
Willow Warbler	5.7	4.5	4.7	3.5	5.6	4.6	A	P6-P8
Eastern Bonelli's Warbler (SS)	5.3	4.3	4.5	3.3	5	4.3	A	(P5 very slightly) P6-P8
Wood Warbler	6.5	4.6	5	3.7	5.6	4.9 (4.4)	B	(P6 ill-defined) P7-P8
Common Chiffchaff and Iberian Chiffchaff	5.4	4.1	4.5	3.5	5.3	4.6	A (wide)	P5-P8
Greenish Warbler (SS)	5.2	4.1	4.3	3.2	4.8	4.6	A	(P5 very slightly) P6-P8

Note: Complete plumages have only been examined for Willow Warbler and Common Chiffchaff. Information for other species is derived from the observation of isolated feathers.

Incomplete descriptions (a non-exhaustive list):

– WOOD and **BONELLI'S WARBLERS:**

Ps, Ss and Rs with well-defined pale margin, more or less bright yellow.

– BONELLI'S WARBLER: outer vane finely bordered with white, more broadly towards the base in the Ss.

– CHIFFCHAFFS and **WILLOW WARBLER:**

very small pale tip to the Ps and Ss.

– GREENISH and **ARCTIC WARBLERS:** inner Ps and Ss with white tip; white tip often also on the inner vane of outer Rs. Some greater coverts with white tip to outer vane.

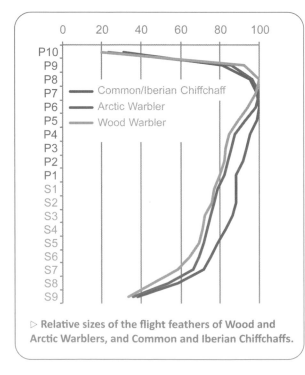

▷ Relative sizes of the flight feathers of Wood and Arctic Warblers, and Common and Iberian Chiffchaffs.

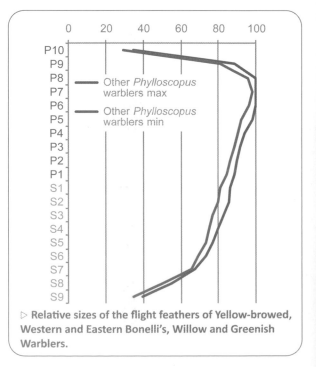

▷ Relative sizes of the flight feathers of Yellow-browed, Western and Eastern Bonelli's, Willow and Greenish Warblers.

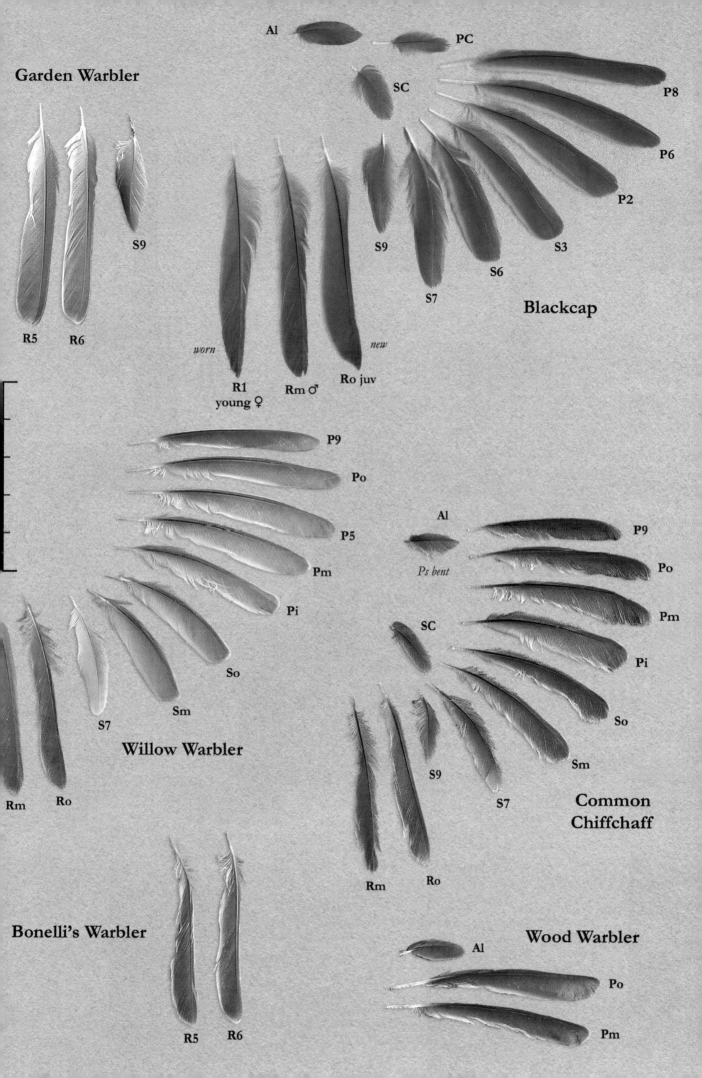

Al

PC

SC

Garden Warbler

S9

R5   R6

worn

new

R1
young ♀

Rm ♂

Ro juv

S9

S7

S6

S3

P2

P6

P8

Blackcap

P9

Po

P5

Pm

Pi

So

Sm

S7

Al

Ps bent

SC

P9

Po

Pm

Pi

So

Sm

S7

Willow Warbler

Rm   Ro

S9

Rm   Ro

S7

Common
Chiffchaff

Bonelli's Warbler

Al

Wood Warbler

R5   R6

Po

Pm

– **YELLOW-BROWED WARBLER:** inner Ps and Ss dark grey with a yellowish-white tip. Inner Ss dark grey with margins and tip yellowish-white. Base of the outer vane of the dark Ss without pale margin. Greater coverts with large yellowish tip.

HIPPOLAIS WARBLERS

The wings are quite triangular, sometimes more rounded or more pointed depending on the species. The tail is rectangular to very slightly forked (R1 ≥ 93% of Rmax).

This group is particularly difficult to identify to species level, especially from isolated feathers. The characteristics of the discovery site can discriminate several species, at least at certain times of the year. For most species moult occurs on the African wintering grounds.

Note: Only complete plumages of Icterine and Melodious Warblers were examined. Information for other species was derived from the examination of poor-quality illustrations and must therefore be used with caution.

The 'large' feathers are brown with a pale margin. Species are difficult to tell apart, except in cases where a fairly complete set of feathers is available, when size and shape of the wing can help. Young and females are smaller than adults and males (10-15%).

▷ **Common Chiffchaff wing.**

▷ **Willow Warbler wing.**

MEASUREMENTS OF THE 'LARGE' FEATHERS OF *IDUNA* AND *HIPPOLAIS* WARBLERS							
Species	Pmax	Pmin	Smax	Smin	Rmax	Rmin	Wing shape
Booted Warbler (SS)	5.3	4.5	4.7	3.4	5	4.5	A
Olive-tree Warbler (SS)	7.6	5.3	5.6	4.1	7.6	6.2	B
Icterine Warbler	7.2 (7.7)	5.1	5.7	4.2	6.5 (7.6)	5.1	B
Western Olivaceous Warbler (SS)	6.2	5.2	5.3	4.2	6.3	5.2	A(b)
Eastern Olivaceous Warbler (SS)	5.7	4.6	4.7	3.5	5.6	4.8	A(b)
Melodious Warbler	5.9	4.4	5.1	3.5	6	5	A(b)

ICTERINE WARBLER *Hippolais icterina:* Ps and Ss brown with white border on inner vane. Outer vane with buff to white margin that reaches the tip. The margin becomes wider and whiter on the median and inner Ss, and is also visible at the tip of the inner vane on the inner Ps and the Ss. Rs brown with white to cream margin on both vanes when fresh, the tip sometimes with a wider pale border. R6 paler in the central part of the outer vane. There is a fairly large variation in size between individuals. Emargination: (P6) P7 and P8.

Possible confusion: with other warblers, especially Garden Warbler, but the rachis is brighter below in this last species.

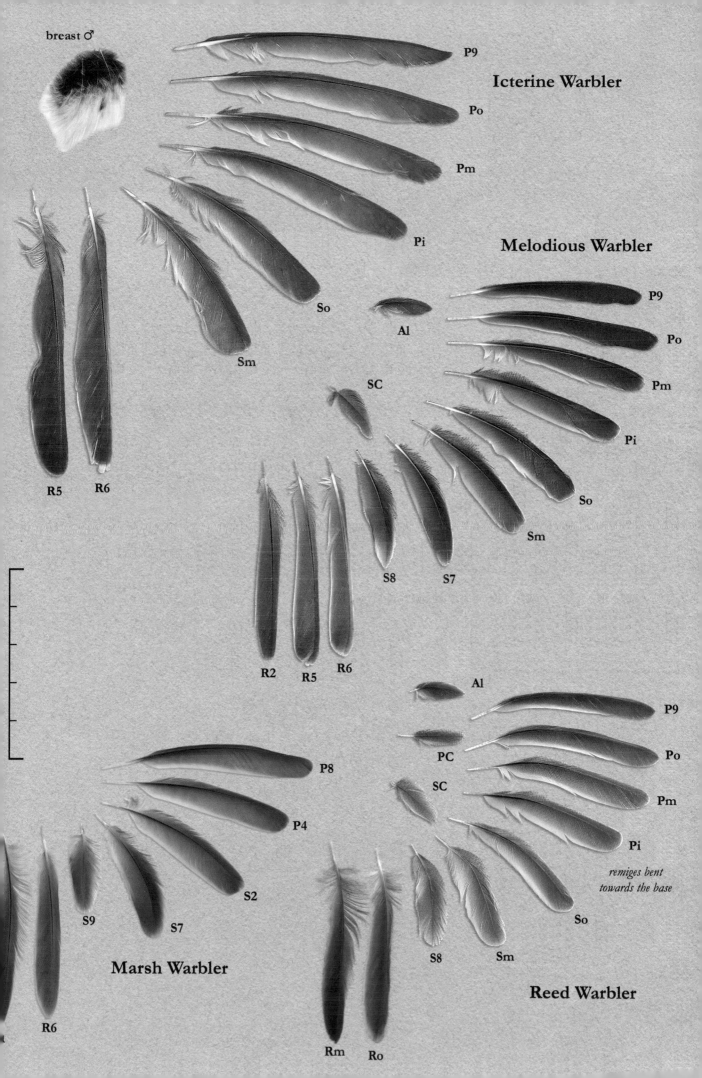

breast ♂

Icterine Warbler

P9

Po

Pm

Pi

So

Al

Sm

SC

Melodious Warbler

P9

Po

Pm

Pi

So

Sm

S8

S7

R5

R6

R2

R5

R6

Al

PC

SC

P9

Po

Pm

Pi

*remiges bent
towards the base*

So

Sm

S8

P8

P4

S2

S9

S7

Marsh Warbler

R6

Rm

Ro

Reed Warbler

MELODIOUS WARBLER *Hipolais polyglotta:* 'large' feathers like those of a small Icterine, but with buff to cream edge; Ss with an edge on the outer vane that is always thin; inner Ps and Ss with a buff to greenish edge. Emargination: (P5) P6-P8.

OLIVE-TREE WARBLER *Hippolais olivetorum* (SS): 'large' feathers as those of Icterine (Rs a little longer) but a darker brown, with a cream or white border to the median and inner Ss. R6 clearly paler on the outer vane. The tip of R6–R4 (R3) can be marked with pale or white, especially on the inner vane.
Possible confusion: of the Rs with Barred Warbler.

EASTERN OLIVACEOUS WARBLER *Iduna pallida* (SS): 'large' feathers very similar to those of the Melodious, with a buff-grey edge on the Ps and outer Ss, becoming wider on the median and inner Ss.

WESTERN OLIVACEOUS WARBLER *Iduna opaca* (SS): 'large' feathers very similar to those of the Melodious, but brown tint paler and duller; outer vanes of primaries and secondaries with buff border and edge without a greenish tinge.

BOOTED WARBLER *Iduna caligata* (SS): 'large' feathers like those of a small Icterine, but with a very wide wing and a generally pale hue. The outer vanes of the remiges have a pale brown edge/border.

ACROCEPHALUS WARBLERS (in part)

Their wings are quite triangular. The tail is rectangular to slightly rounded (R6 ≥ 85% of R1). It is almost impossible to distinguish between species in this group from an isolated feather. The location where the feather was found can discriminate several species, at least at certain times of the year. Moulting usually occurs on the wintering grounds, sometimes during migration.

The 'large' feathers are brown with pale edges; the outer vane is tinted with rufous. The Great Reed Warbler is significantly larger than the others. Small species are very difficult to tell apart; in the case of a complete set of feathers, the colour of the tertials and sometimes the shape of the wing can help.

GREAT REED WARBLER *Acrocephalus arundinaceus:* Ps and Ss a warm brown, outer vane tinted rufous, edged with buff on the Ps. Inner vane bordered with fawn, with a clear limit on the Ps, more blurred on the Ss. Rs warm brown edged with buff. Paler tip sometimes, especially on R6–R4.

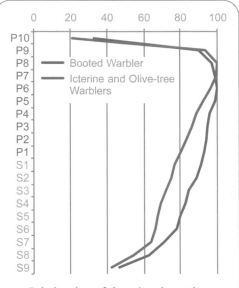

▷ Relative sizes of the primaries and secondaries of Booted, Icterine and Olive-tree Warblers.

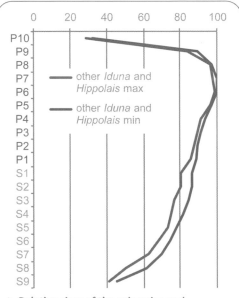

▷ Relative sizes of the primaries and secondaries of Melodious, Eastern Olivaceous and Western Olivaceous Warblers.

MEASUREMENTS OF THE 'LARGE' FEATHERS OF 'REED' WARBLERS

Species	Pmax	Pmin	Smax	Smin	Rmax	Rmin
Blyth's Reed Warbler (SS)	5.8	4.5	4.8	3.5	5.8	4.8
Eurasian Reed Warbler	5.8 (6.3)	4.5	5	3.5	5.8 (6.3)	4.7
Paddyfield Warbler (SS)	5.4	4.2	4.6	3.4	5.7	4.7
Great Reed Warbler	8.3 (9.2)	6.1	7	4.8	8.6	6.9
Marsh Warbler	6	4.5	4.8	3.6	5.8	4.9

Possible confusion, for example with Rufous Bush Robin (but different habitat).

EURASIAN REED WARBLER *Acrocephalus scirpaceus:* Ps and Ss brown with cream border on the inner vane. The outer vane is tinged with rufous, finely edged with cream to buff on the outer and median Ps, wider tawny edge on the inner Ps and Ss. Inner Ss brown bordered with paler tawny. Rs brown, tinged reddish when fresh, edged buff to tawny on both vanes. Tips of R4–R6 sometimes paler (brown to greyish). In this species, R6 can be a paler brown, with rachis paler than vanes.

MARSH WARBLER *Acrocephalus palustris:* like Eurasian Reed Warbler but with R6 as the other Rs, inner Ss sometimes with olive-brown border.

BLYTH'S REED WARBLER *Acrocephalus dumetorum* (SS): like Eurasian Reed Warbler but with uniform inner Ss (ill-defined border when worn). Rs all the same.

PADDYFIED WARBLER *Acrocephalus agricola* (SS): like Eurasian Reed Warbler but with inner Ss more contrasted, with dark centre. Rs all the same.

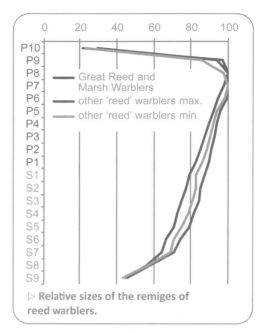

▷ Relative sizes of the remiges of reed warblers.

OTHER WARBLERS, *ACROCEPHALUS*, *LOCUSTELLA*, *CETTIA*, *CISTICOLA*

MEASUREMENTS OF THE 'LARGE' FEATHERS OF OTHER WARBLERS

Species	Pmax	Pmin	Smax	Smin	Rmax	Rmin	Wing shape	Tail shape
Cetti's Warbler	5.6	4.4	5.4	3.5	7	4.9	A	rounded
Zitting Cisticola (SS)	4.8	3.6	4.5	3.2	5	2.9	A	very rounded
Grasshopper Warbler	5.6	4.9	4.6	3.8	6.2	4.1	B	rounded
River Warbler (SS)	6	4.9	4.9	3.8	6.2	5 (4.2)	B	rounded
Savi's Warbler	5.8	4.7	4.7	3.6	6.8	4.3	B	rounded
Moustached Warbler (SS)	5.5	3.9	4.8	3.2	5.4	4.3	A	rounded
Aquatic Warbler (SS)	5.6	4.1	4.5	3.4	5.4	3.8	B	rounded
Sedge Warbler	5.7 (6.8)	4.3	4.6 (5.5)	3.5	5.6	4.4	B	slightly rounded

CETTI'S WARBLER *Cettia cetti:* Ps and Ss warm brown, finely bordered with fawn to cream on the inner vane, rufous-tinted on the outer vane. It has only 10 Rs: reddish-brown, fairly wide, rounded tip, appearing surprisingly large and wide compared to remiges. The wing is rounded and wide; the tail slightly rounded. Individuals from the north of the species' range are larger than those from the south; females are more or less significantly smaller than males (at least 15% difference for the same region).

ZITTING CISTICOLA *Cisticola juncidis:* Ps and Ss dark brown, with a thin buff to white border to the inner vane. Outer vane with a buff edge becoming rufous and wider towards the inner Ss. The tips have a pale edge. Rs brown with a dark subterminal mark and pale tip. R1 brown along the rachis with a wide border from tawny to rufous (sometimes very lightly barred darker). Other Rs with fawn edges on the inner vane and buff to cream on outer vane, with a large, round, black subterminal spot (more contrasting below) and a buff-grey and/or white tip (inner vane). Wide, white tip on R6, whose outer vane is partly edged with white. The wing is very rounded and wide (Ps and Ss almost the same size); the tail is rounded (more or less markedly depending on the individual).

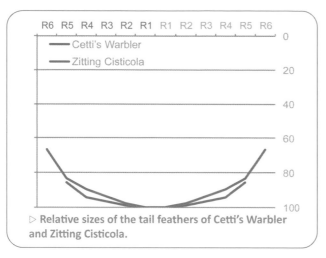

▷ Relative sizes of the tail feathers of Cetti's Warbler and Zitting Cisticola.

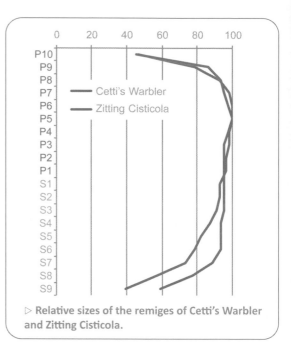

▷ Relative sizes of the remiges of Cetti's Warbler and Zitting Cisticola.

GRASSHOPPER WARBLER *Locustella naevia:* Ps and Ss dark brown, outer vane edged with olive-brown becoming more rufous and wider on the Ss (buff to white on the outer Ps). Inner vane with buff border, edged with rufous on the inner Ss (well contrasted, dark centre and tawny border). Rs brown edged with fawn on both vanes and with paler tip, buff to tawny (especially R4–R6). The tail is clearly rounded in *Locustella* warblers.

RIVER WARBLER *Locustella fluviatilis* (SS): 'large' feathers very similar to those of Savi's Warbler but without the rufous tint. Ps and Ss brown, outer vane tinged with olive-brown, with a paler edge on the outer Ps. Inner vane with buff border, tinged olive on the inner Ss. Rs brown, R1 paler. Small, paler tip, especially on R4–R6.

SAVI'S WARBLER *Locustella luscinioides:* Ps and Ss brown, outer vane tinged with rufous, with buff to tawny edge and white on P9. Inner vane bordered with buff; outer vane tinged with rufous on the inner Ss (uniform). Rs brown tinged with rufous. Small, paler tip, especially on R4–R6.

MOUSTACHED WARBLER *Acrocephalus melanopogon* (SS): Ps and Ss brown, outer vane tinged with rufous, with tawny edge. Inner vane with buff border, larger on the inner Ss (with dark brown centre). Rs brown to dark brown; tawny edge on outer vane, white on R6, and edged with buff on inner vane. The wing is very large and rounded; the tail is slightly rounded.

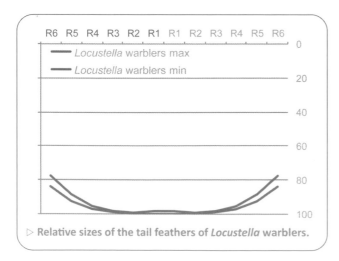

▷ Relative sizes of the tail feathers of *Locustella* warblers.

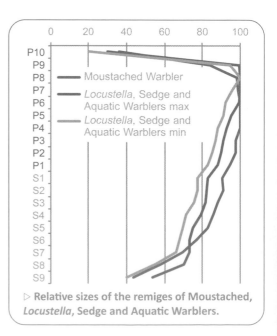

▷ Relative sizes of the remiges of Moustached, *Locustella*, Sedge and Aquatic Warblers.

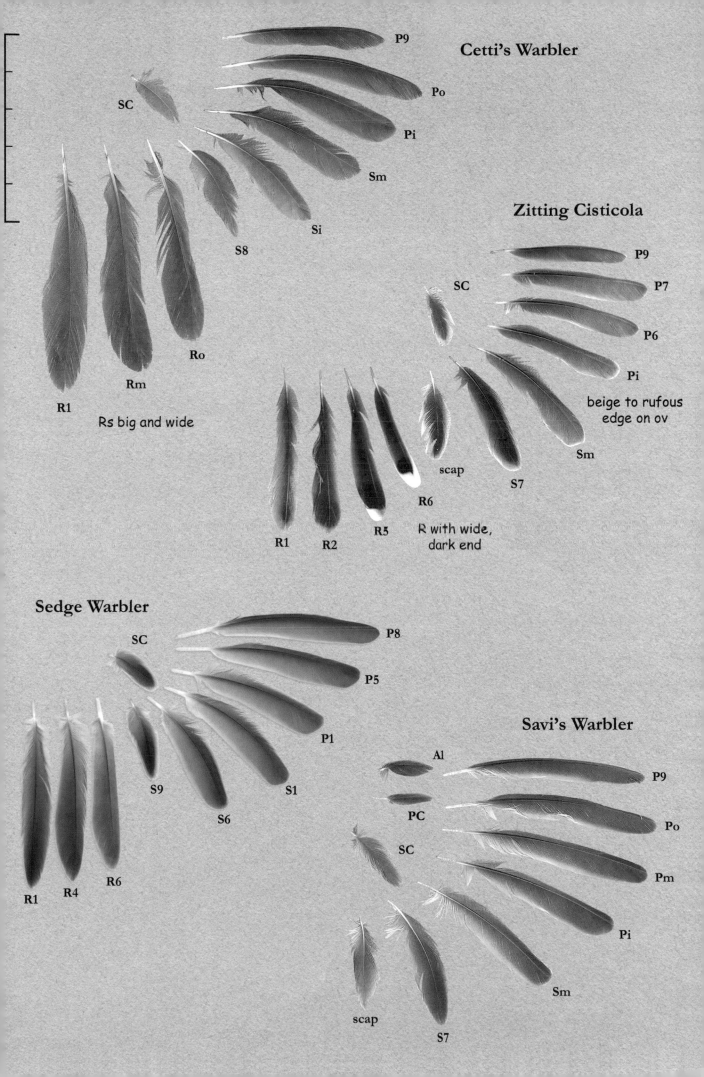

Cetti's Warbler

P9
Po
Pi
Sm
Si
S8
SC

R1
Rm
Ro
Rs big and wide

Zitting Cisticola

SC
P9
P7
P6
Pi
beige to rufous
edge on ov
Sm
S7
scap
R6
R with wide,
dark end
R5
R2
R1

Sedge Warbler

SC
P8
P5
P1
S1
S6
S9
R6
R4
R1

Savi's Warbler

Al
PC
SC
P9
Po
Pm
Pi
Sm
S7
scap

AQUATIC WARBLER *Acrocephalus paludicola* (SS): Ps and Ss brown, outer vane edged with buff, wide and yellowish on the inner Ss (almost black centre). Inner vane with cream to buff border to very tip, yellowish on the inner Ss. Rs dark brown to black, outer vane reddish-brown. R6 paler, with buff to white edge. The tail is slightly rounded.

SEDGE WARBLER *Acrocephalus schoenobaenus:* Ps and Ss brown; outer vane with buff edge, buff border on inner Ss (dark brown centre). Inner vane with cream to buff border to the tip. Rs brown, with tawny to buff edges, tip sometimes a little paler. The tail is slightly rounded.

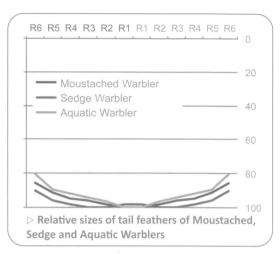

▷ **Relative sizes of tail feathers of Moustached, Sedge and Aquatic Warblers**

• HIRUNDINIDAE: SWALLOWS, MARTINS

Distinctive criteria of group: especially shape of the remiges, colour.

This family has pointed wings, with a fairly long triangular hand and a moderately developed forearm. The tail is often forked, sometimes only slightly (Crag Martin), sometimes with streamers (Barn Swallow). In the swallows and martins, most of which are migratory, moult can be interrupted before migration and completed at the wintering sites, or take place wholly on the wintering sites. Therefore, the frequency with which moulted feathers are found in Europe is relatively low compared to other groups.

Possible confusion: with swifts, and to a lesser degree the auks and storm petrels (on the coast).

Colour of 'large' feathers: quite uniform brown to black; they sometimes have a pale edge (white, buff, brownish) on one or both vanes, rarely a pale tip on the inner Ss. Pale edges wear quite quickly and are not always visible. They have a faint metallic blue, bronze or green iridescence. The Rs may have a pale spot (Sand Martin and Barn Swallow). Leucistic or albino individuals sometimes occur.

Primaries: long, quite narrow and pointed, outer ones slightly curved, scythe-shaped. Their size decreases rapidly towards the inner Ps. They are neither emarginated nor notched.

Secondaries: quite short compared to the Ps. The vanes of the Ss (but not the tertials) are longer than their rachis with a double bump at the tip. This gives a 'slotted' outline to the trailing edge of the wing, which reduces turbulence in flight.

Rectrices: central ones are quite short, outer ones more or less elongated depending on the species.

CRITERIA OF 'LARGE' FEATHERS OF SWALLOWS AND MARTINS								
Species	Pmax	Pmin	Smax	Smin	colour PS and Ss	Rmax	Rmin	colour Rs
Sand Martin	11	5	5.5	4	dark brown with very slight green sheen; inner vane with thin white edge, also on outer vanes of Ss	6	4	dark brown with slight green sheen
Crag Martin	12	5	5.5	5	dark brown; inner vane paler towards the base	7	5	dark brown; round or oval white mark on inner vane (except outer Rs)
Barn Swallow	12	4.5	6	4	black; dark green to blue sheen	12 (13)	4.5	black-brown; green sheen; white mark on inner vane (except central Rs)
House Martin	11	4.5	5.5	4	black-brown with slight green sheen; pale edge, sharp on inner Ss	7	4	black-brown with slight green sheen
Red-rumped Swallow	12	5	6	3	black-brown; green-blue sheen; sometimes pale edge	11	4.5	black-brown; green-blue sheen

SAND MARTIN *Riparia riparia:* 'large' feathers dark brown. Inner Ps, Ss and Rs generally with white edge at tip and at least on inner vane.

CRAG MARTIN *Ptyonoprogne rupestris:* 'large' feathers dark brown; inner vane paler towards the base. Rs (except for R6) with a more or less wide, white oval on inner vane.

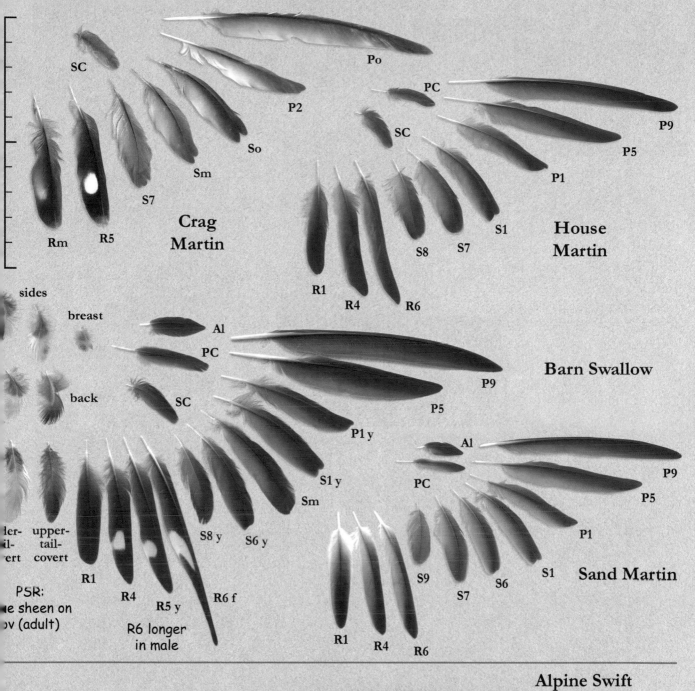

SC

Crag
Martin

Rm   R5   S7   Sm   So   P2

Po

PC

SC

P9

P5

P1

S1

S7

S8

R1

R4

R6

House
Martin

sides

breast

back

Al

PC

SC

P1 y

S1 y

Sm

S8 y   S6 y

R6 longer
in male

Barn Swallow

P9

P5

Al

PC

P9

P5

P1

S1

S6

S7

S9

R1   R4   R6

Sand Martin

PSR:
e sheen on
ov (adult)

upper-
tail-
covert

R1   R4   R5 y   R6 f

der-
il-
ert

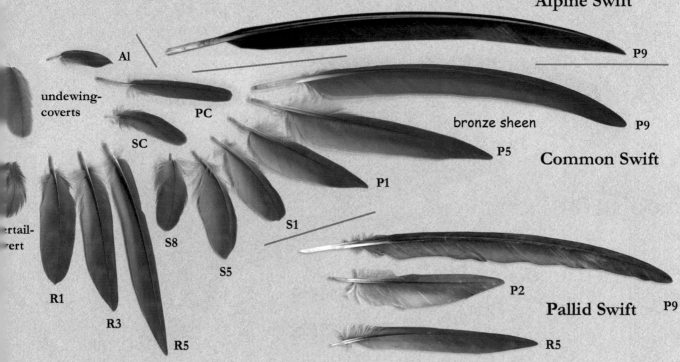

Al

PC

SC

undewing-
coverts

Alpine Swift

P9

bronze sheen

P9

P5

P1

S1

Common Swift

R1

R3

R5

S8

S5

ertail-
ert

P2

Pallid Swift   P9

R5

BARN SWALLOW *Hirundo rustica:* Ps and Ss often with dark blue sheen, slight on the Ps but becoming more obvious on Ss, at least in adults. Outer Rs, so-called 'streamers', of very variable length depending on age and sex of the individual, larger in males and adults. However, the size of the streamers of the same individual may vary from one year to another. Rs (except R1) have a large cream to white oval on the inner vane that extends towards the edge and increases towards the outer Rs, in the shape of a triangle, elongated trapeze or large comma. This mark is smaller in young, often of a rounded shape.

A few measurements of outer R6 (longer or shorter sizes may occur): in juveniles T = 6.5-8; in females T = 8-10; males T = 9.5-13.

HOUSE MARTIN *Delichon urbicum:* remiges black-brown; the inner Ps and Ss are usually finely edged with white to buff at the tip, more broadly on the inner Ss. Rs dark, sometimes edged with pale.

RED-RUMPED SWALLOW *Cecropis daurica:* Ps and Ss black-brown; possibly a buff edge to the tip and inner vane of the inner Ps and Ss. Rs black-brown; sometimes with a grey area on inner vane of outer Rs..

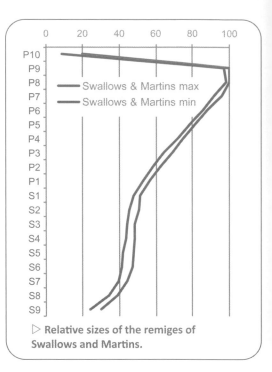

▷ Relative sizes of the remiges of Swallows and Martins.

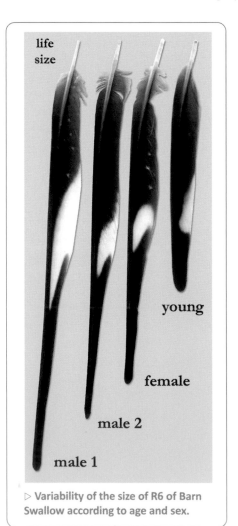

▷ Variability of the size of R6 of Barn Swallow according to age and sex.

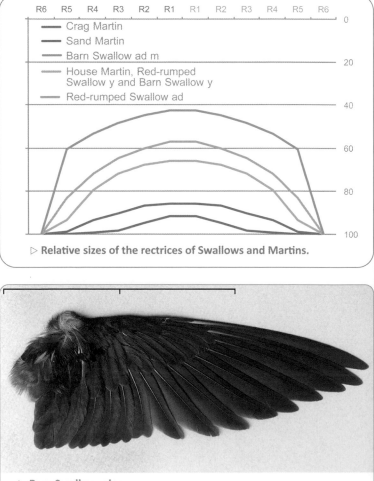

▷ Relative sizes of the rectrices of Swallows and Martins.

▷ Barn Swallow wing.

NEAR PASSERINES

These species are similar to the passerines in the shape and silhouette of their 'large' feathers or their way of life; this similarity is not necessarily phylogenetic.

• APODIDAE: SWIFTS

NP = 10 / NS = 6-9(10) / NR = 10. There are at least six normally developed Ss; however, some inner Ss may be much shorter than the outer ones.

Distinctive criteria of group: shape and colour.

This family has very long wings, with the hand eight times longer than the forearm. High-speed flight, whether flapping or gliding is its specialty. The tail is usually forked. Moulting begins after migration, in the wintering quarters. It is therefore difficult to find isolated feathers of these species and it is more usual to find feathers of plucked birds, and mainly young.

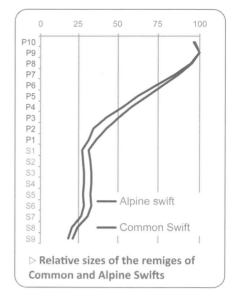
▷ **Relative sizes of the remiges of Common and Alpine Swifts**

Possible confusion: with swallows and martins and, to a lesser extent, auks and storm petrels (on the coast).

Colour of 'large' feathers: black-brown; possibly with slight bronze, blue or green sheen. Paler inner vane and base. The 'large' feathers and body feathers of young have a very narrow pale edge and little sheen, and are more brownish.

Primaries: The outer Ps are excessively elongated, slightly curved and pointed, scythe-shaped. They are neither emarginated nor notched. Their size decreases very rapidly towards the inner Ps (P1 < 35% Pmax). The Ps are quite pointed and therefore wider towards their middle or base.

MEASUREMENTS OF THE 'LARGE' FEATHERS OF SWIFTS						
Species	**Pmax**	**Pmin**	**Smax**	**Smin**	**Rmax**	**Rmin**
Common Swift	15	4.5	5	4	9.5	4.5
Pallid Swift (SS)	15	4.5	5.5	4	9	5
Alpine Swift	22	5	6.5	4	10	5

Secondaries: The Ss are very short compared to outer Ps (about a third of their size); their inner vane may be longer than the outer one. The inner Ss are highly curved inwards.

Rectrices: As with the Ps, they are quite tapered but are straight. The outer ones are even more elongated as the tail is forked.

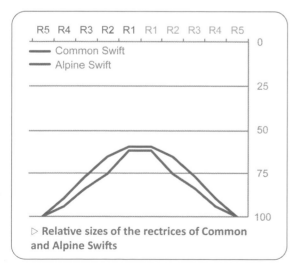
▷ **Relative sizes of the rectrices of Common and Alpine Swifts**

COMMON SWIFT *Apus apus:* Ps dark brown, with outer vane and tip black with blue-green or bronze-green sheen. Inner vane and inner Ps more olive-brown with only a slight sheen. Ss dark brown with very little sheen. Rs olive-brown with slight bronze sheen. 'Large' feathers and coverts of young with a very narrow pale edge.

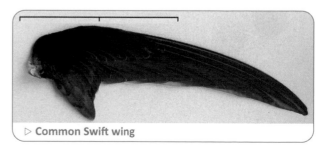
▷ **Common Swift wing**

PALLID SWIFT *Apus pallidus:* Ps and Ss with grey-brown outer vane and tip, inner vane greyer. Rs grey-brown with inner vane paler at base.

ALPINE SWIFT *Tachymarptis melba:* Ps and Ss with olive-black outer vane and tip, inner vane brown-grey. Sometimes a narrow pale edge on the outermost P. Rs olive-black, with slight bronze sheen, the base of the inner vane sometimes paler with a pale edge. Wings slightly narrower than those of Common Swift (P1 = 30% of P10).

• PICIDAE: WOODPECKERS AND WRYNECK

NP = 9 (+1) / NS = 10-12 / NR = 10 (+ 2 very short outer Rs)

Distinctive criteria of group: shape of rectrices (except Wryneck), colour (except Black Woodpecker).

Members of this family have rounded wings and a stiff tail that, except in the Wryneck, serves as a support when the woodpecker climbs trees.

Possible confusion: with Corvidae (Black Woodpecker), Hoopoe (Ss of black and white woodpeckers).

Colour of 'large' feathers: black, black-and-white, or marked with khaki and buff (see below). Wryneck has different shades of brown and rufous.

Primaries and secondaries: in shape they resemble those of passerines of comparable size; it is their colour that makes it possible to distinguish them, except for Black Woodpecker. Sometimes there is a small P (carpal feather) towards the 'wrist'.

Rectrices: except in the Wryneck, the Rs have a very special shape.

The tail serves as a support when climbing trees; the rectrices are rigid. To offer better support, they are almost straight, even the outer ones, and enlarged in their centre. The calamus is not bent upwards as it usually is on the rectrices. The tip is pointed to increase grip on the tree trunk. The tip breaks quite quickly by rubbing against the trunk, so the Rs are often forked, which further improves their grip on the surface. The rachis of the central Rs are thicker than those of the others. The outer Rs are much shorter than the others (about 40% of the adjacent R).

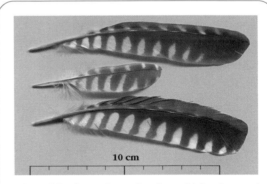

▷ **Carpal feathers of woodpeckers.** P1 (top), carpal feather (centre) and S1 of Green Woodpecker.

SIZE OF THE 'LARGE' FEATHERS OF WOODPECKERS AND WRYNECK							
Species	Pmax	Pmin	Smax	Smin	Rmax	Rmin	R outer
Black Woodpecker	21 (23)	13·5	17 (21)	12	21	11.5	5.5-8 (12)
Green Woodpecker	15	10	12	8	12	8	3-5.5
Grey-headed Woodpecker (SS)	13	9	14	8	12	7	n.d.
Great Spotted Woodpecker	13 (15)	8	10	7.5	10 (11.5)	6.5	3-4
White-backed Woodpecker (SS)	14	10	10.5	8.5	11.5	8.5	4-4.5
Syrian Woodpecker (SS)	11.5	9	10	8	n.d.	n.d.	n.d.
Three-toed Woodpecker	12	7	8.5	7	9	6	n.d.
Middle Spotted Woodpecker	11.5	8.5	9	6.5	9.5	6	n.d.
Lesser Spotted Woodpecker	9·5	6	7	5	7	4	0.5
Eurasian Wryneck	8.5	5.5	7.5	5	8	6	3-3.5 (5)

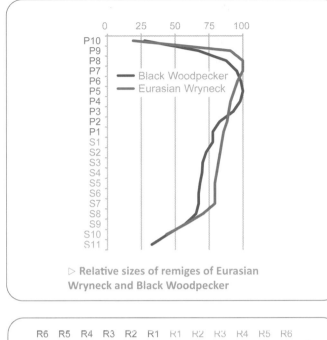

▷ **Relative sizes of remiges of Eurasian Wryneck and Black Woodpecker**

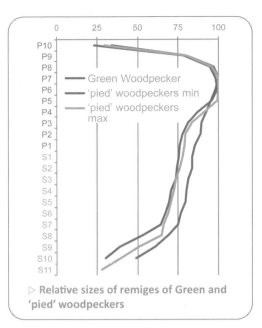

▷ **Relative sizes of remiges of Green and 'pied' woodpeckers**

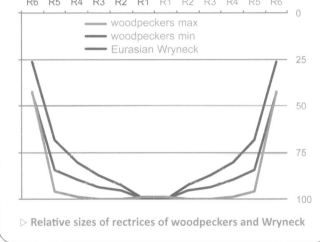

▷ **Relative sizes of rectrices of woodpeckers and Wryneck**

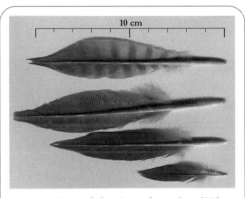

▷ **Comparison of the sizes of rectrices (R2).** Green Woodpecker (left), Black Woodpecker (centre, left), Great Spotted Woodpecker (centre, right), Lesser Spotted Woodpecker (right).

EURASIAN WRYNECK *Jynx torquilla:* Ps and Ss brown with rufous-buff wedge-shaped marks; pale, mottled wedges on the Ss.

Possible confusion: the Rs resemble those of passerines but are rufous-grey barred or marked with dark; those of small owls are downy (see p. 79). For Ps and Ss see also Common Quail, p. 334.

BLACK WOODPECKER *Dryocopus martius:* Ps, Ss and Rs black-brown. Sometimes a paler area at the base of the remiges. A slight bronze sheen on outer vane of Ps and Ss.

Possible confusion: with crows, but the colour is less sustained in the Black Woodpecker. Rs of scoters, but different shape; habitat where feather was found is indicative.

▷ **Comparison of the sizes of secondaries** Lesser Spotted Woodpecker (top), Great Spotted Woodpecker (centre), Hoopoe (bottom)

GREEN WOODPECKER *Picus viridis* and **GREY-HEADED WOODPECKER** *Picus canus:* Ps and Ss brown with buff ovals or wedge-shaped marks, outer vane tinted kaki-green. Rs brown with paler tone in the form of notches or bars with superimposed greenish hue.

		outer vane		inner vane	
Ps and Ss		number of marks	shape of marks	number of marks	shape of marks
Great Spotted Woodpecker	P	(0) 4-6	trapezoid, rectangular; often one at the tip (inner Ps)	3-5	wide, rounded notches
	S	(2) 3-5	rounded notches; 2-3 on the inner Ss	4-5	wide, rounded notches; the marks at the base sometimes connected
Middle Spotted Woodpecker	P	4-6	similar to Great Spotted; marks more angular and larger	3-5	Similar to Great Spotted; marks more angular and larger
	S	(2) 3-5		(2) 3-5	
Three-toed Woodpecker	P	3-8	rectangular, triangular	6-8	oval, elongated notch
	S	0-4	small circle, rectangular, triangular	3-8	oval, elongated notch; more rounded towards inner Ss

Table title: **NUMBER AND SHAPE OF WHITE MARKS ON THE REMIGES OF SOME 'PIED' WOODPECKERS**

These two species are very difficult to distinguish other than by their 'large' feathers.

GREEN WOODPECKER *Picus viridis:* Ps and Ss brown with pale wedges; khaki-green tint on outer vane (sometimes absent on outer Ps). Ps with 7-10 trapeze- or triangle-shaped marks on outer vane and 5-11 oval, rounded or pointed notch-shaped marks on the inner vane. Ss have 8-9 only slightly visible notches on the outer vane and 8-11 oval or elongated notches on the inner vane, tending to become bars on the inner Ss. Rs with pale bars.

GREY-HEADED WOODPECKER *Picus canus* (SS): the Rs are generally less clearly barred than in Green Woodpecker. Sometimes there are no obvious pale bars on outer Rs. In the juvenile, the pale marks of the outer Ps reach the tip.

The 'pied' woodpeckers: Lesser Spotted, Middle Spotted, Great Spotted, Three-toed, White-backed and Syrian.

Without body feathers the species are (very) difficult to differentiate. Ps and Ss are black with rounded notches or white ovals. In some species, the notches may coalesce to form white bars. Central Rs are increasingly marked with white or buff at the tip towards the outer Rs.

The Lesser Spotted Woodpecker is significantly smaller than the others and easily recognisable, but all the others are difficult to distinguish (larger marks in Syrian Woodpecker, smaller in Three-toed Woodpecker, etc.).

GREAT SPOTTED WOODPECKER *Dendrocopos major:* the commonest and most widespread of these woodpeckers.

There is a large variability in size between the subspecies. In the juvenile, the Ps have a pale tip (pale buff to white). Compared to the Great Spotted Woodpecker:

MIDDLE SPOTTED WOODPECKER *Dendrocopos medius:* very similar but the marks are more angular, more like rounded squares, and are very slightly larger in size. In juveniles, the four median Ps have a 2 mm wide white mark at the tip.

THREE-TOED WOODPECKER *Picoides tridactylus:* the marks on the Ps and Ss are more numerous overall and narrower. White tip to the Ps.

SYRIAN WOODPECKER *Dendrocopos syriacus* (SS): on the Ps, the marks on the two vanes may meet, or even form incomplete bars. On the Ss, the marks are wider than in Great Spotted Woodpecker.

WHITE-BACKED WOODPECKER *Dendrocopos leucotos* (SS): the marks are more elongated (similar to those of Syrian). Well-defined white tip on the Ps, whitish end on the Ss. Inner Ss barred. Overall more white on the Rs.

LESSER SPOTTED WOODPECKER *Dendrocopos minor:* Ps and Ss very much smaller. On the inner Ss, the white marks coalesce into bars. In the juvenile, there is often a white border on both vanes of the four median Ps.

Al

underwing-covert

PC

SC

Po

Pm

Pi

R3

R4

Ps and Ss:
pale bars
and notches

R: dark bars on
speckled background

scap

Si

Si

So

Wryneck

rump

belly of young

Al

PC

SC

R5

back

S9

S6

S2

R3

R1

khaki-green

crown
(life size)

P9

P7

P1

**Green
Woodpecker**

plumage black, grey
or dark brown
without a sheen

PC

SC

R6

R5

Al

R3

R1

S9

Sm

So

P10

P8

P5

P2

crown
(life size)

**Black
Woodpecker**

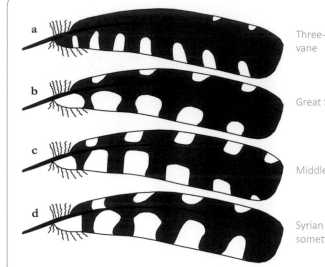

Three-toed Woodpecker (a): many elongated marks on the inner vane

Great Spotted Woodpecker (b): spots

Middle Spotted Woodpecker (c): spots

Syrian Woodpecker (d): wide marks often touching the rachis, sometimes forming a bar

▷ Comparison of the shape of marks on the secondaries of four 'pied' woodpecker species.

COLOUR OF THE RECTRICES IN A FEW SPECIES OF 'PIED' WOODPECKERS	
Species	**Rectrices colour**
Great Spotted Woodpecker	Central Rs black (R1 or R1+R2), sometimes with 1 or 2 white or buff marks. From R3–R5, more extensive pale marks: a single mark at the tip, paler or pale marked with irregular black bars. R6 small, black or with 1 or 2 white marks. Great variability in the markings between individuals.
Middle Spotted Woodpecker	R2 and/or R3 often with a pale edge at tip of outer vane. Overall, markings similar to those of Great Spotted Woodpecker.
Three-toed Woodpecker	Central Rs black (R1), unmarked, or with a small white spot at the base. 2–5 pale marks on the others towards the tip (more numerous on outer vane), forming 1–3 incomplete bars; sometimes also a mark at the base of the inner vane. Overall fewer pale areas than in Great Spotted Woodpecker, bars more numerous and narrower. Often a small white mark at base of inner vane.
Lesser Spotted Woodpecker	Central Rs black (R1 and R2). Other marks similar to those of Great Spotted Woodpecker. Much smaller.

There is great variability in the extent of the white marks on the Rs of individuals within each species. Some individuals have three or four pairs that are entirely black, while others have some white on every R, even the central ones.

▷ Green Woodpecker wing

▷ Great Spotted Woodpecker wing

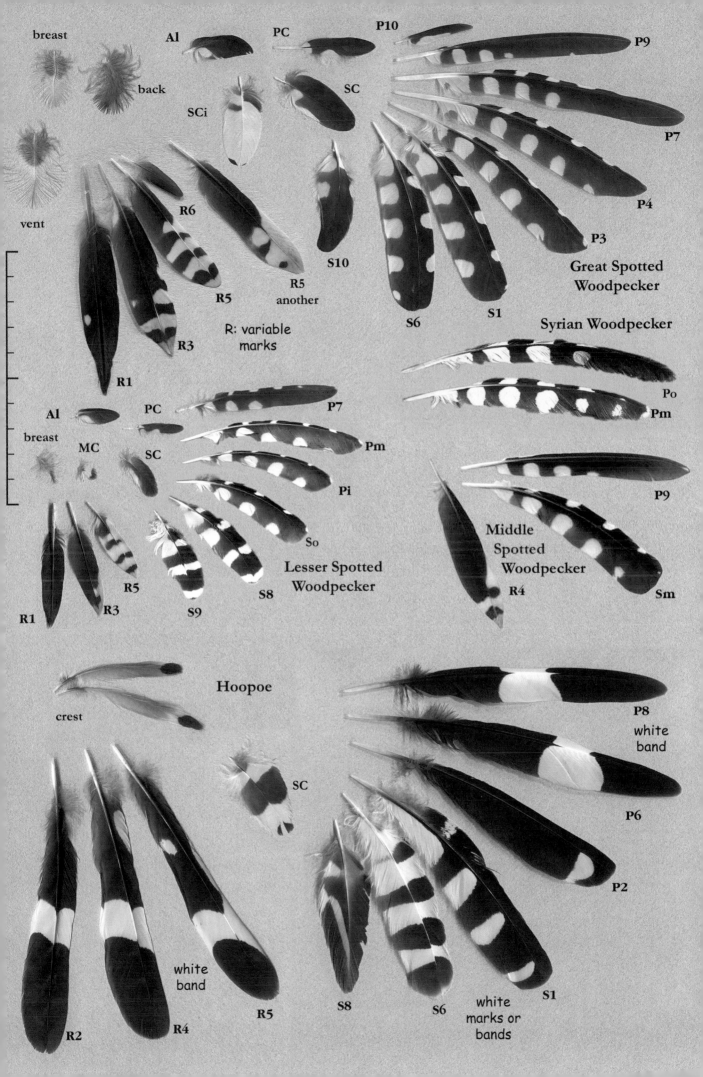

breast

back

SCi

Al

PC

P10

PC

SC

P9

P7

P3

P4

vent

R6

R5

R5
another

S10

R: variable
marks

R3

S6

S1

Great Spotted
Woodpecker

Syrian Woodpecker

R1

Al

PC

P7

breast

MC

SC

Pm

Pi

Po
Pm

R5

So

P9

R1

R3

S9

S8

Lesser Spotted
Woodpecker

Middle
Spotted
Woodpecker

R4

Sm

Hoopoe

crest

SC

P8
white
band

P6

white
band

P2

R2

R4

R5

S8

S6

S1

white
marks or
bands

• UPUPIDAE, ALCEDINIDAE, CORACIIDAE, MEROPIDAE: HOOPOE, KINGFISHER, ROLLER, BEE-EATERS

Distinctive criteria of group: colour.

HOOPOE *Upupa epops*, Upupidae.

NP = 9 (+1) / NS = 10 (11) / NR = 10

It has large rounded wings and a black and white rectangular tail. The general silhouette of the 'large' feathers is reminiscent of that of the passerines.

Possible confusion: with 'pied' woodpeckers (Ps and Ss), passerines (Rs), but the coloured markings are quite characteristic.

Colour of 'large' feathers: all are black and white; the inner Ss may have buff marks.

Primaries: TP = 9.5-14 (17). Black with a white stripe that becomes narrower and closer to the tip going towards the inner Ps, which have only a white spot on the inner vane (often semi-circular). The inner P sometimes has a white spot towards the base of the inner vane.

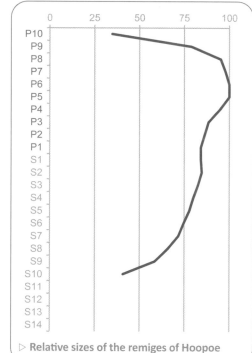

▷ Hoopoe wing

Secondaries: TS = 9-13 (inner ones 6-7.5). More often than not they are striped black and white. White marks may be more rounded, notch-shaped or incomplete bars (outer Ss) or stretch towards the tip (inner Ss).

Rectrices: TR = 9-12.5. The tail is rectangular, with central rectrices measuring between 90% and 100% of the outer ones. They are black with a white stripe towards the middle (central Rs) or the outer two thirds (outer Rs). At the very least, the outer and median feathers may have one or two small white marks towards the base. R6 can have a lot of white on the outer vane.

Body feathers: the pinkish-orange tint of the dorsal body feathers can help identify them. The most characteristic are those of the crest; long, narrow, with a black tip sometimes highlighted with white.

Note: in Europe, some individuals are sedentary (Spain), others are migratory. Migratory birds may moult before or after migration. Most of the feathers found in Europe come from corpses.

▷ **Relative sizes of the remiges of Hoopoe**

COMMON KINGFISHER *Alcedo atthis*, Alcedinidae.

NP = 10 / NS = 12 (11-14) / NR = 12 (14?)

Its wings and tail are quite short; the 'large' feathers have an appearance similar to those of small passerines. Due to a bird's lifestyle and small size, moulted feathers are difficult to find. The relative impermeable nature of the plumage is due to the density and stiffness of the breast and belly feathers.

Possible confusion: with Blue Tit, but larger size and internal rufous-buff edges (Ps and Ss), and Rs more rounded and brighter blue.

Colour of 'large' feathers: on a dark grey background the primaries and secondaries have some blue or green-blue on the outer vane and a rufous to cream edge on the inner vane. The rectrices are black, tinged with dark blue above, becoming lighter when viewed from an angle.

Primaries: TP = 5-7,5.

▷ Common Kingfisher wing

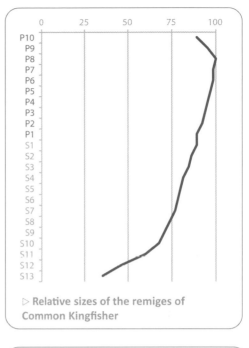

▷ Relative sizes of the remiges of Common Kingfisher

Primaries: TP = 5-7.5.

Secondaries: TS = 4-6.5 (innermost 2-4).

Rectrices: TR = 3.5-4.5 (6), a more compact shape than that of tits. The tail is square, with central and outer rectrices measuring at least 90% of the median ones.

Body feathers: although very difficult to find, some body feathers are quite characteristic. Those of the breast are white with rufous, enlarged tips, and quite rigid. Those of the back are shiny metallic blue. Those of the head and shoulders are dark blue with a pale blue tip.

EUROPEAN ROLLER *Coracias garrulus*, Coraciidae.

NP = 10 / NS = 12 / NR = 12

The wings are rather long with a slightly rounded hand, which facilitates manoeuvring in flight. The tail is rectangular and sturdy. *Possible confusion:* none.

Colour of 'large' feathers: primaries and secondaries black, with some white towards the base and pale blue on the basal part. The blue becomes brighter when viewed from the side. From below, on the black parts visible in flight, the 'large' feathers have a bright purple-blue sheen, but this is duller or even absent in young birds. The rectrices are bluish-grey with pale blue areas and even some sheen.

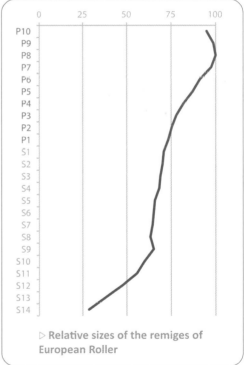

▷ Relative sizes of the remiges of European Roller

In the region considered in this guide, only the Psittacidae (introduced) have 'fluorescent' colours producing the same effect.

Primaries: TP = 11-19. Ps black with a little white at the base (P10) on the inner vane that increases in size and towards the outer vane with a pale blue wash over it, especially on the outer vane. This pale white and blue area covers between one third and one half of P7 to P1.

A small dark blue band is visible towards the base of the black area. The outer P is obviously notched.

Secondaries: TS = 9-12. Ss black with white/blue area on approximately the basal third or half. The dark blue area is better defined.

Rectrices: TR = 12-17. The tail is rectangular, all the rectrices of a similar size. The Rs are blue-grey, with a pale tip on R2, increasing to up to a third of the feather towards the outer ones. In the adult R6 has a small black tip, dull olive in the young. The R1s (central rectrices) are entirely olive-grey tinged with blue. A purple-blue sheen is also present on the dark parts of the underside (except on the central feathers).

Note: some outer Ps may be moulted before migration; the others are usually moulted in the winter quarters. Therefore, it is rare to find these feathers in isolation in Europe.

EUROPEAN BEE-EATER *Merops apiaster*, Meropidae.

NP = 9 (+1) / NS = 14 / NR = 12

Its wings are pointed with a distinctly triangular hand; its tail is rectangular with a central point. Its flight recalls that of swallows and martins with glides and fast flapping.

Possible confusion: with Rose-ringed Parakeet (Rs) and young Great Spotted Cuckoo (Ss).

Colour of 'large' feathers: in the adult, Ps dark grey with green-blue sheen and dark tip. In the young, the Ps are duller, with a medium green to khaki sheen. In the adult, Ss rufous with broad black tip, increasingly tinged with green in the rufous towards the outer Ss. Outer vane sometimes tinged with glossy purple. Inner Ss dark grey heavily tinged with green. In the young, Ss rufous-brown with broad black tip, inner Ss green-grey. From below, the remiges have a rufous-grey tint with a distinct black tip. Rs dark grey with strong green sheen in adults, less marked sheen in the young; a black tip is more or less sharp in all ages.

Primaries: TP = 7.5-13. The outer Ps are quite slender and are reminiscent of those of swallows. The tip (except for outer P) is often split like that of the Ss.

Secondaries: TS = 6-8 (inner ones 3.5-6). The feathers are longer than the rachis at the tip of the Ss and form two bumps (except inner ones).

Rectrices: TR = 9-14. The Rs are quite elongated, the central ones are more tapered and have a pointed tip.

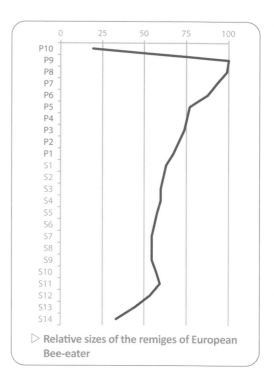

▷ Relative sizes of the remiges of European Bee-eater

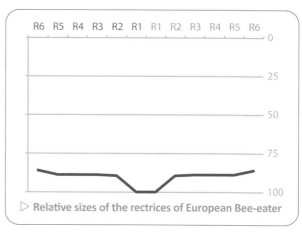

▷ Relative sizes of the rectrices of European Bee-eater

The tail has a central point, the other Rs measure about 90% of the central ones.

Note: Some outer Ps may be moulted before migration; others are usually shed in the winter quarters. These feathers are therefore rarely found in isolation in Europe.

• PSITTACIDAE: ROSE-RINGED PARAKEET AND BUDGERIGAR

NP = 10 / NS = 9-11 / NR = 12

Distinctive criteria of group: shape of rectrices, colour.

Two species predominate among parrot feathers found in the wild; the Budgerigar (individuals escaped from captivity) and the Rose-ringed Parakeet (a feral species). Other species can sometimes be found, but are not described here.

Generally, parrot 'large' feathers have a shape similar to passerines, except for the tail which can be very elongated, especially in parakeets. The bright colours often visible on the 'large' feathers betray their exotic origin. Apart from wild forms, a wide variety of colour mutants may occur. The most common examples are the change of the green areas of a bird to a blue or yellow colour depending on the mutation. Most often the 'large' feathers are grey or black in colour with the bright colours added to this background. Coloured rectrices are usually quite uniform, but the upperside and underside may be tinged with different colours. On the remiges, the trend

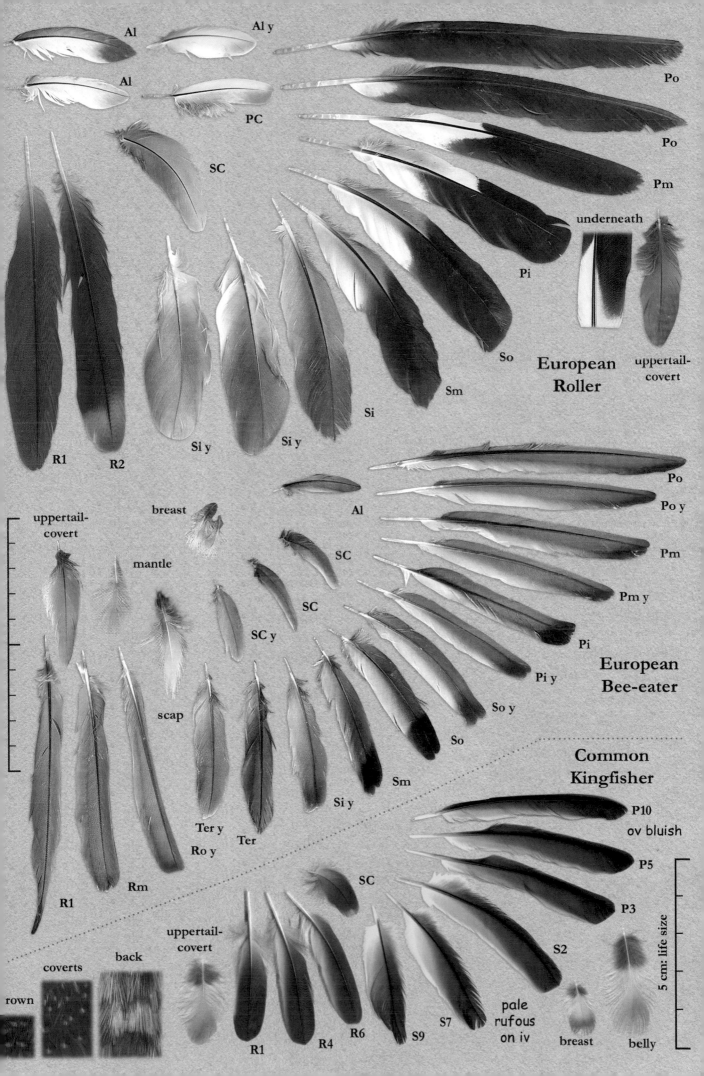

Al

Al y

Al

Al

PC

SC

Po

Po

Pm

underneath

Pi

So

Sm

Si

uppertail-covert

Si y

Si y

**European
Roller**

R1

R2

uppertail-
covert

breast

mantle

Al

SC

SC

Po

Po y

Pm

Pm y

Pi

Pi y

So y

So

Sm

**European
Bee-eater**

scap

SC y

Si y

Ter y

Ter

Ro y

**Common
Kingfisher**

R1

Rm

P10
ov bluish

P5

P3

S2

uppertail-
covert

SC

rown

coverts

back

uppertail-
covert

R1

R4

R6

S9

S7

pale
rufous
on iv

breast

belly

5 cm: life size

is towards 'economy' in the distribution of brightly coloured areas. On the upperside, bright colours cover the outer vane and the central part of the inner vane, which are visible when a bird flies or spreads its wings. The underside is not coloured, or colours only appear on the inner part of the inner vane (the same colour or not as the upperside), because this is the visible part of the underside in flight. This distribution of colour is also found on the coverts and alula.

ROSE-RINGED PARAKEET *Psittacula krameri* (SS): the wings are typical of birds that use flapping flight, with a triangular hand. The tail is very pointed.

Colour of 'large' feathers: the wild form has greenish-coloured 'large' feathers, but blue or yellow mutants are sometimes observed. The remiges are dark grey, with a very narrow yellow to greenish edge; except for the inner part of the inner vane and on the underside where they are khaki-green, which becomes bright green when viewed in profile. The rectrices are pale grey, but this colour is more or less masked by yellow and green. Yellow dominates on the inner vanes and towards the outer Rs. The green areas become brighter when viewed in profile.

Primaries: TP = 10-18. The shape and size are similar to those of doves. The bright colours mean they are easily distinguished.

Secondaries: TS = 5.5-10. They are wider than those of passerines of the same size and quite curved. The tip of the outer and middle Ss is bevelled or quite angular.

Rectrices: TR = 8-23. They are narrow and tapered. The central ones are very long, about 2.5 to 3 times the length of the outer ones.

Possible confusion: with European Bee-eater and other parrots.

BUDGERIGAR *Melopsittacus undulatus:* the wings are slender; the hand is triangular and pointed. The tail is very pointed. Escaped birds often fall victim to predators (Sparrowhawk, falcons, cats, etc.) and feathers from plucked birds are sometimes found.

Colour of 'large' feathers: the base pattern is dark grey and white. There is a round or comma-shaped white mark on the inner vane of the Ps that extends to the base of the feather and touches both vanes. The inner Ps have the basal half white. The Ss have a wide white transverse bar, sometimes with the entire basal half white. The rectrices are dark grey with a wide white band that moves towards the tip towards the centre of the tail. The central Rs are uniformly dark. In the wild form, there are green tints on the dark, and yellow on the white, that add to the pattern. The amount of additional colour varies depending on the mutant concerned: bright yellow, pale yellow, light blue, violet-blue, etc. The density of the dark areas also varies.

Primaries: TP = 4-9. They have a fairly long calamus, only P10 is a little forked at the tip.

Secondaries: TS = 3.5-5. The Ss are slightly curved; the outer vane often longer than the inner one; the calamus is quite long.

Rectrices: TR = 4-11. They are narrow and tapered. The central ones are the longest, about 2.5 to 3 times the length of the outer ones.

Possible confusion: curiously, the remiges of mutants do not have bright colours and are very similar to those of Common Sandpiper!

Other species of this family or this order can be found in the wild, escaped from captivity. Small populations of Fischer's Lovebird *Agapornis fische* can, for example, be found in parts of France and Spain, as well as Alexandrine Parakeet *Psittacula eupatria* in several countries. The latter species is very similar to Rose-ringed Parakeet but is about 1.5 to 1.8 times larger. It is also possible to discover the feathers of various forms of the Cockatiel *Nymphicus hollandicus*, some of which have a mixture of pale yellow and some grey, in pale notches on the remiges and bars on the rectrices. Feathers of other species may also be found at random from escaped birds; examples include the Eastern Rosella *Platycercus eximius*, the Yellow Rosella *P. flaveolus* or various amazon parrots *Amazona* spp., etc. The bright colours of the majority of these birds' 'large' feathers often help to identify this order.

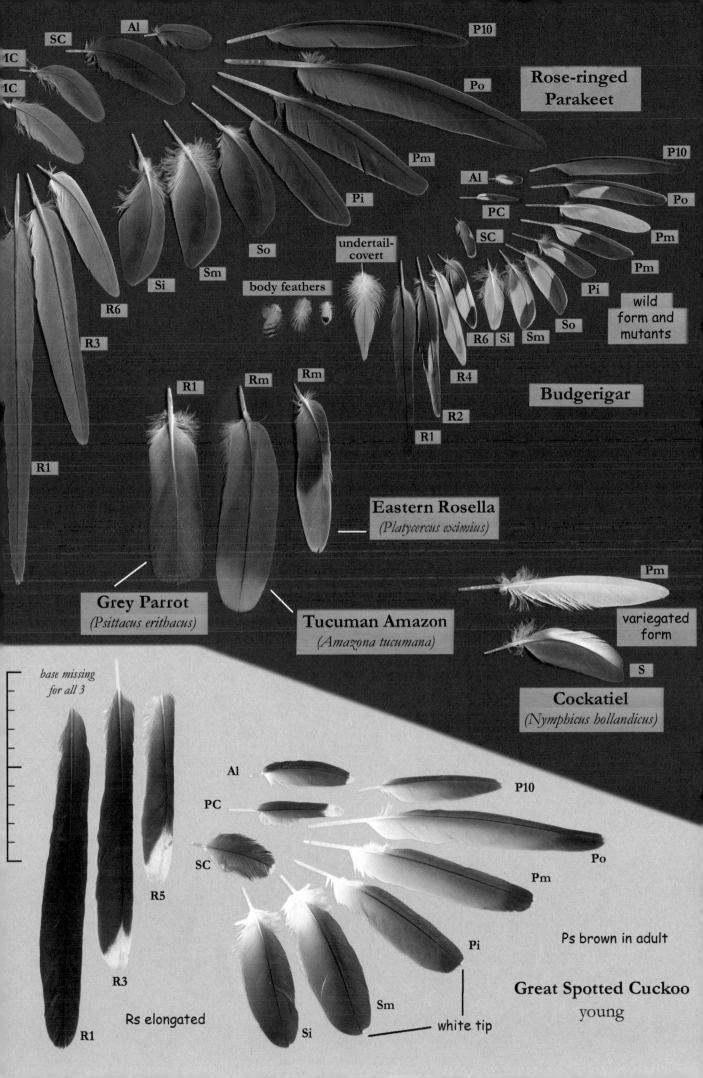

Rose-ringed Parakeet

P10
Po
Pm
Pi
So
Sm
Si
R6
R3
R1
SC
Al
1C
1C

undertail-covert

body feathers

Budgerigar

wild form and mutants

P10
Al
PC
SC
Po
Pm
Pm
Pi
So
R6 Si Sm
R4
R2
R1

Eastern Rosella
(Platycercus eximius)

R1
Rm
Rm

Grey Parrot
(Psittacus erithacus)

Tucuman Amazon
(Amazona tucumana)

Cockatiel
(Nymphicus hollandicus)

Pm
variegated form
S

base missing for all 3

R1
R3
R5

Rs elongated

Al
PC
SC

P10
Po
Pm

Pi
Sm
Si

white tip

Ps brown in adult

Great Spotted Cuckoo
young

• CUCULIDAE: CUCKOOS

NP = 10 / NS = 9-11 / NR = 10

Distinctive criteria of group: colour.

COMMON CUCKOO *Cuculus canorus:* it has pointed wings and a long tail. Its silhouette in flight is similar to the Common Kestrel but its hand is more rounded and its forearm shorter.

Possible confusion: with Common Kestrel (Ps and Ss), and other falcons.

Colour of 'large' feathers: background colour brown-grey, tending towards grey or reddish-brown depending on the morph. Small white tip or edge at tip in young. Ps with large white to rufous notches (outer vane and tip in brown morph and young). Inner Ps and Ss brown-grey with inner vane white at base. In the young, all Ps and Ss have rufous-tinted notches, which can form bars on certain Ss. Rs brown-grey to black-brown with a small white tip. Depending on age and sex, there are a variable number of pale marks: small white spots regularly spaced along the rachis and on the edge of one or both vanes (males more often than not). The white marks can also coalesce forming many narrow bars, remaining white or becoming reddish-brown; some of these marks may remain isolated between two pale bars (especially female and young).

Primaries: TP = 10-19 (22) (outer P 10-12). Compared to falcons, the outer Ps do not have a notch; the tip of the P is much more rounded (sometimes with a white tip). The 'large' feathers have a more supple texture.

Secondaries: TS = 8-12 (inner Ss 4-9). Resemble those of falcons (especially the young) but are quite wide with a square tip, more rounded towards the inner Ss.

Rectrices: TR = 10-19. Long and narrow. The tail is rounded (the outer feathers measuring about 80-85% of the central ones).

Note: in Europe, these feathers are mainly found in the form of feathers plucked by a predator, and are generally from young birds. Moulting usually takes place in the wintering quarters in Africa for adults, but young sometimes moult whilst still in Europe.

GREAT SPOTTED CUCKOO *Clamator glandarius:* its wings are quite long, its hand a little rounded, its tail long and graduated.

Possible confusion: with those of magpies (Rs) but a different colour, or pigeons (Ps and Ss) but much softer texture and thinner rachis in Great Spotted Cuckoo.

Colour of 'large' feathers: remiges are grey-brown with a small white tip, wider on the Ss, and eventually a green or bronze sheen on the Ps and Ss. Sometimes a red tint around rachis of the Ps. The Ps of young birds are reddish-brown with a dark tip; wider on the outer Ps (outermost P completely dark). Rs black-grey with a broad white tip, narrowing towards the central Rs (sometimes without white), olive tint more or less obvious on underside.

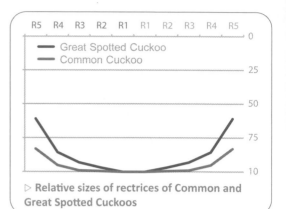

▷ **Relative sizes of rectrices of Common and Great Spotted Cuckoos**

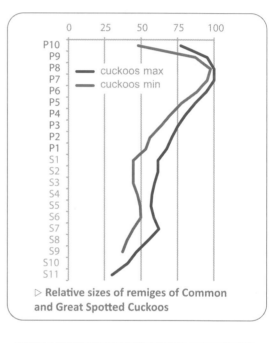

▷ **Relative sizes of remiges of Common and Great Spotted Cuckoos**

▷ **Common Cuckoo wing (juvenile)**

breast

Al

PC

P10 y

Po y

Po ad

Pm y

Pi ad

SC

undertail-covert

uppertail-covert

So ad

Si y

Sm y

Common Cuckoo

Ps and Ss: young with notches or bars
Pi and Ss: adult with inner base white

another pattern of Rs
(reduced size)

R5 y

Rs: variable pattern

R5 ad

R5 y

R3 y

R5 y

10 cm: scale ¹/₂

R1 ♀

R2 ♂

PC

Red-necked Nightjar

Po male: inner
white mark

Po

Ps of ♂

Po ♀

C

So

Pm ♀

Pi ♀

back

uppertail-covert

undertail-covert

Si ♀

Sm ♀

So ♀

European Nightjar

Ro male:
white tip

R6 y ♀

R6 ad ♂

R5 y ♂

R4 y ♂

R1 y

LC

MC

SC

scap

Primaries: TP = 10-20 (22) (outer P 8-10). Ps with rounded tip; slight notch on P10 to P7.

Secondaries: TS = 7-11 (13) (inner Ss 5-8). Ss broad with rounded tip; sometimes more square.

Rectrices: TR = 10-23. Rs quite narrow; longer towards the centre of the tail. Their shape is reminiscent of those of the Common Magpie. The tail is graduated (outer feathers measuring about 60-65% of central ones).

Note: in Europe, it is unusual to find isolated feathers. Moult usually occurs in the African wintering quarters between July and February.

• CAPRIMULGIDAE: NIGHTJARS

NP = 10 / NS = 13 / NR = 10

Distinctive criteria of group: colour.

Their wings are pointed, quite narrow. The tail is long, rectangular. The silhouette is reminiscent of that of falcons. The base of the 'large' feathers (especially the inner vane) and the body feathers have, when freshly shed, a texture reminiscent of the 'velvety' surface of the feathers of owls. However, this 'hairiness' is much less developed in the nightjars, and there is little risk of confusion with those of owls. The body feathers are more similar, but the surface structure generally makes it possible to differentiate between them.

▷ **European Nightjar wing (male)**

Possible confusion: with other species with camouflage colours, resembling bark or dead leaves, such as pheasants (especially females), diurnal raptors (speckling and narrower bars in nightjars), owls ('large' feathers more elongate in nightjars), etc.

Colour of 'large' feathers: the background is brown-black, with pale barring: cream, buff, tawny, buff or grey. The proportions of dark and light are very variable according to the feather. The dark colour forms wide or narrow bars, partially broken triangular bars, speckling, etc.

The marks are fairly regular on the same 'large' feather. On the terminal third of the outer Ps and at

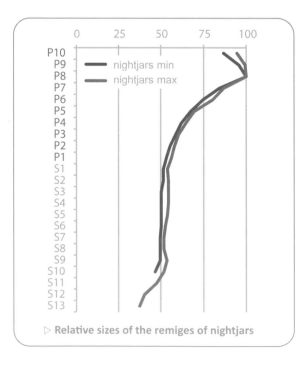

▷ **Relative sizes of the remiges of nightjars**

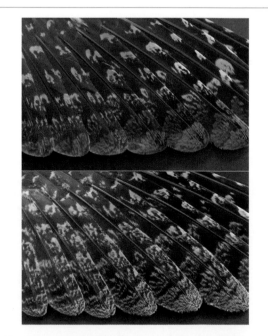

▷ **Comparison between the sexes of the tips of the primaries of European Nightjar.**

In the male, the background is rufous and becomes clearly grey towards the tip. In the female the contrast is less clear.

the edge of R4 and R5 there is a large, paler, un-vermiculated area in some individuals. In adult males, this spot is pure white (European Nightjar) or white or cream (Red-necked Nightjar).

Primaries: the edge of the inner vane of the outer primaries is slightly serrated at the tip. There is a slight emargination on P9 and P8 and a slight notch on P10 and P9.

Secondaries: normal shape.

Rectrices: long and narrow. Tail rectangular

Note: The rectrices are moulted in winter in Africa.

EUROPEAN NIGHTJAR *Caprimulgus europaeus:* TP = 9-18, TS = 8-11, TR = 12-16. The mark on the Ps and Rs is pure white in adult males. In females, this mark is yellowish-brown or absent. In most juveniles, the mark on the Ps is absent, but some young males have a large buff spot on one or two outer Ps.

RED-NECKED NIGHTJAR *Caprimulgus ruficollis:* TP = 8-18 (21), TS = 8-12, TR = 15-18 (20). The mark on the three Ps and the two Rs is creamy-white in adult males. It is generally rather buff in young and females. The mark on the Ps in Red-necked Nightjar is bigger than that on European Nightjar.

• COLUMBIDAE: PIGEONS AND DOVES

NP = 10 (+1) / NS = 11-12 (10-15) / NR = 12

Distinctive criteria of group: shape of the remiges and calamus, the colour and shape of the rectrices. These birds can fly fast and for long periods. They have reinforced 'large' feathers, with a fairly long calamus, a wide rachis and rigid vanes. Their body feathers are also different and quite recognisable. Species that are sedentary migrate little and can moult throughout the year. This explains in part why their feathers are very frequently found, and even in cities (especially Feral and Wood Pigeons and Collared Dove).

▷ **Body feathers of pigeons.**
Note the very thick rachis in the downy part that narrows at the level of the free barbs.

Possible confusion: with other white species for domesticated varieties.

Colour of 'large' feathers: Ps and Ss of pigeons and larger doves: black to grey, possibly an edge or pale area on the outer vane. Ps and Ss in smaller doves: grey to brownish, more brown to buff for inner Ss. Rs of all species with two or three grey, brown or black bands or marks, showing a sharper contrast below; often a transverse band; the central Rs sometimes uniform. In the young, 'large' and body feathers generally have slight brown to buff edge. Beware of domestic variants: the structure and shape are identical, but the colour is very variable: pure white, grey, brown, cinnamon, etc.

Primaries: the base of the outer vane is slightly enlarged. The outer Ps have a slightly convex outline. Slight emargination on two or three outer Ps and notched on the one or two outermost.

Secondaries: the rachis is thick. The inner Ss are almost straight and do not have the classic silhouette of Ss. On which side of a bird they grew is sometimes difficult to determine.

Rectrices: the base of the rachis is thick. They have a rectangular silhouette in pigeons, more rounded at the tip in doves (the outer Rs measure 85 to 95% of the central ones depending on the species). The Rs are quite elongated (Wood Pigeon, small doves) or quite short (Rock and Stock Doves).

Body feathers: very unusually, *the body feathers in this group have a characteristic structure.* They have a wide, rigid rachis

▷ **Feral Pigeon wing (pattern similar to wild Rock Dove).**

MEASUREMENTS OF THE 'LARGE' FEATHERS OF PIGEONS AND DOVES						
Species	Pmax	Pmin	Smax	Smin	Rmax	Rmin
Rock Dove	20	12	13	8	14	11
Feral Pigeon	20.5	11	13	8	14	12
Stock Dove	19.5	11	12	9	14	10
Wood Pigeon	21.5	13.5	14.5	9.5	19	15.5
Collared Dove	15	9	10	9	16	12
Turtle Dove	15	7.5	10	8	14	10

Note: there are no measurement data for Laughing Dove

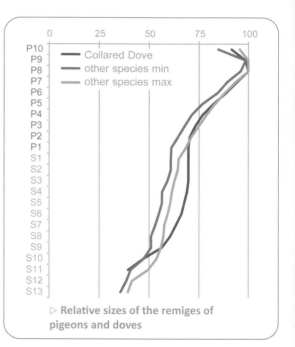

▷ Relative sizes of the remiges of pigeons and doves

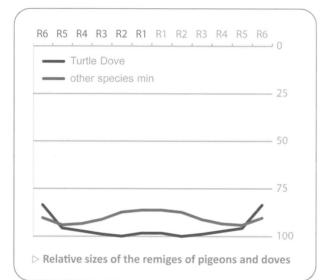

▷ Relative sizes of the remiges of pigeons and doves

▷ Wood Pigeon wing (young)

that narrows abruptly. The free barbs at the base form a voluminous and dense down. Even the smaller down feathers of these species are recognisable. Generally, the body feathers are white, grey, or buff on the back; grey or pinkish on the belly.

The neck feathers are short and broad, and have iridescent colours or white and/or black markings forming a collar. Some of the secondaries are sometimes identifiable: there is a black mark in Rock and Stock Doves, a white band in Wood Pigeon, a black centre and a rufous border in Turtle Dove.

ROCK DOVE *Columba livia:* Ps grey, with the tip becoming darker, even black. Outer Ss grey with dark ill-defined tip. Other Ss mid grey with a transverse black band (often asymmetric). The grey tip decreases in size towards the inner Ss. Rs dark grey with black tip, possibly with a terminal grey spot. Outer Rs with white in the grey of outer vane.

FERAL PIGEON *Columba livia* form *domestica*): birds with wild-type plumage: Ps grey to dark grey, with darker to black tip; Ss mid grey, tip ill-defined black area or a black mark; Rs dark grey with blurred black tip, sometimes with a lighter terminal spot; outer vane of R6 partially white.

Other colour forms: Ps, Ss and Rs of very variable colour; grey, brown, chocolate, white, marbled, etc. Grey Ss often have black marks. Colours can be evenly distributed or mixed.

WOOD PIGEON *Columba palumbus:* Ps dark grey, with a white edge on the outer vane, usually well-defined. Pale edge often less well-defined on median and inner Ps. Ss dark grey with blurred edge or pale border on the inner vane of the outer Ss, becoming less obvious or disappearing towards the inner Ss. No well-defined dark tip.

Al

Al

P9

sharp
white edge

Po

LC

PC

SC

SC

SC

Pm

Pi

Wood Pigeon

Si

Sm

So

S: no dark tip

Rs: 3 grey
bands

neck

belly

back

axillary

uppertail-
covert

R1

R4

R6

PC

uppertail-covert

SCi

SC

P9

Po

pale
base ov

Si

Sm

So

pale
base

dark tip

pale
base

Pi

Pm

R1

R4

R6

Rs: 3 grey bands

bases broken by predator

Stock Dove

COLOUR OF REMIGES OF PIGEONS AND DOVES. Criteria present on at least one remige						
	P			S		
Species	main colour	pale edge (at least on outer vane)	pale zone	colour	black tip	other mark
Rock Dove	pale grey to black			pale grey	yes	black mark
Feral Pigeon: wild type	pale grey to black			pale grey	yes	black mark
Feral Pigeon: domestic	very variable	possible	possible	very variable	possible	possible
Stock Dove	dark grey	very thin; possible	outer vane	dark grey	yes	pale grey
Wood Pigeon	dark grey	yes; sharp		dark grey		pale border
Collared Dove	dark brown-grey	outer and medians		brown-grey		pale border
Turtle Dove	brown	eventually		grey-brown		pale border

COLOUR OF RECTRICES OF PIGEONS AND DOVES. Criteria present on at least one rectrix				
Species	colour	uniform	2 zones	3 zones
Rock Dove	dark grey, black, white		(R1–R5)	yes; 4 for R6
Feral Pigeon: wild type	dark grey, black, white		(R1–R5)	yes; 4 for R6
Feral Pigeon: domestic	variable; often black and grey	possible	often	sometimes
Stock Dove	dark grey, pale grey, black		sometimes	often
Wood Pigeon	dark grey, pale grey, black		central sometimes	yes
Collared Dove	black, white, grey, buff	central	outer	median
Turtle Dove	black, white, grey, buff	central	median and outer	

The remiges of immature birds have buff edges at the tip (as well as the wing-coverts). Rs are characteristic, quite long, with three sharper bands underneath. From the base: mid grey, pale grey, black-grey. The pale band is sometimes reduced or absent (especially central Rs and in immature birds), but the size is diagnostic. **Secondary coverts are identifiable:** the outer ones have outer vane (and part of the inner vane) white; the medians ones have a wide white edge on the outer vane. The wing-coverts of young, especially medium and lesser, are tinged with light brown and have a buff edge.

STOCK DOVE *Columba oenas:* Ps dark grey, becoming darker towards the tip (darker grey than Rock Dove and Wood Pigeon). There is a pale grey area on the outer vane of the median and inner Ps (on about the basal two thirds). They sometimes have a very thin pale to white edge. Ss dark grey, with an ill-defined black tip; the outer vane is paler grey on the basal two thirds. Inner Ss mid grey, often with a black mark on the outer vane. Rs similar to those of the Wood Pigeon, grey with a black tip. There is normally a paler narrow band between the two (broken line or double curve). The outer vane of the R6 has a white base.

▷ **Reconstructed Stock Dove wing; feathers from a plucked corpse.**

uppertail-covert

Al

PC

PC

SC

P9

Po

Pm

Pi

Si   Sm   So

R6

R1   R4

Collared Dove

Al

LC

SC

flank

P9*

Po*

= different individual

ov white

uppertail-covert

Pm

neck

Si

Pi

breast

R1   white tip   R4

R6

Sm   So

vague reddish border

Turtle Dove

SC

P9

pale base and dark tip

Pm

scap

dark tip

neck

Sm

black mark

Si

R6

ov white base

R1   R4

grey tip

Feral Pigeon
wild-type plumage

The outer primary coverts are dark grey, and rapidly become light grey towards the wrist. The inner secondary coverts (especially the greater) have a black mark on the outer vane.

COLLARED DOVE *Streptopelia decaocto:* Ps dark brown-grey, with a pale blurred edge on the outer and median Ps. Ss brown-grey, with a vague blurred edge on the outer vane; inner Ss browner. Rs with black, white, grey, buff: central Rs uniform brown-grey with a dark area below about one third or half from the base. The following Rs are similar; the dark area is more obvious underneath. Outer Rs black with a white tip; tinted grey or buff underneath.

▷ Collared Dove wing (some SCs missing)

TURTLE DOVE *Streptopelia turtur:* Ps brown to grey, with a possible blurred rufous edge, and a pale blurred base. Ss browner, with a vague rufous or greyish border. Rs with some black, white, grey-brown. Central Rs uniform grey-brown (sometimes with ill-defined white tip). Median and outer Rs black-grey with a white tip. Outer vane of R6 white. It is the only species in the group to have black feathers with a rufous border: scapulars, secondary coverts, and sometimes inner Ss.

Note: the inner Ps are sometimes moulted in Europe before migration, the rest of the moult occurs in Africa.

grey tip

▷ Turtle Dove wing (feathers growing); reconstructed from a plucked bird

LAUGHING DOVE *Streptopelia senegalensis:* species at the edge of its range in the area considered here. 'Large' feathers similar in size to those of Turtle Dove. The colouring of the 'large' feathers is similar to Collared Dove, but overall darker; there is more contrast in colours on the Rs (the dark parts even darker). On the Rs the limit between white and dark is clearly curved (it is straight or nearly so in Turtle and Collared Doves).

• PTEROCLIDAE: SANDGROUSE

NP = 10 (+1) / NS = 18 / NR = 16-18

Distinctive criteria of group: shape, colour.

This family has pointed wings with an elongated hand. The tail is rounded or with elongated central feathers. *Possible confusion:* between the Rs of Pin-tailed Sandgrouse and the Rs of Woodcock.

Colour of 'large' feathers: grey to brown, with more or less yellow-rufous to greenish-yellow.

Primaries and secondaries: shape reminiscent of those of waders, but a little broader. Calamus quite long.

Rectrices: quite stout, slightly pointed at the end, central ones elongated in Pin-tailed Sandgrouse.

Body feathers: they have easily recognisable patterns and colours, at least for those on the back.

PIN-TAILED SANDGROUSE *Pterocles alchata* (SS): TP = 8-17, TS = 7-9, TR = 6.5-18. Ps brown-grey, darker towards the tip, ash-grey outer vane (black on P10), tip with buff edge (except outer Ps); the ill-defined dark tip is bigger towards the inner Ps.

Male: Ss dark brown on the outer vane and the tip of the inner; the rest white to buff with white edge. Towards the medians, the dark is tinted olive-green. The inner Ss (S10 and following) have an increasingly large olive-yellow mark on the outer vane; inner vane brown-grey. The tip becomes yellower and wider towards the tertials.

secondaries

3 SCs

primaries

Feral Pigeon

colour variants

primaries

3 tail
feathers

10 cm: scale ¹/₂

Pigeon and dove tail feathers: underside

Feral Pigeon

Stock Dove

Note the more marked contrast
below; shape either short or long;
the number and position of the
dark and pale zones.

Turtle
Dove

Wood
Pigeon

Collared
Dove

Female: Ss grey with a broad, blurred, black mark at tip, descending further along the outer vane. Pale edge along the black mark. Towards the medians, the base of the black mark is speckled on the outer vane and then has buff to red vermiculations, which quickly form long, pale, wavy longitudinal lines. The inner Ss become pointed and are golden-red with brown bars and then grey bars towards the tip.

Rs of both sexes golden-rufous with thin black-brown bars. Towards the outer Rs, the pale areas become more buff and the pale barring fades on the inner vane (sometimes uniform brown). Small white tip, descending a little on the outer vane of outer Rs. Rs slightly pointed; tapered central feathers in both sexes.

BLACK-BELLIED SANDGROUSE *Pterocles orientalis* (SS): no data on isolated feathers. Ps and Ss black-grey. Rs with golden-rufous and black barring.

▷ Black-bellied Sandgrouse rectrices

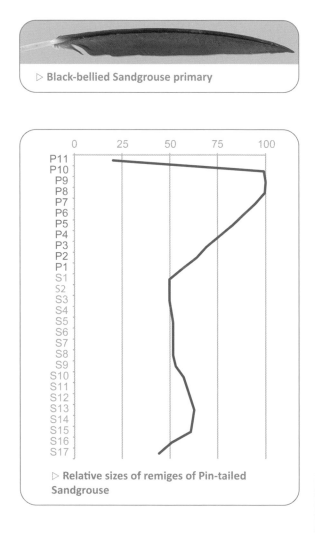

▷ Black-bellied Sandgrouse primary

▷ Relative sizes of remiges of Pin-tailed Sandgrouse

▷ Black-bellied Sandgrouse greater coverts

AQUATIC BIRDS OF THE COAST AND WETLANDS

• CHARADRIIFORMES: GENERAL CRITERIA

Distinctive criteria of group: shape of remiges, size and colour for some species, where they were found.

This order includes a wide variety of species, most of them occurring in aquatic environments: waders, gulls, terns, skuas, auks, etc.

The wings are pointed, more or less elongated. The tail is often short and rectangular, but can be more rounded (many waders), pointed (skuas) or forked (terns).

Possible confusion: with gannet, petrels, divers.

Colour of 'large' feathers: shades of white, grey, black or brownish, often with two associated colours (sometimes only grey or three colours).

Primaries: neither emarginated nor notched (except in Lapwings, which have the tip of the outer Ps slightly narrowed), calamus very long (C = 20-25 (30)%). Inner Ps wedge-shaped (pointed hand). Outer Ps are tapered in terns and related species.

Secondaries: calamus very long (C = 20-25 (30)%), curved in shape (except the innermost secondaries that are elongated and lanceolate); inner vane sometimes longer than the outer one (median and inner Ss especially); almost identical shape and size throughout the forearm.

Possible confusion: with the Ss of some Anatidae.

Rectrices: usually short, rounded, or rectangular at the tip (except for some of the terns and skuas).

Some examples:

– barred and quite short: waders, many narrow bars or a single (sub)terminal bar of variable width
– irregularly marked or with a (sub)terminal bar: young of gulls, snipes, Oystercatcher, Lapwing
– white: adult gulls, some waders
– greyish and pointed: small waders (sandpipers, etc.)
– outer feathers form streamers: terns
– dark, short and rigid: auks

• LARIDAE: GULLS

NP = 10 (+1) / NS = 20-25 / NR = 12

Distinctive criteria of group: shape, colour, where they were found.

Possible confusion: with Gannet, Fulmar, immature diurnal birds of prey, skuas, waders with black-and-white tails, etc.

Colour of 'large' feathers: the 'large' feathers of adults are usually coloured grey and/or white and/or black. Outer and median Ps are often darker towards the tip (melanin strengthens the structure of the feathers and thus limits wear of the terminal part).

The feathers of immatures are quite difficult (if not impossible) to identify to species when they are found in isolation, especially in medium-sized species. The size and shape of their 'large' feathers are as in adults. On the other hand, the colour differs more or less markedly between birds of different ages. They are generally darker than those of adults (rarely paler). Immatures undergo at least two and sometimes up to four different plumage changes before acquiring the adult plumage. The 'large' feathers are mostly brown or strongly tinged with dark in the first year. Over the years the juvenile markings progressively disappear. In immatures (especially in the early years), the 'large' feathers frequently differ from those of adults by the absence of white marks at the end of the Ps for species with white spots at the primary tips, the absence of sharp contrast on the Ps, the absence of a white terminal bar on the Ss (or a narrower bar), etc. In addition to the strong resemblance between some species, the variation between individuals and different subspecies and the gradation from immature to adult plumage often make it (very) difficult to identify isolated feathers. However, some species can be distinguished from the rest of family, either by their size (Great Black-backed Gull, Little Gull, etc.), or by their very particular coloration (Kittiwake, etc.).

Primaries: they are neither emarginated nor notched. The calamus is quite long (C = 20-30%). There is often an area on the underside of the outer Ps around the rachis where the barbs are thicker. This surface is more or less shiny, but still significantly less so than in the Anatidae (that have a real tegmen). The inner Ps are often uniform grey; black appears on their tip on the medians and increases in size towards the outermost Ps. A 'mirror' is present in many species (large white spot at the end of one or more outer Ps). Within the same species, and especially in the genus *Larus*, the coloration of the adult Ps (especially outer Ps) varies more or less clearly and there may be large variations in coloration between different subspecies, both in the shade of grey and the extent of the black and/or white markings.

Secondaries: they are curved and quite wide. The calamus is very long (C = 20-35%). The adult Ss are pale grey to grey-black, often with a white tip, rarely entirely white. The Ss of immatures are brown, usually speckled with pale or with a large dark mark and become greyer with each moult.

Rectrices: the tail is rectangular, with a size difference of no more than 10% between Rmin and Rmax. The Rs are rectangular with a square tip. The central ones and those of the smaller species have a rounded tip.

They are white and matt in adults. On the other hand, in immatures they have a dark terminal band and/or irregular spots (which can form bars); the tip is sometimes edged with white. These marks vary greatly depending on age but also between individuals, making identification often tricky.

Possible confusion: between the rectrices of immature gulls with black tips and those of waders with a terminal bar on the tail. Generally in waders, the width of the terminal bar increases markedly towards the centre of the tail, the shape of the rectrices is generally more rounded (except for Lapwing, with square tip), but these criteria are not necessarily sufficient. However, apart from a few large waders, size quickly distinguishes the gulls. Compared to other aquatic birds, the inner vanes of the rectrices are thin and flexible. As a result, the rectrices have a more fragile appearance. In order to distinguish Rs of immatures from those of diurnal raptors, it is enough to compare their texture, the vanes being much more rigid in raptors than in gulls. The place where the feather was found may also provide a good clue in most cases. For all gull species, size is an important criterion, but it does not allow one to distinguish all species from each other.

Wing colour in Gulls

The wing pattern of adult gulls can be distinguished according to the distribution of the three colours (white, grey and black) on the upper wing: a uniform wing (grey, with white border or not), grey wing with a black tip (including or not white spots towards the tip of the remiges), or tricoloured (three areas of distinct colours). Size (wingspan is indicated here) also allows for separation of species' groups, although there is some overlap.

For the 'tricoloured' wing pattern and in Little Gull, the wing appears darker from below. In the Little Gull, the underside of the 'large' feathers is really darker, but in the others, this aspect is due to the darker coloration of the inner vanes, that are more visible on the underside of the wing.

Wing pattern / Wing size	uniform	with black tip	tricoloured
very small (Ws = 60-80 cm)	Little Gull		
small (Ws = 85-120 cm)	Mediterranean Gull	Common Gull, Kittiwake	Slender-billed and Black-headed Gulls
medium (Ws = 115-130 cm)		Audouin's Gull	
large (Ws = 115-150 cm)	Iceland Gull	Lesser Black-backed, Herring, Yellow-legged, and Caspian Gulls	
very large (Ws = 135-170 cm)	Glaucous Gull	Great Black-backed Gull	

For 'black-tipped' gulls, the criteria are the size, the colour of the inner vane (tone of grey), and the white spot on the outer P (position and size).

| Species | Pmax | Pmin | Primaries: colours present | | | | | Longitudinal black band | Dark tip | White mirror/ white tip |
			white	pale grey	mid grey	dark grey	black			
SIZE AND COLOUR OF THE PRIMARIES OF GULLS										
Little Gull	(10.5) 18.5 (21)	8.5 (5)	imm		ad, imm	ad	imm	imm: ov (P5) P6-P10	imm: (P5) P6-P10	ad, imm: end P1-P4; tip P5-P6
Black-headed Gull	25	11	outer	inner	x	median	outer-median	outer	median	tip sometimes (especially imm)
Mediterranean Gull	30	12	P10	x		Po im	(P9) P10	P10 ov; (P9)		
Slender-billed Gull	25 (30)	12	outer	inner	x	median	outer-median	outer	median	tip sometimes (especially imm)
Kittiwake	26	12	imm	x			P5-P10	P10 ov	P5-P10	tip, especially P5-P6
Common Gull	29	13	x	x			P5-P10		P5-P10	x
Iceland Gull (SS)	32	18	x	x						x
Audouin's Gull (SS)	31	16	x	x			P5-P10		P5-P10	tip (P5-P10); narrow tip (P1-P4)
Lesser Black-backed Gull	33	16	x			x	x		(P4) P5-P10	x
Herring Gull	34	18	x	x			x		(P5) P6-P10	x
Caspian Gull (SS)	n.d.	n.d.	x	x			x		(P4) P5-P10	x
Yellow-legged Gull	35	18	x		x		x		(P4) P5-P10	x
Glaucous Gull	40	24	x	x						x

Note: the colours indicated are not necessarily present on all the feathers. An 'x' indicates that this is often the case; if not the position of the colour is given.

| Species | Smax | Smin | Main tint | | | | Marks | |
			pale grey	mid grey	dark grey/ black	black	white tip	black mark
SIZE AND COLOUR OF THE SECONDARIES OF GULLS								
Little Gull	10 (11.5)	7.5			ad, imm		ad, imm	ov especially
Black-headed Gull	13	10	x				inner Ss ad (blurred); Ss imm often	imm: tip darker or dark mark
Mediterranean Gull	15	10	x				blurred ad/ sharp in imm	imm
Slender-billed Gull	15	12	x				inner Ss ad (blurred); S imm often	imm: tip darker or dark mark
Kittiwake	14	10	x				x	imm: inner Ss
Common Gull	14	9	x				large	-
Iceland Gull (SS)	19	17	x				x	-
Audouin's Gull (SS)	16	13	x				narrow and blurred	-
Lesser Black-backed Gull	18	12			x		wide and neat	-
Herring gull	20	13	x				wide and neat	-
Caspian Gull (SS)	n.d.	n.d.	x				wide and neat	-
Yellow-legged Gull	19	15 (12)		x			wide and neat	imm (dark mark)
Glaucous Gull	24	17	x				x	-
Great Black-backed Gull	24	18			x		large	-

Note: immature gulls often have a dark mark on their Ss, the shape recalling that on the Ss of certain waders, such as Dunlin or Oystercatcher.

SIZE OF RECTRICES OF GULLS (adults and immature) AND COLOUR OF THE RECTRICES OF IMMATURES (whose rectrices are white in adults)					
Size	species	Rmax	Rmin	terminal bar	irregular marks
small to medium 8-16 cm	Little Gull	11	8	black	-
	Common Gull	16	13	black	sometimes dark edge
	Slender-billed Gull	14	12	black	-
	Mediterranean Gull	14	11	black	sometimes dark edge
	Black-headed Gull	13	11	black	-
	Kittiwake	15	13	black	-
medium 15-20 cm	Audouin's Gull	16	14	brown	large brown zone (juvenile)
	Iceland Gull	18	17	-	pale
	Herring Gull	19	16	brown	brown
	Lesser Black-backed Gull	18	15	brown	brown, quite extensive
	Yellow-legged Gull	20	17	brown	brown
	Caspian Gull	20?	17?	brown	brown
large > 18 cm	Glaucous Gull	22	18	-	pale
	Great Black-backed Gull	23	18	brown	brown

Note: Tails of gulls are rectangular to square; all rectrices measure between 90 and 100% of Rmax.

LITTLE GULL *Hydrocoloeus minutus*: it is distinguished in particular by the small size of its 'large' feathers. Rs of adult white, or with a little ash-grey at the base (especially outer vane and along rachis). Rs of immature black on the tip; about a quarter on central feathers, lessening outwards (a small mark on the outers, sometimes all white). Ps and Ss of adult dark grey, ashy-looking, with white tip, underside blackish. Inner Ps and Ss often paler. Remiges of immatures with a lot of white. Outer Ps white with black on the outer vane, along the rachis and at the tip. The dark mark becomes grey towards the median Ps, which also have a white tip. The internal Ps are white with a broad grey mark in the middle of the outer vane, extending a little onto the inner vane. The Ss are white at the tip and base, with a dark grey or black mark in the centre (especially outer vane).

BLACK-HEADED GULL *Chroicocephalus ridibundus:* Rs of the adult white. Rs of immatures with black subterminal band, often with irregular edge, narrower towards outer Rs (R6 sometimes without black mark); small white tip (quickly worn).

Ps with a very variable pattern depending on their location. Outer Ps white with black borders (P10, P9, often P8); black

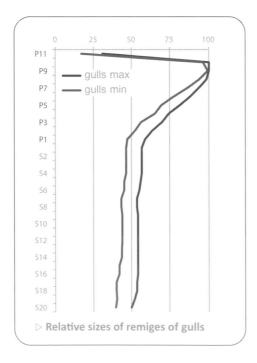

▷ Relative sizes of remiges of gulls

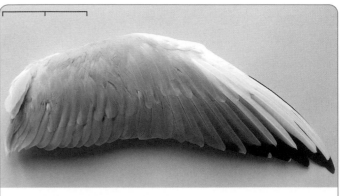

▷ Black-headed Gull wing (adult).

tip; outer vane black, bicoloured or white; wide black border on the inner vane (becoming grey towards the base). Ps mid grey with black tip; white still neat on P8 fading towards P6; the inner black border becomes paler from P8 to P5. P4 often grey with darker or black tip (very tip or two marks). P1 to P3 mid grey. Ss mid grey, a little paler towards the inner Ss, which often have a vague white tip.

The remiges of juveniles and immatures much more variable, generally darker than those of adults. Outer Ps have a higher proportion of black and are sometimes entirely dark; the black can also sometimes form irregular patterns. The dark border on the outer vane

▷ Black-headed Gull wing (first-year).

of the median Ps is wider. The inner Ps from P1 to P4-P5 are mid grey, darker towards the tip, with the grey outer vane paler on (P2) P3-P5. The Ps often have a variably-sized white tip. The Ss are mid grey, with a dark mark bordered with light at the end, dark grey to brown-grey, which becomes smaller towards the inner Ss. This mark varies in extent depending on the individual.

SLENDER-BILLED GULL *Chroicocephalus genei:* 'large' feathers very similar to those of Black-headed Gull; difficult to differentiate. Outer Ps overall wider and with more contrasting white and a darker inner vane. Immatures often have a little more dark on their Ss; inner Ps sometimes have more black. Juveniles have remiges sometimes entirely dark except for a small white tip.

▷ Comparison of the shape of the white zone in the outer primaries of Black-headed (upper) and Slender-billed (lower) Gulls.

P7 and P8 can be differentiated in some cases. The tip of the white central area is quite pointed in Black-headed Gull, while it appears wider in Slender-billed Gull.

MEDITERRANEAN GULL *Larus melanocephalus:* Rs of adults white. Rs of immatures white with a wide black subterminal bar; sometimes white along the rachis in the black.

Ps and Ss pale grey, often with white tip. The P10 is paler or white, with black edge on the outer vane (but tip and base often white). P9 can have a little black on the outer vane in its distal third. The tip of the Ss is white but ill-defined; quite narrow on the external Ss and widening markedly towards the inner ones. The remiges of immatures are marked with grey-black, especially on the outer vane and along the rachis of the Ps. The Ss have a dark distal mark and a white tip. In the second juvenile plumage the remiges are pale grey, only (P5) P6-P9 have a large black subterminal mark, more elongated towards P9.

KITTIWAKE *Rissa tridactyla:* Rs of adults white. Rs of immatures white with a black terminal bar, decreasing in size towards outer Rs (Rs sometimes white or with a small black spot on the inner vane).

Outer Ps pale grey with black tip, sharply defined, as if dipped in ink. Black tip decreases

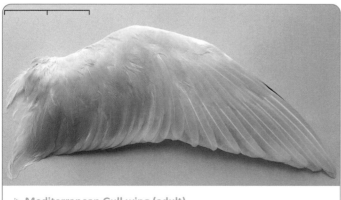

▷ Mediterranean Gull wing (adult).

in size from P10 to P5. P10 has outer vane with black border (sometimes totally black). Often a small white spot on P5 and P6. Border of inner vane is paler; sometimes some white before the black towards the tip. P1 to P4 pale grey; border of the inner vane paler to white. Ss pale grey, with a small white tip becoming wider towards the inner Ss.

Immatures with outer Ps black with inner vane largely white towards the edge. Median Ps pale grey with black at end (tip or subterminal bar); often dark towards the base of the outer vane. Ss and inner Ps pale grey with white tip, sometimes all white. Tertials grey with a distal black mark along the rachis on the outer vane. Second plumage similar to the adult, but on the outer Ps the black descends on the outer vane from P10 to P8.

▷ **Kittiwake wing (adult).**

COMMON GULL *Larus canus:* Rs of adults white. Rs of immatures with a large, terminal, or subterminal black band sometimes reduced to a small mark on R6.

Ps of adults pale grey with white tip. P10 to P8 with a lot of black, covering (almost) all the outer vane and at least half of the inner vane on P8; P10 is sometimes all black. P7 to P5 have only a little black at the tip. P4 to P1 are usually pale grey (rarely still a little black at the tip). All Ps have a white tip: large subterminal mark on P10 (sometimes terminal, sometimes like P9); subterminal mark and tip on P9; tip on P8–P5; end on P4-P1. Ps of young pale grey

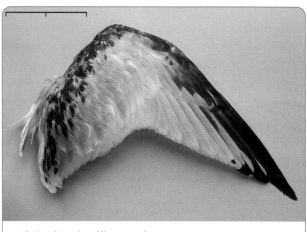

▷ **Kittiwake wing (first-year).**

with dark grey on outer vane and the tip. The contrasts are less sharp, the dark can form more random marks towards the tip and outer edge of the Ps. Some dark also on the inner vane of the outer Ps, less often on the others. Ps of second-year with a pattern similar to those of adult, but with more black and less white at the tip (sometimes without white).

Ss of adults pale grey with a well-defined white tip, wider towards the median and inner Ss.

Ss of the young pale grey with an ill-defined broad dark tip, more or less bordered with white at the tip. The following year, the dark is a more irregular mark, sometimes split into several marks.

The 'large' feathers of the adult differ from those of the Herring, Yellow-legged and Caspian Gull group in their slightly smaller size (about 20% less).

AUDOUIN'S GULL *Larus audouinii* (SS): Rs of adults white. Rs of young largely brown, then white with black subterminal bar.

Ps of adult: P1-P4 pale grey with narrow, ill-defined white tip. Other Ps with white tip. P5-P7 grey with black subterminal band; P8 and P9 black with grey base to inner vane; P10 often all black. P10 has a white tip, sometimes with another larger subterminal white spot on the inner vane. Ss of adult pale grey with a fairly small, ill-defined white tip.

Ps of young with a lot of dark brown-grey, sometimes with a small white tip. Ss largely dark brown-grey with white base and a well-defined white tip. In the following plumages, a little pale grey appears towards the inner Ps, the others becoming paler (except black areas).

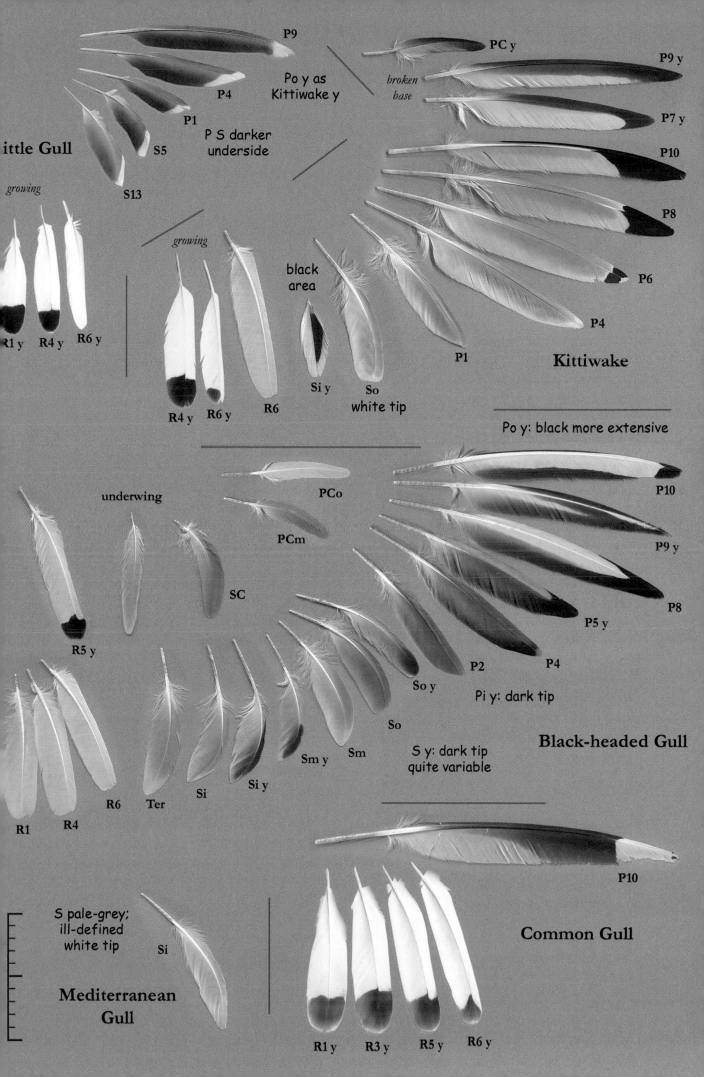

Little Gull

P9
P4
P1
S5
S13

Po y as
Kittiwake y

P S darker
underside

growing

R1 y   R4 y   R6 y

growing

black
area

R4 y   R6 y   R6

Si y   So
white tip

PC y

*broken
base*

P9 y

P7 y

P10

P8

P6

P4

P1

Kittiwake

Po y: black more extensive

underwing

PCo

PCm

SC

R5 y

R6   R1   R4

Ter   Si   Si y   Sm y   Sm

So

So y

Sm y

Si y

P10

P9 y

P8

P5 y

P4

P2

Pi y: dark tip

S y: dark tip
quite variable

Black-headed Gull

P10

Common Gull

S pale-grey;
ill-defined
white tip

Si

**Mediterranean
Gull**

R1 y   R3 y   R5 y   R6 y

The isolated moulted feathers of the following three species are difficult to identify to species. The grey areas are generally a little darker in Yellow-legged Gull, but several feathers must be compared together to appreciate the subtle difference. In adults, the Rs are white and Ss pale grey (Herring Gull *argentatus*) to deeper grey with a white tip (Yellow-legged Gull). Some black and white occurs on the tip of the median Ps, the amount of black increasing towards the outer Ps (wing pattern with black tip).

Herring Gull *Larus argentatus:* the white mark on the Ps is quite large, especially in the subspecies *argentatus*; the white spots are a little less extensive in *argenteus*. P10 has a large white spot at the tip, possibly with a broken bar or subterminal black marks (particularly *argenteus*). P9 has white tip, then a wide black bar and a white mark. The other Ps have a white tip. The black band on P5 is quite narrow, sometimes incomplete or absent (*argentatus*).

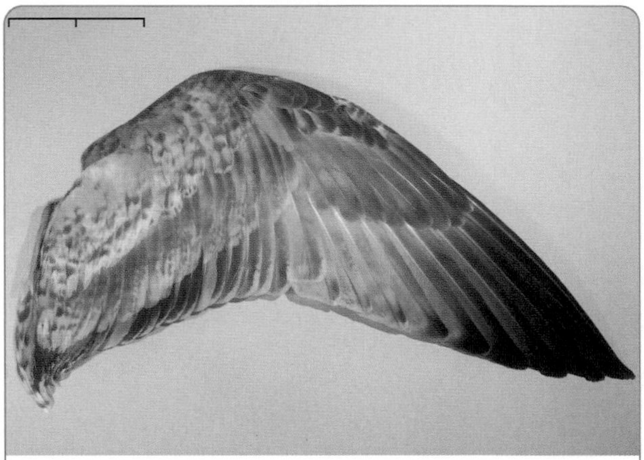

▷ Herring Gull wing (2nd winter immature)

Yellow-legged Gull *Larus michahellis*: the black wing-tip is generally larger than in the other two species and the grey is more intense. The white spots at the tip of the Ps are generally quite small (they disappear quickly with wear). P10 has a large white spot with a clear subterminal black stripe within it (sometimes the tip after the white area is totally black). P9 usually has a white tip and a second inner white spot. The other Ps have a white tip. P5 normally has a wide black band at its tip; P4 sometimes has a small black mark or band.

▷ Yellow-legged Gull (adult)

Caspian Gull *Larus cachinnans:* the white marks on the outer Ps resemble those of the Herring Gull. P5 normally has a well-defined black bar (P4 sometimes also with black). On the other hand, there is proportionately slightly less black on the outer Ps; the pale grey is more extensive on the inner vane towards the tip.

Lesser Black-backed Gull *Larus fuscus:* the remiges are a darker grey than those of the three preceding species, but that varies markedly according to subspecies, from ash-grey to almost black.

Rs of adults white. Rs of immatures white, more or less marked from dark brown to black depending on the individual and age, often with a largely dark tip and the base marked with irregular marks or bars.

Ps of adults (very) dark grey, with black end (except P1-P3) and white tip. P10-P8 with much black; P7 still marked with black on the outer vane; P7-P5 (P4) with black tip. P4 with white tip, often a small black mark before that (outer vane). P10 with a large subterminal white mark, often partially with a grey wash. P9 to P6 with a small white tip only in darker subspecies. P9 also has a subterminal white mark (less extensive than on P10) in *graelsii*, but generally none or very small in *fuscus*. The white on the tip of the Ps decreases in size according to subspecies: *graelsii* > *intermedius* > *fuscus*. Ss of adults (very) dark grey with a large white tip; sometimes the base of the inner vane also mainly white. Ps and Ss of young dark brown with paler inner vane, Ss with light tip; vane sometimes marked with more or less well-defined pale spots.

Great Black-backed Gull *Larus marinus:* it differs from other gulls by its very large size and the dark colour of its 'large' feathers. Rs of adults white. Rs of young white with more or less black-brown markings depending on

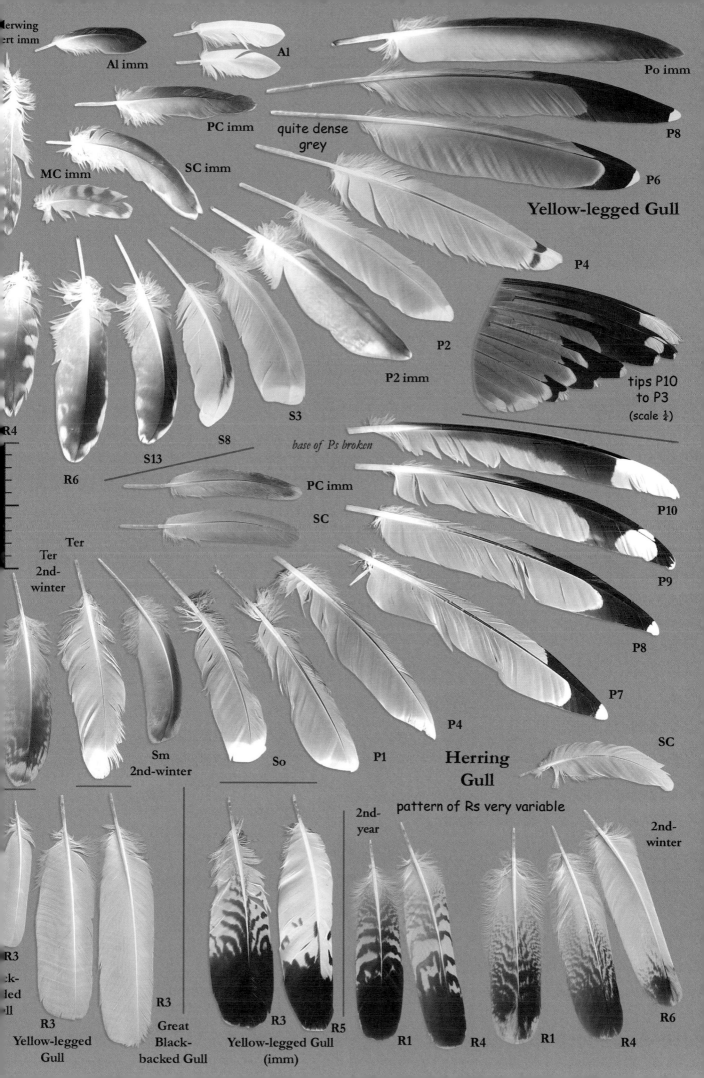

Underwing covert imm

Al imm

Al

PC imm

quite dense grey

Po imm

P8

P6

MC imm

SC imm

Yellow-legged Gull

P4

P2

P2 imm

tips P10 to P3 (scale ¼)

R4

S3

S8

S13

base of Ps broken

R6

PC imm

SC

P10

Ter

Ter 2nd-winter

P9

P8

P7

Sm 2nd-winter

So

P1

P4

Herring Gull

SC

pattern of Rs very variable

2nd-year

2nd-winter

R3

R3
Yellow-legged Gull

R3

R3
Great Black-backed Gull

R3

R5
Yellow-legged Gull (imm)

R1

R4

R1

R4

R6

individual and age; usually with one or two large marks towards the tip; then irregular spots towards the base, which can form bars.

Ps of adults dark grey. P10 is mainly black (outer vane and much of the inner vane), with a large white tip. P9 and especially P8 have less black on vanes. P9 has a large white mark at the tip, crossed by a subterminal black bar (on one or both vanes). Often grey marks the edge of the white area on P9 and P10. P8 has a small white tip and often another white mark (with black between the two).

P6 and P7, and sometimes P5 and P8 have white tip and a black subterminal bar, often bordered with a little white towards their base. P1-P5 have a well-defined white tip. Ss dark grey with a large white tip.

Ps of young dark brown-grey; pale appearing on the inner vane and towards the inner Ps (inner Ps paler with dark tip). Ss of young fairly dark brown-grey, with a white tip, and possibly irregular pale markings. The colour evolves towards that of adults as a bird ages, with the appearance of a more contrasting pattern at the tip of the wing.

GLAUCOUS GULL *Larus hyperboreus* and **ICELAND GULL** *Larus glaucoides:* Iceland Gull is rarer and a little smaller than the Glaucous Gull.

The Ps differ from those of other gulls by the combination of large size and the absence of black at the tip. The 'large' feathers of juveniles are distinguished by their large size and the paler colour of markings.

Rs of adults white. Rs of young pale brown-grey, with small buff to white markings, more or less regularly spaced. Ps and Ss of young and immature white to paler grey washed with pale brown-grey, with ill-defined markings, but more contrasting towards the tip. Ps and Ss pale to vary pale grey in adults, with a white tip.

• STERNIDAE: TERNS

NP = 10 (+1) / NS = 15-20 / NR = 12

Distinctive criteria of group: shape, colour, where they were found.

This family has narrow, pointed wings; the tail is more or less forked.

Possible confusion: with smaller gulls (outer Ps), waders (Ss, Rs).

Colour of 'large' feathers: colours range from white to black, through all shades of grey. The rachis may be light or dark.

– In these species, the young generally have darker 'large' feathers than adults.

– In the marsh terns, feather shape resembles that of other terns. However, the colour tone is more uniform on each feather, from light grey (Rs especially) to mid grey (Ps and Ss) to dark grey (outer Ps).

– In other terns, the wing is fairly uniform in colour, but the tip and trailing edge of the hand are often darker. The Ps and Ss are pale to mid grey; the inner vane is largely white. The outer Ps are darker, with grey or black forming a band on the outer vane and often on the tip of the inner vane. The Rs are quite pale, but grey colours the outer vane of the outer Rs in many species. All the 'large' feathers of the terns are marked (at least in a fresh state) by a silvery-grey hue superimposed on the ground colour, which differentiates this group from gulls and waders.

Primaries: the outer Ps are elongated, narrow and pointed. The central Ps are also more pointed than those of the gulls. The calamus is not very long (C = 15-20%).

Secondaries: Ss resemble those of gulls (greyish colour) and waders (oval spot), but the silvery coloration is diagnostic. The calamus is more or less long (C = 15-25%).

Rectrices: the central Rs are oval, the median ones more rectangular, and the outer ones elongated. The more the outer Rs are elongated the more the tail is forked.

In most cases there is a difference in coloration between immatures and adults. Thus the rectrices of immatures are generally darker (a deeper grey) and the grey edges are wider. The outer rectrices are also shorter than in adults.

The rectrices of adults are white, but mostly tinged with grey or with a silvery wash.

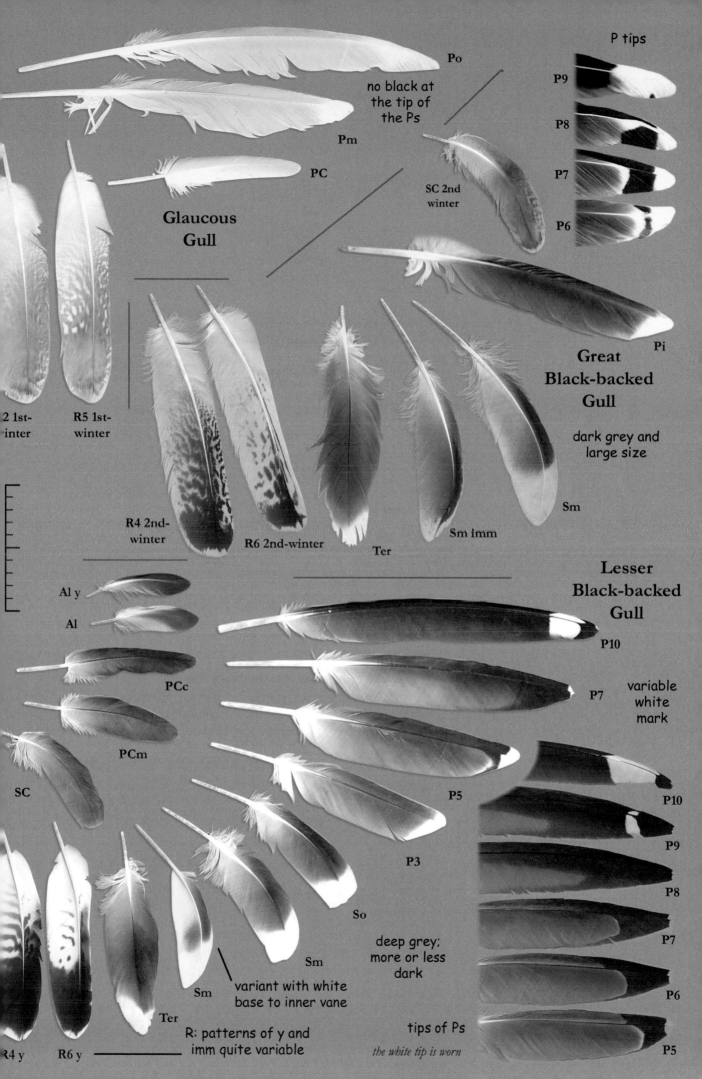

P tips

P9

P8

P7

P6

Po

no black at
the tip of
the Ps

PC

Pm

SC 2nd
winter

**Glaucous
Gull**

2 1st-
inter

R5 1st-
winter

**Great
Black-backed
Gull**

Pi

dark grey and
large size

R4 2nd-
winter

R6 2nd-winter

Ter

Sm imm

Sm

Al y

Al

PCc

PCm

SC

R4 y

R6 y

Sm

Ter

Sm

variant with white
base to inner vane

R: patterns of y and
imm quite variable

So

P3

P5

P7

P10

**Lesser
Black-backed
Gull**

variable
white
mark

deep grey;
more or less
dark

tips of Ps

the white tip is worn

P10

P9

P8

P7

P6

P5

The tails of these species are forked to very forked (tail with streamers). The central Rs are oval and quite short, the median ones more rectangular. The outer ones are elongated to very elongated depending on the species.

▷ Roseate Tern wing, adult

SOME EXAMPLES OF TAIL CHARACTERISTICS IN TERNS

Species	Rmax	Rmin	shape of outer feathers in comparison to central ones	colour of Rs
Little Tern	9	4	elongated (155%)	white; some outers tinted grey
Sandwich Tern	14 (18)	7.5	very elongated (190%)	white with slight grey tint
Common Tern	17 (18.5)	7.5	very elongated (195%)	white; grey on outer vane
Arctic Tern	20	7.5	very elongated (220%)	white; grey on outer vane
Gull-billed Tern	14.5	9	elongated (140%)	pale grey
Black Tern	11	7	arrow-shaped, but not particularly long (115%)	mid grey
Collared Pratincole	11.5	5.5	very elongated (200%)	white base; distal half black

Note: Collared Pratincole is a wader with Rs similar in shape to those of terns

MEASUREMENTS OF THE 'LARGE' FEATHERS OF TERNS

Species	Scientific name	Pmax	Pmin	Smax	Smin	Rmax	Rmin
Little Tern	*Sternula albifrons*	14 (16.5)	5.5	6.5	4.5	9	4
Gull-billed Tern (SS)	*Gelochelidon nilotica*	28	10.5	11	9.5	14.5	9
Caspian Tern	*Hydroprogne caspia*	32	14	14.5	11	17.5	12
Whiskered Tern (SS)	*Chlidonias hybrida*	21	8.5	10	8	10	7
Black Tern (SS)	*Chlidonias niger*	> 13	10	n.d.	n.d.	11	7
White-winged Tern (SS)	*Chlidonias leucopterus*	n.d.	n.d.	n.d.	n.d.	8.5	7
Sandwich Tern	*Sterna sandvicensis*	23 (30)	9	9	7.5	14 (18)	7.5
Common Tern	*Sterna hirundo*	22	8	8.5	7	17 (18.5)	7.5
Arctic Tern	*Sterna paradisaea*	22	8	8.5	7	20	7.5
Roseate Tern (SS)	*Sterna dougallii*	20	7	7	5	n.d.	n.d.

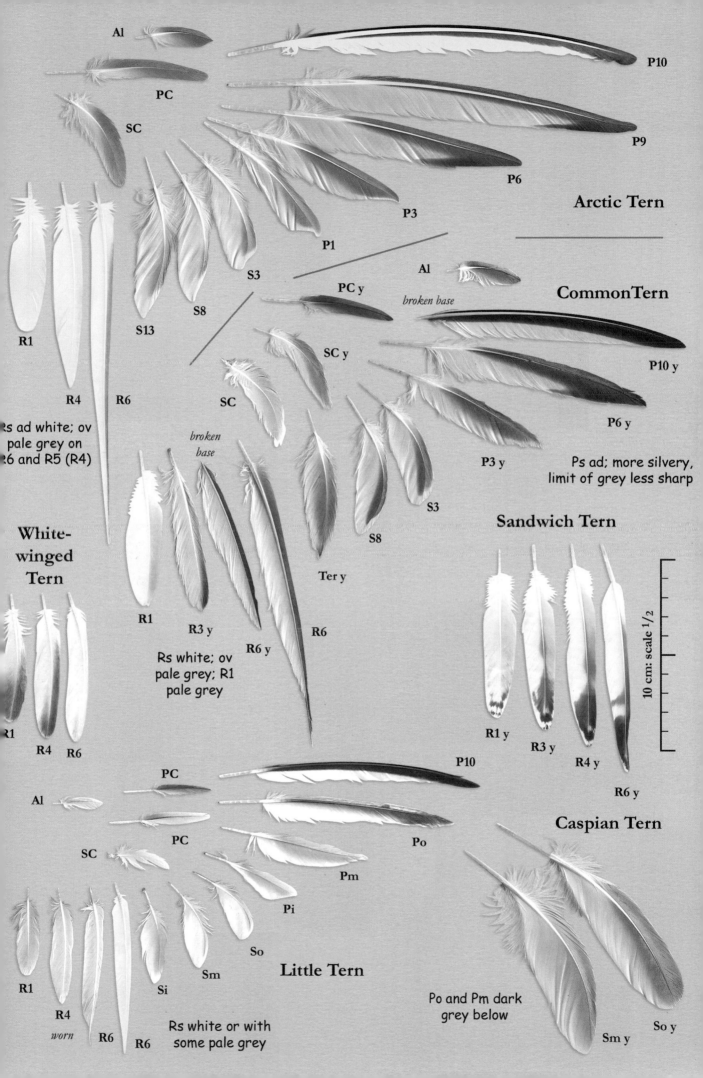

Al

PC

SC

R1

R4

R6

Rs ad white; ov
pale grey on
R6 and R5 (R4)

White-
winged
Tern

R1

R4   R6

Al

SC

PC

PC

SC

R1

R4
R6   R6
worn

R1

S13

S8

S3

P1

P3

P6

P9

P10

Arctic Tern

Al
broken base

PC y

SC y

SC
broken base

P3 y

P6 y

P10 y

CommonTern

Ps ad; more silvery,
limit of grey less sharp

R1

R3 y

R6 y   R6

Ter y

S3

S8

Rs white; ov
pale grey; R1
pale grey

Sandwich Tern

R1 y

R3 y

R4 y

R6 y

10 cm: scale ¹/₂

Caspian Tern

P10

Po

Pm

Pi

So

Sm

Si

Rs white or with
some pale grey

Little Tern

Po and Pm dark
grey below

Sm y

So y

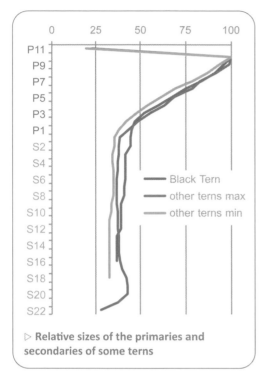

▷ **Relative sizes of the primaries and secondaries of some terns**

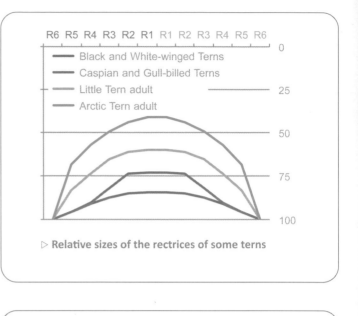

▷ **Relative sizes of the rectrices of some terns**

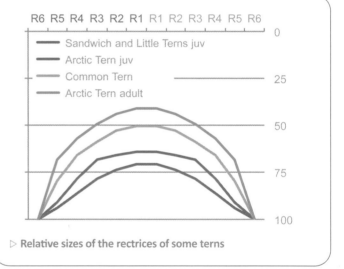

▷ **Relative sizes of the rectrices of some terns**

• STERCORARIIDAE: SKUAS

NP = 10 (+1) / NS = > 18 / NR = 12

Distinctive criteria of group: size, colour, where the feather was found.

This family has pointed wings, more or less long; the tail is variable, either short and square or with elongated central feathers.
Possible confusion: with gulls and diurnal birds of prey.

Colour of 'large' feathers: brown with a more or less extensive white base, often larger on the Ps (contributes to the hand having a white patch). In the skuas the brown is warmer than the brown-grey of young gulls.

Primaries and secondaries: similar in shape and size to those of the gulls (see p. 228). The boundary between white and brown is rather ill-defined and irregular. Ps and Ss of young may have a pale tip and a very speckled brown/white boundary. Rachis pale on outer Ps at least, sometimes on all Ps.

Rectrices: rectangular except for the central ones in several species. Brown with a white base, sometimes greyer; the white spreads more along the inner vane and on the outer Rs (the central ones have only a small white base). It should be noted that the rectrices of the skuas closely resemble those of the gulls. Their colour makes it possible to differentiate them: they are grey or brown, often with a white base; the rachis is pale (except for central Rs); they do not have well-defined bars or marks (see immature gulls, p. 228).

P10

P8

Pi

PC

SC

Great Skua

PC

So

Sm

R5

Pııı

pale spots at tip
in young

Arctic Skua

young

young

S17

S12

S3

Long-tailed Skua

young
(different individual)

small
undertail-
covert

young

R1   **R4**

R6

R1

R5

R1   **R4**

R6

CHARACTERISTICS OF 'LARGE' FEATHERS OF SKUAS								
Species	Pmax	Pmin	Smax	Smin	Rmax	Rmin	colour	shape of central feathers*
Pomarine Skua (SS)	27 (30)	13	15	13	18	15	dark brown; white base	longer and wider; twisted
Arctic Skua (SS)	31	12	14	11	17	12	brown-grey; white base	longer and wider; twisted
Long-tailed Skua (SS)	32	13	n.d.	n.d.	33	12	slate-grey to white; white base	very elongated
Great Skua (SS)	31 (34)	18	20	16	18	15 (14)	brown; white base	rectangular

* only slightly developed in young. Note: small sample sizes

POMARINE SKUA *Stercorarius pomarinus* (SS): rachis white on the Ps; wholly on the outer Ps, only at the base on the inner ones. Proportion of white on Ps quite variable depending on the individual. Inner Ss with white tip.

ARCTIC SKUA *Stercorarius parasiticus* (SS): the remiges of adults have a little less white than those of Great Skua (much less in the dark morph). Rachis mostly white on at least P8-P10 (dark morph), often up to P5.

LONG-TAILED SKUA *Stercorarius longicaudus* (SS): the Ps of adults have much less white than in other skuas but those of young are similar. The rachis is mainly white on at least P10 and P9, sometimes also P8 and P7.

▷ **Pomarine Skua wing (1st-year)**

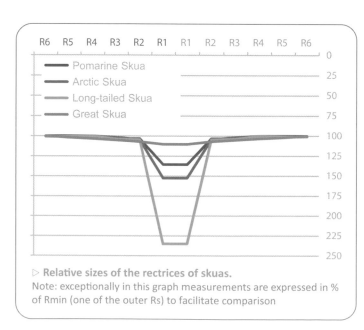

▷ **Relative sizes of the rectrices of skuas.**
Note: exceptionally in this graph measurements are expressed in % of Rmin (one of the outer Rs) to facilitate comparison

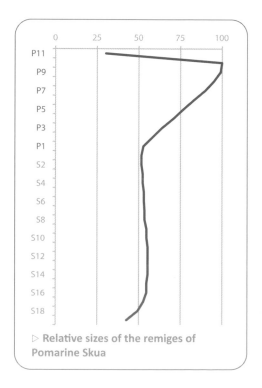

▷ **Relative sizes of the remiges of Pomarine Skua**

GREAT SKUA *Stercorarius skua:* rachis is mainly white (Ps, Ss and Rs). Slightly larger than the other three species. Ss sometimes have a reddish tint to the outer vane. The Rs can have pale at the tip. The tail is rectangular.

• ALCIDAE: AUKS

NP = 10 (+1) / NS ≥ 15-21 / NR = 12.

Distinctive criteria of group: shape of the remiges, colour, where feather was found.

This family has very small wings relative to body size, with a rather pointed hand, a little curved. The tail is short to very short and rigid. Their feathers are found on the coastline. Due to the pelagic lifestyle of these species, feathers are only rarely found in isolation.

Possible confusion: with storm petrels (on the coast), swifts and swallows, but the 'large' feathers are stiffer in the auks.

Colour of 'large' feathers: brown-grey-black, sometimes with white markings. All 'large' feathers have a dark rachis.

Primaries: rigid, curved in profile to compensate for the heavy wing-loading. Long calamus (C − 15-30%), pointed tip, inner Ps with a wedge-shaped tip. Black-brown colour, with lighter base (white in the Black Guillemot).

Secondaries: long calamus (C = 20-30%); shape and size very similar to those of other seabirds, but black-grey colour, with or without white tip. S1-S3 can look a lot like inner Ps (almost straight rachis). The greater coverts are almost as long as the remiges.

Rectrices: short and stiff, longer and more pointed in the Razorbill; black-brown; with white at the tip in the Little Auk.

The differences between species are quite small, and all the more difficult to distinguish as the feathers fade with wear. The Little Auk is very small and is distinguishable by size; the Puffin is intermediate and the other species are about the same size as each other.

CHARACTERISTICS OF PRIMARIES AND SECONDARIES OF AUKS							
Species	Pmax	Pmin	colour of Ps	Smax	Smin	main colour of Ss	white tip
Black Guillemot	14.5 (16)	9	brown black; white base on the underside	10	6.5	brown black; internal vale paler brun-noir, base interne blanche	sometimes, small
Common Guillemot	13.5 (15)	8	grey-brown-black; very pale at the base	9	5	grey-brown-black	x
Razorbill	15	7	brown-black; very pale base	9	5	brown-black; base very pale	x
Atlantic Puffin	13	6.5	grey-brown-black; inner vane paler	8	5	grey-brown on the outer vane; grey on the inner vane	-
Little Auk	10	5.5	brown-black; inner vane paler	7	4.5	brown-black; inner vane very pale	x

Note: no data could be found for Brünnich's Guillemot *Uria lomvia*

CHARACTERISTICS OF RECTRICES OF AUKS				
Species	**Rmax**	**Rmin**	**shape and main colour of Rs**	**white tip**
Black Guillemot	8	5	copper-brown (sometimes white edge to tip of outer Rs); base paler	-
Common Guillemot	7	5	black-brown	-
Razorbill	9.5	6.5	black; slightly pointed; central feathers pointed and arrowhead-shaped	-
Atlantic Puffin	7	5	black; slightly elongated	-
Little Auk	5	4	black	edge

Note: no data could be found for Brünnich's Guillemot *Uria lomvia*

BLACK GUILLEMOT *Cepphus grylle:* Ps and Ss dark brown-grey, with a rounded white mark at the base of the inner vane, on the side of the vane (except sometimes outer Ps). On the Ss, the mark is more rounded and touches the rachis at the base. Possibly a pale edge at the tip of the Ss. Inner Ss (almost) without white. Rs copper-brown with a paler base, sometimes a white edge at the tip of the outer Rs.

Greater coverts with a white spot on the outer vane, sometimes on both.

COMMON GUILLEMOT *Uria aalge:* Ps and Ss dark brown-grey. Border of the inner vane of Ps paler to white on the basal half (except P1, and sometimes P2). Ss with white tip, S1-S3 (S4) often with less white, reduced to the edge or absent (S1). Inner Ss without white. Rs dark grey-brown with a paler base.

RAZORBILL *Alca torda:* Ps and Ss black-brown; rachis dark. Border of the inner vane of the Ps paler on the basal half. Ss paler towards the base of the inner vane; a white tip, smaller on S2 and S3 and absent on S1. Inner Ss without white. The white tip is quite large in adults (> 6 mm), smaller in young. Rs black, quite pointed. Central Rs more pointed and arrowhead-shaped.

Possible confusion: with 'pied' woodpeckers (Rs), but different habitat.

ATLANTIC PUFFIN *Fratercula arctica:* Ps and Ss dark brown-grey, paler on inner vane. Ps with a dark tip also on the inner vane. Ss much paler on the inner

▷ **Common Guillemot wing**

▷ **Little Auk wing**

vane; quite grey, sometimes with a thin white edge at the tip. Rs black; the same size but a little more elongated than those of the Common Guillemot.

LITTLE AUK *Alle alle:* Ps and Ss dark brown; paler on the inner vane. Ps with dark tip. Ss paler on the inner vane; grey, with a sharp white tip, becoming larger towards the inner Ss. S3 and S4 often with less white, reduced to an edge on S2 and absent on S1. Inner Ss with a white tip or no white. Rs black, paler towards the base, with a very narrow white edge towards the tip. Inner greater coverts with white border towards the tip of both vanes, but not covering the tip.

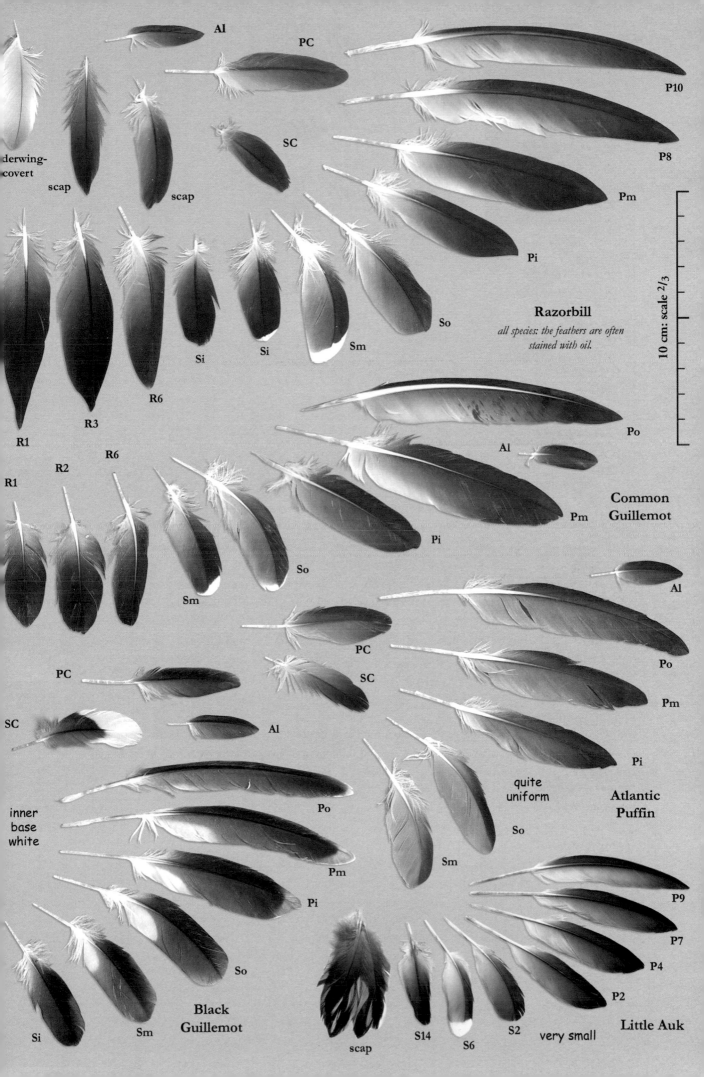

Al

PC

P10

P8

Pm

SC

Pi

derwing-covert

scap

scap

Si

Si

Sm

So

Razorbill

all species: the feathers are often stained with oil.

10 cm: scale 2/3

R6

R3

R1

Po

Al

Common Guillemot

R6

R2

Pm

R1

Pi

Sm

So

Al

PC

Po

SC

Pm

PC

SC

Al

Pi

inner base white

Po

quite uniform

Atlantic Puffin

So

Pm

Sm

Pi

P9

So

P7

Si

Sm

Black Guillemot

scap

S14

S6

S2

very small

P4

P2

Little Auk

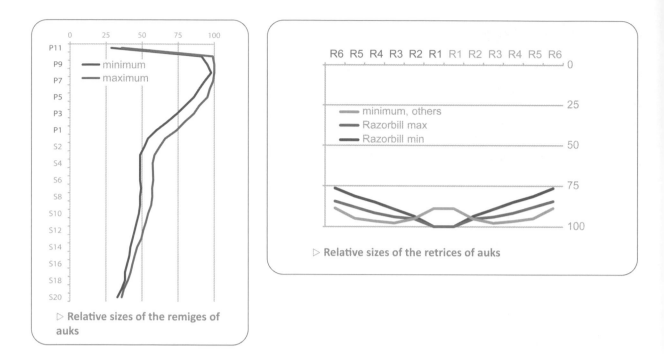

▷ **Relative sizes of the remiges of auks**

▷ **Relative sizes of the retrices of auks**

• CHARADRIIFORMES (WADERS): CURLEWS, GODWITS, SANDPIPERS, PLOVERS, ETC.

NP = 10 (+1) / NS = 15 (21) / NR = 12 (18)

Distinctive criteria of group: shape, colour for some species, where the feather was found.

These feathers are found almost exclusively near water, especially on coastlines, sometimes in marshes or along streams and lake shores. Certain species are more common in meadows and fields, such as Lapwing or Stone Curlew, or in wooded areas, such as Woodcock. It should be noted that the shape of the 'large' feathers of Woodcock and Lapwing are quite different and contrast with the relative homogeneity within this group.

Certain waders moult most of their 'large' feathers after the post-nuptial migration. These are therefore mainly found in their wintering quarters, outside Europe (Africa for the most part). This means that it is very rare to find moulted 'large' feathers of these species in Europe. Plucked corpses or other remains are generally found. The following species are exceptions to the general rule: Little Ringed Plover (a few outer Ps and inner Ss are sometimes moulted before migration); Woodcock ('large' feathers sometimes found during its migratory stopover in Spain); Wood Sandpiper (sometimes some of the 'large' feathers are moulted before migration).

Possible confusion: mainly with gulls, terns, auks, small waterfowl (Ss) and Sandgrouse (Ss).

Colour of 'large' feathers: very frequently wing and tail feathers have markings, often in two colours; sometimes tricoloured or uniform. Often, there are longitudinal marks on the Ps; oval-shaped marks or at the tip on the Ss; regular bars or a terminal bar on the Rs. See the key to identifying 'large' feathers by colour on p. 253 .

Primaries: Ps narrow; fairly straight, inner ones wedge-shaped; fairly long calamus (C = 20-25 (30)%), outer Ps not emarginated, except in three particular species: Lapwing (outer Ps narrow at tip); Stone Curlew (outer Ps slightly notched); Woodcock (Ps curved).

Secondaries: for Ss particular, shape is an important criterion (curved with long calamus: C = 20-25 (30)%), which places them in the same shape group as the gulls, terns, and auks.

Rectrices: generally quite short and not robust. Square or rounded tip, more pointed in sandpipers and snipes. Only a few common or highly recognisable wader species are described individually below. Descriptions by feather category are given in the identification keys. A summary of the patterns observed for each species is also presented in the table on p. 257 . The sizes of the 'large' feathers are given in that table as well as in the descriptions included in the keys.

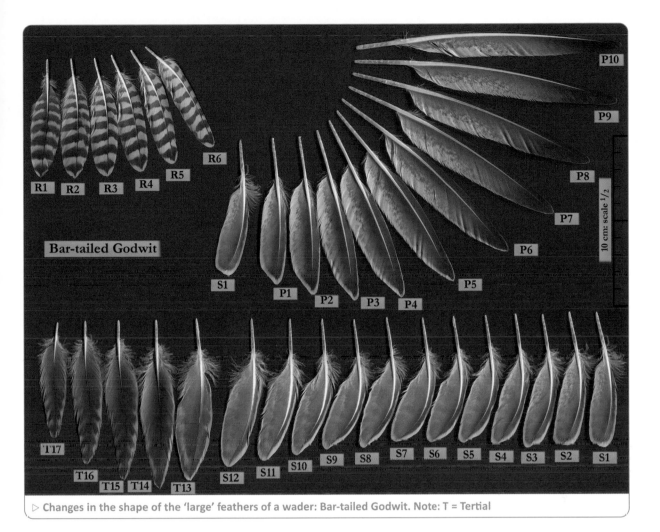

▷ Changes in the shape of the 'large' feathers of a wader: Bar-tailed Godwit. Note: T = Tertial

Large waders

Northern Lapwing *Vanellus vanellus:* the coloration of Ps and Ss is fairly constant. The Ps are black, the outer ones with a characteristic shape; the distal part is narrowed. In addition, the same part of the feather is pale (white to grey or buff). The Ss are black with the base of the inner vane white. The colour of the Rs is much more variable. They are usually white with a third to half of the distal part black, with a narrow pale tip; there is less black towards the outer Rs.

Spur-winged Lapwing *Vanellus spinosus:* the 'large' feathers are generally very similar to those of Northern Lapwing but the outer Ps are not marked with pale at the end and have a pale white base; the Ss have white on both vanes and the inner Ss are buff; the Rs have more black.

Stone Curlew *Burhinus oedicnemus:* the patterns of the 'large' feathers vary greatly depending on the position of the feather and the individual, especially on Ps and Rs. The Ps are a mix of dark and pale in various proportions; the Ss are dark with a pale inner base. The Rs are quite large, with a narrower rounded tip, most often mixing a white-buff background with bars or dark patterning, with a well-defined black tip.

▷ Stone Curlew wing

EURASIAN OYSTERCATCHER *Haematopus ostralegus:* the 'large' feathers are black and white, with interposed areas of colour on the Ps. The median and inner Ss sometimes white. Juvenile: sharp white tips on the Ss, white edge towards the tip of the inner 4 or 5 Ps.

PIED AVOCET *Recurvirostra avosetta:* outer Ps black-brown with small white base; increasing amount of white towards the inner Ps (often P1 and P2); P3 generally white with a little black at the tip. Ss white, inner ones tinged with dark brown. Rs white to pale grey. Young with more or less white at the tip of their dark Ps.

BLACK-WINGED STILT *Himantopus himantopus:* Ps and Ss black-brown, with a pale edge towards the tip (except outer Ps) in the young. Rs white to light grey; darker in the young with a pale edge at the tip.

EURASIAN CURLEW *Numenius arquata* and **WHIMBREL** *Numenius phaeopus:* Ps and Ss are brown with white or off-white notches, sometimes connected in a comb-shape. The Rs are white to brown-grey with dark brown barring. In the Whimbrel, the Ps and Ss are smaller; Rs are smaller and more heavily tinged with brown-grey between the darker bars.

In both species there is variability in the coloration, mainly of outer Ps, this is related to the geographic origin of birds and also potentially to their age. Pale notches are sometimes restricted to spots or flecks; the outer two or three Ps are sometimes entirely dark (Eurasian Curlew, ssp. *orientalis*).

BAR-TAILED GODWIT *Limosa lapponica:* Ps and Ss brown; generally P7-P10 without a pale edge. Pale area covers much of the inner vane, speckled with irregular margins. Ss sometimes with faint pale bars. Brown tertials with pale notches, resembling those of curlews.

Rs white to buff with brown bars; wider and more regular in the young. Female with thinner bars or similar to the male. Rs of male largely tinged with brown, more towards the central ones; barred with pale or pale notches on the edges, sometimes with more irregular brown and pale patterns. Outer feathers usually well barred.

▷ Avocet wing

▷ Whimbrel wing

▷ Bar-tailed Godwit wing, reconstructed from a tide-line corpse

BLACK-TAILED GODWIT *Limosa limosa:* Ps black-brown, white towards the base. Partially white inner vane with ill-defined limit on P10 and P9 (P8), the mark becoming sharp and rounded; often brown along the rachis within the white area. Outer vane of P1-P6 (P7) white on basal third to two thirds. P7 possibly with white edge at the base. Ss white with about the distal third black. The shape of the black is rounded as on the Ps; the white continues along the rachis in the black part. Fine white edge on the Ss, sharper towards the inner Ss, on which the black mark becomes paler and decreases on the outer vane. Inner Ss brown, sometimes with white base. Rs with white base; black on about two thirds to three quarters (central feathers), the black gradually decreasing towards the outer ones (less than half). Outer Rs (R5 sometimes) with very asymmetrical black mark, narrower on the outer vane. Tip often paler on R1–R3 or R4.

Woodcock and Snipes

WOODCOCK *Scolopax rusticola*: confusion impossible: dark 'large' feathers with triangular wedges or narrow buff, rufous or warm brown bars. Rectrices with grey tip, white underneath. Ps are wider than those of other waders, with a more rounded end; they are more curved when viewed from above or in profile. Ss of more elongated shape. Rs quite curved and wide. Vague resemblance to the Ps and Ss of partridges.

GREAT SNIPE *Gallinago media* (SS): Ps and Ss brown. Tip of Ps with very thin pale border, more obvious on Ss. Inner vane of Ps paler; that of the Ss with pale speckling. Tip of Ss usually rounded, sometimes slightly pointed. R1 black-brown at the base and spotted with rufous towards the tip. Following ones with more and more rufous and incomplete dark bars; the tip paler. Outer Rs (R5 and especially R6–R8) with lots of white and with dark barring. Inner vane and tip often white, with more dark markings in the young.

COMMON SNIPE *Gallinago gallinago*: Ps and Ss brown, paler on the inner vane. Ss more clearly marked with pale (mottling, etc.) on the inner vane. Tip of the inner Ps edged with pale. Ss with broad white tip (tip rounded).

Rs characteristic black-brown, rufous and with a white tip, but with a very variable pattern depending on the individual. Rs have little white at the tip of outer pairs (R5–R8). Outer Rs paler than the others. Outer Rs of the male with strong calamus, narrower at the tip and slightly twisted in profile (the Rs produce the characteristic drumming during display flights).

JACK SNIPE *Lymnocryptes minimus*: Ps and Ss brown. Ps vaguely edged with pale towards the tip. Ss with pointed tip and an ill-defined white mark. Rs brown with pale (buff to rufous) markings on the outer vane and the tip; edged with pale on the inner vane. Central pairs spotted on both vanes. Rs quite narrow; increasingly pointed towards central pair, which are significantly longer than the other pairs.

▷ Woodcock wing

▷ Common Snipe wing

▷ Jack Snipe wing

Sandpipers and 'shanks

COMMON SANDPIPER *Actitis hypoleucos*: it is the only wader to have a white spot of this shape on the remiges. Ps dark brown with a large white mark on the inner vane, more so in the centre (often absent on P10 and sometimes on P9). The white spot is wide towards the edge of the vane and narrows into a triangle or oval towards the rachis, with the white mark wider towards the base. Ss dark brown, crossed by a wide white band. Median Ss also have more white and a little white at the tip. Rs brown; central feathers with the beginning of black bars, often with a green iridescence. A small white tip appears on R2 and R3 and widens towards R6. The brown background

▷ Common Redshank wing, reconstructed

▷ Sandpiper wings: Sanderling (upper), Dunlin (middle) and Little Stint (lower).

becomes paler and the bars become sharper as the white increases in proportion. R6 often white with more or less sharp bars on the inner vane; sometimes one or two on the outer vane.

SPOTTED REDSHANK *Tringa erythropus:* the pattern of the 'large' feathers is reminiscent of that of curlews, but they are distinctly smaller in size.

Small waders (*Calidris* spp)

Some species are very similar so that isolated feathers are difficult to identify. Certain details can sometimes help with identification, such as rachis colour, size or colour contrasts.

Included here are three examples that illustrate such differences.

Graphs of the relative sizes of 'large' feathers

The following graphs give the relative sizes of the 'large' feathers by species groups (minimum and maximum), or in isolation for atypical species.

Note: No data could be found for sets of 'large' feathers of the following species: Black-winged Pratincole, Broad-billed Sandpiper, and Red-necked Phalarope.

For the remiges, six groups by shape have been identified and four species are discussed individually.

These species are:

Grey Phalarope: has a 'classic' wing shape; pointed hand; narrow forearm; long tertials.

Collared Pratincole: has a narrow forearm (short Ss).

Woodcock: has a very broad wing (Ss very big); only a slightly pointed hand (slight difference of Pmax to P1).

Northern Lapwing: has a very round hand.

▷ Spotted Redshank wing

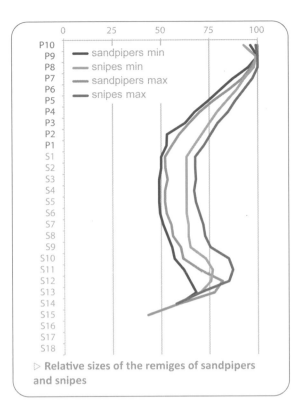

▷ Relative sizes of the remiges of sandpipers and snipes

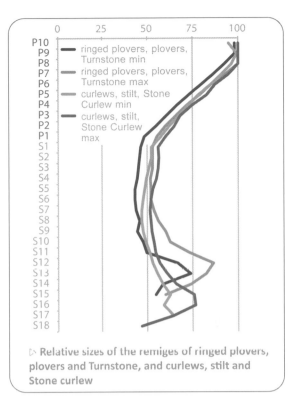

▷ Relative sizes of the remiges of ringed plovers, plovers and Turnstone, and curlews, stilt and Stone curlew

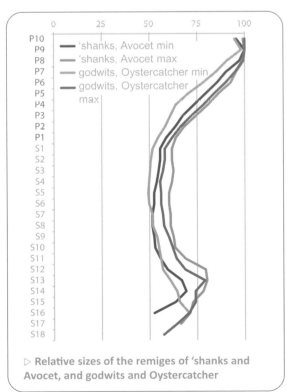

▷ Relative sizes of the remiges of 'shanks and Avocet, and godwits and Oystercatcher

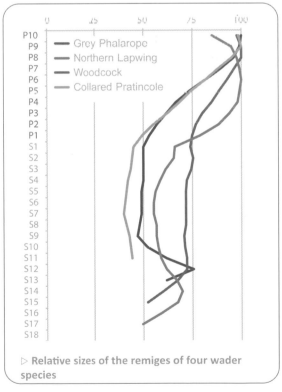

▷ Relative sizes of the remiges of four wader species

For the rectrices, most species have a square or rectangular tail, i.e. the difference in size between the largest and smallest rectrices of the same bird does not exceed 10% (i.e. Rmin ≥ 90% Rmax). As a result, no graphs are given for the following species:

Bar-tailed Godwit	Dunlin	Little Ringed Plover	Ruff
Black-tailed Godwit	Eurasian Curlew	Little Stint	Sanderling
Black-winged Stilt	Eurasian Dotterel	Marsh Sandpiper	Spotted Redshank
Common Greenshank	Eurasian Oystercatcher	Northern Lapwing	Spur-winged Lapwing
Common Redshank	European Golden Plover	Pied Avocet	Temminck's Stint
Common Ringed Plover	Green Sandpiper	Purple Sandpiper	Whimbrel
Common Snipe	Grey Plover	Red Knot	Wood Sandpiper
Curlew Sandpiper	Kentish Plover	Ruddy Turnstone	

The other species are presented in three graphs.

Note: only one individual per species is presented, values may vary slightly between individuals.

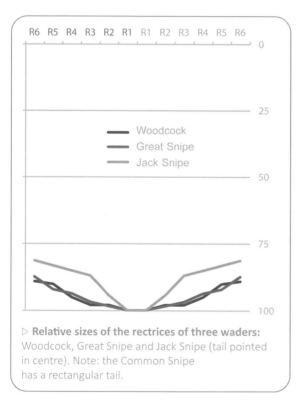

▷ **Relative sizes of the rectrices of three waders:** Woodcock, Great Snipe and Jack Snipe (tail pointed in centre). Note: the Common Snipe has a rectangular tail.

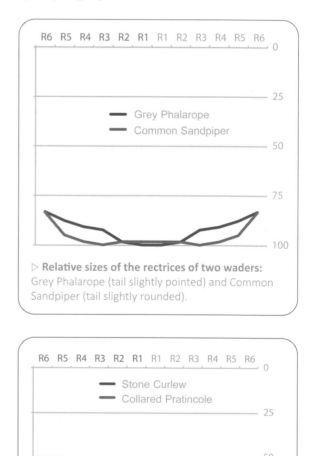

▷ **Relative sizes of the rectrices of two waders:** Grey Phalarope (tail slightly pointed) and Common Sandpiper (tail slightly rounded).

▷ **Relative sizes of the rectrices of two waders:** Stone Curlew (rounded tail) and Collared Pratincole.

• KEYS TO IDENTIFICATION BY THE COLOUR OF THE 'LARGE' FEATHERS OF WADERS (AND ASSOCIATED SPECIES)

The table below shows the main colour criteria found on 'large' feathers (not including tertials). Each colour category is detailed in the following keys, for one of the three feather types: primaries, secondaries and rectrices.

Summary of colour patterns found in waders

Refer to the descriptions in the keys for more detail.

Species	Pmax	Pmin	Smax	Smin	Rmax	Rmin	uniform: white	uniform: silver-grey / pale grey markings clear	uniform: dull grey	uniform: pale wedges on dark background	uniformly dark: entirely dark	uniformly dark: small pale mark / pale tip	black and white	dark and white: pale base to inner vane	dark and white: several patches of colour	dark and white: dark on a single distal patch	dark and white: black/grey/brown; irregular pale area or pale tip	dark and white: white with one or more dark marks	dark and white: regular bars; not black and white	diverse contrasting marks	diffuse marks	other pattern or hues
Large waders																						
Lapwing	20	13	13.5	9	13	10	(Ro)	(Ro)			P	P	R	S								
Spur-winged Lapwing	20?	13?	13.5?	9?	13?	10?					P	R	PS		S							S
Stone Curlew	20	5	12	8.5	14	9					P	R	PS	P	P					R		
Oystercatcher	20.5	11	12	9.5	13	9.5	S				P	P	R	P	P	PS						
Avocet	18	10	10.5	9	10.5	9.5	PSR				P				P							
Black-winged Stilt	20	10.5	11	9	9.5	8	R				PS (Rc)	S								(R)		
Curlew	24	13	13.5	11	14	11.5				PS	Po							P	R			
Whimbrel	19.5	8.5	10.5	7	12	10				PS								P	R			
Bar-tailed Godwit	17.5	8.5	9.5	7	10	7						R		R			PS		R	R		
Black-tailed godwit	18	8.5	10	7.5	10	7.5								R	P	P	PS					
Woodcock and snipes																						
Woodcock	15.5	10.5	12	10	10.5	7.5				PSR												S
Great Snipe	11.5	6	7.5	6	6.5	5.5					P	S	R					S	Ro	R		
Common Snipe	11	6	7.5	6	7 (9)	5				Ro	P	S						PS		R		
Jack Snipe	9	6	7.5	6	6.5	5					P Ro	S						S		R		
Ringed plovers, plovers, and Turnstone																						
Little Ringed Plover	9.5	5	6	4.5	7	5.5					PSRc	S	Ro							R		
Ringed Plover	11	5	7	4.5	8	5.5	S Ro					Rc		R	P	P	PS			R		

Species																				
Kentish Plover	10	5	6.5	4.5	7	5.5	S Ro				PR		Ro	P	P	PS			R	R
Dotterel *	13	5.5	7.5	5.5	8.5	7					PSR	R								
Golden Plover	15.5	7.5	9	7.5	9.5	7				S				P	P	PS		R	R	
Grey Plover	16	8	9	6.5	9.5	7.5				S			R	P	P	PS		R		
Turnstone	13.5	6.5	7.5	5.5	8.5	6							R	P		PS			R	
Calidris species																				
Knot	13.5	6.5	7.5	6	7.5	6					Pe Rc			PS			PS		R	R
Sanderling	10.5	4.5	6	4.5	6	5	S Ro	R			Rc			P		PS			R	R
Little Stint	8.5	4.5	5	4	5	4	Ro	Ro			P Rc			PS			PS			R
Temminck's Stint*	8.5 ?	4.5 ?	5 ?	4 ?	5 ?	4 ?	Ro				P Rc			PS						R
Curlew Sandpiper	11	5	6.5	4.5	6	5	S				P Rc			PS			PS		R	R
Purple Sandpiper*	11	5	6.5	4	6	5	S				Pe R			PS		S		S		Re
Dunlin	9.5	4.5	5.5	4.5	6	4	S				P Rc			P		PS			R	R
Tringa and allied species																				
Ruff	14.5	7	8.5	6.5	8.5	5.5					PR			PS		S			R	R
Spotted Redshank	13.5	7	8	6.5	8.5	7				PSR					P	S	R			S
Redshank	13.5	7.5	8	6.5	8	6	S							P				P	PS	R
Marsh Sandpiper*	13	6.5	7.5	5	7	6							R	P				PS	Ro	
Greenshank	15.5	6	8.5	5.5	9	8	(Ro)			S	P		R				PS	R	R	Rc
Green Sandpiper	13.5	7	8	5.5	7.5	6	(Ro)			(S)	PS		R							
Wood Sandpiper	11.5	5.5	6.5	5	6.5	5					PS		R				PS	R	R	R
Common Sandpiper	9.5	5	6	4.5	7	5					PR	P(Rc)	P					S	R	R
Phalaropes and pratincoles																				
Red-necked Phalarope*	n.d.	n.d.	n.d.	n.d.	n.d.	n.d.								P		PS			R	
Grey Phalarope*	10.5	5	6	4.5	7	5.5								P		PS			R	
Collared Pratincole	18	7	8	5.5	11.5	5.5					PS	S	R							
Black-winged Pratincole	19 ?	7 ?	8.5 ?	5.5 ?	9 ?	5.5 ?					PS		R							
Species with which waders can be confused																				
gulls									PS			S								
terns								PS												
marsh terns											PS									
auks											PS	S		P						
wildfowl												S								

The sign* indicates a small sample size, for measurements and/or for descriptions.

Note: no data were available for Broad-billed Sandpiper Calidris falcinellus.

The following key organises species according to the colour patterns of their 'large' feathers. First, it is necessary to assign the 'large' feather to be identified to group type (primary, secondary or rectrice). Descriptive summaries illustrated with diagrams allow a choice to be made at subsequent stages. The diagrams are representative of the group described, but not necessarily of all the secondary descriptions, which must be referred to afterwards. For each pattern, one or more subcategories are presented. The ranges of size presented and descriptions by species can be used to find the one that best suits the observed 'large' feather. Illustrations of several species are also available on the full-page plates.

The 'large' feathers of the wader species described here include sizes for primaries with a length between 4 cm and 24 cm, and for secondaries and rectrices between 4 cm and 14 cm. The size of individuals, and therefore that of their 'large' feathers, varies in particular according to their sex and geographical origin. For example, individuals of the same species are on average larger the further north they occur in Europe.

The proportions indicated for the coloured areas relate to the length of the vanes from their base (not the total length of the feather).

Notation: Species name in English (T = total size range in cm of the 'large' feather type - P, S or R concerned (possibly T for primaries).

For measurements: 'T = x-y' indicates 'total size between x and y cm'. These sizes consider all the individuals examined to produce this key. Some decorative 'large' feathers may have a total size outside this range. In cases where feathers are significantly larger than the description (P > 25 cm, S > 14 cm and R > 15 cm), refer in particular to gulls, and possibly to skuas and petrels. Species marked with an asterisk* are not waders but their 'large' feathers are found in the same locations and their shapes and sizes are very similar to those of waders. These species are therefore included when necessary in this key.

Note on colour: on worn feathers, the original black may appear brown or dark grey, so try these descriptions as well. By default, the adjective 'dark' refers to the darkest colour of the description (e.g. brown if the feather is buff and brown). The edges and tips wear quite quickly and the colours that might be visible there can disappear completely on worn feathers. There is sometimes marked colour variability between individuals of the same species; several descriptions may correspond to the feathers from the same location on a bird. A species can therefore be found included for several different patterns.

In some cases rachis colour can be important to differentiate between species.

In the case of plucked corpses, 'large' feathers of a different category and pattern can be identified separately and then verified by comparing the results obtained.

KEY TO THE PRIMARIES OF WADERS

• List of the colour criteria used in the key

I. Uniform pale (white, pale grey)		p. 257
a) entirely white		p. 257
b) with a silvery-grey mark		p. 257
c) entirely dull grey (eventually with darker tip)		p. 258
II. With pale wedge(s) (white, buff, rufous, etc.) on dark background (grey, brown, black)		p. 258
a) a single triangular wedge; on the inner vane		p. 258
b) several wedges; at least on the inner vane		p. 258
III. Uniform dark black, grey, or brown: (eventually with a small, pale area)		p. 258
a) entirely dark		p. 258

a-1 dark rachis		p. 258
a-2 entirely pale rachis		p. 260
a-3 partly pale rachis (a few cm)		p. 260
b) small pale mark		p. 260
b-1 mark at feather base		p. 260
b-2 mark towards the middle or tip		p. 262
IV. Dark and white		p. 262
a) dark with white or pale at base on inner vane; limit with a clear edge		p. 262
a-1 rachis entirely dark		p. 262
a-2 rachis entirely pale		p. 262
a-3 rachis pale for a few centimetres of its distal (or median) part		p. 262
b) several zones of white and dark		p. 264
c) the dark on the vanes in a single distal zone		p. 266
V. Black, grey, or brown; inner vane paler with irregular margin; speckled or vermiculated		p. 266
VI. White with one or more brown-grey marks		p. 268

KEY TO THE SECONDARIES OF WADERS

• List of the colour criteria used in the key

I. Uniform pale (white, pale grey)		p. 268
a) entirely white or pale grey		p. 268
b) marked with silvery-grey; sometimes very pale		p. 268
c) dull grey; with or without a white tip		p. 268
II. With pale wedge(s) (white, buff, rufous, etc.) on a dark background (grey, brown, black)		p. 270
III. Uniform dark (black, brown, grey)		p. 270
a) entirely dark		p. 270
b) dark with pale tip (white or grey)		p. 270
IV. Dark and white		p. 272
a) dark with pale base to inner vane		p. 272
b) dark on the vanes in a single distal zone (one or both vanes); white edge present or not		p. 272
V. Grey or brown, inner vane with pale markings (border, spots, etc.)		p. 274
VI. White with one or more grey, brown, or black marks		p. 274
VII. Other patterns or colours		p. 275

KEY TO THE RECTRICES OF WADERS

• List of colour criteria used in the key

I. Uniform pale (white, pale grey)		p. 275
a) entirely pale grey to white		p. 275
b) white with pale grey markings		p. 275
II. Uniform dark		p. 275
a) tip darker with white border/white tip		p. 275
b) uniform brown-grey without pale tip		p. 276
III. Black and white		p. 276
a) a single black area towards the tip		p. 276
b) black distal mark with white tip		p. 277
c) black mark(s) on inner vane		p. 277
d) black mark(s) on outer vane		p. 277
e) black tip + other black marks (a few bars or spots)		p. 277
f) several black bars; complete or not		p. 278
IV. Regular barring (black, brown, grey / white, pale grey, buff)		p. 278
a) dark bars on a paler background (area of dark ≤ area of pale)		p. 278
b) pale bars on a darker background (area of dark ≥ area of pale)		p. 279
V. Pale wedges on dark background		p. 279
VI. Regular bars reduced to marks on a white background		p. 279
VII. Grey to brown, with contrasting paler or darker marks		p. 279
a) regular marks	 	p. 279
b) irregular marks		p. 280
VIII. Grey to brown; with diffuse paler or darker areas		p. 281

KEY TO THE PRIMARIES OF WADERS

Note: The Ps of the lapwings and Woodcock have an atypical shape (but size and colour are easily recognisable).

The outermost primary is excluded from the descriptions because of its small size and the great difficulty of its identification. 'Outer P' therefore refers to the first 'large' primary from the outside (P10), and not to this small P11.

Observed colour of primary:

I. UNIFORM PALE (WHITE, PALE GREY)

Ia) ENTIRELY WHITE

Pied Avocet T = 10-18): 2 (3) inner Ps (T = 10-13). Rachis white.

Ib) MARKED WITH SILVERY-GREY

See terns p. 236 (T = 5,5-28).

Ic) ENTIRELY DULL GREY

(eventually a darker tip)

See gulls p. 227 (T > 11).

II. PALE WEDGE(S) (white, buff, rufous, etc.) on a dark background (grey, brown, black)

IIa) A SINGLE TRIANGULAR WEDGE,
ON INNER VANE

Common Sandpiper (T = 5-9.5): Ps black-brown. A white oval (P9-P8) quickly becomes an asymmetrical triangular white notch on the inner vane, extending more towards the base and slightly touching the rachis towards the median Ps. On inner Ps, white rarely passes on to the outer vane (small spot). Brown rachis.

IIb) SEVERAL WEDGES,
AT LEAST ON INNER VANE

Eurasian Curlew (T = 13-24): Ps with brown-grey background; very pale wedges sometimes linked together at feather edge (comb shape). Wedges sometimes stained with dark or reduced to small spots. Pale edge towards the tip of the inner Ps. Rachis same colour as the vanes, sometimes a little lighter; pale on the outer P.

Whimbrel (T = 8.5-19.5): Ps with brown-grey background; very pale wedges sometimes reduced to ovals. Pale edge towards tip of inner Ps. Rachis same colour as the vanes, sometimes a little paler; pale on the outer P.

Eurasian Woodcock (T = 10.5-15.5): Ps with brown background; numerous triangular wedges (often more than 10 per vane on inner P, more than 15 on other Ps). Rufous wedges on the outer vane; buff on the inner vane; paler on the outer Ps. Tip with pale border. Dark rachis.

Spotted Redshank (T = 7-13.5): Ps brown-grey; very pale wedges on the inner vane, with variable margin (notches, comb, and pseudo-bars); sometimes speckled or vermiculated. Pale edge at tip of the median and inner Ps. Rachis dark; pale on the outer P or Ps.

III. UNIFORM DARK: BLACK, GREY OR DARK BROWN (SOMETIMES A SMALL PALE ZONE)

IIIa) ENTIRELY DARK

IIIa-1 dark rachis

Northern Lapwing (T = 13-20): Ps entirely black; except outer Ps, marked with white (see IIIb-2).

Eurasian Oystercatcher (T = 11-20.5): outer Ps sometimes entirely black (T = 18-20.5).

Black-winged Stilt (T = 10.5-20): Ps black-brown. Pale edge towards tip in young.

Common Greenshank (T = 6-15.5): outer and median Ps dark brown-grey (T = 7-15.5). Inner vane paler towards the edge, sometimes with darker speckles. Outer P has a pale rachis. Rachis of the second outermost P sometimes partly pale on the distal half.

Green Sandpiper (T = 7-13.5): Ps brown to black. Sometimes a pale edge to the median and inner Ps.

Eurasian Dotterel (SS) (T = 5.5-13): Ps dark brown. Inner vane sometimes a little paler towards the base and edge. Often a vague reddish-brown edge on the outer vane of the median and inner Ps.

Wood Sandpiper (T = 5.5-11.5): Ps black-brown. Pale edge at tip of median and inner Ps.

Great Snipe (T = 6-11.5): Ps brown. Light edge (white to buff) at their tip (except outer Ps). Outer vane of the outer P sometimes pale. Rachis quite dark.

Common Snipe (T = 6-11): PS dark brown. Pale tip to the end of the inner Ps. Outer vane of the outer P is partially pale.

Jack Snipe (T = 6-9): Ps black-brown. Tip with white edge except for one or more outer Ps. Outer vane of outer P paler.

Common Sandpiper (T = 5-9.5): a few outer Ps (often 1 or 2) black-brown (T = 8-9.5). Rachis brown; paler on distal third.

Little Ringed Plover (T = 5-9.5): Ps brown-grey. Base of the inner vane a little paler. Thin pale edge on the median and inner Ps. Outer P with pale rachis.

Temminck's Stint (SS) (T = 4.5-8.5?): 3-4 outer Ps dark brown (T = 7.5-8.5?). Outer P with pale rachis.

ollared Pratincole

R6 and R4 spots
in the white

R1

R4

R6

S7

S3

P1

P5

P9

Black-winged Stilt

Po

Pm

So

no, or very
little, white
at base

Pied Avocet

Po

Pm

R5

Sm

the white at base
increases towards
the Pi

SC (in situ)

crest

Northern Lapwing

Po

pale and
narrowed

Pm

Pi

scap

base of
iv white

So

Si

R6

Rm

Rm

R1

Rm

R6

Si

So

white
base

Po

Pm

Pi

Spur-winged Lapwing

Collared Pratincole (T = 7-18): Ps black. Base and edge of the inner vane sometimes dark grey. Black rachis (except outer P, white).

Black-winged Pratincole (T = 7-19?): as Collared Pratincole

Black Tern* (T = 8-20?): slate-grey.

Auks* (T = 5.5-16): Ps brown to black, base paler. Ps more rigid, more curved from above and more pointed than those of the waders. See p. 243 .

IIIa-2 entirely pale rachis

Note: In many species, one or two outer Ps have a pale rachis.

Eurasian Curlew (T = 13-24): outer P brown-grey (T = 21-24), with little colour; usually with an irregular pale edge on the inner vane. Diagnostic size T > 20 cm.

Common Greenshank (T = 6-15.5): outer P dark brown-grey (T = 9-15.5). Inner vane paler near the edge.

Eurasian Dotterel (SS) (T = 5.5-13): outer P dark brown (T = 10-13).

Wood Sandpiper (T = 5.5-11.5): outer P black-brown (T = 8.5-11.5).

Kentish Plover (T = 5-10): outer P brown-grey to black-brown (T = 9-10). Inner vane paler near the edge.

Little Ringed Plover (T = 5-9.5): outer P brown-grey (T = 8.5-9.5). Rachis white; often with a small brown zone towards the base.

Red Knot (T = 6.5-13.5): outer Ps brown-grey (T = 11.5-13.5). Border of inner vane paler.

Purple Sandpiper (SS) (T = 5-11): outer and median Ps black-brown (T = 6-11). The base and edge of the inner vane are paler. Fine white edge on median Ps. Inner Ps have border of basal part of inner vane white and a white edge.

Dunlin (T = 4.5-9.5): outer Ps black-brown (T = 8-9.5). Base and border of inner vane paler.

Little Stint (T = 4.5-8.5): outer P dark brown-grey; rachis sometimes partly dark (T = 7.5-8.5).

Temminck's Stint (SS) (T = 4.5-8.5?): outer P dark brown (T = 8-8.5?).

Collared Pratincole (T = 7-19): outer P black (T = 14-18).

Black-winged Pratincole (T = 7-19?): outer P black (T = 14-19?).

Black Tern* (T = 15-20?): outer Ps dark, silvery-grey.

IIIa-3 rachis only partly pale (a few cm)

Eurasian Oystercatcher (T = 11-20.5): sometimes 1 or 2 outer Ps black (T = 18-20.5). Rachis pale on its distal part. Sometimes a little white at the base of the inner vane and/or a vague pale halo on the inner vane.

Ruff (T = 7-14.5): Ps brown to black-brown. Base of the inner vane paler. Possible pale edge towards the tip of the inner vane of inner Ps. Rachis paler than vanes, brown hue in the basal third (sometimes completely pale on the inner 2 or 3 Ps).

Curlew Sandpiper (T = 5-11): 2-4 outer Ps brown-grey (T = 8-11); paler towards the base and edge of the inner vane (very faded IVa pattern). Rachis pale but darker on the basal third or half.

Kentish Plover (T = 5-10): 4-5 outer Ps brown-grey to black-brown (T = 7.5-10). Inner vane paler towards the edge. Rachis pale on the distal half or third.

Little Ringed Plover (T = 5-9.5): Ps brown-grey. Fine pale edge on the median and inner Ps. Rachis dark; sometimes with a lighter zone in the distal half.

Dunlin (T = 4.5-9.5): 3-5 outer Ps black-brown (T = 6.5-9.5). Tip darker. The base and edge of the inner vane are paler. Rachis dark; pale on the distal third to two thirds.

Little Stint (T = 4.5-8.5): 3-4 outer Ps dark brown-grey (T = 7-8.5). Base and border of the inner vane a little paler. Rachis pale, with a brown zone towards the basal third.

IIIb) SMALL PALE MARK
IIIb-1 mark at the base

Eurasian Oystercatcher (T = 11-20.5): sometimes 1 to 3 outer Ps (T = 18-20.5). Rachis pale in its distal part.

Spur-winged Lapwing (T = 13-20?): Ps black with a little white at base. Rachis the same colour as the vanes.

Northern Lapwing (T = 13-20): Ps sometimes have a little white at their base, the white continuing up the rachis a little. Rachis same the colour as the vanes.

Pied Avocet (T = 10-18): outer Ps (often from P4 onwards) black-brown (T = 12-18). Border of inner vane and tip sometimes paler. The dark/white limit is more or less well-defined. Proportion of white at the base very variable depending on the individual (from a few mm to 5 cm). Rachis dark in the dark area.

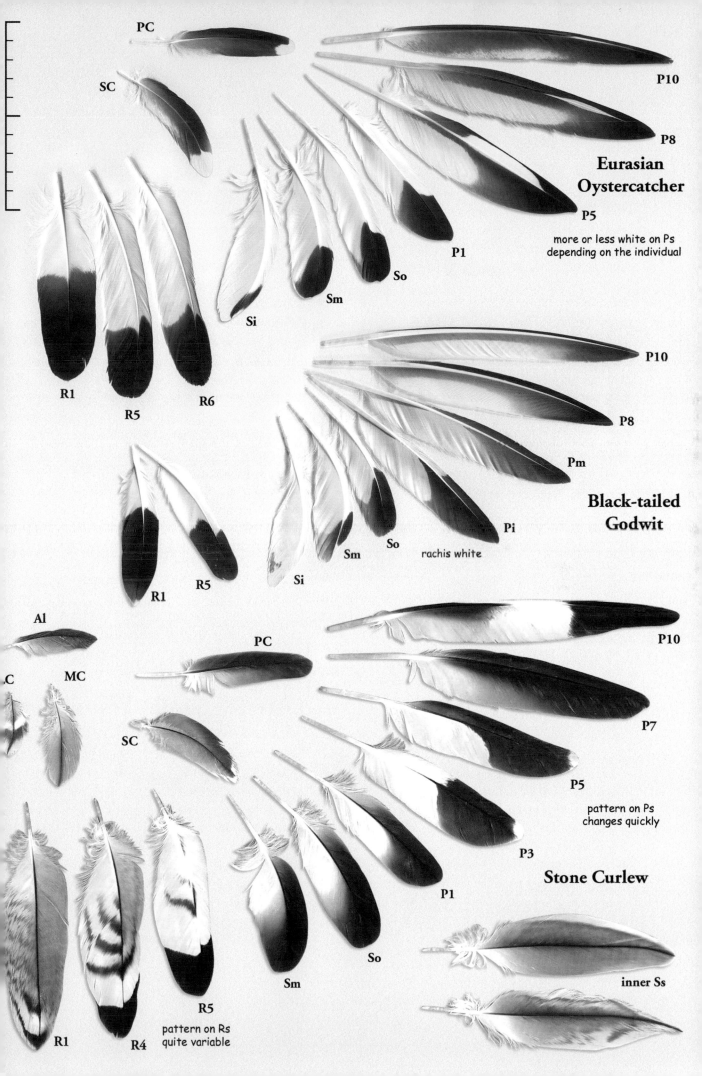

PC

SC

P10

P8

Eurasian Oystercatcher

P5

more or less white on Ps
depending on the individual

Si

Sm

So

P1

R1

R5

R6

P10

P8

Pm

Black-tailed Godwit

R1

R5

Si

Sm

So

Pi

rachis white

Al

C

MC

PC

P10

SC

P7

P5

pattern on Ps
changes quickly

P3

Stone Curlew

P1

So

Sm

R1

R4

R5

pattern on Rs
quite variable

inner Ss

IIIb-2 mark towards the middle or at tip

In the case of a short pale edge, refer to paragraph IIIa (completely dark primary).

Northern Lapwing (T = 13-20): 1-4 outer Ps (T = 16-20) with an ill-defined subterminal white spot (white to greyish or buff), sometimes also with a small white base. Tip narrower. Rachis the same colour as the vanes.

Stone Curlew (T = 10-20): 2 (3) outer Ps (T = 16.5-20) white (sometimes grey) mark touching the edge of the inner vane, located near the centre; very asymmetrical. Rachis the same colour as the vanes, but darker than them in the dark zones.

Common Sandpiper (T = 5-9.5): outer and median Ps (T = 6-9·5) black-brown with a triangular or oval white mark on the inner vane, touching the edge of the vane, located near the centre, with a more or less regular margin. On the inner Ps the white sometimes extends onto the outer vane (small spot). Rachis brown.

IV. DARK AND WHITE

IVa DARK WITH WHITE OR PALE AT BASE ON INNER VANE, LIMIT WITH A SMOOTH EDGE

If it is narrow and the dark mark is elongated, parallel to the edges, and with a silvery reflection: see terns p. 236 .

If the area around the pale zone is vermiculated or irregular, see V.

IVa-1 rachis entirely dark

Spur-winged Lapwing (T = 13-20?): Ps black. Base of the outer vane sometimes also white. Rachis the same colour as the vanes.

Stone Curlew (T = 10-20): median Ps (T = 13.5-19.5) and innermost Ps (T = 10-13), ill-defined limit; the pale zone white to grey. Rachis same colour as the vanes, but darker in the dark areas.

Common Redshank (T = 7.5-13.5): outer and median Ps black-brown (T = 9-13.5).

Auks* (T = 6-16): see especially Common Guillemot*.

IVa-2 rachis entirely pale

Black-tailed Godwit (T = 8.5-18): four outer Ps (T = 13-18). Rachis white, black at tip.

Ruff (T = 7-14.5): outer P (T = 11-14.5), sometimes the others also, brown to black-brown. Ill-defined limit. Paler edge at tip.

Common Redshank (T = 7.5-13·5): outer Ps black-brown (T = 11·5-13·5). White base to inner vane.

Marsh Sandpiper (SS) (T = 6.5-13): outer P black-brown (T = 10-13). Base of inner vane white, sometimes speckled with dark.

Ruddy Turnstone (T = 6.5-13.5): outer and median Ps black-brown (T = 7.5-13.5). Limit quite ill-defined. Rachis white, second and third outer Ps sometimes with a small area of brown towards the base.

Red Knot (T = 6.5-13.5): outer Ps brown-grey with off-white zone (T = 10-13.5). Fine white edge at the tip. Rachis pale but sometimes with brown towards the base.

Purple Sandpiper (SS) (T = 5-11): inner Ps black-brown (T = 5-7.5). Small white border on the inner vane. White edge.

Red-necked Phalarope (SS) (T = 4-10): outer and median Ps brown-grey. Limit ill-defined.

Grey Phalarope (SS) (T = 5-10.5): Ps black-brown. Limit ill-defined. Rachis white; sometimes a small brown area towards the base on the second and third outer Ps.

Sanderling (T = 4.5-10.5): outer P (T = 8-10.5); more rarely the other outer and the median Ps black-brown (T = 5.5-10.5). Limit ill-defined. Median Ps with white edge on base of outer vane.

IVa-3 rachis pale for a few centimetres of its distal (or median) part

Eurasian Oystercatcher (T = 11-20.5): outer Ps (T = 17-20.5).

Grey Plover (T = 8-16): 4-5 outer Ps (T = 12-16). Contrast of the black and white has well-defined limit, but less so at the tip of the white.

Golden Plover (T = 7.5-15.5): 5-7 outer and median Ps (T = 9.5-15.5) brown-grey, sometimes blacker. Pale to white area on the inner vane covering half or less of the feather length. Blurred boundary.

Ruff (T = 7-14.5): Ps brown to black-brown. Boundaries ill-defined. Pale edge at tip of inner and median Ps. Rachis pale on the distal third or half.

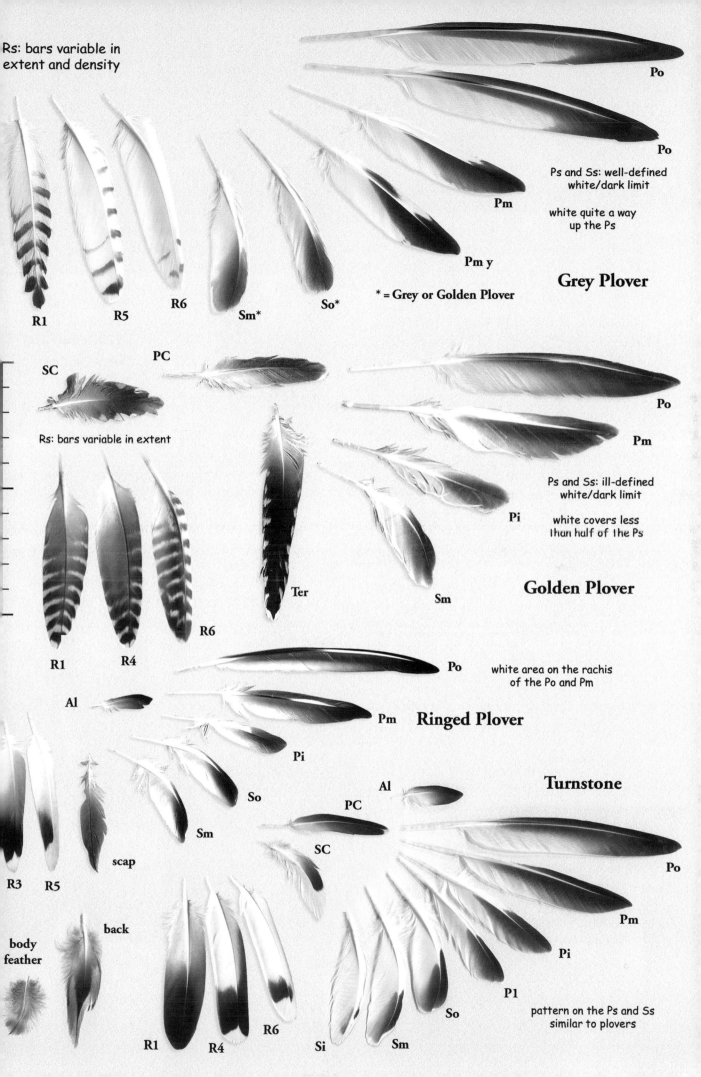

Rs: bars variable in
extent and density

R1 R5 R6 Sm* So*

Po
Po
Pm
Pm y

Ps and Ss: well-defined
white/dark limit

white quite a way
up the Ps

* = Grey or Golden Plover

Grey Plover

SC PC

Rs: bars variable in extent

Po
Pm

Ps and Ss: ill-defined
white/dark limit

white covers less
than half of the Ps

Pi

R1 R4 R6

Ter Sm

Golden Plover

Po

white area on the rachis
of the Po and Pm

Al

Pm **Ringed Plover**

Pi

So

Sm

Turnstone

Al

PC

SC

R3 R5

scap

Po

Pm

body
feather

back

Pi

P1

R1 R4 R6

Si So

Sm

pattern on the Ps and Ss
similar to plovers

Ringed Plover (T = 5-11): 4-5 outer Ps black-brown (T = 7.5-11). Rachis brown; white along a few centimetres in centre.

Kentish Plover (T = 5-10): 4-6 outer Ps brown-grey to black-brown (T = 7-10). White edge on the outer vane of at least one median P. Rachis white at the base and on a large distal part.

Curlew Sandpiper (T = 5-11): median Ps brown-grey (T = 6-10.5). White edge on part of the outer vane. White rachis with a brown area towards the basal third.

Sanderling (T = 4.5-10.5): outer and median Ps black-brown (T = 5.5-10.5). Ill-defined limit. Rachis white but with a brown area on the basal third of the outer 3-4 Ps.

Dunlin (T = 4.5-9.5): outer and median Ps black-brown (T = 5.5-9.5). Ill-defined limit. Fine white edge at the tip, wider on the outer vane. Inner Ps with much white towards the base (see IVc). Rachis white in the centre, with a brown area on the basal half.

Little Stint (T = 4.5-8.5): median and inner Ps dark brown-grey (T = 5.5-7.5). Ill-defined limit. Well-defined white edge on the outer vane (on at least the basal two thirds). Rachis pale, with a dark area on the basal third.

Temminck's Stint (SS) (T = 4.5-8.5?): median and inner Ps black-brown (T = 4.5-7.5?). Base paler to white. Ill-defined limit. White edge on about the basal two thirds of the outer vane. Possibly a pale edge at the tip or pale tip. Rachis brownish, paler in the middle.

IVb) SEVERAL ZONES OF WHITE AND BLACK

Black-tailed Godwit (T = 8.5-18): 2-3 median Ps (T = 10.5-15.5). Pattern rarer, with some dark near the rachis towards the base of the vanes. Rachis almost completely white. Usual pattern IVc.

Stone Curlew (T = 10-20): 2 (3) outer Ps (T =16.5-20) (see also IIIb2) and sometimes the inner Ps (T = 11.5-14.5). Outer Ps black with area of white in the centre, which may extend towards the base, especially on the inner vane; rachis black (except sometimes on the outer P); pattern with the most white

having a black tip and black at the base around the rachis. Sometimes inner Ps (P3 and P4) patterned like IVc with black at the base. Rachis usually black.

Grey Plover (T = 8-16): four or five median and inner Ps black and white (T = 8-12.5). White spot on the outer vane more or less extensive. Fine white edge at the tip. Rachis usually the colour of the outer vane, or white from the white area of the outer vane as far as the base.

Golden Plover (T = 7.5-15.5): inner and median Ps (T = 7.5-14) brown-grey to black and off-white. Ill-defined limit. The dark base of the outer vane is often reduced to spots or a border along the rachis and/or at the edge of the vane. Rachis dark at the tip, white at the level of the white area of the outer vane and often as far as the base.

Ringed Plover (T = 5-11): median and inner Ps black-brown (T = 5-10). White zone on the outer vane located at about one half or two thirds of its length. White descending to the base of the outer vane on the innermost Ps (1 to 3). Rachis usually the colour of the outer vane.

Kentish Plover (T = 5-10): inner Ps brown-grey to black-brown (T = 5-7). The white on the outer vane becomes closer to the base towards the inner Ps. The white extends a little up the rachis towards the tip. Rachis white, sometimes a small brown area towards the base.

Classical pattern or distal white spot sometimes reduced to a narrow white area along the rachis:

Eurasian Oystercatcher (T = 11-20.5): outer and median Ps (P3-P8) (T = 12.5-20). Rachis brown, white when one or both vanes are white next to it.

Grey Plover (T = 8-16): median Ps (T = 10-14.5), rarely inner Ps (T = 8-10.5). Rachis generally the colour of the outer vane.

IVc) THE DARK OF THE VANES IN A SINGLE DISTAL ZONE

The length of the black zone often differs between the two vanes. The white of the base of the outer vane is sometimes limited to a border.

If a silvery aspect is apparent, see terns, p. 236.

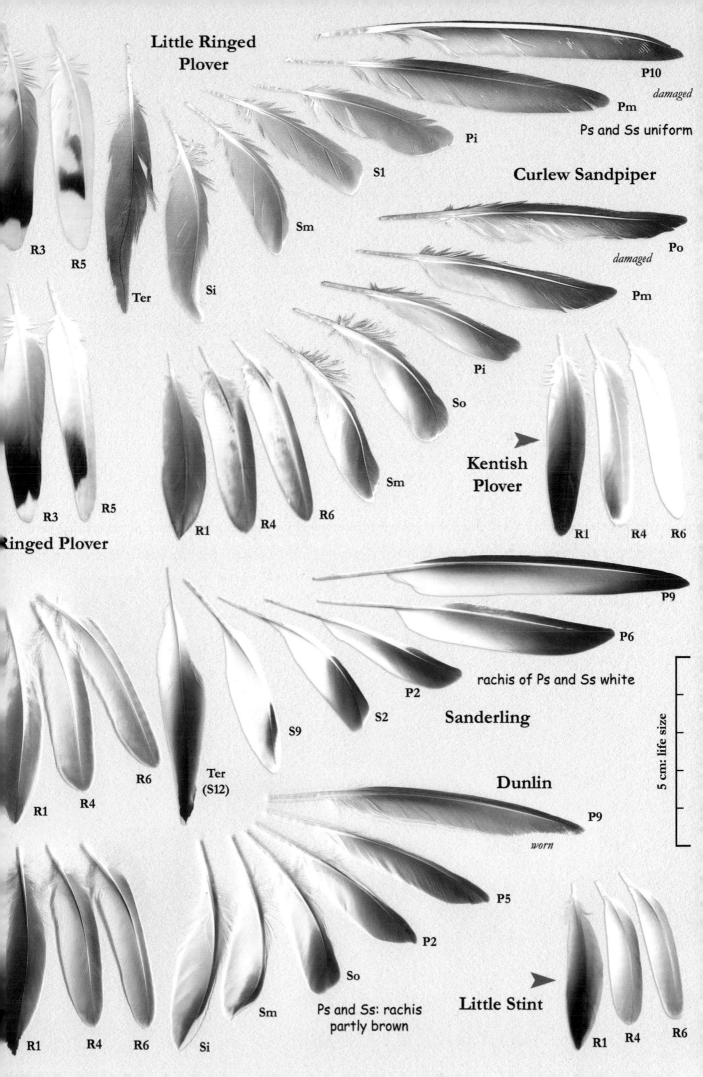

Little Ringed Plover

P10
damaged
Pm
Ps and Ss uniform

Pi

S1

Sm

R3

R5

Ter

Si

Curlew Sandpiper

Po
damaged

Pm

Pi

So

Sm

R3

R5

R1

R4

R6

Sm

Kentish Plover

R1

R4

R6

Ringed Plover

P9

P6

P2

rachis of Ps and Ss white

S2

S9

Sanderling

R1

R4

R6

Ter
(S12)

Dunlin

P9

worn

5 cm: life size

P5

P2

So

R1

R4

R6

Si

Sm

So

Ps and Ss: rachis
partly brown

Little Stint

R1

R4

R6

Eurasian Oystercatcher (T = 11-20.5): 1-3(4) inner Ps (T =11-15). Rachis white, dark in the black zone.

Stone Curlew (T = 10-20): inner Ps (P2-P5, not P1) (T = 10.5-16). Rachis generally black, possibly with white tip.

Pied Avocet (T = 10-18): inner Ps, P3 and P4 (T = 11-14). Often a pale edge on the inner vane. Very small mark on P3 (on one or both vanes). Rachis dark in the black zone.

Black-tailed Godwit (T = 8.5-18): median and inner Ps (T = 8.5-15.5). Rachis mainly white in the black zone, black at tip.

Grey Plover (T = 8-16): inner Ps (T = 8-10.5). Limit of black on the outer vane sometimes blurred (in the immature, for example), but often sharp or sawtooth-shaped. Fine white edge at tip. Black of the inner vane tip descends in a point along the rachis, covering up to half or more of the vane. Rachis usually white at the base and also partly so in the black zone.

Golden Plover (T = 7.5-15.5): inner Ps brown-grey to black (T = 7.5-10). Dark limit ill-defined. Fine white edge at the tip. Rachis partly white in the black zone.

Ruddy Turnstone (T = 6.5-13.5): median and inner Ps black towards the tip (T = 6.5-12). Black of the inner vane descends more than on outer, in a triangle along the rachis. Fine white edge at the tip. Rachis partly white in black zone.

Ringed Plover (T = 5-11): 1-3 inner Ps (T = 5-7). Dark often descends along the rachis on the inner vane, as far as the base. Rachis the same colour as the vanes.

Kentish Plover (T = 5-10): often the three inner Ps (T = 5-6.5) brown-grey to black-brown, larger on the inner vane. The white goes up along the white rachis a little towards the tip.

Curlew Sandpiper (T = 5-11): inner Ps brown-grey (T = 5-8). Dark outer vane edged white on the basal half or two thirds. Inner vane white on basal two thirds except near the spine, with a dark border. Rachis white with a brown zone towards the basal quarter.

Sanderling (T = 4.5-10.5): median and inner Ps black-brown (T = 4.5-9.5). Ill-defined limit. The dark descends lower on the inner vane. Sometimes a thin white edge at the tip (immature). White

extends a little along the rachis in the dark area. Rachis white.

Dunlin (T = 4.5-9.5): inner Ps sometimes black-brown (T = 4.5-6.5). A rather ill-defined limit. Dark descending to the base along the inner vane, covering at least half of the vane. Rachis white, brown at both ends.

Red-necked Phalarope (SS) (T = 4-10): inner Ps with dark brown-grey zone (T = 4-7). Rachis white on at least the basal half.

Grey Phalarope (SS) (T = 5-10.5): median and inner Ps (T = 5-9.5). The black-brown covers from two thirds (median) to half (inner) of the Ps. Fine white edge at tip. Rachis white.

V. BLACK, GREY OR BROWN, INNER VANE PALER WITH IRREGULAR MARGIN, SPECKLED OR VERMICULATED

Eurasian Curlew (T = 13-24): 1 to 4 outer Ps (T = 20-24) in the ssp. *orientalis* (ssp. *arquata* with wedges). Rachis white on the outer Ps (or the two outermost). On the others, rachis the same colour as the vanes, sometimes a little paler.

Whimbrel (T = 8.5-19.5): 1 to 4 outer Ps (T = 13-19.5), less colour than wedges. White rachis on the outer Ps (or the two outermost). On the others, spine the same colour as the vanes, sometimes a little paler.

Bar-tailed Godwit (T = 8.5-17.5): all the Ps. Pale edge to the median and inner Ps. Rachis paler than the vane (rachis white on the outer Ps).

Common Greenshank (T = 6-15.5): median and inner Ps dark brown-grey (T = 6-14.5). Inner vane paler, spotted or vermiculated. Possibly paler edge at the tip. Rachis dark.

Spotted Redshank (T = 7-13.5): Ps brown-grey. Largely pale inner vane, with dark vermiculations. Pale edge at the tip of the median and inner Ps. Rachis dark (light on the outer P or Ps).

Common Redshank (T = 7.5-13.5): outer and median Ps grey-brown. (T = 9-13.5). White base to inner vane. Irregular limit slightly vermiculated. Rachis dark (white on the outermost P).

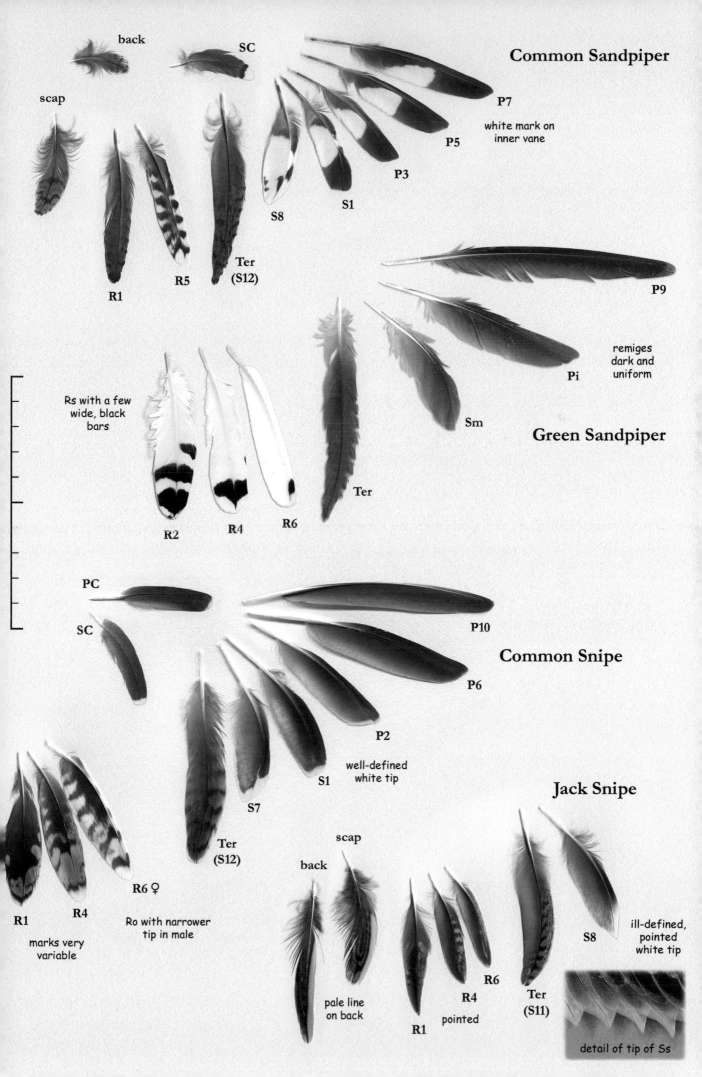

back

SC

Common Sandpiper

scap

P7

white mark on
inner vane

P5

P3

S1

S8

Ter
(S12)

R1

R5

P9

Rs with a few
wide, black
bars

remiges
dark and
uniform

Pi

Sm

Green Sandpiper

R2

R4

R6

Ter

PC

SC

P10

Common Snipe

P6

P2

well-defined
white tip

S1

S7

Jack Snipe

Ter
(S12)

R6 ♀

R1

R4

Ro with narrower
tip in male

marks very
variable

scap

back

pale line
on back

R1

R4

pointed

R6

Ter
(S11)

S8

ill-defined,
pointed
white tip

detail of tip of Ss

Wood Sandpiper (T = 5.5-11.5): inner Ps (T = 5.5-8.5) black-brown with a small, irregular pale area along the edge of the inner vane. Pale edge at the tip. Rachis the same colour as the vanes.

Marsh Sandpiper (SS) (T = 6.5-13): Ps dark brown-grey. Inner vane paler and speckled or vermiculated. Possibly a paler edge at the tip. Rachis dark (white on the outer Ps and partly paler in the distal third of the second outermost P).

Red Knot (T = 6.5-13.5): Ps brown-grey. Pale area more or less speckled or vermiculated. White edge. Outer vane of the median and inner Ps often with a wide white edge and/or white base. Rachis pale (sometimes brown at the base of outer Ps).

Common Snipe (T = 6-11): inner Ps, and sometimes median Ps, brown to black-brown (T = 6-10). Pale edge at the tip. Small pale zone on inner vane. Rachis dark.

VI. WHITE WITH ONE OR MORE BROWN-GREY MARKS

Common Redshank (T = 7.5-13.5): inner and median Ps (T = 7.5-12.5). Large dark mark around the rachis and on the outer vane, with blurred limit, extending along two thirds to three quarters of the length. Tip paler to white, vermiculated with dark. Often a thin, blurred, subterminal, parallel-sided bar at the tip. Rachis same colour as the vanes. The inner Ps may have a single large mark along the rachis, with no other dark mark.

KEY TO THE SECONDARIES OF WADERS

The inner secondary feathers (tertials) are not considered in this key (except where indicated). Their size is significantly larger than other secondaries, their coloration is close to that of the scapulars; they are elongated and pointed. The similarity of their coloration in several species means it is often difficult to distinguish between them.

Possible confusion: with ducks that have pale grey to black Ss (T = 7-13 generally). Often with a paler, or darker tip. See Anatidae p. 290 .
Observed colours of secondaries:

I. UNIFORM PALE (WHITE, PALE GREY)

Ia) ENTIRELY WHITE OR PALE GREY

Eurasian Oystercatcher (T = 9.5-12): median and inner Ss white. Rachis white.

Pied Avocet (T = 9-10.5): Ss white. Rachis white.

Common Redshank (T = 6.5-8): inner Ss sometimes all white, but usually with dark towards the base touching the rachis. Rachis the same colour as the vanes.

Ringed Plover (T = 4.5-7): inner Ss sometimes; in general the base is grey. Rachis white.

Kentish Plover (T = 4.5-6.5): inner Ss sometimes; in general a greyish zone in the distal third. Rachis white.

Curlew Sandpiper (T = 4.5-6.5): inner Ss pale brown-grey around the rachis. Rachis pale.

Purple Sandpiper (SS) (T = 4-6.5): inner Ss white or with some dark marks towards the tip.

Sanderling (T = 4.5-6): inner Ss white with or without a small dark mark at the tip, along the rachis. Rachis totally white.

Dunlin (T = 4.5-5.5): inner Ss white with or without a small dark mark at tip, along the rachis. Rachis white, generally with a small area of brown near the base.

Ib) MARKED WITH SILVERY-GREY, SOMETIMES VERY PALE

See terns, T = 4.5-14.5. See p. 236.

Ic) DULL GREY, WITH OR WITHOUT A WHITE TIP

See gulls, T = 10-24. See p. 227.

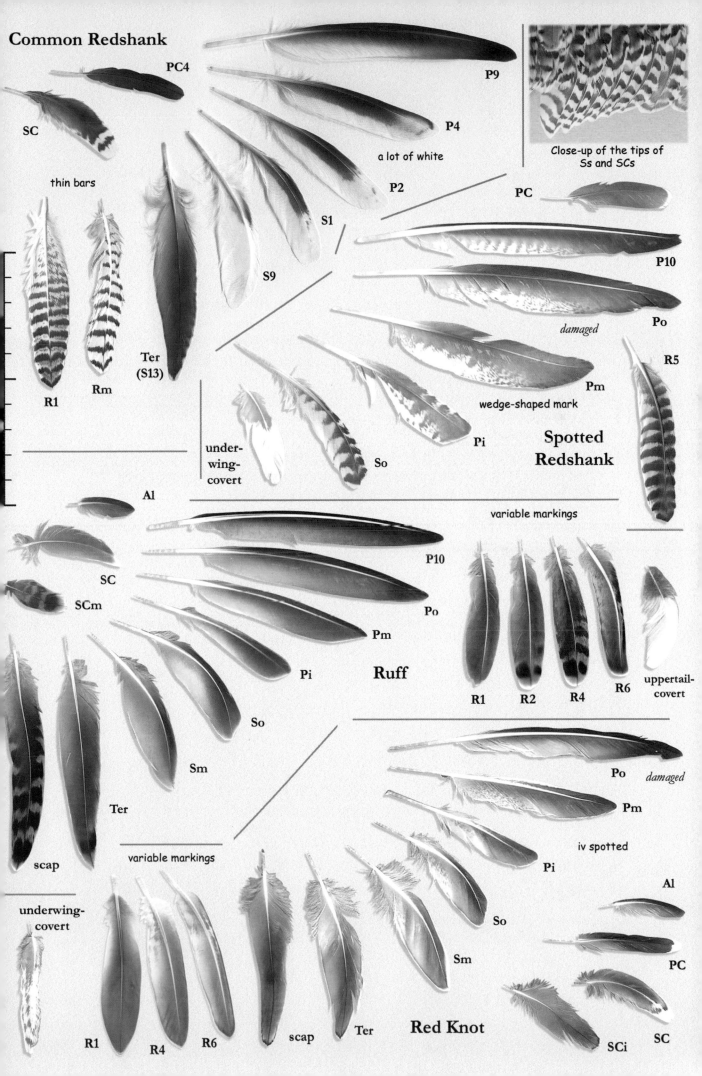

Common Redshank

SC

PC4

thin bars

R1

Rm

Ter
(S13)

S9

S1

P9

a lot of white

P4

P2

Close-up of the tips of
Ss and SCs

PC

P10

Po

damaged

Pm

wedge-shaped mark

Pi

So

R5

Spotted
Redshank

under-
wing-
covert

Al

variable markings

P10

Po

Pm

Pi

So

Ruff

R1 R2 R4 R6

uppertail-
covert

SC

SCm

Ter

scap

variable markings

Sm

Po damaged

Pm

iv spotted

Pi

So

Sm

Al

PC

underwing-
covert

R1 R4 R6

scap Ter

Red Knot

SCi

SC

II. WITH PALE WEDGE(S) (white, buff, rufous...) on a dark background (grey, brown, black)

Eurasian Curlew (T = 11-13.5): Ss, background brown-grey, wedges very pale (often four to seven) forming incomplete bars on the inner Ss. Tip with pale edge. Rachis same colour as the vanes, sometimes a little paler.

Whimbrel (T = 7-10.5): Ss, background brown-grey, wedges very pale (often three to six) forming incomplete bars on the inner Ss. Tip with pale edge. Wedges smaller than in Eurasian Curlew. Rachis same colour as the vanes, sometimes a little paler.

Woodcock (T = 10-12): Ss, background black-brown, numerous triangular wedges, rufous on the outer vane, buff on the inner vane. Wedges towards the tip may coalesce to form narrow bars (especially towards the inner Ss). Tip with pale edge. The background may be marked with pale, more so towards the inner Ss. Longer than the typical shape. Rachis dark.

Golden Plover (T = 7.5-9): inner Ss grey-brown; white towards the base and inner part. Sometimes slight golden wedges (better defined on the tertials). Rachis paler than the vanes.

Grey Plover (T = 6.5-9): inner Ss grey-brown; white towards the base and inner part. Sometimes inconspicuous buff wedges (better defined on tertials). Rachis same colour as the vanes.

Common Greenshank (T = 5.5-8.5): Ss brown-grey; more or less well-defined pale wedges and spotted. Pale edge especially at tip. Rachis same colour as the vanes.

Spotted Redshank (T = 6.5-8): Ss brown-grey, very pale wedges. The dark is sometimes reduced to triangles or bars. Rachis the same colour as the vanes.

Green Sandpiper (T = 5.5-8): inner Ss sometimes marked with short buff wedges towards the tip on a brown to black background. Rachis dark.

III. UNIFORM DARK (BLACK, BROWN, GREY)

IIIa) ENTIRELY DARK

Black-winged Stilt (T = 9-11): in adult all Ss black-brown. Rachis dark like the vanes.

Eurasian Dotterel (SS) (T = 5.5-7.5): Ss dark brown-grey; a little paler towards the base and on the inner vane.

Reddish-brown edge more or less sharp at the tip and on outer vane. Rachis the same colour as the vanes, paler towards the base.

Green Sandpiper (T = 5.5-8): Ss brown to black. Rachis dark.

Wood Sandpiper (T = 5-6.5): outer and median Ss black-brown. White edge at tip. Rachis the same colour as the vanes.

Little Ringed Plover (T = 4.5-6): Ss brown-grey. Base of inner vane a little paler. Pale edge at tip, wider and longer at tip of outer vanes of inner Ss. Rachis the same colour as the vanes.

Collared Pratincole (T = 5.5-8): outer Ss black-grey. Blurred tip paler. Rachis dark.

Black-winged Pratincole (T = 5.5-8.5?): Ss black-grey. Rachis dark.

Black Tern* (T = 7-9?): slate-grey.

Auks* (T = 4.5-10): Ss brown to black-brown. See especially Atlantic Puffin* and Black Guillemot* (p. 244).

IIIb) DARK WITH PALE TIP (white or grey)

Black-winged Stilt (T = 9-11): Ss black-brown, pale edge at tip in young. Rachis the same colour as the vanes.

Common Snipe (T = 6-7.5): Ss brown grey. Small speckled white area at the base of the inner vane. Rachis dark.

Jack Snipe (T = 6-7.5): Ss brown-grey. Blurred white tip. Pointed tip. Rachis dark.

Great Snipe (T = 6-7.5): Ss brown-grey, inner vane paler. White to buff edge at tip. Tip rounded to triangular. Rachis dark.

Little Ringed Plover (T = 4.5-6): inner Ss brown-grey with an irregular, small white tip descending on the border of the outer vane. Rachis same colour as the vanes.

Collared Pratincole (T = 5.5-8): Ss black-grey. Blurred paler tip on the outer Ss. Wide white to buff edge at tip of median and inner Ss. Rachis dark.

Auks* (T = 4.5-10): Ss black-brown. See especially Razorbill*, guillemots* and Little Auk* (p. 244).

See also gulls* (T>7.5) p. 227 .

See also ducks* Ss without a shiny speculum (T > 6), see p. 290.

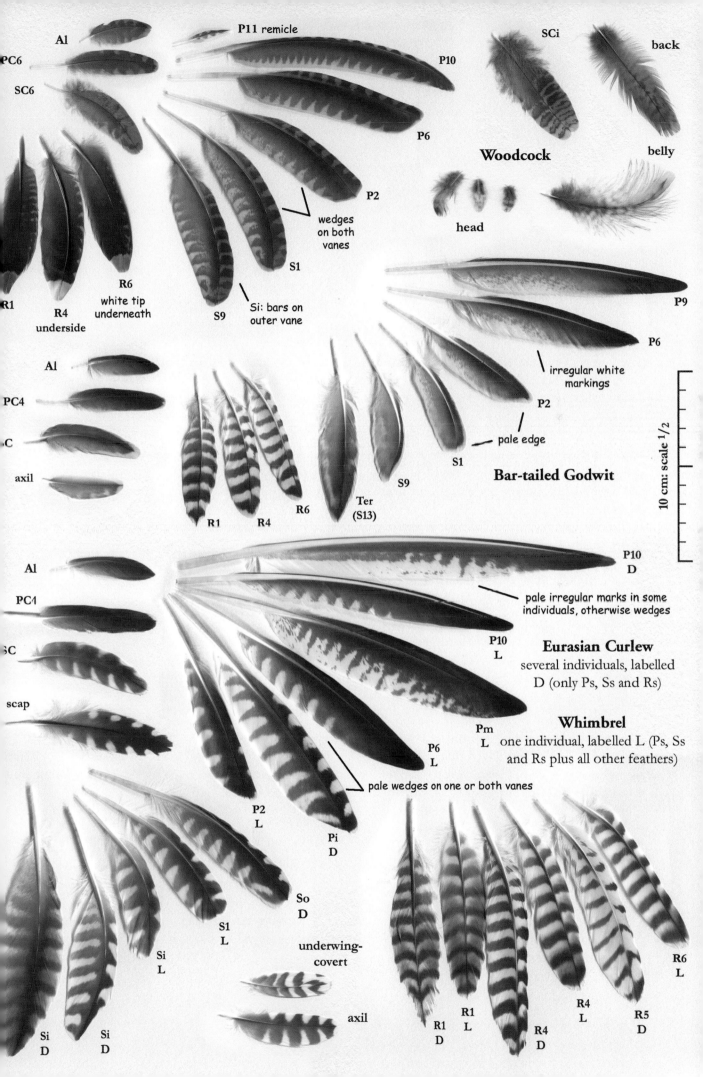

Al

PC6

SC6

P11 remicle

P10

P6

SCi

back

belly

Woodcock

head

R1

R4
underside

R6

white tip
underneath

S9

wedges
on both
vanes

P2

S1

Si: bars on
outer vane

Al

PC4

C

axil

R1

R4

R6

Ter
(S13)

S9

S1

pale edge

P9

P6

irregular white
markings

P2

Bar-tailed Godwit

10 cm: scale $^1/_2$

Al

PC4

SC

scap

P2
L

Pi
D

P6
L

pale wedges on one or both vanes

P10
D

pale irregular marks in some
individuals, otherwise wedges

P10
L

Eurasian Curlew
several individuals, labelled
D (only Ps, Ss and Rs)

Pm
L

Whimbrel
one individual, labelled L (Ps, Ss
and Rs plus all other feathers)

Si
L

S1
L

So
D

underwing-
covert

axil

Si
D

Si
D

R1
D

R1
L

R4
D

R4
L

R5
D

R6
L

IV. DARK AND WHITE

IVa) DARK WITH WHITE BASE TO INNER VANE

Northern Lapwing (T = 9-13.5): Ss black; no or little white at the base of the outer vane. Boundary more or less blurred and speckled. Sometimes a short pale or white edge at the end of the inner and median Ss (juvenile). Rachis the same colour as the vanes.

Spur-winged Lapwing (T = 9-13.5?): outer Ss white at the base of the outer vane. Rachis the same colour as the vanes.

Stone Curlew (T = 8.5-12): Ss black-brown with some white; sometimes also white at the base of the outer vane. Blurred boundary. Rachis the same colour as the vanes but darker than them in the dark zones.

Ruff (T = 6.5-8.5): Ss brown to black-brown. White area with blurred boundary. White edge especially at the tip. Rachis paler than vanes.

Red Knot (T = 6-7.5): Ss brown-grey. White-edged. Pale area with blurred boundary, partly speckled with brown. Inner Ss more uniform pale brown, with a wider white edge on the tip of the outer vane and speckling sometimes weak or absent. Rachis white.

Curlew Sandpiper (T = 4.5-6.5): Ss brown-grey, inner Ss paler. Boundary blurred. White edge at the tip, narrower on the inner vane (sometimes absent), wider on the outer vane of the inner Ss (becomes a white tip on worn Ss). Rachis pale.

Purple Sandpiper (SS) (T = 4-6.5): Outer Ss black-brown. White at base of the inner vane with an irregular outline. White edge. Rachis white.

Little Stint (T = 4-5): outer Ss black-brown. Blurred boundary. Rachis white in the middle, brown at both ends (rarely without brown towards the base).

Temminck's Stint (SS) (T = 4-5?): outer and median Ss black-brown (white does not reach the rachis). Blurred boundary. Fine white edge on the outer vane, and on both vanes at the tip (wider on the outer vane). Towards the inner Ss, the dark becomes paler from the base and becomes more diffuse. S10 is brown-grey with a white tip. Rachis brownish, paler in the middle.

IVb) DARK ON THE VANE IN A SINGLE DISTAL ZONE (ONE OR BOTH VANES), WHITE EDGES PRESENT OR NOT

Particularly look to see if the rachis is white or dark in the dark zone.

For T > 9 cm, see also young gulls. Their Ss generally have less contrast than those of waders of the same size. See p. 227.

Spur-winged Lapwing (T = 9-13.5?): outer and median Ss; two thirds black; on the outer Ss the white of the outer vane is less extensive than on the inner vane but becomes more extensive towards the median Ss. White of the outer vane tinged with buff towards the inner Ss replaces the black on most of the feather on the inner Ss. Rachis the same colour as the vanes, paler on the inner Ss.

Eurasian Oystercatcher (T = 9.5-12): Ss with a black mark, and a thin white edge (quickly worn) on the outer Ss; wider on the inner Ss. The black mark decreases in size towards the inner Ss. Rachis white, black in the black zone. On the inner Ss, the white can extend a little up the rachis in the black zone.

Black-tailed Godwit (T = 7.5-10): Ss sometimes with a fine white edge. Inner Ss with totally white outer vane and little black on the inner vane. The white extends up the rachis into the black zone (the black zone appears to be 'split').

Golden Plover (T = 7.5-9): Ss with brown-grey to black mark. Blurred boundary. Fine white edge becoming buff on the inner Ss. The dark expands on the inner Ss (almost entirely grey-brown). Rachis partly pale in the dark zone.

Grey Plover (T = 6.5-9): a black mark on the outer Ss becomes paler, wider, and more blurred towards the inner Ss. On the outer Ss the dark descends lower on the inner vane, equalises on the median Ss and then descends even lower on the outer vane of the inner Ss. White edge at the tip. White rachis at the base; usually brown in the dark area (but sometimes white on a small part).

Ruddy Turnstone (T = 5.5-7.5): Ss with a black mark, decreasing towards the inner Ss. Black at the same level or lower on the outer vane. White extends on the rachis into the black zone. On the

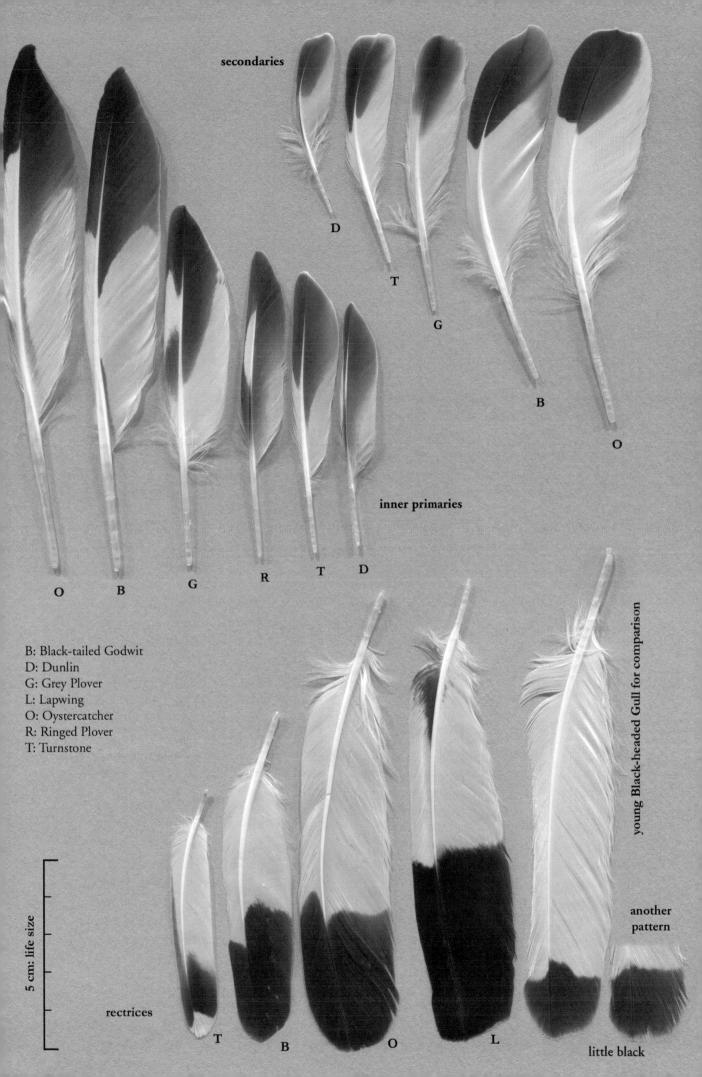

secondaries

D

T

G

B

O

inner primaries

O   B   G   R   T   D

B: Black-tailed Godwit
D: Dunlin
G: Grey Plover
L: Lapwing
O: Oystercatcher
R: Ringed Plover
T: Turnstone

young Black-headed Gull for comparison

5 cm: life size

rectrices

T   B   O   L

another
pattern

little black

inner Ss, the black becomes grey and decreases in extent. On the outer vane, it decreases to become non-existent or shows only as a small dark spot. Rachis white.

Ringed Plover (T = 4.5-7): Ss with black mark, decreasing in size towards the inner Ss with blurred boundary on the inner vane. Fine white edge at the tip. Rachis white at the base, dark at tip. White extends along the rachis and at the tip next to the rachis on the outer vane.

Kentish Plover (T = 4.5-6.5): Ss with a brown-grey to black-brown mark, with ill-defined border. White edge at the tip, wider and longer on the outer vane. Greyish and reduced mark on the inner Ss. Rachis white, darker on the inner Ss.

Curlew Sandpiper (T = 4.-6.5): Ss with a brown-grey mark. Ill-defined boundary. The dark can reach the base by descending along the rachis. Fine white edge, wider on the outer vane of the inner Ss. Rachis pale. Inner Ss almost entirely brown-grey, paler towards the base, sometimes almost white.

Purple Sandpiper (SS) (T = 4-6.5): Ss black-brown. White area with irregular margin. White edge, wider towards the tip. White at base covers at least half of the surface of the inner vane. Rachis white.

Sanderling (T = 4.5-6): Ss with a black-brown mark. A rather ill-defined boundary. The dark may descend a little along the edge of the inner vane. Mark smaller and paler on inner Ss. Rachis white at the base; white extends a little along the rachis in the dark area.

Dunlin (T = 4.5-5.5): Ss with black-brown mark, decreasing in size towards the inner Ss. White edge; wider at the tip of the outer vane of the inner Ss. Rachis white in its centre, brown at both ends.

Little Stint (T = 4-5): median and inner Ss with brown to black-brown mark. Blurred border, very ill-defined on inner Ss. White edge, wider at the tip on the outer vane. Rachis white in the middle, brown at both ends (rarely without brown towards the base).

Red-necked Phalarope (SS) (T = 4-6?): Ss with brown-grey mark, smaller towards the inner Ss. White edge at tip. Rachis pale.

Grey Phalarope (SS) (T = 4.5-6): Ss with black-brown mark descending on the inner vane, smaller towards the inner Ss. White edge at the tip, wider towards the inner Ss. Rachis white.

V. GREY OR BROWN, INNER VANE WITH PALE MARKINGS (BORDER, SPOTS, ETC).

Bar-tailed Godwit (T = 7-9.5): Ss brown; brown covering a greater area. Pale edge. Wider and paler towards the inner Ss. Rachis pale, browner on the inner Ss.

Ruff (T = 6.5-8.5): Ss brown. Rachis paler to white. White edge, wider on inner Ss. Pale area vermiculated with dark only towards the base or all over.

Common Greenshank (T = 5.5-8.5): Ss dark brown-grey. More or less regular pale markings. Pale area often vermiculated or speckled. Pale edge, especially at the tip. Rachis same colour as the vanes.

Marsh Sandpiper (SS) (T = 5-7.5): Ss brown-grey, inner vane with paler border. Pale area vermiculated with dark. Pale edge at the tip. Rachis the same colour as the vanes.

Wood Sandpiper (T = 5-6.5): Ss brown to black-brown. Border of the inner vane marked with more or less regular pale spots. Pale edge to the tip. Rachis the same colour as the vanes.

Great Snipe (T = 6-7.5): Ss brown-grey. Inner vane sometimes slightly speckled with pale. White to buff edge at the tip. Rachis the same colour as the vanes.

Common Snipe (T = 6-7.5): Ss dark brown-grey. Large white tip. Pale marks more or less extensive, sometimes in very ill-defined bars. Rachis the same colour as the vanes.

Jack Snipe (T = 6-7.5): Ss sometimes brown-grey with white tip and inner vane with pale marbling. Rachis the same colour as the vanes.

Red Knot (T = 6-7.5): Ss brown-grey. Tip fairly uniform; base vermiculated with grey. White edge, wider on the outer vane of the inner Ss. Rachis white.

VI. WHITE WITH ONE OR MORE GREY, BROWN OR BLACK MARKS

Common Redshank (T = 6.5-8): Ss with at least one brown-grey mark at the base around the rachis, extending over up to two thirds of the vanes. Main mark often accompanied by small, regular dark spots, sometimes with irregular spots. Rachis the same colour as the vanes.

Spotted Redshank (T = 6.5-8): Ss with fairly regular marks, in bars or triangles. Rachis the same colour as the vanes.

Common Sandpiper (T = 4.5-6): Ss have a mixture of black-brown or brown-grey and white. On all Ss there is dark at the base (more extensive on the outer vane) and a white edge at the tip, more or less wide. Rachis the same colour as the vanes. For the outer and median Ss: dark area towards the tip quite wide. For inner Ss: dark area of the tip with more white (only a few small dark spots remain). For the inner S (S10 or S9): dark area more or less uniform; tip white; broad white rounded mark in the centre of the inner vane; white in the middle and brown at the ends. Tertials brown with black markings.

Purple Sandpiper (SS) (T = 4-6.5): inner Ss white with only a few dark spots towards the tip. Rachis white.

VII. OTHER PATTERNS OR COLOURS

Spur-winged Lapwing (T = 9-13.5?): inner Ss buff with white background; darker distal mark on the inner vane appearing towards the median Ss (see IVb). Rachis paler than the vanes.

Woodcock (T = 10-12): inner Ss with rufous-buff bars on a dark, non-uniform background (with pale markings). Rachis dark.

Spotted Redshank (T = 6.5-8): Ss with brown-grey bars on a very pale background (rarer variant of pattern II). Rachis the same colour as the vanes.

KEY TO THE RECTRICES OF WADERS

The rectrices have two main shapes:
1) **rectangular with rounded or square tip;**

2) **more oval; rounded or pointed (small size).**
This latter form is generally observed in sandpipers and smaller snipes.

The pratincoles have tapered outer rectrices; the shape resembles that of swallows.

The colour patterns are described regardless of the shape of the tail feather.

The number of bars indicated, such as x-y, can be read as 'x to y bars'. That is, depending on the exact position of the feather, the observed vane and the individual bird, the number of bars most commonly observed is between these two values (inclusive).

An additional number in brackets indicates that this number is more rarely observed.

Observed colours of rectrices:

I. UNIFORM PALE (WHITE, PALE GREY)

Ia) ENTIRELY PALE GREY TO WHITE

Pied Avocet (T = 9.5-10.5): Rs white. Rachis white.

Black-winged Stilt (T = 8-9.5): Rs pale grey (rachis paler); outer Rs whiter, especially on the inner vane (rachis white).

Common Greenshank (T = 8-9): Outer Rs sometimes but usually with small dark marks at the base. Rachis the same colour as the vanes.

Green Sandpiper (T = 6-7.5): outer Rs sometimes, but usually with at least one black mark on the outer vane. Rachis white.

Ringed Plover (T = 5.5-8): outer Rs. Sometimes a diffuse grey-buff area or a black mark some two thirds along the inner vane. Rachis white; sometimes dark near the base.

Kentish Plover (T = 5.5-7): outer Rs white or with grey tint. Rachis the same colour as the vanes.

Sanderling (T = 5-6): outer Rs. Rachis white.

Temminck's Stint (SS) (T = 4-5?): outer Rs grey-white, usually with dark mark along the rachis. Rachis brown, pale at the tip.

See terns (T = 4-20), p. 236.

Ib) WHITE WITH PALE-GREY MARKINGS

Northern lapwing (T = 10-13): outer Rs sometimes. Rachis pale.

Sanderling (T = 5-6): outer and median Rs white; pale grey border towards tip and on outer vane. White edge. Rachis white.

Little Stint (T = 4-5): outer Rs. White edge. Rachis white with a dark zone towards the base.

See terns (T = 4-20). See p. 236.

II. UNIFORM DARK

IIa) TIP DARKER WITH WHITE BORDER/ WHITE TIP

Bar-tailed Godwit (T = 7-10): central Rs brown. Inner vane sometimes with small pale zone (patterned in male?). Rachis same colour as or darker than the vanes.

Eurasian Dotterel (SS) (T = 7-8.5): Rs dark brown-grey, darker at the tip. Large white to buff tip. Rachis same colour as the vanes.

IIb) UNIFORM BROWN-GREY WITHOUT PALE TIP

Black-winged Stilt (T = 8-9.5): central Rs of young brown-grey. Rachis paler.

Ruff (T = 5.5-8.5): Rs brown to dark brown. Base of the inner vane can be paler. Pale edge, blurred at tip. Sometimes a darker mark at the tip. Rachis paler colour than the vanes.

Common Sandpiper (T = 5-7): central Rs; rarely brown without marks. Tip often darker. Rachis same colour as the vanes.

Ringed Plover (T = 5.5-8): central Rs brown-grey, becoming black towards the tip. Tip sometimes with buff edge. Rachis same colour as the vanes.

Little Ringed Plover (T = 5.5-7): central Rs brown-grey becoming black towards the tip. Tip sometimes with buff edge. Rachis brown.

Kentish Plover (T = 5.5-7): central Rs brown-grey, becoming black towards the tip. Tip sometimes with buff edge. Rachis same colour as the vanes.

Jack Snipe (T = 5-6.5): outer Rs brown; sometimes uniform but usually marked with pale on the outer vane. Rachis brown.

Curlew Sandpiper (T = 5-6): central Rs brown-grey. Pale edge, more or less regular on the inner vane. Rachis pale, tinged with brown towards the base.

Red Knot (T = 6-7.5): central Rs brown-grey. Pale edge, often underlined with dark towards the tip. Possible pale edge or paler zone towards the base of the inner vane. Rachis paler colour than the vanes, but brown towards the base.

Sanderling (T = 5-6): central Rs brown-grey, darker at the tip. Rachis paler. Pale edge. Often a pale area around the rachis.

Dunlin (T = 4-6): 1 or 2 pairs of central Rs brown-grey; darker at tip and on inner vane. Rufous-buff edge. Rachis dark. R2 brown-grey with a darker subterminal, curved band. Pale edge. The base of the inner vane lighter towards the edge. Rachis dark; even darker towards the base.

Purple Sandpiper (SS) (T = 5-6): 1–3 pairs of central Rs black-brown. Pale edge on outer vane. Rachis pale at base for about two thirds of its length. Other pairs becoming paler towards the exterior; pale edge; rachis paler colour than the vanes. 1 or 2 outer pairs with basal border of inner vane white, with a blurred boundary.

Little Stint (T = 4-5): central Rs dark brown-grey. Large rufous-buff edge on the outer vane. Rachis brown.

Temminck's Stint (SS) (T = 4-5?): 1 or 2 pairs of central Rs brown-grey. Eventually a rufous-brown edge. Rachis dark..

III. BLACK AND WHITE

IIIa) A SINGLE BLACK AREA TOWARDS THE TIP

For T = 8.5-17 cm, see also young gulls.

Stone Curlew (T = 9-14): outer Rs, black lower on outer vane. Well-defined border. Rachis same colour as the vanes, but darker than them in dark areas.

Northern Lapwing (T = 10-13): one third to one half black, more extensive on central Rs; sometimes bordered with buff. Border of black more or less sharp. Rachis the same colour as the vanes. Rarely, white extends slightly along the rachis in the black zone.

Spur-winged Lapwing (T = 10-13?): one third to two thirds black, more extensive on central Rs. Tip sometimes buff. Rachis the same colour as the vanes; white extends slightly on the rachis in the black area.

Eurasian Oystercatcher (T = 9.5-13): black covers one quarter (outer Rs) to two thirds (central Rs). Border more or less well-defined. Black on the vanes at the same level or separated. Black rachis in the black zone, but white can extend slightly on and around the rachis.

Black-tailed Godwit (T = 7.5-10): Rs sometimes have a paler or white edge. Rachis the same colour as the vanes. White may sometimes extend a little on the rachis in the black zone.

Ruddy Turnstone (T = 6-8.5): central and median Rs of immature. Blurred margins. Possibly a pale edge at the tip. Rachis the same colour as the vanes; white continues a little in the dark zone.

Collared Pratincole (T = 5.5-11.5): Rs with white base and black extremity. White covers less than the basal quarter on central Rs and up to two thirds on outer Rs. On the outer Rs, the white extends further on the outer vane and can go up to the tip. Rachis the same colour as the vanes but white extends a little into the dark. On the outer Rs, black can extend a little along the rachis. The outer Rs have an atypical shape (see Barn Swallow); they are obviously elongated, a little narrower towards the tip.

Black-winged Pratincole (T = 5.5-9?): as in Collared Pratincole.

IIIb) BLACK DISTAL MARK WITH WHITE TIP

For T = 8.5-17 cm, see also young gulls.

Northern Lapwing (T = 10-13): white to buff tip. Possibly green iridescence. Limit of black more or less well-defined. Rachis the same colour as the vanes. Rarely, white extends slightly along the rachis into the black zone.

Ruddy Turnstone (T = 6-8.5): Rs with diamond-shaped white tip. Black mark more blurred at its basal limit, usually extending as a blurred black border on the outer vane (more or less wide, as far as the base or not). Dark distal mark covering 20% (outer Rs) to 50% (central Rs) of vanes. Rachis the same colour as the vanes; white goes into the dark area a little. Immature: white tip less well-defined; sometimes very small; border of dark area sometimes irregular.

Green Sandpiper (T = 6-7.5): Rs sometimes with this unusual type of mark. Mark non-rectangular but pointed near tip of the rachis. Rachis the same colour as the vanes.

Ringed Plover (T = 5.5-8): mark sometimes visible on the second outermost pair of Rs. Large white tip. Asymmetrical black mark. Margin of the black sharp towards the tip and blurred towards the base. White rachis; sometimes a small brown area towards the base.

Collared Pratincole (T = 5.5-11.5): 1 or 2(3) pairs of central Rs black or dark grey (T = 5.5-7). Small white base and white tip or wide white edge at the tip. Rachis the same colour as the vanes.

Black-winged Pratincole (T = 5.5-9?): as Collared Pratincole

IIIc) BLACK MARK(S) ON THE INNER VANE

Northern Lapwing (T = 10-13): Outer Rs. Triangular, oval, or non-geometric mark. Rachis pale.

Ruddy Turnstone (T = 6-8.5): outer Rs. White tip. Mark quite large, sometimes accompanied by a black or grey spot on the outer vane. In general, this pattern is found in immatures. Rachis the same colour as the vanes.

Ringed Plover (T = 5.5-8): 1 or 2 pairs of outer Rs. Mark more or less large. Always white. Sometimes some dark on the rachis; usually all white.

Little Ringed Plover (T = 5.5-7): 1 or 2 pairs of outer Rs. Mark more or less large and regular. Tip always white. Rachis the same colour as the vanes.

Kentish Plover (T = 5.5-7): outer and median Rs with white to buff-grey background. Ill-defined black-grey mark towards the tip of the inner vane. Rachis the same colour as the vanes.

IIId) BLACK MARK(S) ON OUTER VANE

Northern Lapwing (T = 10-13): outer Rs, quite rare. Rachis pale.

Green Sandpiper (T = 6-7.5): outer Rs, 1 or 2 marks on the border of the vanes. Sometimes the back touches the rachis, in which case the black continues along the white rachis a little.

IIIe) TIP BLACK + OTHER BLACK MARKS (A FEW BARS OR SPOTS)

Stone Curlew (T = 9-14): outer and median Rs. White background on outer Rs; white or grey to buff on the others. Well-defined black tip, lower on the outer vane. 1 to 4 blurred black bars; sometimes marks parallel to the rachis or as three

sides of a rectangle. Rachis the same colour as the vanes.

Northern Lapwing (T = 10-13): Rs with black distal mark. Possibly with white or buff tip. Base marked with black; often a triangle against the rachis on the outer vane, sometimes on both vanes; rarely an ill-defined oblique bar at the base of the outer vane. Rachis the same colour as the vanes.

Green Sandpiper (T = 6-7.5): less common pattern (see IIIf) with last bar wider or as an arrowhead-shaped mark, accompanied by black spots, more numerous on the outer vane. Usually a white edge at the tip. Rachis the same colour as the vanes.

IIIf) SEVERAL BLACK BARS, COMPLETE OR NOT

Grey Plover (T = 7.5-9.5): Rs with white to very pale-buff background; sometimes more brown-grey towards the tip of the central ones. Black to brown-grey bars, increasingly complete towards central Rs.

Often 2-10 bars on the outer vane, 2-7 on the inner vane. The bars often appear a little 'faded', especially on the inner vane. Towards the outer Rs, the bars gradually disappear from the base, more so on the inner vane. Last bar sometimes large and arrowhead-shaped. Tip pale. Juvenile: wider bars. Rachis same colour as the vanes.

Common Greenshank (T = 8-9): Rs white (sometimes pale buff on the outer vane) to pale brown-grey (central Rs). Generally 9-13 bars with a 'shaken' look. Last one or two often arrowhead-shaped. Rachis same colour as the vanes (sometimes pale on central Rs).

Green Sandpiper (T = 6-7.5): Rs with white base without any bar. Few bars, wide, sometimes fused (outer vane: 1-4(5) bars, inner vane: 0-3 bars). Rachis same colour as the vanes.

Marsh Sandpiper (SS) (T = 6-7): Rs with 8-11 narrow bars, complete or not (present at least on the outer vane). Rachis same colour as the vanes.

Wood Sandpiper (T = 5-6.5): Rs with a white background, slightly tinged with rufous-buff at the tip towards the central Rs. Black or dark brown bars quite wide; small in number (often 7-9 on the outer vane, 1-6 on the outer web); sometimes only on the outer vane (except terminal bar, always present).

Bars staggered between the two vanes (except sometimes on central Rs). Rachis same colour as the vanes.

Great Snipe (T = 5.5-6.5): Rs with a white background, maybe washed with rufous. Black-brown bars, sometimes only on the outer vane; base darker. Rachis same colour as the vanes.

IV. REGULAR BARRING (BLACK, BROWN, GREY/WHITE, PALE GREY, BUFF)

IVa) DARK BARS ON A PALER BACKGROUND (AREA OF DARK ≤ AREA OF PALE)

Eurasian Curlew (T = 11.5-14): Rs with 8-11 dark brown bars on white (outer Rs) to grey-buff (median and central Rs) background; paler tip. Last mark sometimes larger and diamond-shaped. In general, area of pale ≤ area of dark. Rachis often brown, sometimes paler in the middle and/or at the base.

Whimbrel (T = 10-12): R with 6-8 brown-black bars on buff to greyish background; lighter tip; background darker towards the central Rs. Last dark mark sometimes larger; diamond- or triangular-shaped. Rachis often pale in the middle and brown at the ends; sometimes all brown.

Bar-tailed Godwit (T = 7-10): Rs with a white to buff-grey background. Brown-grey bars. Last mark often larger and diamond-shaped. Bars sometimes partially merged. Tip pale or with pale edge. Rachis often pale at the base and then dark in the distal half or two thirds; pale for up to three quarters of outer Rs. A pattern usually found in immatures and adult females. Other rectrices largely brown, see VIIb and V.

Grey Plover (T = 7.5-9.5): 1 or 2 pairs of central Rs with very pale white to buff background, sometimes greyer towards the tip. 8-10 black to brown-grey bars. Towards the outer Rs, the bars fade quickly and become smaller towards the inner vane commencing at the base (see IIIc). Pale tip. 2 or 3 last bars sometimes merged partly around the rachis. Juvenile: wider bars. Rachis same colour as the vanes.

Common Greenshank (T = 8-9): Rs white (sometimes pale-buff on the outer vane) to light brown-grey (central Rs). Brown-grey bars, often 9-13; narrow; 'blurred'; complete or not. Last one(s) often arrowhead-shaped. Rachis same colour as the vanes, (sometimes paler on central Rs).

Common Redshank (T = 6-8): Rs with white (outer Rs) to grey-buff (central Rs) background. Brown-grey bars; numerous (often 10-17); 'blurred'; usually complete; sometimes staggered between the two vanes. Last bar parallel to others or arrowhead-shaped. Rachis same colour as the vanes.

Common Sandpiper (T = 5-7): Central and median Rs. Grey-brown background. Bars darker. Subterminal bar wider; tip paler to white. Rachis the same colour as the vanes.

Wood Sandpiper (T = 5-6.5): central Rs with white to rufous background. 7-8 wide black-brown bars. Rachis the same colour as the vanes.

IVb) PALE BARS ON A DARKER BACKGROUND (AREA OF DARK ≥ AREA OF PALE)

Golden Plover (T = 7-9.5): Rs with a more or less dark, brown-grey background. About 8-10 white (outer Rs) to golden-buff (central Rs) bars. Tip pale. Rachis brown, darker in the basal half.

Spotted Redshank (T = 7-8.5): Rs with a brown-grey background, darker at the tip. Narrow bars, very pale, numerous (often 9-16). Rachis brown.

V. PALE WEDGES ON A DARK BACKGROUND

Woodcock (T = 7.5-10.5): Rs with black-brown background. Rufous-yellow to rufous-brown wedges on the outer vane, sometimes on the inner vane. Tip grey above and bright white below. Wedges sometimes elongated into narrow bars. Rachis the same colour as the vanes.

Bar-tailed Godwit (T = 7-10): Rs of the male in particular. Brown background; darker tip with white border. Pale wedges. Rachis often pale at the base then brown-black on distal half or two thirds, pale for up to three quarters of outer Rs.

Common Snipe (T = 5-7[9]): 1 or 2 pairs of outer Rs. Few wedges and more towards the base, sometimes touching the rachis (like bars). Distal third variable, often buff to rufous with an irregular subterminal dark bar and white tip (see VIIb). Rachis the same colour as the vanes.

Spotted Redshank (T = 7-8.5): Rs with a brown-grey background, many pale, narrow wedges (often 9 to 16), less visible on central Rs, may form narrow bars. Rachis brown.

Common Sandpiper (T = 5-7): median and outer Rs. Brown background. White notches. Black subterminal bar. White tip. Central Rs sometimes brown with small buff wedges separated by darker colour. Rachis the same colour as the vanes.

VI. REGULAR BARS REDUCED TO MARKS ON A WHITE BACKGROUND

Common Greenshank (T = 8-9): especially outer Rs, sometimes median. Marks often mainly only on the inner vane, but on both vanes towards the tip. Last mark usually following the outline of the tip. Rachis pale.

Marsh Sandpiper (SS) (T = 6-7): Outer Rs. About 3-5 dark grey marks on the outer vane; 0-3 on the inner vane. Rachis the same colour as the vanes.

Wood Sandpiper (T = 5-6.5): Outer and median Rs. About 7-9 black-brown marks on the outer vane; 0-4 on the inner vane. Terminal bar often complete. Rachis the same colour as the vanes.

Great Snipe (T = 5.5-6.5): outer Rs white with a black-grey base. Inner vane with 3-5 black-grey bars, sometimes attached. Rachis the same colour as the vanes.

VII. GREY TO BROWN, WITH CONTRASTING PALER OR DARKER MARKS

VIIa) REGULAR MARKS

Stone Curlew (T = 9-14): Rs with a white, greyish, buff, or yellowish background. Black to brown-grey marks rather blurry; well-defined black tip larger (except sometimes on the two pairs of central Rs, without black tip). Variable marks: parallel, wavy, in three sides of rectangle, in bands of flecking, etc. Rachis the same colour as the vanes, but darker than them in dark areas.

Golden Plover (T = 7-9.5): Outer Rs (1-2[4] pairs) with a more or less dark brown-grey background. About 8-10 white bars. Tip pale. The bars may be indistinct or absent, especially towards the base of the inner vane, in which case they are brown-grey. The inner vane can be uniform, with only 1 or 2 pale bars towards the tip and a fine pale edge at the tip. The base of the inner vane is often white. Rachis brown; darker in the basal half.

Common Greenshank (T = 8-9): central Rs. White to buff-grey background; darker tip. Blurred, brown-grey bars. Last bar(s) often parallel to the curve of the tip. Rachis the same colour as the vanes, sometimes paler at the tip.

Ruff (T = 5.5-8.5): Rs brown. Complete or incomplete bars towards the tip; more extensive on the outer vane, alternating from paler brown to rufous-brown with some black-brown. Often 1 or 2 bars, but depending on the individual as many as 4 or 5 dark bars. Rachis paler than the vanes.

Common Sandpiper (T = 5-7): Median and central Rs with brown background. Darker marks or the beginnings of bars. Black subterminal bar. Tip paler brown (central Rs) to white (median Rs). Dark area sometimes ill-defined around the rachis in a large brown zone. Rachis the same colour as the vanes.

Common Sandpiper (T = 5-7): outer Rs with 5-7 brown-grey bars on white background on the inner vane. Subterminal bar darker brown to black. White tip. Often only 1 or 2 dark marks on the outer vane of outer Rs, with dark triangle at the base touching the rachis. Rachis the same colour as the vanes.

Wood Sandpiper (T = 5-6.5): 1 or 2 pairs of outer Rs white, with about 6-9 black-brown marks on the edge of the outer vane, more or less connected. Rachis the same colour as the vanes.

Common Snipe (T = 5-7 [9]): Rs with dark base, brown-grey to black. Rufous tip with dark bars. Central Rs almost entirely dark with rufous bars, complete or not, towards the tip and with a white to buff tip. Rachis the same colour as the vanes.

Great Snipe (T = 5.5-6.5): Median Rs buff to rufous with a few black-grey bars. Grey-black base; often white tip. Towards the outer Rs the rufous areas are replaced by white. Rachis the same colour as the vanes.

Jack Snipe (T = 5-6.5): Rs becoming more pointed and longer towards the central ones. Rs brown with rufous-buff marks on the outer vane and the tip of the inner vane. Sometimes a buff edge on one or both vanes. Rachis dark.

Sanderling (T = 5-6): outer Rs (mostly) light brown-grey. Rachis pale. Pale edges. More or less white at the base of the inner vane and around the rachis. Slightly darker bars, especially on the outer vane.

VIIb) IRREGULAR MARKS

Stone Curlew (T = 9-14): Rs sometimes with irregular marks, but similar to the regular pattern, see VIIa.

Bar-tailed Godwit (T = 7-10): Rs of the adult male particularly: brown-grey background marked with more or less pale spots or stripes. Rachis pale at the base and then dark towards the tip.

Common Sandpiper (T = 5-7): Rs sometimes with irregular marks, similar to the regular pattern, see VIIa.

Wood Sandpiper (T = 5-6.5): Rs with a white background. Black-brown marks on the outer vane, more or less elongated. Incomplete terminal bar often present. Rachis the same colour as the vanes.

Ruff (T = 5.5-8.5): outer Rs brown with paler rachis. Inner vane marked with pale and/or outer vane marked with dark; sometimes also rufous. Sometimes a wide black subterminal bar.

Ruddy Turnstone (T = 6-8.5): median Rs black-brown (blacker towards the tip). Well-defined white tip. The white at the base of the inner vane also extends over part of the outer vane. White extends a little on the rachis (white at the base) into the dark area. Immature: similar pattern but the limits are much more blurred and irregular; the white to buff tip is sometimes reduced to a thin edge. See also IIIb.

Ringed Plover (T = 5.5-8): central Rs brown becoming black at the tip. A black mark (more or less diamond-shaped) appears at the tip on the second central pair of Rs and increases towards the outer pairs. At the same time, the white increases from the base to half of the inner vane and the outer web also becomes paler from the base. Rachis white to brown depending on the colour of the vanes; sometimes a brown tint towards the base. For all the Rs involved, the dark end contrasts sharply with the white tip. The dark area of the inner vane may be irregular with black and white smudges mingling with brown from the edges of the vanes.

Kentish Plover (T = 5.5-7): central and adjacent Rs more or less brown-grey, becoming black at the tip. Rufous-buff edge, wider at the tip; sometimes absent. Rachis the same colour as the vanes.

Little Ringed Plover (T = 5.5-7): central Rs (1 or 2 pairs) brown-grey, becoming black at the tip. Buff edge at the tip. A white mark (diamond-shaped or irregular) appears at the tip on the second or third pair and increases in size towards the outer Rs. At the same time, white increases from the base to half of the inner vane and also lightens the outer vane from the base upwards. Rachis white to brown depending on the colour of the vanes. For all those Rs involved, the dark end contrasts sharply with the white tip. The 2 or 3 outer pairs have a white tip; the zone next to it is black and pales towards the base forming smudges, mixing brown with white.

Common Snipe (T = 5-7[9]): 1 to 3 pairs of outer Rs with a dark centre on basal one to two thirds. Outer vane white, buff or barred brown and rufous. Inner vane dark speckled buff. Incomplete dark bars or pale wedges. Tip white. Rachis the same colour as the vanes.

Common Snipe (T = 5-7[9]): central and medium Rs with brown-grey to black base with more and more broad reddish marks, speckled or not dark, towards the tip. Tip can be paler (white to buff). Rachis the same colour as the vanes.

Great Snipe (T = 5.5-6.5): central and medium Rs with a brown-grey base, colour increasing towards the tip due to large rufous marks; with or without dark speckling. Tip generally white. Rachis the same colour as the vanes.

Jack Snipe (T – 5-6.5): sometimes irregular marks, similar to regular pattern, see VIIa.

Grey Phalarope (SS) (T = 5.5-7): Rs brown-grey; pale edge at the tip. 1 or 2 central pairs darker at the tip, edged buff. Towards the outer Rs, white border becomes more extensive especially on the outer vane with more and more white on the edge of the inner vane. Rachis paler than vanes, brown on central Rs.

Red Knot (T = 6-7.5): Rs brown-grey. Fine pale edge, wider at the tip, often highlighted with dark. Pale marks especially towards the edges but closer to the rachis on the outer Rs. Pale rachis, brown at the base (sometimes only a few mm).

Curlew Sandpiper (T = 5-6): Rs brown-grey. Fine pale edge especially on the inner vane and at tip. A fairly sharp contrast between dark and white, but with a very irregular margin. More white on the inner vane. Rachis pale; brown towards the base.

Sanderling (T = 5-6): central Rs sometimes brown; marked with pale along the rachis and/or at the tip and/or at the base. Rachis paler than the vanes.

Dunlin (T = 4-6): Rs brown-grey and white. White edge. Limit of the dark more or less regular. The amount of white increases towards outer Rs from the base of the inner vane and the centre, around the rachis. Rachis white towards the tip; brown on the basal third to one half. The outer Rs are sometimes white with a little blurred brown towards the tip.

VIII. GREY TO BROWN, WITH DIFFUSE PALER OR DARKER AREAS

Black-winged Stilt (T = 8-9.5): in young, Rs greyish with the base of the inner vane paler; outer Rs paler. Rachis paler than the vanes.

Ruff (T = 5.5-8.5): Rs brown with a paler zone. Rachis paler than the vanes.

Kentish Plover (T = 5.5-7): central and near Rs black-brown; base of the inner vane paler; often a rufous-buff edge, wider at the tip. Rachis the same colour as the vanes. Median Rs paler, brown-grey, becoming darker at the tip. Tip white. Rachis pale.

Red-necked Phalarope (SS) (T – 5-7?): central Rs brown-grey with pale edge. Rachis paler than the vanes.

Red Knot (T = 6-7.5): Rs brown-grey. Inner vane paler towards the edge and base. Possible pale edge. Rachis pale-brown towards the base.

Curlew Sandpiper (T = 5-6): see VIIb.

Purple Sandpiper (SS) (T = 5-6): outer Rs brown-grey. White edge. Basal border of the inner vane white. Rachis pale.

Sanderling (T = 5-6): Rs brown-grey, paler towards the outer Rs. Pale edge. Base of the inner vane and centre of outer vane often white. Rachis white.

Dunlin (T = 4-6): Rs brown-grey. Pale edge. Border and base of inner vane paler to white. Rachis white for about one half its length, darker brown towards the base.

Little Stint (T = 4-5): Rs paler brown-grey towards outer Rs. Fine white to buff edge, wider at the tip. Inner vane white towards the edge. Rachis pale-brownish at the tip, with a brown area on the basal half or third.

Temminck's Stint (SS) (T = 4-5?): median and outer Rs brown-grey, becoming increasingly pale towards outer Rs. Buff edge towards tip; wider towards the outer Rs. Dark edge to outer vane. Rachis dark, at least at the base.

• PROCELLARIIDAE, HYDROBATIDAE: SHEARWATERS, PETRELS, STORM PETRELS

NP = 10 (+1) / NS = 14-22 / NR = 12. Northern Fulmar: NR = 14.

Distinctive criteria of group: shape, colour, where found.

These families have narrow wings with a long and pointed hand. The tail is usually rounded or square, and sometimes more pointed or forked. Their feathers are found on the coastline. Due to the pelagic lifestyle of these species, feathers are rarely found in isolation. Those of Northern Fulmar are the most commonly found (large feathers); other species are rarely found, except at nesting sites.

Possible confusion: with gulls for large species; for storm petrels with auks, swifts, hirundines (different habitat for these last two groups). The Ps (except those of Northern Fulmar) resemble those of auks but their tip is less pointed. The Ss are similar to those of gulls and waders, but are squatter. The Rs are similar to those of gulls and waders (rectangular to round). The elongated outer Rs of some storm petrels are reminiscent of the Rs of Blackbird.

Colour of 'large' feathers: white, grey, black, uniform, or mixed tones (e.g. Ps dark grey with a white base in Cory's Shearwater).

Primaries: quite long and pointed (except Fulmar: rounded), quickly shortening towards the inner Rs

Secondaries: curved and squat.

Rectrices: rectangular to rounded, of a grey to black colour, without bars.

MEASUREMENTS OF THE LARGE FEATHERS OF SHEARWATERS, PETRELS AND STORM PETRELS						
Species	**Pmax**	**Pmin**	**Smax**	**Smin**	**Rmax**	**Rmin**
Northern Fulmar	26	11.5	12.5	9	15	10
Cory's Shearwater	26	11.5	12.5	10	15	12
Manx Shearwater	18.5	9	9.5	7	10 (11)	8
European Storm Petrel	10.5	5	5	4	6.5	5.5
Leach's Storm Petrel	13.5	6.5	6.5	6	9.5	6.5

Note: small sample size for all these species. No data were obtained for Sooty Shearwater *Puffinus griseus*, Yelkouan/Balearic Shearwaters *Puffinus yelkouan/mauretanicus*, Wilson's Storm Petrel *Oceanites oceanicus* or Madeiran Strom Petrel *Oceanodroma castro*.

NORTHERN FULMAR *Fulmarus glacialis*: moulted feathers of this species are the most commonly found of the group. The hand is quite short.

Possible confusion: mainly with gulls.

Colour of 'large' feathers: Ps and Ss are dark grey becoming white towards the inner side and base. Ps and Ss are more or less dark depending on the plumage morph, but the difference in colour between individuals is mostly visible on the Rs which range from light grey bordered with white to deep grey.

Remiges: comparable in size to those of gulls. They have a more compact shape, the tip rounded and the vanes broader.

Primaries: neither emarginated nor notched; calamus long (C = 20-25%).

▷ **Northern Fulmar wing**

Secondaries: short and broad; calamus long (C = 20-25%).

Rectrices: rectangular with rounded tip.

CORY'S/SCOPOLI'S SHEARWATERS *Calonectris borealis/diomedea* (SS): Ps black-grey with a white mark on the inner vane, starting from the base to feather centre, leaving grey along the rachis and a little on the border of the vane. Ss black grey with a white mark on the inner vane, diagonally from the base, and with a speckled edge.

282

Al

P10

Northern
Fulmar

P7

Sm

So

PCo

C

PC

R4   R7

Rs of variable
shades from mouse
grey to white

Pi

'large' feather broad
with rounded tip

uppertail-
covert

Cory's Shearwater

P10

Po

Pm

Pi

certain feathers
soiled or damaged

So

wing-
coverts

Si

Sm

So

PC

P10

Po

Pm

SC

Pi

So

Ter   R1   Rm   Ro

R1   R4   R6

Leach's Storm Petrel

10 cm : scale ¹/₂

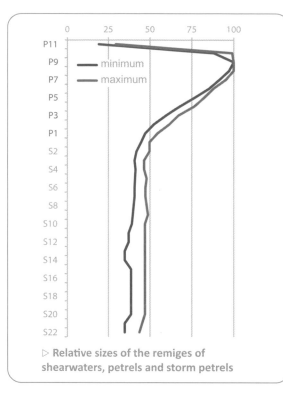

▷ **Relative sizes of the remiges of shearwaters, petrels and storm petrels**

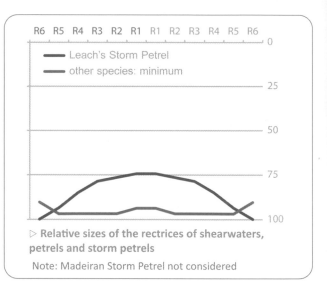

▷ **Relative sizes of the rectrices of shearwaters, petrels and storm petrels**

Note: Madeiran Storm Petrel not considered

▷ **Leach's Storm Petrel wing**

Rs like the Ss but with the tip becoming black. R1 without white.

Manx Shearwater *Puffinus puffinus* (SS): Ps, black-grey; paler inner vane, especially towards the base. Rs black.

Leach's Storm Petrel *Oceanodroma leucorhoa*: 'large' feathers black. Rs with grey and white base. Outer Ps very elongated and pointed, others with bevelled tip. Rapidly decreasing in size towards the wrist. Ss curved and quite short. Rs rectangular with rounded tip; increasing in size towards outer Rs.

Possible confusion: mainly with auks, swifts and hirundines.

European Storm Petrel *Hydrobates pelagicus* (SS): Ps black. Ss black with white base and ill-defined edge, a white edge at the tip of inner Ss. Rs black. R1 uniform, the others with a small white base more extensive towards the outer Rs (a quarter to a third), R6 with more white on the outer vane.

• SULIDAE: GANNETS

NP = 10 (+1) / NS ≥ 25 / NR = 12

Distinctive criteria of group: shape, size, colour, where found.

These birds are very good gliders, using the air currents at the sea surface. Their wings are disproportionately elongated. Their hands are pointed; they also have remiges on their arms, just like albatrosses and frigatebirds. They therefore have more than thirty secondaries and tertials. The tail is pointed and used during diving. Only one species normally occurs in Europe, the Northern Gannet.

Possible confusion: with Ps and Ss of skuas and gulls, Ss of ducks and divers.

Northern Gannet *Morus bassanus*

'Large' feather colour: the young are black-brown with pale speckling, become pearly-white with age. The 'large' feathers can remain on a bird for more than a year, so they are sometimes very worn when moulted. The brown

adult	immature	subadult
Ps bronze-brown; white base with graded border; rachis paler than the vanes		
Ss pearly-white	Ss brown; white graded base	pearly-white: pure white or with half black-grey (along the length), or other black-grey mark (line, band, or longitudinal triangle)
Rs pearly-white; rachis white to yellowish	Rs bronze-brown; white graded base; rachis white to yellowish	most pearly-white adult-type, except central ones which are immature-type; possibly others brown with white tip

feathers of juveniles are progressively replaced by white ones, except on the hand where they will remain brown throughout a bird's life.

Primaries: TP = 18-37. The three outer Ps are clearly notched and have long emarginations; calamus long (C = 20-25 [30]%); bevelled end on the median and inner Ps. The size decreases very quickly from the outer to the inner Ps.

Secondaries: TP = 14-19. Quite curved, similar to those of gulls; long calamus (C = 20-30%). Size and shape fairly homogeneous on the forearm and arm.

Rectrices: TR - 12(11)-29. Typical, spearhead-shape; increasingly elongated towards the centre of the tail. Calamus not upturned and very long (C = 20-25%). Patterns quite variable, but white, brown, or bicoloured (white from the tip).

Body feathers: those of the young are brown with a small white drop-shaped mark.

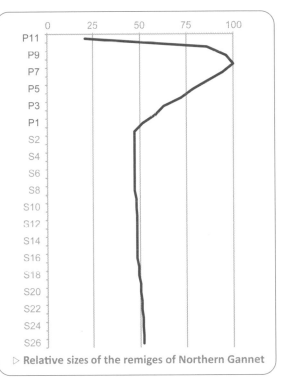

▷ Relative sizes of the remiges of Northern Gannet

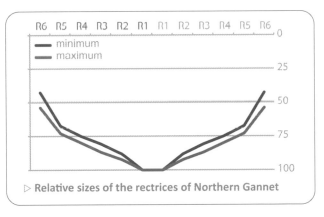

▷ Relative sizes of the rectrices of Northern Gannet

• PHALACROCORACIDAE: CORMORANTS AND SHAG

NP = 10 (+1) / NS = 18-22 (25) / NR = 12-16

Distinctive criteria of group: colour, where found.

Cormorants use flapping flight; the hand is slightly pointed, but much less so than in most birds that use this type of flight. The tail is suited for diving.

Possible confusion: with ducks (remiges, but Ps with a tegmen), black corvids (but truly black), herons (but shape different), Gannet (Rs, see shape and colour).

'Large' feather colour: black-brown, with bronze tint; sometimes dark green iridescence on the outer vane.

Primaries and secondaries: the Ps and Ss are not very different from the basic remige shape. Ss and Ps are similar apart from the Ps having a longer calamus and the outer Ps having notches/emarginations. The size of the Ps decreases very slightly towards the wrist; the 7 or 8 outer Ss measure at least 80% Pmax; their size decreases a little afterwards. Calamus quite long, C = 20-25%.

Rectrices : their shape and structure are characteristic. The Rs are adapted for diving during which cormorants use their tails as rudders, thrust being provided mainly by the legs. The tail is rounded; all Rs are about the same size. Their tips are bevelled when fresh but wear quickly. The Rs have a modified shape to improve their water resistance during swimming: very long calamus (C = 20-25%) that is straight not upturned; rachis thick; Rs almost straight, even the outer ones.

Similar adaptive shape: Northern Gannet, woodpeckers.

Body feathers: they are often recognisable. The majority of the feather are black-brown with bronze or greenish reflections; often black borders. This contrast makes for a scaly pattern on the back and wings, visible on birds in good light.

The 'large' feathers of Shag are slightly smaller than, and not as dark as, those of Great Cormorant.

Species	Pmax	Pmin	Smax	Smin	Rmax	Rmin
Great Cormorant	27 (31)	19	20	16	20 (23)	14
European Shag	21	17	18	13	17	12
Pygmy Cormorant (SS)	16	12	14	11	15?	9 ?

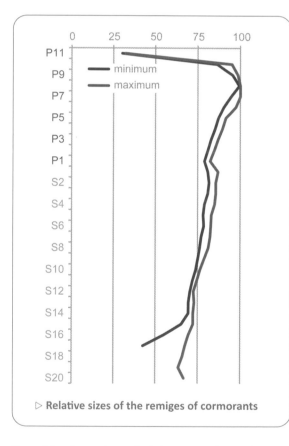

▷ **Relative sizes of the remiges of cormorants**

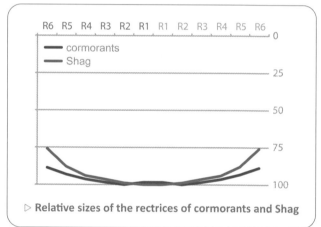

▷ **Relative sizes of the rectrices of cormorants and Shag**

• GAVIIDAE: DIVERS

NP = 10 (+1) / NS = 18-23 / NR = 12

Distinctive criteria of group: shape, colour, where found. This family has relatively narrow wings for their size and a very short tail.

Possible confusion: with the auks, ducks (Ss) and even gulls and cormorants, but the calamus is very long in all the 'large' feathers. Further, compared to the Ss of ducks and gulls, those of divers have a less curved rachis and the dark brown hues of this family are absent in the gulls.

'Large' feather colour: dark brown, paler underneath and towards the base. Sometimes white on the inner vane of the remiges or at the tip of the rectrices. All the 'large' feathers have a dark brown rachis.

Primaries: the calamus is exceptionally long (C = 30-35%), which gives them a very recognisable shape

Secondaries: they resemble those of ducks and geese but with a very long calamus (C = 25-35%), often with some white at the tip. They are quite narrow. The inner Ss viewed from above are bent outwards (look at the width and curvature of the vanes in profile to identify the side of the tail they come from).

Rectrices: they have a fairly long calamus for rectrices (C = (15) 20-25%) and are quite rigid. They often have a white edge or a small tip, sometimes in two separate marks.

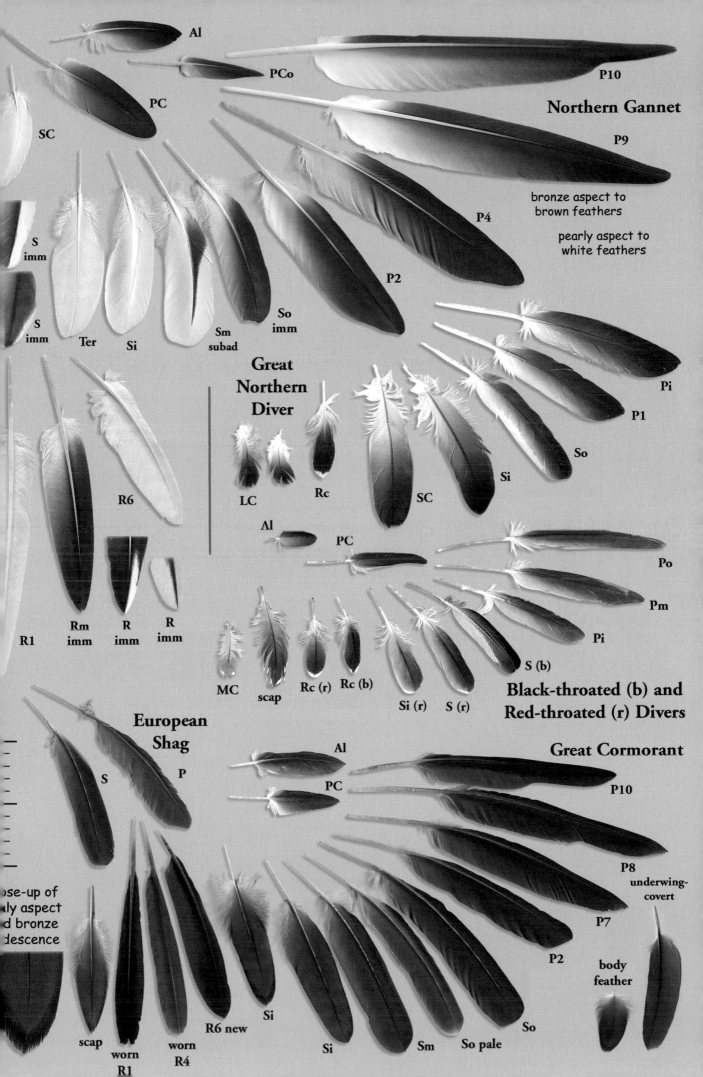

Al

PCo

PC

SC

P10

Northern Gannet

P9

P4

bronze aspect to
brown feathers

pearly aspect to
white feathers

S
imm

S
imm

Ter

Si

Sm
subad

So
imm

**Great
Northern
Diver**

R6

LC

Rc

SC

Al

PC

Pi

P1

So

Si

Po

Pm

Pi

S (b)

R1

Rm
imm

R
imm

R
imm

MC

scap

Rc (r)

Rc (b)

Si (r)

S (r)

**Black-throated (b) and
Red-throated (r) Divers**

**European
Shag**

S

P

Al

PC

Great Cormorant

P10

P8
underwing-
covert

P7

ose-up of
ly aspect
d bronze
descence

scap

worn
R1

worn
R4

R6 new

Si

Si

Sm

So pale

So

P2

body
feather

Species	Pmax	Pmin	Smax	Smin	Marks on the Ss	Rmax	Rmin
Red-throated Diver	21 (23)	11	13	8	a white edge at the tip, or a small mark	8	6
Black-throated Diver	22 (23)	11	13	9	white border on the inner vane	8	5
Great Northern Diver	30	16	16 (21)	11	white border on the inner vane; the innermost with one or two white marks	9	7

The 'large' feathers of Great Northern Diver are much larger than those of Black-throated and Red-throated Divers, which is usually sufficient to differentiate them. The White-billed Diver *Gavia adamsii* is similar to Great Northern in size.

The differences between Black-throated and Red-throated Divers are subtle. For less damaged 'large' feathers, the Red-throated's Ps and Ss appear paler, brown-grey, while those of Black-throated appear darker. However, this criterion is difficult to see in worn feathers.

No data found for White-billed Diver.

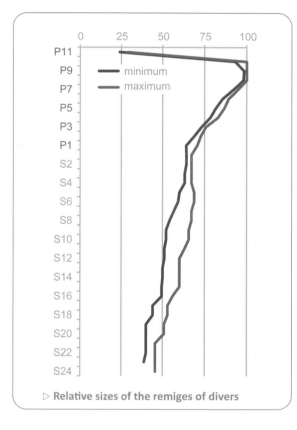

▷ Relative sizes of the remiges of divers

▷ Red-throated Diver wing

▷ Relative sizes of the rectrices of divers

• PODICIPEDIDAE: GREBES

NP = 11 (+1) / NS = 17-22

Distinctive criteria of group: shape, colour, where found.

The grebes have small wings relative to their size, so they are very concave. The tail is barely visible; the vestigial rectrices look like body feathers (so are not described here).

Possible confusion: with thrushes (but Ps curved and rigid in grebes), ducks including 'sawbills' (Ss), but are generally distinguished by size or different colour.

'Large' feather colour: Ps and Ss are brown and/or white. The extent of white varies depending on the species, the age of a bird, and the location of the feather on a bird. The brown may sometimes be quite reddish on the Ps and tends towards grey on the inner Ss. The Ps often have a little white at the base, sometimes also a white tip. Certain species have white secondaries (often median ones), possibly with brown on the rachis. The tertials are turned towards the outside of the wing; they are often two-tone, sometimes with rufous as well. The shape and extent of these marks vary greatly within each species. The rachis is brown or white depending on the species and the position of the feather. Remiges are relatively short; the Great Crested Grebe has remiges that are only slightly larger than those of the Mistle Thrush. The Ps especially, and the Ss to a lesser extent, are curved in profile and rigid enough to support the high wing-loading.

Primaries: the 3 or 4 outer Ps have a long notch on the inner vane (Not = 75-90%).

Secondaries: the calamus is quite long (C = 15-25%).

CHARACTERISTICS OF THE REMIGES OF GREBES						
Species	Pmax	Pmin	Colour of Ps	Smax	Smin	Colour of Ss
Great Crested Grebe	14 (18)	10	brown; white base on inner vane; inner Ps sometimes with white tip (especially immatures)	11	8	white; often with some brown along the rachis and below. Outer Ss more marked with brown on outer vane; inner Ss with brown tip and outer vane
Red-necked Grebe	14 (18)	9	brown; no or very little white at base of inner vane	11	7.5	brown; the white quickly increases towards the base from the tip; rachis brown at the base. Inner Ss partially brown (tip and outer vane)
Slavonian Grebe	11 (14)	7.5	brown; a little white at base of inner vane	8.5	6.5	white; white rachis. Outer Ss (1 to 3) as Ps or with white more or less marked with brown at the tip
Black-necked Grebe	10.5 (14)	7	brown with white at least 2/3 along the edge of the inner vane (outer Ps) and then on all the inner vane (inner Ps)	7.5	5.5	white; white rachis. Inner Ss partially brown (tip and outer vane)
Little Grebe	8 (8.5)	6	brown; with a little white at the base on the inner vane, extending along the edge	7.5	5.5	brown with more or less white on the inner vane and a little on the outer vane

LITTLE GREBE *Tachybaptus ruficollis:* much smaller than the others. Ps brown (outer Ps often) then with white at the base of the inner vane and running along its edge. Ss quite variable; brown with more or less white from the base of the inner vane. The white can cover the whole of the inner vane and the tip of the outer vane in a fairly irregular zone. Ss sometimes white with a little brown on the rachis and at the tip. Often grey-buff marks on the tip of the Ss. Inner Ss mainly brown.

GREAT CRESTED GREBE *Podiceps cristatus:* Ps brown with a white base on the inner vane. Inner Ps sometimes have a pale tip (especially immature). Some individuals have much more white on the Ps which can cover the totality of the inner vane. Ss quite variable in the proportions of brown or white (depending on age and origin). Often all white with white rachis, but also with brown on the rachis, and often with brown around the rachis. Sometimes brown over almost the whole of outer vane or over nearly all the feather (outer Ss especially). Inner Ss white with more or less brown on the outer vane and at the tip; sometimes rufous.

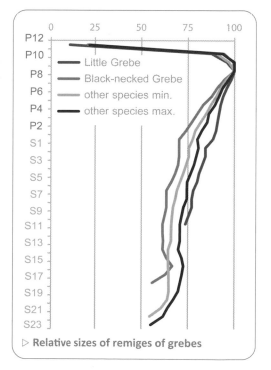

▷ Relative sizes of remiges of grebes

▷ Great Crested Grebe wing

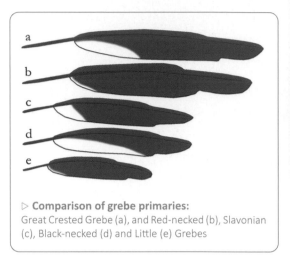

▷ **Comparison of grebe primaries:**
Great Crested Grebe (a), and Red-necked (b), Slavonian
(c), Black-necked (d) and Little (e) Grebes

• ANATIDAE: SWANS, GEESE, DUCKS

Swans: NP = 10 (+1) / NS = 20-28 / NR = 20-24

Geese and Shelducks: NP = 10 (+1) / NS = 12-20 / NR = 14-20

Ducks: NP = 10 (+1) / NS = 12-18 / NR = 14-18 (20)

Distinctive criteria of group: shape of rectrices, tegmen of Ps, size for swans and geese, colour for some species, where found.

This family has pointed wings and a short tail. They have a high wing-loading; their flight is fast and can be maintained for long periods. Therefore, especially the Ps, less so the Ss, have a strengthened structure. In the Ps this is achieved by an increased thickness of the barbs and a thickened and elongated calamus. The inner Ss are elongated and reinforce the aerodynamic cohesion between the body and the wings.

Possible confusion: no confusion possible for the Ps, due to the presence of the tegmen (see below). The Ps of geese may resemble those of raptors or other gliding birds, but the tegmen distinguishes them. The same is true for the smaller diving species whose 'large' feathers resemble those of grebes. The gulls also have a paler area on underside of the Ps but it is much less distinct. As for the galliforms, the silhouette of the 'large' feathers and their curvature easily differentiates them, see pp. 79-80.

For the Ss, confusion possible with divers, waders, and gulls. For the Rs, possible confusion: with herons or large waterbirds, and for diving species with divers p. 286, auks p. 243, rallids p. 321. See also gulls, other seabirds, herons, storks, cranes, pelicans and flamingos, etc.

'Large' feather colour: very variable. Swans have grey-brown 'large' feathers when immature and dull white when adult. In other species, Ps are generally brown, black, or grey. In connection with the frequent presence of a speculum, many species have secondaries of contrasting colours; they often have a metallic area on the outer vane and a different coloured tip. Apart from swans and geese, the rectrices are frequently of different colours depending on the gender, and sometimes on age. Females tend to have brown hues (dark brown with lighter markings); males with rather grey, black or white hues.

Primaries: they have a particular structure, the tegmen. The barbs are thicker around the rachis, which is noticeable to the touch and results in a bright area on the underside of the feather. The overall look of the Ps is also characteristic.

In surface-feeding (dabbling) ducks and other species the calamus is long (C = 20-35%, C = 25-35% in swans, geese, shelducks). The outer Ps are well-emarginated and have a fairly short and well-defined notch (Not = 65-80% in swans, geese, shelducks; Not = 75-90% in the others). The median Ps have a rather pointed tip and a rather triangular outline, with the inner vane widening towards the base and the wider outer vane towards the basal third. The inner Ps have a bevelled tip and fairly parallel sides.

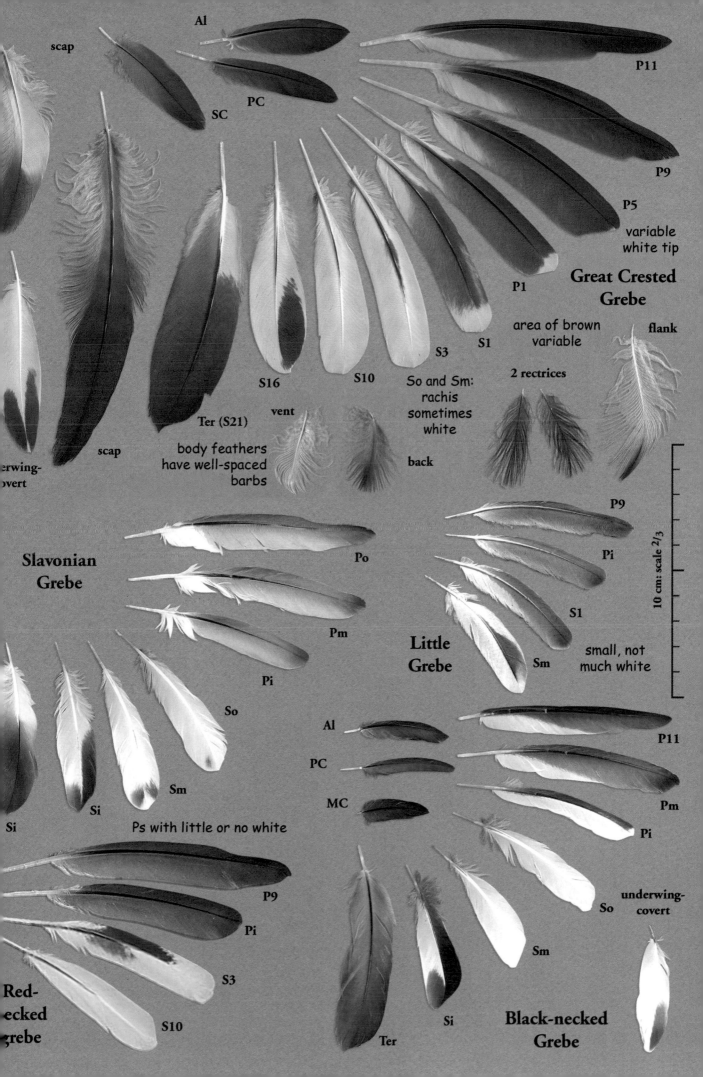

scap

Al

SC

PC

P11

P9

P5

variable
white tip

**Great Crested
Grebe**

P1

S1

area of brown
variable

flank

S3

So and Sm:
rachis
sometimes
white

S10

S16

Ter (S21)

vent

body feathers
have well-spaced
barbs

back

2 rectrices

scap

erwing-
covert

**Slavonian
Grebe**

Po

Pm

Pi

So

Sm

Si

Si

Ps with little or no white

**Little
Grebe**

P9

Pi

S1

Sm

small, not
much white

10 cm: scale $^2/_3$

Al

PC

MC

P11

Pm

Pi

So

underwing-
covert

Sm

Ter

Si

**Black-necked
Grebe**

**Red-
ecked
grebe**

P9

Pi

S3

S10

In the diving ducks, the narrowing of the Ps is often less pronounced. The notch of the outer Ps is not very marked and very high up the feather (Not = 75%). The tip of the Ps is more pointed and elongated than in the dabbling species. **Secondaries:** there is no tegmen on the Ss. The calamus is quite long in some species, but is not always very obvious (C = 20-30%, C = 25-30% in swans and geese).

The colouring of the Ss is quite indicative as to species. They often have a contrasting edge. Indeed, many medium and small-sized species have a speculum on the forearm, which results in strongly contrasting colours on at least part of the Ss. Examples of coloration: white base + black tip, or white tip highlighted black + outer vane with a metallic colour.

Inner Ss have a characteristic, elongated and leaf-like shape. They are longer than the medians and effectively fill the space between the wing and the body. The scapulars completely cover the wing in ducks and most other Anatidae at rest. The colour of the scapulars and inner Ss is often identical or at least very similar. Some have a distinctive pattern; for example, contrasting longitudinal stripes (Wigeon, Pintail, Teal, Shelduck, etc.).

Rectrices: The Rs are short relative to the size of the bird. Except for a few species with ornamental central Rs (Pintail, Long-tailed Duck, Mallard, etc.), the Rs are curved and arrowhead-shaped. In diving ducks they are even shorter, but also often less curved, narrower, and more pointed. The tail serves little purpose in the movements of dabbling species. The difference in the width of vanes depends on the position of the Rs and is not as marked as in the Ps, so it is sometimes difficult to position them in relation to each other. On the other hand, the Rs are more curved inwards towards the centre of the tail the more they are located on the outside of the tail, which helps to clarify their position.

Body feathers: they are often recognisable by their marked curvature when viewed in profile and by the dense down at the base. The scapulars are quite tapered in ducks. The greater coverts are sometimes identifiable (see descriptions).

Size is an important criterion for distinguishing the main groups of Anatidae: T swans > T geese > T shelducks > T dabbling and diving ducks.

On the other hand, many species have similar measurements. For the same body size, the 'large' feathers of diving ducks are a little shorter than those of dabbling ducks.

SWANS

The three European species (genus *Cygnus*): 'large' feathers of adult dull white, more or less mixed with light grey to dirty brownish in young. Somewhat rounded tail; external Rs measuring 60-65% of central ones; less rounded in Bewick's Swan (80% of central feathers).

BLACK SWAN *Cygnus atratus*: feral; breeds in the wild; 'large' feathers black and white. The primaries are white with black tips (part of the rachis is black in the area around the white vanes). Black increases towards the base in the secondaries (more on the outer vane). The rectrices are black with a small white base. Possible confusion: with Sacred Ibis, but the swan has a tegmen on the underside of the primaries and the black tip of the secondaries is not well-defined, while it appears dipped in ink in the Sacred Ibis.

SIZE OF 'LARGE' FEATHERS OF SWANS						
Species	Pmax	Pmin	Smax	Smin	Rmax	Rmin
Mute Swan	43	32	36	29	24	14
Bewick's Swan (SS)	35	22	23	21	21	16
Whooper Swan (SS)	47	32	33	31	22	13
Black Swan (SS)	38	26	27	23	14	12

Note: small sample sizes, except for Mute Swan

GREY GEESE (genus *Anser*):

The rachis is white on the 'large' feathers (dark in *Branta* geese), although the 'large' feathers of young have a dark rachis. Ps and Ss grey; base of the inner vane paler. Centre and base of Ps often paler than the rest. Ss often with fine white edge. Central Rs brown with white tips; the white increases towards the base and from the innermost feathers outwards. Ventral body feathers brown-grey to buff with a small pale tip.

Al y

Al y

P10

PC y

SC y

P7

uppertail-
covert

P5 y

y: tip
tinted with
grey or brown

back

P1

S4

S14

Mute Swan

Rm y

Ro y

Black Swan

Al

Rm

P10

ses broken

P3

S5

R4

Rm

Ro

Po

Al

Pi

PC

Sm

SC

Greylag Goose

10 cm: scale 1/3

Po

PC

SC

Rm

vent

Sm

Canada Goose

Si

Al

Po

Rm

MC

Po

**Barnacle
Goose**

Rm

Ro

Sm

Po

**Brent
Goose**

Si

Pi

flank

flank

GREYLAG GOOSE *Anser anser:* TP = 23-40 (45), TS = 20-24, TR = 13-19.

Ps dark grey and Ss browner; paler towards the base. Ss with white edge on the outer vane. Rs brown with large white tip; edged with white on the outer vane. Compared to other geese, the white on the Rs is quite extensive and entirely covers the outer Rs; white with small brown base.

BEAN GOOSE *Anser fabalis* (SS): Ps and Ss dark grey with ash-grey base. Ss edged with white on the outer vane and at the tip. Rs brown with large white tip, narrower towards the central feathers; edged with white on the outer vane.

PINK-FOOTED GOOSE *Anser brachyrhynchus* (SS): Ps dark grey; silvery-grey around the rachis. Rs very similar to those of the Greylag Goose.

WHITE-FRONTED GOOSE *Anser albifrons* (SS): Rs brown with large white tip, narrower towards the central feathers; edged with white on the outer vane.

LESSER WHITE-FRONTED GOOSE *Anser erythropus:* No data.

THE BRANTA GEESE

The rachis is dark on the 'large' feathers (white in adult grey geese). Ps and Ss are black-brown; browner in Canada Goose, with grey on the outer vane in Barnacle Goose.

CANADA GOOSE *Branta canadensis:* Ps and Ss dark brown; base a little paler. Ss more or less tinged with reddish-brown on the outer vane. Rs almost black.

BARNACLE GOOSE *Branta leucopsis:* Ps and Ss dark grey, tinged with grey on the outer vane and lighter on the base of the inner vane. Rs dark grey to black; inner base sometimes paler or with a little white.

BRENT GOOSE *Branta bernicla:* Ps, Ss and Rs black-grey with a slightly lighter base. In young, pale tip on several outer pairs of Rs and white border at the tip of Ss (wider on the median and inner Rs).

Size of 'large' feathers of *Branta* geese					
Species	Pmax	Pmin	Smax	Smin	Rmax	Rmin
Canada Goose	48	23	29	20	20	16
Barnacle Goose	39	18	22	18	18	11
Brent Goose	32 (44)	14	18 (24)	12	18	9

▷ Brent Goose wing (young)

Introduced species

These species are mainly found near ornamental ponds, in cities or in parks. However, some individuals become mixed with wild populations in other environments and may leave feathers in the same places as them. The probability of discovering them is very low, but their possible presence should not be forgotten, especially when faced with feathers difficult to associate with locally occurring species. Apart from Wood and Mandarin Ducks, these species are not included in the description tables.

BAR-HEADED GOOSE *Anser indicus* (SS): remiges quite similar to those of European *Anser* geese; black-grey; more ash-grey towards the base (especially the Ps which are quite pale, except at the tip). Rs silvery brown-grey with white tip and edge.

SNOW GOOSE *Anser caerulescens* (SS): Ps black-grey with white rachis, except in generally white P1; sometimes dark or bicoloured. White Ss and Rs. In the dark morph, Ss are like the brown-grey Ps and Rs, bordered with white and with a white tip.

EGYPTIAN GOOSE *Alopochen aegyptiacus* (SS): 'large' feathers similar to those of Ruddy Shelduck but a little bigger. Generally, speculum a little darker (sometimes purple) and tertials a little more 'brick'-coloured. Secondary coverts white with a thin black subterminal bar

RUDDY SHELDUCK *Tadorna ferruginea* (SS): Ps black; possibly a small pale base. Ss black with the wide speculum (on outer vane) green except at the tip (less extensive on the inner Ss); sometimes a white tip. Tertials dark red on the outer vane, brown-grey on the inner. Rs black.

WOOD *Aix sponsa* (SS) and **MANDARIN DUCKS** *Aix galericulata:* see tables for descriptions. Specially shaped tertials, extended in hourglass-shape in Wood Duck, and with a very enlarged rufous vane in Mandarin Duck.

Swans and geese are distinguished from other species by their large size and colour. They are not included in the following tables.

Species	Sex	Pmax	Pmin	Main colour	Uniform	Tip	Pale edge	Base of inner vane	Other marks
MAIN COLOUR CRITERIA OF THE PRIMARIES OF DUCKS Reminder: all Ps of the Anatidae have a tegmen, very visible from below.									
Dabbling ducks									
Common Shelduck	mf	26	12	black-brown	X	irregular; white on Pi of young	-	small; white	-
Eurasian Wigeon	mf	22	11	brown	X	darker	fine; especially visible on Pi	white or paler with dark mottling	-
Northern Pintail	mf	22 (25)	11	brown	X	darker	very fine on ov and tip; wider and white at tip of Pi	-	-
Northern Shoveler	mf	19 (22)	10	brown	X	darker	-	paler but not well defined	-
Garganey	mf	16	9	brown	X	darker	pale, fine, on ov and tip	paler to white	-
Eurasian Teal	mf	16 (18)	8	brown	X	darker	fine, on ov and tip	-	-
Gadwall	mf	21	10	brown	X	darker	eventually on tip of Pi	white or paler with dark mottling	-
Mallard	mf	22	13	brown	X	darker	very fine, on ov and tip	most often pale brown; sometimes pure white or mottled	-
Domestic Duck	mf	> 25	12	often brown; sometimes pure white or spotted	variable	variable	variable	variable	-
Mandarin Duck	m	21	11	brown-grey	X	-	wide white edge on ov (except P1 and P2)	-	-
Wood Duck	mf	n.d.	n.d.	brown-grey	X	tinted with dark blue (m+i)	blurred white zone on ov, towards tip (m+o)	-	-
Diving Ducks									
Smew	mf	15	8	black-dark grey	X	-		-	-
Common Pochard	mf	16	11	grey	X	dark grey		-	Po with ov darker
Tufted Duck	mf	16	9	dark brown-grey	o	black		-	ov grey in centre on P1-P5 (P6)
Greater Scaup	mf	17	9	dark brown-grey	o	black		-	ov grey to white in centre on P1-P6 (P7)
Long-tailed Duck	mf	18	9	brown to black	X	-	white edge in some birds	-	-

Species	Sex	Pmax	Pmin	Main colour	Uniform	Tip	Pale edge	Base of inner vane	Other marks
Red-crested Pochard	mf	19	11	white to brownish-grey	-	black		-	Po with ov black, at least in part, on P5-P10
Common Scoter	mf	19	10	black to dark brown	X	darker		ov paler except tip	-
Velvet Scoter	mf	21	11	black; sometimes dark brown	X	-	-	-	-
Goldeneye	mf	22	10	black	X	-	-	-	-
Goosander	mf	22	12	dark grey-black	X	rarely a white tip P1-P2	-	-	-
Common Eider	mf	23	14	black (m) to brown (f)	X	A little darker	-	m: sometimes white tip (Pi)	
Red-breasted Merganser	mf	19 (21)	10	dark grey-black	X	-	-	-	-

MAIN COLOUR CRITERIA OF THE SECONDARIES OF DUCKS

Species	Sex	Smax	Smin	Main colour	Speculum	Pale tip	Dark tip	Black subterminal bar on outer vane	White base to vanes	Other marks	Inner Ss	Colour of greater coverts
						Dabbling ducks						
Common Shelduck	mf	16	10	black-brown	bottle green (slight in y.); often ov of S1-S9	white and wide in young; sometimes white tip in ad.	black	-	ov: wide; iv: wide and rounded, speckled border, larger in y	-	ad: white with large dark rufous mark bordered with black on the ov	-
Eurasian Wigeon	mf	13	9	Brown; ov black under the speculum	bottle green; small and towards base; larger and brighter in m, especially on S4-S9	more or less wide	-	very wide	small on iv and ov	-	S10 with ov nearly all white and eventual black edge; others with grey, black and white longitudinal bands	m: black tip; iv grey; ov white; f: brown with pale tip or spotted
Northern Pintail	mf	12	9	dark brown	bottle green tending towards bronze-brown; sometimes purple; not well defined, especially S3-S10	wide and sharply defined; white to buff depending on position (border sometimes mottled)	-	-	-	-	contrasting, longitudinal bands; brown or grey and black	m: grey with orange tip ; f: grey-brown with buff tip
Northern Shoveler	mf	11	7	dark brown	green with blurred edge; especially S5-S10	fine and blurred; sometimes absent	-	wide	border of vane white towards the base	-	brown with pale edge; m: with pale band along the rachis; f: with irregular pale marks	brown with white end; eventual black tip

Species	Sex	Smax	Smin	Main colour	Speculum	Pale tip	Dark tip	Black subterminal bar on outer vane	White base to vanes	Other marks	Inner Ss	Colour of greater coverts
Garganey	mf	10	7	brown-grey; ov black-grey	dark green; slightly marked; towards the tip on S2 increasing towards S10	wide and well defined	-	-	-	-	brown with pale edges; sometimes pale line along the rachis	brown-grey with white tip; larger in male
Eurasian Teal	mf	9	6	dark brown; ov black	bright green; very shiny; low and small on S3 or S4; complete S8 - S10	white; finer towards Si (and in f)	-	-	-	f: eventual pale vermiculations on tip of ov	grey-brown; ov paler with black border and white edge	brown with large white tip becoming rufous-buff towards inners
Gadwall	mf	12	8	brown-grey	white, limited to ov of S9 - S11	more or less wide and well defined	-	-	-	ov uniform or with pale marks	brown; pale edge on ov, with pale markings in f	m: brown-grey with white tip or small black tip slightly underlined with white
Mallard	mf	15	11	brown; ov darker	dark blue (S2 small, S3-S10 well defined)	large; very large on ov	-	wide	-	S1 may have pale marks or vermiculations	brown; ov darker; often pale edge; well-formed vermiculations in m	brown-grey with black tip and white subterminal band
Domestic Duck	mf	>20	10	variable; often brown to white	often blue; sometimes white or absent	often	-	slight	-	variable	variable	variable
Mandarin Duck	m	13	11	brown-grey	dark blue; green border (m+o); especially visible on S6-S11	white, except Si	-	x	paler	-	one with iv rufous and very wide	dark brown-grey
Wood Duck	mf	n.d.	n.d.	brown-grey	dark blue; extensive; m: on S2-S9; f: on S5-S9	white; m: thin; f: wide	-	m: well defined; f: slight	-	-	S10 and following larger and differently marked	brown-grey; bluish tip towards inners
Diving ducks												
Red-crested Pochard	mf	13	11	grey	pale	thin; white	dark grey tip	-	-	-	brown-grey	brown-grey with darker tip
Common Pochard	mf	11	9	pale grey	-	small white band on So and Sm	dark grey; blurred on So	-	-	-	grey; mottled (m)	grey
Tufted Duck	mf	10	7	black, grey, white	white	-	black	-	-	base grey; centre white; tip black	black	dark grey with black tip
Greater Scaup	mf	11	8	black, grey, white	white	-	black	-	-	base grey; centre white; tip black	black	dark grey; sometimes paler marks at tip
Common Eider	m	15	11	black (dark brown)	-	eventually white or buff; small	-	-	-	triangular white tip appears towards S8-S9; very large on S11	S12 and following white	-

Species	Sex	Smax	Smin	Main colour	Speculum	Pale tip	Dark tip	Black subterminal bar on outer vane	White base to vanes	Other marks	Inner Ss	Colour of greater coverts
Common Eider	f	15	11	dark brown	-	often a buff edge on ov and tip; wider towards Si	-	-	-	-	-	brown with white tip
Long-tailed Duck	mf	11	8	dark brown	-	-	-	-	-	ov tinted with brown to rufous	brown to black	-
Common Scoter	mf	12	8	black/dark brown	-	female sometimes	-	-	-	-	black/dark brown	black/dark brown
Velvet Scoter	mf	13	9	white / black	white	-	-	-	-	S1 black and white	black	black with white tip

MAIN COLOUR CRITERIA OF RECTRICES OF DABBLING DUCKS

Rectrices quite large, curved (especially outer ones) and slightly pointed or bevelled.

Species	Sex	Rmax	Rmin	Main Colour	Uniform	Tip	Edge	Marks	Particular shape
Common Shelduck	mf	14	10 (8.5)	white/black	o	black; with blurred border and/or mottled	-	-	tip often bevelled
Common Shelduck	y	14	10 (8.5)	white/brown	o	possibly white	-	diffuse brown; except base of iv	tip often bevelled
Wood Duck	mf	n.d.	n.d.	black-brown	x	-	-	-	rounded tip
Mandarin Duck	m	14	10	black-brown	x	-	-	-	-
Northern Pintail	m	22	8	buff-grey; R1 and R2 black-brown	X	-	white to buff on ov and tip	y: fine undulating, buff lines	ad: outer ones slightly pointed; central ones pointed; R1 elongated
Northern Pintail	f	14	8	brown	-	-	buff on ov and tip	mottled buff wedges on ov; also in R1 and R2 on iv (almost pale bars)	outer ones slightly pointed; central ones pointed
Eurasian Wigeon	m	12 (16)	9	grey-brown	c	pale border	buff (ov)	base of iv sometimes white	tip slightly pointed; c longer and more pointed
Eurasian Wigeon	f/ym	12 (16)	9	dark grey-brown	X	pale border (m + o)	buff (ov)	-	tip slightly pointed
Northern Shoveler	m	13	8	white or black	-	-	white on ov of central Rs	white with grey base; spotted and mottled with brown-grey. Rc black with white border and mottling at base	tip slightly pointed
Northern Shoveler	f	13	8	buff/brown	-	often a pale border	-	increasing number of brown marks towards c; brown with small pale marks	tip slightly pointed

Species	Sex	Rmax	Rmin	Main Colour	Uniform	Tip	Edge	Marks	Particular shape
Gadwall	m	10	8	pale brown-grey	X	-	buff (ov)	-	-
Gadwall	f	10	8	buff to greyish	(c)	dark arrowhead-shaped mark	-	brown on pale background, more or less regular	-
Mallard	m	12	8	black (c); white to grey	c	white (other than c)	-	pale brown-grey; nearly uniform towards c and increasingly mottled and dispersed towards outers	central ones curved upwards (up curled)
Mallard	f/y	12	8	buff/brown	-	often a pale border	-	brown; arrowhead mark towards tip; irregular or bars (c)	central ones slightly pointed
Domestic Duck	mfy	20	8	variable	some-times	sometimes pale	sometimes	often (f)	see Mallard
Garganey	f	9	7	grey-brown to brown (c)		-	pale; ov and tip	pale marks and mottling, especially towards tip	tip slightly pointed
Eurasian Teal	m/y	9 (10)	6	grey-brown to grey (c)	X	-	buff; ov and tip	-	tip slightly pointed
Eurasian Teal	f	9 (10)	6	dark brown	-	-	pale; ov and tip	at least one pale; often several	tip slightly pointed

MAIN COLOUR CRITERIA FOR RECTRICES OF DIVING DUCKS
Short rectrices and spearhead, bevelled.

Species	Sex	Rmax	Rmin	Main colour	Uniform	Tip	Edge	Marks	Particular shape
Common Eider	m	14	8	black to dark brown	X	-	-	-	pointed tip
Common Eider	f	14	8	dark brown	X	-	pale; thin to wide on ov	sometimes pale marks towards tip	pointed tip
Velvet Scoter	mf	9 (11)	6	black	X	-	-	-	quite pointed tip
Common Scoter	mf	13	6	black	X	-	sometimes thin and white at tip	-	pointed tip; Rc very pointed
Long-tailed Duck	m	28	8	black, white	c	sometimes white on iv	white	-	tip very pointed; Rc very elongated
Long-tailed Duck	f	14	8	brown, white	c	white tip	white on ov, sometimes partially	-	tip very pointed; Rc a little longer
Goldeneye	mf	13	5	dark brown-grey with nuance of silvery-grey	X	silver more pronounced	-	-	pointed tip

Red-breasted Merganser	mf	10 (12)	7	grey	X	-	eventually, buff on ov	possibly pale vermiculations; ov of R10 partly white	slightly pointed
Smew	mf	8 (10)	6	grey	X	-	-	possibly pale vermiculations; ov of R10 partly white	slightly pointed
Tufted Duck	mf	8	6	black	X	-	-	-	pointed tip
Greater Scaup	mf	8	6	black-brown with a grey tint	X	-	-	-	pointed tip
Common Pochard	mf	8	5	grey to brown	X	-	-	-	pointed tip
Red-crested Pochard	mf	n.d.	n.d.	brown-grey		-	large pale, ill-defined border	-	-

Nota: the *Aythya* ducks quite often hybridise; generally hybrid 'large' feathers are similar to the parent species and not identifiable

Dabbling (surface feeding) ducks

The Ps in this group are mostly brown and very similar. They can sometimes be distinguished by size (small for EurasianTeal and Garganey; about 75% of the size of other *Anas* ducks), and possibly by the clear base for some, but several species are almost indistinguishable without another associated feather.

The Ss frequently have a metallic bright blue or green speculum, and the tertials also have particular markings, at least in the males.

Rs are often uniform in males and mottled in females, but several species are very close in their coloration.

The greater coverts and scapulars sometimes have identifiable patterns and field guides can be useful, especially in identifying scapulars (photographs are preferable).

COMMON SHELDUCK *Tadorna tadorna:* TP = 12-26, TS = 10-16, TR = 10 (8.5)-14; females are smaller than males.

Possible confusion: with other ducks (Ps), Northern Lapwing (Ss). Ps and Ss black (rachis dark); white base more extensive on the inner vane of the secondaries. Bronze-green iridescence on the inner vane of the Ss. In the juvenile, there is a white tip on the inner Ps and the Ss, while the entire inner vane of the Ss is white (only the base in adults).

Possible confusion: with Ps of Brent Goose (also exotic species), but small white base in the Shelduck. Rs of juvenile with grey geese, but much smaller size.

▷ **Common Shelduck wing (young)**

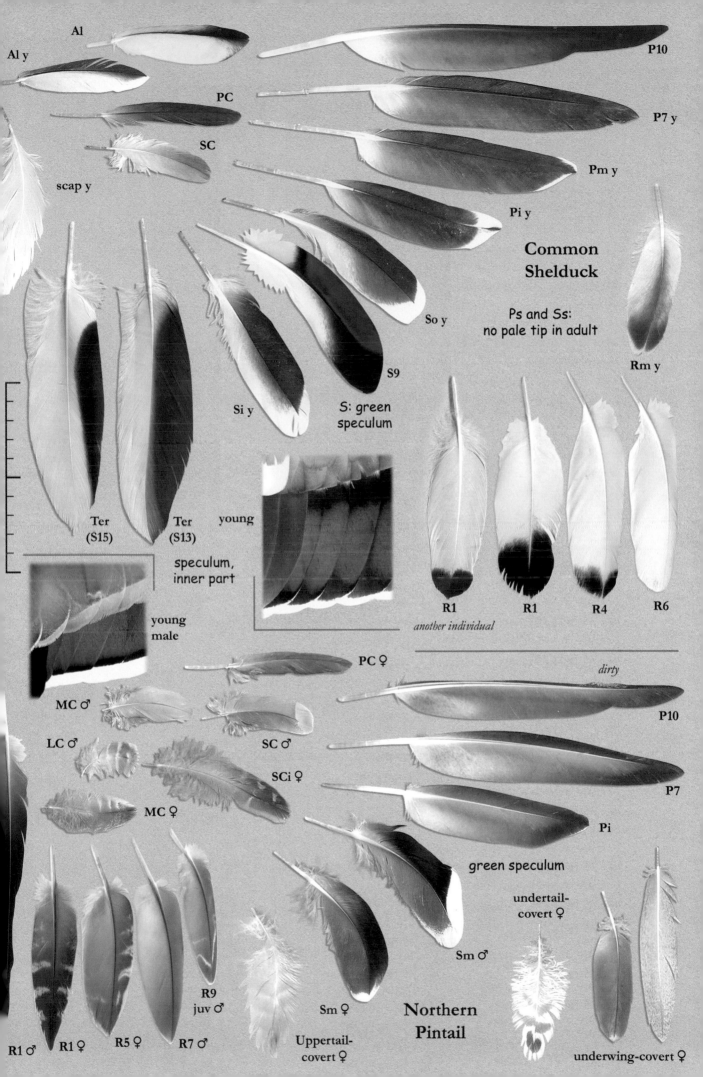

Al

Al y

PC

SC

scap y

P10

P7 y

Pm y

Pi y

Common Shelduck

Ps and Ss:
no pale tip in adult

Rm y

So y

S9

Si y

S: green
speculum

young

Ter
(S15)

Ter
(S13)

speculum,
inner part

R1 R1 R4 R6

another individual

young

young
male

MC ♂

LC ♂

SC ♂

SCi ♀

MC ♀

PC ♀

dirty

P10

P7

Pi

green speculum

undertail-
covert ♀

Sm ♂

R1 ♂ R1 ♀ R5 ♀ R7 ♂ R9
juv ♂

Sm ♀

Uppertail-
covert ♀

**Northern
Pintail**

underwing-covert ♀

EURASIAN TEAL *Anas crecca:* Ss of males with a narrower white tip. The bright green speculum starts at S3, small and basal, and extends to S7. Green usually covers the whole of the outer vane towards (S7) S8-S10 (often without a white tip). More contrasting tertials in males: they are grey-brown with the outer vane paler towards the rachis, then a wide longitudinal black band and a white to buff edge (wider in the male).

GARGANEY *Anas querquedula:* compared to Eurasian Teal, the Ps are more contrasting, with a paler base and a tip that looks darker (difficult to see on heavily worn feathers). The Ss have a wider white tip; the speculum is less bright and less contrasting (appears towards the tip of S2 or S3 and then extends towards the base). Rs of the female have more pale markings. Scapulars of male are grey to black, with a white line along the rachis.

MARBLED DUCK *Marmaronetta angustirostris:* very rare, found in Spain: no data of isolated 'large' feathers. Ps darker brown at the tip. Ss buff to greyish. Rs brown-grey at the tip with a pale edge.

EURASIAN WIGEON *Anas penelope:* Ss with fine white tip, not very prominent dark green speculum, mainly visible on S4-S9. More vivid on more Ss and more extensive in the male. Rs pale brown-grey, paler in the male. Greater coverts of male with black tip, grey inner vane and white outer one.

GADWALL *Anas strepera:* Ss without shiny speculum but one or more of S9-S11 have white outer vane. Rs of male pale brown-grey, those of female buff with brown markings. Outer greater coverts partly dark rufous (male).

NORTHERN PINTAIL *Anas acuta:* Ss with a bottle green speculum tending towards bronze-brown, sometimes purple. Ill-defined green mark, especially S3-S10, paler in female. Tertials with contrasting longitudinal bands, brown or grey and black. Rounded tail with central Rs slightly longer in the female, and very elongated in the male (R1 measures about 1.5 x R2). Rs of the male brown-grey, black-brown towards central ones. Rs of the female brown with pale wedges mottled with dark, almost forming bars on the central Rs. Greater coverts brown-grey with orange (male) or buff (female) tip.

NORTHERN SHOVELER *Anas clypeata:* Ss with ill-defined green speculum, especially on S5-S10, sometimes on S2-S4, and thin or absent white tip. Rs of male white, grey towards base and mottled, black central feathers bordered with white, possibly mottled white. Rs of female various beiges with brown spots; more numerous towards the central Rs, almost brown. The small and median secondary coverts of the male are pale blue.

▷ **Eurasian Teal wing (young female)**

▷ **Eurasian Wigeon wing (young male)**

▷ **Northern Pintail wing (young male)**

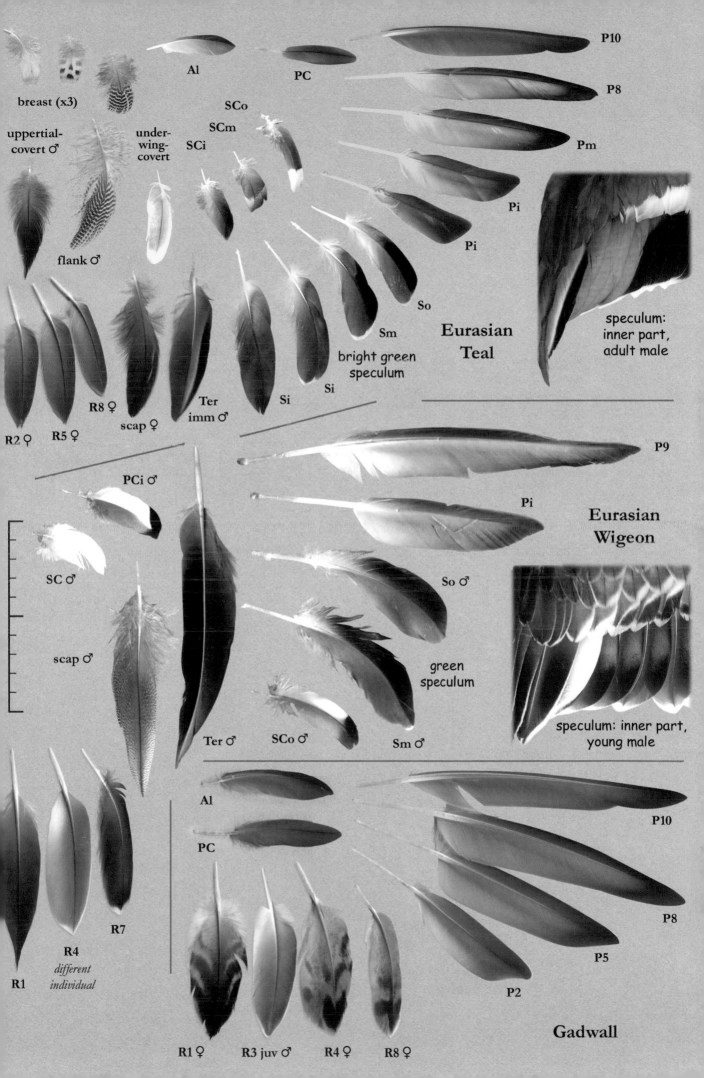

breast (x3)

uppertial-covert ♂

under-wing-covert

flank ♂

Al

PC

SCo

SCm

SCi

P10

P8

Pm

Pi

Pi

So

Sm

bright green speculum

Si

Si

Ter imm ♂

scap ♀

R8 ♀

R5 ♀

R2 ♀

Eurasian Teal

speculum: inner part, adult male

PCi ♂

SC ♂

scap ♂

P9

Pi

Eurasian Wigeon

So ♂

green speculum

speculum: inner part, young male

Ter ♂

SCo ♂

Sm ♂

R1

R4

different individual

R7

Al

PC

R1 ♀

R3 juv ♂

R4 ♀

R8 ♀

P10

P8

P5

P2

Gadwall

MALLARD *Anas platyrhynchos:* its feathers are mainly found on the shores of lakes or ponds, even on small city-centre ponds. This duck has readily identifiable Ss with a large blue speculum, but the rectrices show a wide variety of coloration.

Possible confusion: with Northern Shoveler (male Rs), and other female ducks (female Rs).

Rs of male: black and central feathers curved upward in a 'hook.' Other Rs more variable with white tip, almost uniform brown-grey (broad pale edge on outer vane) to white of outer Rs in certain forms. Generally white more or less more heavily mottled with brown-grey. Sometimes white with a small dark mark towards the tip around the rachis.

▷ **Northern Shoveler wing (male)**

Rs of female: creamy buff-red background, with dark brown markings. Often an arrowhead-shaped mark towards the tip and irregular rounded spots. The outer Rs may have only a little brown around the rachis; the central Rs can have almost regular large bars (sometimes almost entirely brown).

Greater coverts with brown-grey base, then a white band and a black tip.

DOMESTIC DUCKS *Anas* sp.: this 'species' includes various hybrids, most of which originate in part from the wild Mallard, often crossed with the Barbary Duck or sometimes other wild species of duck also kept in captivity. Their feathers are usually found near farms or villages, sometimes in cities on ornamental ponds.

That a bird is of the dabbling duck group is fairly easy to determine from the shape of the 'large' feathers, but their specific identification is much more complicated. Some of the most common variants arise from a change in the 'large' feathers of the Mallard, which reduces the colour or includes white patches on the feathers (including body feathers). The speculum is sometimes missing. Some 'large' feathers are therefore entirely white, and this may come as a surprise. Again, structure and form will be the determining criteria.

Diving ducks

Compared to dabbling ducks of a similar body size, the 'large' feathers are generally smaller and stiffer. The tips of the Ps and Rs can be more pointed. Metallic speculum is replaced by a black/grey/white contrasting wing-bar.

RED-CRESTED POCHARD *Netta rufina:* Ps and Ss white or light grey with dark tip. Ps of the male white with black tip; black also on the outer vane of outer Ps (P10-P8), then decreasing towards the median Ps and absent on P4-P1. Ps of the female similar, but the white replaced by brownish-grey. Ss of both sexes grey with darker tips, bordered with white to pale grey (Ss of the male more contrasting). Often some white ('blurred') in the centre of the inner vane, less on the outer vane. Tertials brown-grey. Rs (SS) brown-grey with a wide pale border, greyer in males, browner in females. Greater coverts brown-grey with darker tips.

Possible confusion: Ps and Ss, with the *Aythya* ducks, but those of Red-crested Pochard are much larger (on average 2 to 3 cm); their calamus is slightly larger.

▷ **Red-crested Pochard wing (female)**

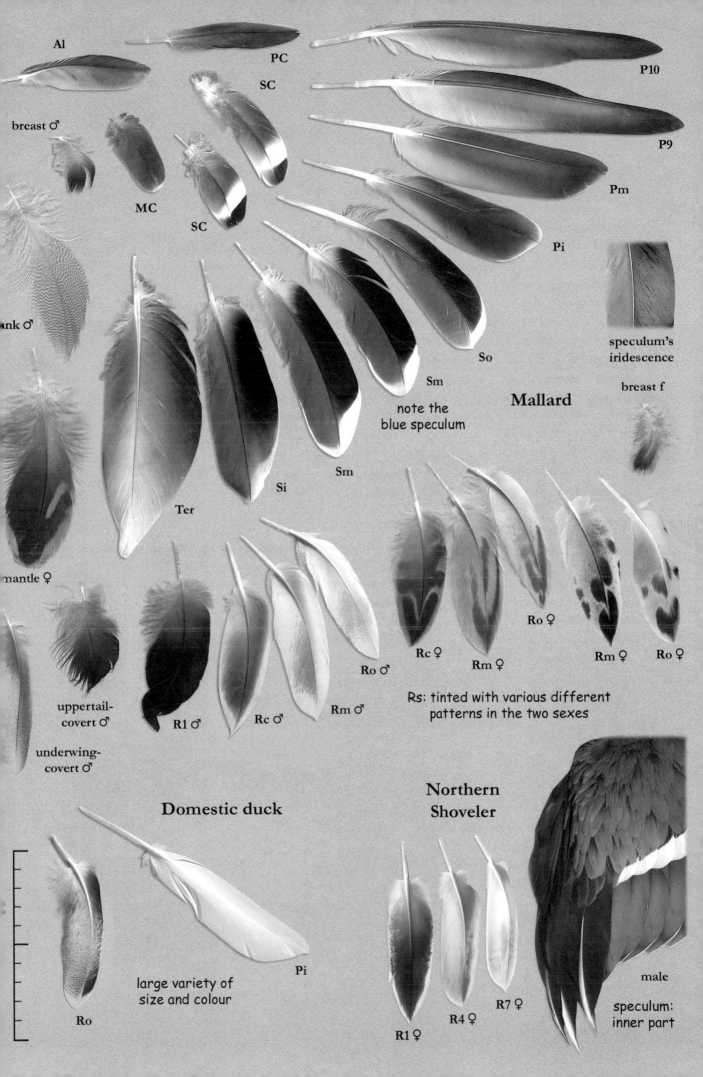

Al

PC

SC

breast ♂

MC

SC

P10

P9

Pm

Pi

So

Sm

speculum's
iridescence

breast f

Mallard

note the
blue speculum

Sm

nk ♂

Si

Ter

mantle ♀

Rc ♀

Rm ♀

Ro ♀

Rm ♀

Ro ♀

Ro ♂

uppertail-
covert ♂

R1 ♂

Rc ♂

Rm ♂

Rs: tinted with various different
patterns in the two sexes

underwing-
covert ♂

Domestic duck

Northern
Shoveler

large variety of
size and colour

Pi

Ro

R1 ♀

R4 ♀

R7 ♀

male

speculum:
inner part

COMMON POCHARD *Aythya ferina:* Ps mid grey with dark grey tip with blurred border, wider on the inner vane, crescent-shaped on the outer Ps but quite straight on the others. Outer vane of the outer 4–5 Ps with a silvery shade and vague pale edge at the darker tip. In the female, slight brownish tint to the Ps. Ss grey with tip darker on the outer Ss (wider on the inner vane), this mark disappearing towards the median Ss. Neat white edge at the end of the Ss, disappearing towards S10, which is totally grey. Tertials grey in males, brownish in females. Pale vermiculations appear towards the median Ss in the male, accentuating towards the tertials. Rs dark grey in males, sometimes browner in females.

TUFTED DUCK *Aythya fuligula:* 'large' feathers with black, grey, and white, the black being sometimes dark brown in the female. Outer Ps black-brown, darker on tip and outer vane. Towards the median Ps (P5-P6) the tint is rather dark grey with a black tip; a long grey zone (from mouse-grey to very light grey) appears towards the middle of the outer vane. The majority of Ss are tricoloured: grey base, white centre, black tip, usually with grey along the rachis. Black is denser and wider on the outer vane. Towards S9 the basal grey increases from the base upwards and the white diminishes. S10 is often dark grey with a black tip and a small white spot in the centre of the outer vane. Tertials black. Rs black-brown to black often with a bronze hue (especially outer vane). Greater coverts dark grey with a black tip.

GREATER SCAUP *Aythya marila* (SS): 'large' feathers very similar to those of the Tufted Duck (slightly larger on average) but there are some differences in coloration. On the Ps, the mark on the outer vane is wider and whiter on P1 to P7. On the Ss, the black tip is more diffuse and wider on the inner vane. There is sometimes grey in the black band at the tip. S10 has more white and S11 often also has a white mark (reduced). Rs have a slight grey hue (appear paler). Greater coverts dark grey, often with a paler area at the tip.

COMMON EIDER *Somateria mollissima:* Ps, Ss and Rs of male black or dark brown (worn); browner on the female. Sometimes a white tip on the inner Ps of the male. Inner Ss black in the male, with a white mark towards S8-S9: a white tip then the white increases on the inner vane towards the base, with S11 often half white. S12 and following totally white. Ss of female all brown or with a buff tip and sometimes a broad pale edge (possibly mottled) on the outer vane. Ss of young male like those of female but fewer pale marks. Rs of the males brown to black. Rs of female dark brown with pale (buff-red to greyish) more or less wide edge on the outer vane; sometimes paler spots on the distal part. Greater coverts of female brown with white tip.

LONG-TAILED DUCK *Clangula hyemalis:* Ps and Ss black-brown; base a little paler (especially inner vane). Ps black (male) to dark brown, with white edges in some individuals. Ss browner; outer vane often tinged with reddish-brown to olive (becoming paler with wear). All Rs quite pointed and narrow towards the tip, R1 of the male very long and tapered, R2 also quite long and narrow. R1 generally measures between 20 and 23 cm, sometimes more (27 cm). Rs of male black and white: R1 often black, R2 black with a narrow white edge on both vanes, R3 almost totally white inner vane and the others with decreasing black. R6 and R7 white with a little black along the rachis, but R7 sometimes all white. Rs of female dark brown with a small white tip and a wide white edge to the outer vane, sometimes incomplete, and the base of the inner vane is paler (sometimes white). In winter plumage, scapulars of male white, narrow and very elongated.

COMMON SCOTER *Melanitta nigra:* Ps and Ss black to dark brown, Rs black. Ps and especially Ss a little browner or dark grey in the female. Inner vane of Ps paler (except at the tip). Outer Ps slightly pointed. Rs slightly pointed, the central ones obviously; tail slightly more pointed in the male. In females the tip of Rs sometimes very finely edged

▷ Common Scoter wing (male)

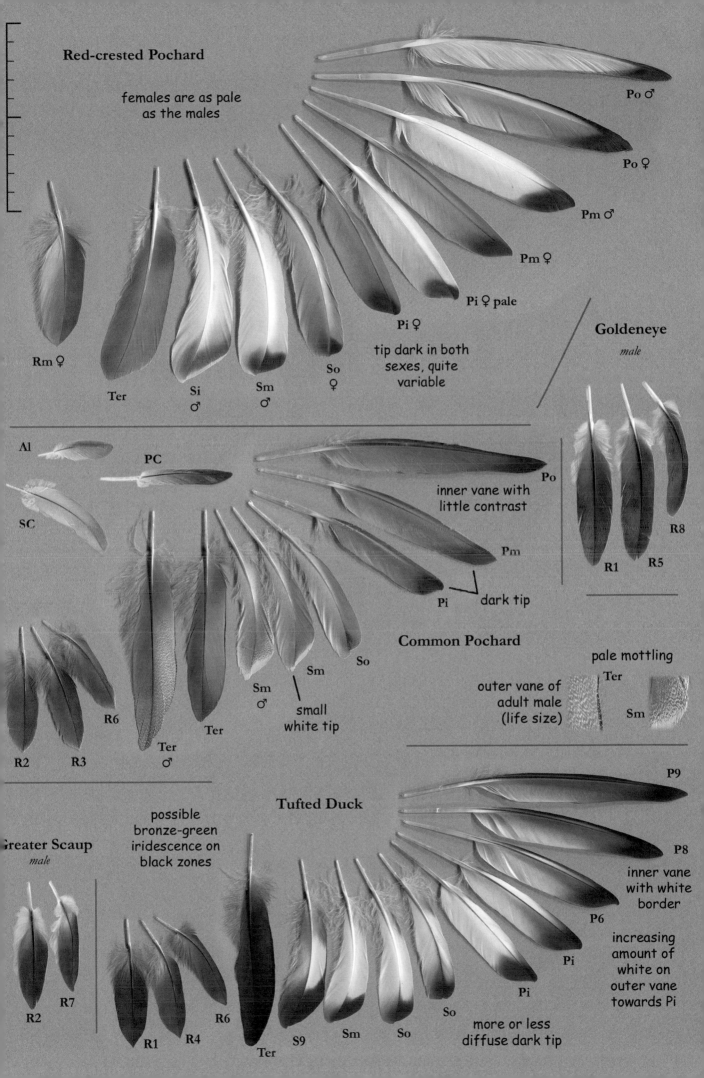

Red-crested Pochard

females are as pale
as the males

Po ♂

Po ♀

Pm ♂

Pm ♀

Pi ♀ pale

Pi ♀

tip dark in both
sexes, quite
variable

Rm ♀

Ter

Si
♂

Sm
♂

So
♀

Goldeneye
male

Al

PC

SC

Po

inner vane with
little contrast

Pm

Pi

dark tip

R8

R1

R5

Common Pochard

pale mottling

Ter

outer vane of
adult male
(life size)

Sm

Sm

Sm
♂

So

Ter

small
white tip

R6

R2

R3

Ter
♂

Tufted Duck

P9

P8

inner vane
with white
border

P6

increasing
amount of
white on
outer vane
towards Pi

Pi

Pi

Greater Scaup
male

possible
bronze-green
iridescence on
black zones

R2

R7

R6

So

R1

R4

Ter

S9

Sm

So

more or less
diffuse dark tip

with white and Ss with a small white tip (outer vane especially). With wear, Ps especially become brown and pale on the inner vane (almost white) and may resemble those of other species. Greater coverts black.

Possible confusion: with Rs of Black Woodpecker, but generally found in a different habitat. In Black Woodpecker: Rs larger (11-21 cm) with narrowed tip and tip often split.

VELVET SCOTER *Melanitta fusca:* Ps black to dark brown, Rs black. Ss white with a little grey at the base. S1 (possibly S2) black and white: inner vane fully or partially white; black outer vane possibly with a small white mark. Often a little black at the tip of S2 and S3 (others sometimes also). Sometimes a black mark at the tip towards S10 and S11. Tertials black. Outer Ps obviously pointed. Rs slightly pointed; shorter and less pointed than those of Common Scoter. Ps also fade and become brown. Ps with less colour contrast between the two vanes than Common Scoter. Greater coverts black with a white tip.

Possible confusion: see Common Scoter, p. 306

COMMON GOLDENEYE *Bucephala clangula:* Ps black. Outer Ss (S1-S4) black with or without a white tip on the inner vane. The white mark increases from S1 to S4, often with a large white mark at the tip and sometimes towards the base. S5-S10 white with a small grey base (sometimes absent). S10 sometimes with black at the tip of the inner vane. S11 and following black, with dark green iridescence in male. Rs dark grey to black, tinged with silvery-grey, especially at the tip. Greater coverts two-toned with black basal half, the rest white usually with an ill-defined black tip. Male has white scapulars, with broad, black edge, at least on the outer vane.

Sawbills

Ps black-grey. Ss black or white and black, distinguished by the detail of their pattern. Rs grey. Size of 'large' feathers is important in distinguishing between the three species (especially for Ps and Rs).

SMEW *Mergellus albellus* (SS): 'large' feathers slightly smaller than those of Red-breasted Merganser. Ps black to dark grey. S1-S10 black with white tip (small on S1). S11 dark grey on the inner vane, black on the outer one, which has a large oval mark that does not touch the base. S12 and following ash-grey to black depending on age and sex. Rs ash-grey, outer vane often mottled with white, very much so on outer Rs. Greater coverts black-brown with small white tip.

RED-BREASTED MERGANSER *Mergus serrator:* 'large' feathers slightly smaller than those of Goosander. Ps dark grey to black. Ss with on average less white than Goosander. S1-S4 grey-black with white tip and the distal part of the inner vane white (poorly defined). S5-S10 white with a black base (one third to one half), neat and black on the outer vane, greyer and blurred on the inner vane. Grey sometimes extends along the rachis. S11 quite variable; rather grey on the inner vane and black on the outer vane, with a more or less extensive white mark on the outer one (sometimes only a black edge on the outer vane, otherwise wholly white). S12-S15 of male white with small dark base with black edge on the outer vane. S16 black. Those of the female and young male are grey with a black edge. Rs ash-grey; outer Rs with outer vane completely marbled with white. Other Rs with pale edge on the outer vane, or with a large border of white markings. Greater coverts black-brown near the wrist, others with basal-half black and the rest white, usually with a black tip.

GOOSANDER *Mergus merganser:* Ps dark grey to black; sometimes a small ill-defined white tip on P1 (P2). Ss black and white: S1 to S4 grey-black with a blurred white tip; white can extend over a large part of the inner vane in a diffuse zone, the outer vane remaining dark. S5-S10 white with the base of the outer vane marked with a well-defined black or grey zone; the inner vane is more or less marked with dark grey, especially near the rachis and at the tip. S11 and the following white with small grey base and black edge, at least on the outer vane. Rs ash-grey, the outer (R10) with a largely white outer vane. Other outer Rs sometimes marbled with pale on the outer vane. Greater coverts black-brown towards the wrist (sometimes speckled with white near the tip), others with a black base (one half to two thirds) and the rest white.

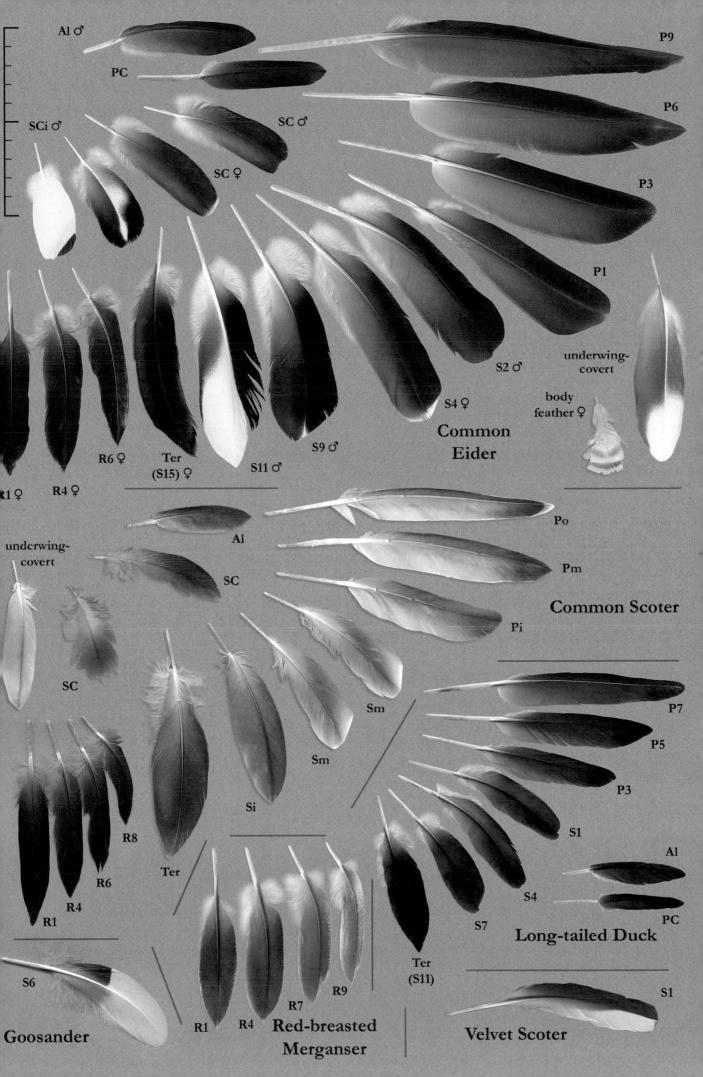

Al ♂
PC
SCi ♂
SC ♂
SC ♀
P9
P6
P3
P1

R6 ♀
Ter (S15) ♀
S11 ♂
S9 ♂
S4 ♀
S2 ♂
underwing-covert
body feather ♀

Common Eider

R1 ♀
R4 ♀

underwing-covert
SC
Al
SC
Po
Pm
Pi

Common Scoter

R8
R6
R4
R1
S6

Goosander

Sm
Sm
Si
Ter
P7
P5
P3
S1

R1
R4
R7
R9
Ter (S11)

Red-breasted Merganser

S4
S7
Al
PC

Long-tailed Duck

S1

Velvet Scoter

For the following species, no 'large' feather data could be found; the following comments are derived from the observation of whole birds (so that the bases of the 'large' feathers cannot be described):

Ferruginous Duck *Aythya nyroca:* Ps and Ss white with black tip (pattern of Ps similar to those of Red-crested Pochard; pattern of Ss similar to those of Tufted Duck). Inner Ss black-brown with greenish iridescence. Rs black-brown.

King Eider *Somateria spectabilis:* Ps, Ss and Rs brown tending towards black in the male. Ss often edged with white at the tip (wider in the female). Scapulars developed into 'sails'.

Steller's Eider *Polysticta stelleri:* Ps and Ss black-brown. Ss with a wide white tip, and possibly a bluish speculum.

Harlequin Duck *Histrionicus histrionicus:* Ps, Ss and Rs dark brown. Bluish speculum on Ss of male.

Barrow's Goldeneye *Bucephala islandica:* Ps and outer Ss black. One or two white and black median Ss (black towards tip and outer vane); the others white. Rs black in the male, dark brown-grey in the female.

Ruddy Duck *Oxyura jamaicensis:* Ps and Ss brown with lighter tip, outer vane speckled with pale at the tip. Rs dark brown. Wedge-shaped tail, pointed rectrices. Confusion possible with cormorants.

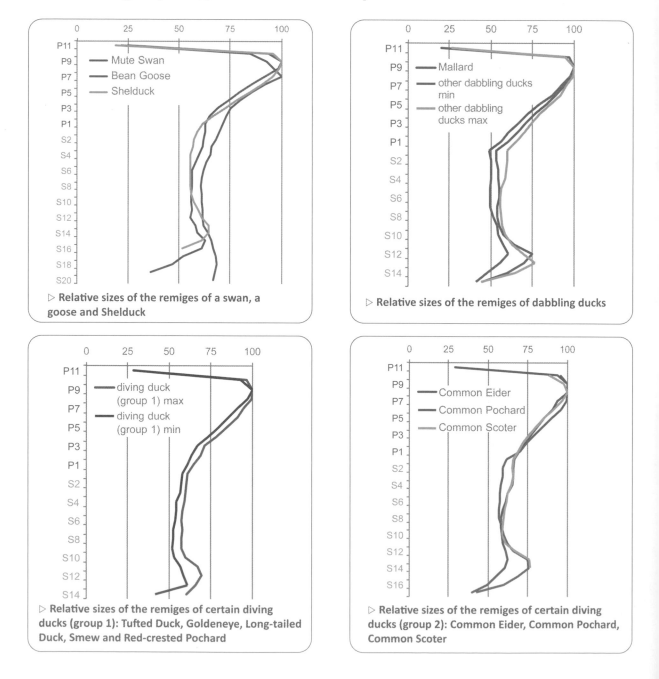

▷ **Relative sizes of the remiges of a swan, a goose and Shelduck**

▷ **Relative sizes of the remiges of dabbling ducks**

▷ **Relative sizes of the remiges of certain diving ducks (group 1): Tufted Duck, Goldeneye, Long-tailed Duck, Smew and Red-crested Pochard**

▷ **Relative sizes of the remiges of certain diving ducks (group 2): Common Eider, Common Pochard, Common Scoter**

WHITE-HEADED DUCK *Oxyura leucocephala:* Ps dark brown. Ss brown marked with pale on the outer vane and tip. Rs dark brown. Tail wedge-shaped, pointed rectrices. Confusion possible with cormorants.

Tail shapes

The Anatidae generally have rounded tails, but with variations. The proportions between the rectrices may not only vary within a species depending on age and sex, but also individually.

A square to slightly rounded tail is found when the rectrices are similar in size (Rmin > 90% Rmax), for example in Bean Goose, Brent Goose, Common Shelduck, Northern Shoveler, Eurasian Teal and Tufted Duck. Eurasian Wigeon also has a slightly rounded tail, but its central rectrices can measure 15% more than the ones next to them.

The tail is more obviously rounded (Rmin between 80 and 90% of Rmax) in other species, such as Bewick's Swan, Mallard, Common Eider, the sawbills, etc.

The tail is clearly rounded to slightly wedge-shaped in several species, such as Mute Swan, Whooper Swan, female Northern Pintail, Long-tailed Duck, Velvet Scoter, Common Goldeneye, etc. It is clearly wedge-shaped in White-headed Duck, for example.

Ornamental tail shapes are found in the males of some species, in which the central rectrices may be a little elongated (Eurasian Wigeon) or the tail very elongated (Northern Pintail and Long-tailed Duck).

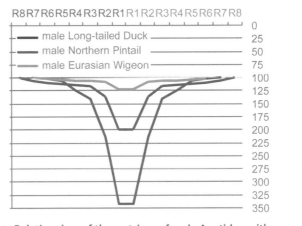

▷ Relative sizes of the rectrices of male Anatidae with a pointed tail: Long-tailed Duck and Northern Pintail (Eurasian Wigeon for comparison).
Note: for easier interpretation the relative sizes are given as a percentage of the shortest rectrices (outer rectrix = 100%)

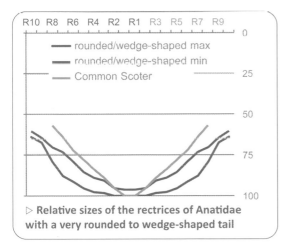

▷ Relative sizes of the rectrices of Anatidae with a very rounded to wedge-shaped tail

• ARDEIDAE: HERONS, EGRETS, ETC.

NP = 10 (+1) / NS = 15-20 / NR = 12

Distinctive criteria of group: shape of outer Ps and rectrices, size and colour for some species, where found.

This family has rectangular or rounded wings. The tail is short and square or rounded.

Possible confusion: with diurnal birds of prey (large, non-white species), white pigeons (white species), Rallidae (Little Bittern), gulls and cranes (rectrices), etc.

MEASUREMENTS OF THE 'LARGE' FEATHERS OF HERONS						
Species	Pmax	Pmin	Smax	Smin	Rmax	Rmin
Little Bittern (SS)	12	9	9	8	6	5
Cattle Egret (SS)	22	14	16 (17)	12	11	8
Squacco Heron (SS)	n.d.	n.d.	n.d.	n.d.	12	8
Little Egret	24	17	19	14	13,5	9
Night Heron	26	19	20	14	13	10
Eurasian Bittern	30	20	22	17	12	8
Great Egret	35 (38)	25 (22)	28 (30)	21	20	16
Purple Heron	34	24	26	19	15 (20)	12
Grey Heron	37	25	28	20	20	15

Primaries: the shape of the outer 2-3 Ps is characteristic for some species; they are quite wide and clearly notched high up the feather. The Little and Eurasian Bitterns have no notch or emargination.

Secondaries: quite wide and long (about 70 to 80% of Pmax).

Rectrices: short, rounded, sometimes curved inwards (especially outer ones). Square or slightly rounded tail (all Rs of an individual are of a similar size).

Body feathers: most species have 'plumes' and/or ornate scapulars, or other ornamental feathers during the breeding season. The 'plumes' growing on the head (two per bird in general) are very elongated with short, attached barbs. The barbs of the scapulars are mostly free and very long. Certain species were almost hunted to extinction due to unregulated collection of their ornamental feathers for use in the millinery industry.

Grey to black 'large' feathers:

GREY HERON *Ardea cinerea:* Ps and Ss black-grey with a paler base. Median and inner Ps with inner vane dark grey towards the edge. Ss dark grey becoming black at the tip and on the distal half of the outer vane. Inner Ss ash-grey. Rs ash-grey, slightly darker towards the tip (especially median and central Rs). Greater coverts ash-grey on the outer vane, darker on the inner. Grey ornamental scapulars; 'plumes' black; black and white neck feathers.

▷ Grey Heron wing

PURPLE HERON *Ardea purpurea:* Ps and Ss black-brown; Ps darker, inner Ss browner. Ss sometimes more contrasting, with the outer vane more greyish. Rs dark brown-grey. Greater coverts like those of the Grey Heron but tinged with grey-brown on the border of the outer vane. Ornamental scapulars with some rufous; 'plumes' black; black and white neck feathers most of which have red or buff markings.

BLACK-CROWNED NIGHT HERON *Nycticorax nycticorax:* Ps, Ss and Rs mid grey, without any contrast. The 'large' feathers of the young are brown-grey, the remiges with a neat white tip. Inner Ss of adult black. 'Plumes' white.

▷ Purple Heron wing

▷ Eurasian Bittern wing

▷ Details of primary of Eurasian Bittern

▷ Details of secondary of Eurasian Bittern

Al

P10

PC

SC

P8

Grey Heron

P4

So

plume

body feathers scale 1/4

Sm

Rm

R6

Si

body feathers   underwing-covert   scapulars

R1

underwing-covert

P10

PC

Black-crowned
Night Heron

SC

10 cm: scale 1/3

P8

P4

P10

P1

P8 imm

S1

P7

S11

S7

Purple Heron

urasian Bittern

So

Sm imm

R1
imm

Ro

scap imm

Si imm

MC imm

SC imm

breast

R3

R5

'Large' feathers rufous-buff with black markings:

EURASIAN BITTERN *Botaurus stellaris:* its 'large' feathers are easily identifiable, although at first glance they resemble those of a pheasant or raptor. However, because the outer Ps are neither notched nor emarginated identification can be confirmed by the shape and size of the 'large' feathers. It is, however, rare to find isolated Bittern feathers, given its relatively inaccessible favoured habitat. Ps and Ss are barred but the outline of the bars is quite irregular. Inner Rs and Ss are marked more haphazardly, with marks sometimes forming incomplete bars. Depending on the individual and the position of the feather, the background colour can be buff to very rufous, with dark brown to black marks. The wing is rounded, with the Ss measuring about 80% Pmax.

Small 'large' feathers, brown to black:

LITTLE BITTERN *Ixobrychus minutus:* its very small 'large' feathers can be confused with those of Water Rail or Moorhen. The Ps have no notch or emargination. Ps and Ss brown, inner Ss rufous in the male. Rs black or dark grey. Moulting begins before migration and is completed in the winter quarters.

▷ Little Bittern wing (male)

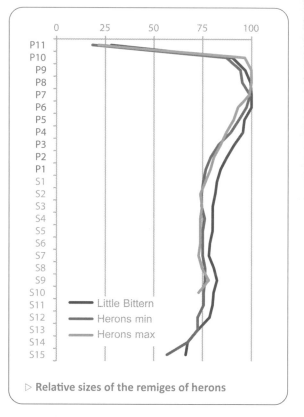

▷ **Relative sizes of the remiges of herons**

White 'large' feathers:

See also p. 55 for other species with white 'large' feathers.

The details to note are the general lack of any distinctive character except the high emargination/notch on the outer Ps; the Rs are quite curved and rounded, rather short.

After contact with mud and water, the feathers of these species may appear greyish.

GREAT EGRET *Casmerodius albus:* size of Grey Heron. No 'plumes' but very long ornamental scapulars, exceeding 20-25 cm and sometimes 35 cm.

LITTLE EGRET *Egretta garzetta:* pure white 'large' feathers. Remiges of similar size to those of pigeons (the latter have a thicker rachis, especially at their base). Body feathers and 'plumes' white; scapulars fluffy (up to 20-22 cm at least).

CATTLE EGRET *Bubulcus ibis:* 'large' feathers similar in size to those of the little Egret; apparently not separable. Sometimes off-white and not pure white. Many yellow to orange body feathers in breeding plumage.

SQUACCO HERON *Ardeola ralloides:* slightly smaller than the previous two species. 'Large' feathers pure white in adult, sometimes with yellowish wash; those of young may be washed grey-brown. Body feathers yellowish to purple-brown; ornate nape feathers with black edges.

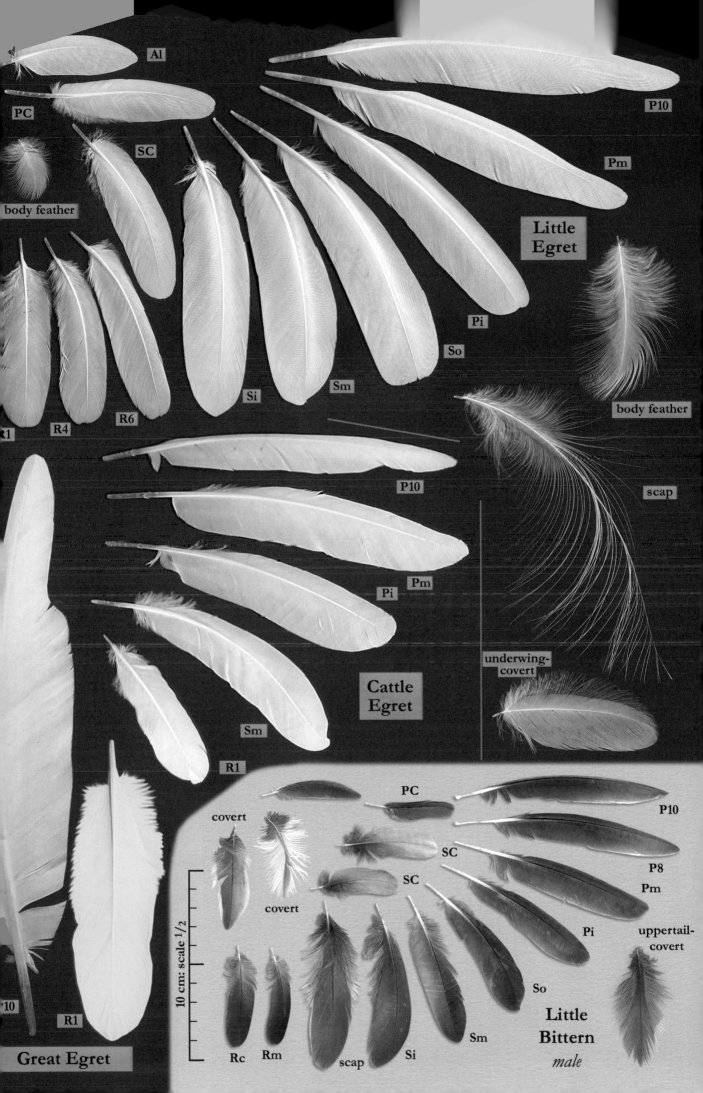

Al

PC

SC

body feather

R1

R4

R6

Si

Sm

So

Pi

Little
Egret

P10

Pm

body feather

scap

underwing-
covert

P10

Pi

Pm

Cattle
Egret

Sm

R1

Great Egret

R1

P10

covert

covert

10 cm: scale 1/2

Rc

Rm

scap

Si

Sm

PC

SC

SC

So

P10

P8

Pm

Pi

uppertail-
covert

Little
Bittern
male

• PELECANIDAE: PELICANS

White Pelican: NP = 11 / NS = 30 / NR = 24

Distinctive criteria of group: size, colour, where found.

Pelicans have broad wings and are quite capable of soaring. The hand is rectangular with 'fingers', the forearm is very wide. The tail is proportionally quite small and rounded.

Possible confusion: with storks, other large wetland birds with 'fingered' primaries (Ps) or short tails (Rs).

Colour of 'large' feathers: the Rs are pearly-white, sometimes slightly pink (White Pelican). Those of the young are tinted grey to brown towards the tip around the rachis and on the outer vane. The Ps are black-brown becoming greyer towards the wrist. The base of the Ps is much paler. A 'frosty' appearance covers the remiges, especially on the inner vane of the Ps; it becomes very pronounced on the Ss. The Ss have distinctly white edges to the outer vane, with the frosty appearance on their surface that

▷ **Base of a White Pelican primary.**
The frosted appearance is due to the particular placing of certain barbs

becomes more diffuse towards the rachis. From below the white band is quite narrow and is not diffuse. The Ss are brown and grey, with an extensive frosty appearance on the outer vane. The tertials are brown-grey and well developed, and the inner ones are white. Ss of young are darker, less white and less 'frosted' than those of adults. The 'frosty' appearance can disappear with wear.

Primaries: the 5 outer Ps are notched and emarginated very low on the feather (about 35-60%). Their calamus is quite long (C = 20-25%) and robust.

Secondaries: the Ss have a rather elongated silhouette and a long, robust calamus (C = 25-30%). The tertials are fairly straight, or even slightly bent towards the outer wing and appear to be from the opposite wing (observe the curvature of the vanes in profile to make sure).

Rectrices: the Rs are robust but quite short, a little curved, with a long calamus (C =20-25%).

WHITE PELICAN *Pelecanus onocrotalus:* TP = 33-50 (60), TS = 30-38, TR = 15-22 (SS).

The rachis of the Ps is pale, at least at the base of the inner Ps. The underside of the Ps and Ss appears quite dark. Their upperside, however, has a silvery or frosty appearance, especially on the outer vane of the Ss (which appears white) and the inner vane of the Ps (which appears pale grey). The upperside and underside of the feather are of different colours. The Ss have a white edge (except sometimes the outer ones), narrow underneath but wide on top, due to the extent of the frosty appearance.

DALMATIAN PELICAN *Pelecanus crispus* (SS): TP = 32-50. TS = 30-38. TR = 20-28.

The rachis of the Ps is dark. The inner vane of Ps and Ss is mainly white, as is the base of the outer vane. The underside of the remiges appears pale.

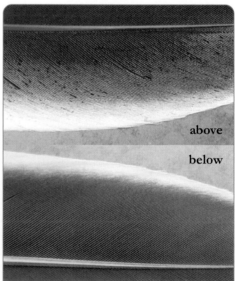

above

below

▷ **Border of the outer vane of a secondary of White Pelican.**
The white edge is quite narrow and well-defined below, wide and diffuse above.

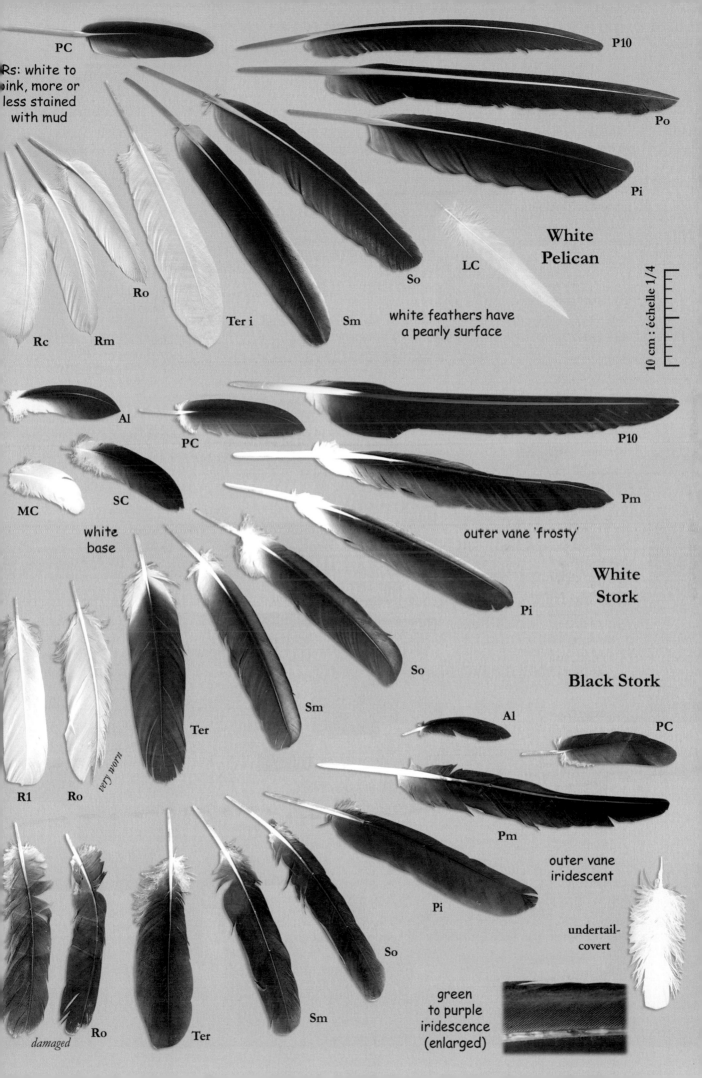

PC

Rs: white to
pink, more or
less stained
with mud

P10

Po

Pi

Ro

Rc

Rm

Ter i

Sm

So

LC

**White
Pelican**

white feathers have
a pearly surface

10 cm : échelle 1/4

Al

PC

P10

MC

SC

Pm

white
base

outer vane 'frosty'

**White
Stork**

R1

Ro

very worn

Ter

Sm

So

Al

PC

Black Stork

Pm

outer vane
iridescent

Pi

So

undertail-
covert

damaged

Ro

Ter

Sm

So

green
to purple
iridescence
(enlarged)

• OTHER LARGE, LONG-LEGGED WETLAND BIRDS: STORKS, IBISES, SPOONBILL, FLAMINGO, CRANE

Distinctive criteria: size, colour.

Several criteria are common to these species from different orders: large size, outer Ps clearly notched, relatively short tail, dominant black and/or white colours on 'large' feathers. Confusion may occur between species in this group; mainly between pelicans, large herons, large waterfowl and diurnal birds of prey.

WHITE STORK *Ciconia ciconia*, Ciconiidae):

NP = 11 (+1) / NS = 22 / NR = 12

TP = 28-49. TS = 26-31. TR = 18-27.

P9-P5 emarginated and P10-P6 with a low notch: Not = 40-70%, Em = 35-65%. Ps and Ss black with white base, with green or purple iridescence and a grey to white area on the outer vane giving a 'frosted' appearance when fresh (wider on the Ss). This frosted appearance varies in size and is more or less apparent depending on the individual; it can also partly disappear with wear. Rachis white at the base (maximum basal one third), then black. Rs white.

Possible confusion: Ps and Ss with pelicans, Common Crane, Black Stork and Egyptian Vulture. Rs with White-tailed Eagle (but it has larger and more rigid Rs).

BLACK STORK *Ciconia nigra* (Ciconiidae):

NP = 11 (+1) / NS = 22 / NR = 12

TP = 30-50 (54). TS = 26-33. TR = 20-26 (32).

Ps and Ss black-brown with strong purple and green iridescence, with no (or very little) white at the base. Rachis quickly becomes brown. Rs similar to remiges and tinged with brown. Rectangular tail; very slightly rounded (Rmin > 88% Rmax).

Possible confusion: with Glossy Ibis (smaller), White Stork, pelicans, Common Crane.

COMMON CRANE *Grus grus* (Gruidae):

NP = 10 (+1) / NS ≥ 16 + 7 tertials / NR = 12

TP = 26-45. TS = 26-32. TR = 21-24.

Ps black with grey to white base. P6-P9 emarginated and P7-P10 with very low notch. Not = 35-50% and Em = 30-55%. Calamus quite long (C = 20%). Ss black becoming grey towards the base, greyer on the inner vane. Rs grey becoming dark grey to black towards the tip. Tail square. Extended, ruffled, often black-edged tertials (reminiscent of the scapulars of Grey Heron). Greater coverts with ill-defined, more or less extensive, black at the tip.

Possible confusion: with other large wetland species, diurnal birds of prey and Grey Heron (Rs).

GREATER FLAMINGO *Phoenicopterus roseus* (Phoenicopteridae):

NP = 11 / NS ≥ 25 / NR = 14

TP = 20-33. TS = 16-22. TR = 11-16.

▷ **Common Crane wing**

▷ **Greater Flamingo wing (young)**

▷ **Detail of the tertials on the wing of a Common Crane.**

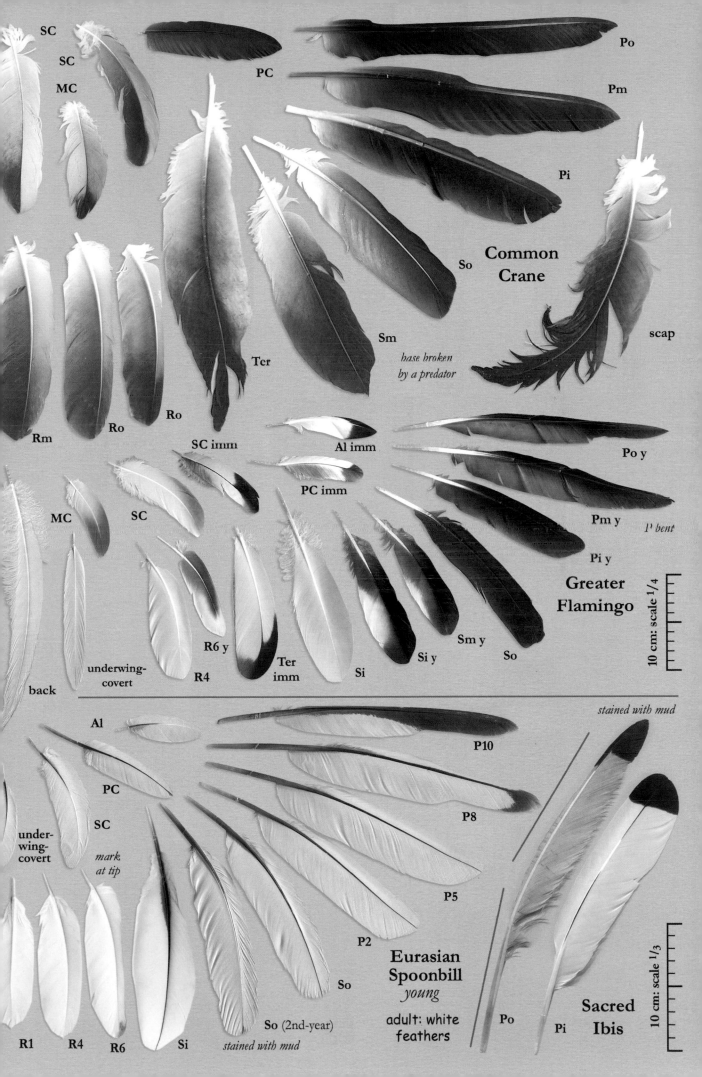

SC

SC

MC

PC

Po

Pm

Pi

So

**Common
Crane**

Sm

*base broken
by a predator*

scap

Ter

Ro

Ro

Rm

Ro

SC imm

Al imm

Po y

PC imm

Pm y

P¹ bent

Pi y

MC

SC

**Greater
Flamingo**

R6 y

Ter
imm

Si

Sm y

Si y

So

underwing-
covert

R4

back

10 cm: scale ¹/₄

stained with mud

Al

P10

PC

SC

P8

under-
wing-
covert

*mark
at tip*

P5

**Sacred
Ibis**

R1

R4

R6

Si

So (2nd-year)

stained with mud

P2

So

**Eurasian
Spoonbill**
young

adult: white
feathers

Po

Pi

10 cm: scale ¹/₃

Ps and Ss black. Ps with emarginations (P8-P9) and notched (P9-P10, very prominent) but high on the feather (Not - 70%). Inner Ss white in young and then pale pink in adults. In the young, the Ss have a blurred white spot along the edge of the inner vane, more or less extensive (and perhaps absent). Rs of the adult pale pink, especially visible on the outer vane (very pale pink to white on the inner vane). Rs in young brown-grey with broad, ill-defined white tip; the white descends a long way on the edge of the vanes towards the outer Rs. Square tail, very slightly rounded (Rmin > 88% Rmax).

Most body feathers are tinged with pale pink in adults, the under- and upperwing-coverts are pale pink to bright red-pink. In the young, the alula and many of the wing-coverts are white with a black-brown tip.

Possible confusion: Ps and Ss with Grey Heron, Purple Heron; the Rs with those of pelicans, swans, geese (imm).

GLOSSY IBIS *Plegadis falcinellus* (Threskiornithidae):
NP = 10 (+1) / NS ≥14 / NR = 12
TP = 15-24. TS = 14-17. TR = 10-13.

Ps, Ss and Rs black-brown with bronze, purple or green iridescence. Ps with green iridescence (often also on underside). Ss with green iridescence and then also purple towards the inner Ss. Rs with purple and dark green iridescence at the edges. Square tail.

Coverts and body feathers black, most with some iridescence.

SACRED IBIS *Threskiornis aethiopicus* (Threskiornithidae):
NP = 10 (+1) / NS ≥ 14 / NR = 12
TP = 24-30. TS = 20-28. TR = ?

Ps and Ss white with tip 'dipped in black ink'. The rachis can be marked with black or be completely white; the calamus is white. Rs white, at least in their distal half. (Some of the illustrations consulted show Rs with red on the outer vane, at the base, or on the border at the base).

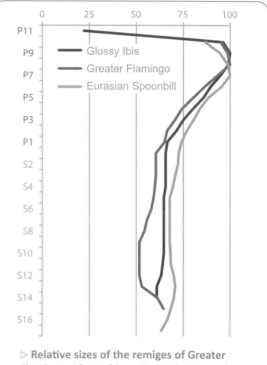

▷ **Relative sizes of the remiges of Greater Flamingo, Glossy Ibis and Eurasian Spoonbill**

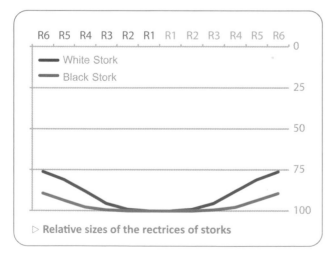

▷ **Relative sizes of the rectrices of storks**

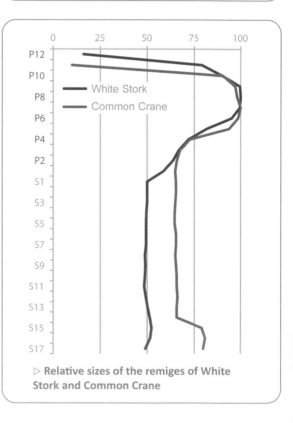

▷ **Relative sizes of the remiges of White Stork and Common Crane**

Possible confusion: with Black Swan, but there's a tegmen on the underside of the Ps in the swan and they have a different shape, young of Eurasian Spoonbill (see below).

EURASIAN SPOONBILL *Platalea leucorodia* (Threskiornithidae):
NP = 10 (+1) / NS >17 / NR = 12
TP = 21-31. TS = 17-22. TR = 12-15.

Ps and Ss white in adults. In young the tip of the remiges and the rachis are black. The outer Ps have a large black tip with irregular edge. Towards the inner Ps, the black on the vanes disappears, but the Ps and Ss have a black rachis and often a small black tip. The black of the rachis appears to spread a little onto the vanes towards the base of the remiges. Some individuals have more black at the tip on all Ps and Ss. The dark colour can extend to the calamus. In subadults, the feather is white and only the base of the rachis is dark. Rs white. The species frequents mudflats so the Rs are sometimes grey due to staining by mud. Tail square, very slightly forked (Rmin [central] > 86% Rmax). *Possible confusion:* Ps and Ss of young, compared to the Sacred Ibis, the black/white border is less well defined; the black descends to the base of the rachis in the Spoonbill.

• RALLIDAE: RAILS, CRAKES, GALLINULES

NP = 10 (+1) / NS = 10-20 / NR = 10-14 (16)
Distinctive criteria of group: shape, where found.
This family has short, broad wings, with a slightly rounded triangular hand. The tail is short.
Possible confusion: remiges: with passerines, pigeons, Little Bittern (but less curved in profile).
Colour of 'large' feathers: Ps and Ss grey in coots, black with blue iridescence in Purple Swamphen, brown more or less mixed with rufous in other species. Rs black to brown, often with some rufous.
Primaries: they are curved in profile to compensate for the heavy wing-loading; often a little hooked with a fairly long calamus (C = up to 25%). The curvature is less obvious in small species.

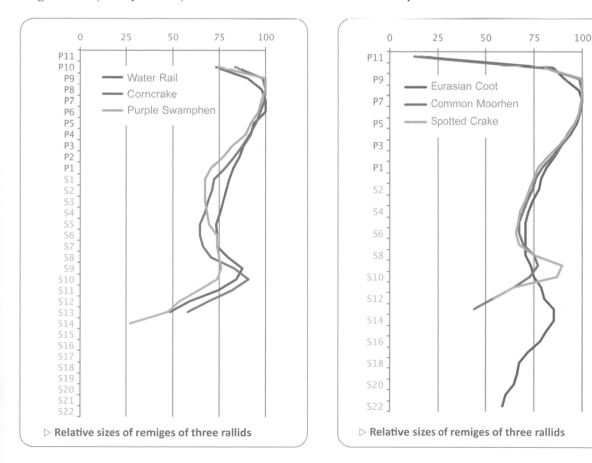

▷ **Relative sizes of remiges of three rallids**   ▷ **Relative sizes of remiges of three rallids**

Secondaries: tip wedge-shaped (not rounded), except in the coots. Tertials more elongated.

Rectrices: they are generally (very) short and appear fragile, as they are hardly used when moving.

MEASUREMENTS OF THE 'LARGE' FEATHERS OF RAILS, CRAKES AND GALLINULES							
Species	Pmax	Pmin	Smax	Smin	S inner	Rmax	Rmin
Water Rail	10.5	7	8.5	7	5.5-10	7	4
Spotted Crake	9.5	7	7.5	6	5-8.5	6	4
Little Crake (SS)	9	6	6.5	5	4-6	6	4.5
Baillon's Crake (SS)	7,5	5	5.5	4	4.5-6	?	?
Corncrake	12 (14)	7.5	9	6.5	6.5-10	7.5	5
Common Moorhen	15	10	11	8.5	6-10.5	9	6
Purple Swamphen (SS)	20 (25)	16	16.5 (20)	14.5	11-16	9.5	5.5
Eurasian Coot	16.5	10.5	12.5	9	8-12	8	5
Red-knobbed Coot (SS)	20	14?	16?	12?	n.d.	10?	6?

WATER RAIL *Rallus aquaticus:* Ps and Ss like those of a small Common Moorhen. Ps and Ss brown, slightly tinged with rufous at the base on the edge of the vanes, sometimes also at the tip of the Ss. P10 edged with grey on the outer vane. Tertials reddish-brown with darker centre. Rs brown with reddish tint, becoming more contrasting towards the central feathers; the central feathers almost black with rufous. Tail square very slightly rounded (Rmin > 85% Rmax). Secondary coverts and the majority of dorsal body feathers black-brown in their centre, largely bordered with reddish-brown to buff. Flank and belly feathers and underwing-coverts black with several white bars.

▷ **Water Rail wing**

Body feathers of vent with grey down, then with a thin white bar, a black bar, and a light rufous bar.

SPOTTED CRAKE *Porzana porzana:* Rs and Ss very similar to those of Water Rail, though slightly smaller. Fine white bars appear on the outer vane of inner Ss, becoming well-defined (often 4 bars) on the tertials, darker in the centre with a large reddish border on the inner vane. Rs black-brown with reddish-brown border (mostly outer vane), some with small white longitudinal marks or lines towards the edge of the vanes (one or both vanes with marks). Tail rounded (Rmin > 75% Rmax).

Secondary coverts and the majority of dorsal body feathers black-brown in the centre and with wide reddish-brown to buff border and thin white lines along the borders, or in 2-4 incomplete bars. Some ventral body feathers brown mixed with black with well-defined white to buff bars.

LITTLE CRAKE *Porzana parva* (SS): Ps and Ss brown. Outer Ps with an ill-defined rufous edge on the outer vane and buff on the tip and the inner vane. Tertials darker in the centre and bordered with buff to rufous. Rs brown, more or less clearly bordered with rufous on the outer vane and the tip, and also on the inner vane of the central feathers.

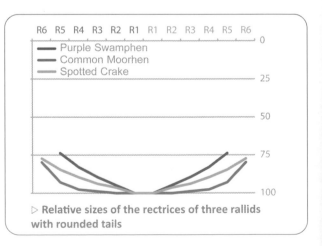

▷ **Relative sizes of the rectrices of three rallids with rounded tails**

Al

PC

Po

Water Rail

SC

Pm

underwing-covert

Pi

So

Sm

Si

scap

Ro

R1

Rm

underwing-covert

P10

Al

PC

Spotted Crake

SC

Po

Pm

Pi

fine white bars
towards the Si

Ro

Sm

R1

Rm

Si

scap

scap

more or less white stripes

Al

P9

PC

SC

P6

MC

Pi

Pi

back

Si

S1

Sm

Rm

rufous
border

Si

Corncrake

two individuals of different colour

BAILLON'S CRAKE *Porzana pusilla* (SS):): 'large' feathers very similar to those of Spotted Crake, but a little smaller. Generally fewer white marks on inner Ss and on the Rs.

CORNCRAKE *Crex crex:* Ps and Ss rufous or brown largely tinged with reddish on the outer vane. P10 with buff edge on the outer vane. Tertials black in their centre with rufous border. Rs black in their centre with a wide reddish border; outer Rs sometimes without black. Tail square, very slightly rounded (Rmin > 80-85% Rmax).

COMMON MOORHEN *Gallinula chloropus:* Ps and Ss dark brown, sometimes with a very thin buff to red edge to the whole feather. P11 and P10 have a white edge on the outer vane. Ss with wedge-shaped tip. Tertials elongated and almost black. Rs black, longer than those of coots. Rounded tail (Rmin > 75% Rmax).

Possible confusion: with pigeons (Ps), but more supple.

PURPLE SWAMPHEN *Porphyrio porphyrio:* 'large' feathers black (or dark grey), with a blue to purple iridescence on the outer vane; rather grey on P10. Tail rounded (Rmin > 75% Rmax). Alula, wing-coverts and the majority of body feathers similar. White undertail-coverts. Depending on age and sex, blue iridescence and black background are more or less evident.

▷ **Common Moorhen wing**

EURASIAN COOT *Fulica atra:* Ps and Ss grey. Ps slightly darker towards the tip, P10 with white edge on the outer vane. Sometimes P1 and P2 with small amount of white at tip. Ss with a white mark at the tip, larger towards the inner Ss; rachis black right to the tip. White may spread more widely and diffusely on the inner vane (as much as to the distal two thirds, usually in young birds). Tertials elongated and black-grey. Rs black, short and curved. Tail square (Rmin > 90% Rmax).

Possible confusion: with pigeons (but Ps more flexible, Ss with irregular white mark), ducks for the Ss (but underside of rachis dark grey, usually pale in ducks).

RED-KNOBBED COOT *Fulica cristata:* 'large' feathers slightly larger than those of Eurasian Coot, without a pale mark at the tip of the Ps and Ss; darker below.

TERRESTRIAL BIRDS OF DRY HABITATS

• OTIDIDAE: BUSTARDS

NP = 10 / NS ≥ 15 / NR = 18-20

Distinctive criteria of group: shape, size, colour, where found.

This family has wide, rectangular wings, with a fingered hand. The tail is rectangular to rounded. The size of the 'large' feathers is comparable to

Species	Pmax	Pmin	Smax	Smin	Rmax	Rmin
Little Bustard	18	12.5	13.5	11.5	12	10
Great Bustard	61	33	43	21	29	21

that of Common Pheasant for Little Bustard and of the storks for Great Bustard.

Colour of 'large' feathers: Ps and Ss brown to black and white, or whitish. Rs with white and/or rufous and black bars.

Primaries: outer Ps strongly emarginated/notched.

Secondaries: oval, quite curved. Tertials slightly pointed.

Rectrices: quite wide, with rounded tip.

LITTLE BUSTARD *Tetrax tetrax* (SS): Ps and Ss white and black (or dark brown). P7-P10 notched quite low, brown on the narrow part and white at the base. The black is rarely extensive and restricted to the tip.

Male: other Ps with black only at the tip in the form of a terminal bar to P6 and then a subterminal bar towards the inner Ss. The black mark is often staggered and lower on the outer vane. Ss usually white (possibly with buff

Al

PC

P9

P5

scap

P1

S1

S6

S10

**Eurasian
Coot**

ill-defined white
tip; dark rachis

rip of Ss: variable placing
and extent of white

R1

R4 underside

R6

Rs: wide ashy-grey border

Common Moorhen

broken base in some feathers

Al

white edge

flank

belly

SC

P10

back

scap

Pm

Pi

Si

So

R6

Sm

R1

R4 underside

brownish zone

dertail-
covert

Rs: wide greyish border

P10

Al

P6

PC

SC

Ps and Ss:
outer vane
tinted blue

underwing-
covert

P1

Si

S6

S1

**Purple
Swamphen**

base); inner ones sometimes with black tip. Tertials buff, yellow, or reddish with black V-shaped bars and small black marks. Greater coverts black with large white base and small white tip.

Female: black mark at the tip of the Ps less extensive; possibly other black marks; often another incomplete bar on the outer vane (especially inner Ps). Ss white (possibly with buff base) with some wavy black bars, complete or not, and often other small black marks. Tertials as those of male.

Rs of both sexes buff-red for the central feathers (R1 at least, R2 and R3 often), white for the outer ones (2 pairs at least) and a mixture of the two shades for the others. All have several wavy black bars, with a variable number of small black marks between each bar, especially in the distal part, but usually the tip is without marks if the background is white.

Possible confusion: with diurnal birds of prey (outer Ps and Rs) but their pattern is quite characteristic.

GREAT BUSTARD *Otis tarda* (SS): males are much larger than females; they are the heaviest flying birds in Europe. Ps dark brown to black becoming white at the base, more white on the inner vane. Pale rachis on at least P4-P10, a lot of white around the rachis on the underside. P6-P10 emarginated and notched quite low (Not = 45% for P9). Ss black with about half the basal part white; speckled border, a little more extensive on the inner vane. The tip of the Ss and inner Ps is sometimes marked irregularly with white to buff. Some tertials white with small black tip, the others rufous with black barring. Complete wide black bars usually alternate with thin, wavy and incomplete bars.

Rs rufous becoming white at the base, with black bars and white tip. Females often have less white than males, more black bars and the rufous colour is paler. Several black, wide bars (often 1 to 3 in the male), alternate with thinner or incomplete bars or lines of flecking. R1 rufous and black; the others with white tip after a wide black subterminal bar. From the central feathers outwards; more white starting at the base (outer Rs sometimes black and white), fewer black bars (sometimes only one) and small marks. Tail rounded (Rmin > 80% Rmax).

Most of the coverts and body feathers are recognisable from their upperside, with colour and pattern similar to that of the central tail feathers.

Possible confusion: with the Ps and Rs of diurnal birds of prey and large waterbirds.

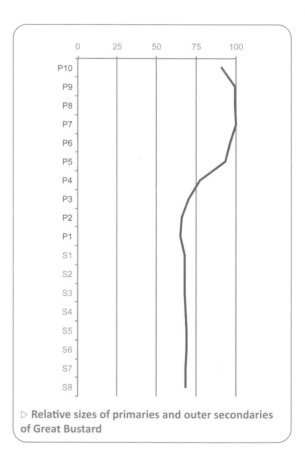

▷ Relative sizes of primaries and outer secondaries of Great Bustard

• PHASIANIDAE: PARTRIDGES, PHEASANTS, GROUSE, ETC.

NP = 10 / NS = ≥ 12-20 / NR = 10-24 (overall 10-16 in smaller species and more in large species).
Domestic and introduced species: for some NP = 10 (+1)
Distinctive criteria of group: shape, colour, sometimes size.

This group includes two subfamilies, the Phasianinae (pheasants, partridges, quail) and the Tetraoninae (capercaillie, grouse, ptarmigan). Common Quail is a special case; it has wings with a pointed hand and is migratory. The other species fly little and are sedentary.

The wings are short and rounded, the wing-loading is high. The tail is generally short or square. It is developed for courtship in grouse and pheasants. In this group young birds are able to fly when very young. While they have not yet reached adult size their 'large' feathers are already rigid enough to carry their weight

body
feathers

growing

soiled

soiled

Po

Pm

SC

Pi

uppertail-
covert

S

body feathers

Si

R

R

body feathers

Little Bustard

Great Bustard

Po ♂

Pm ♂

SC   SC

male much bigger than
female

underwing-
covert

S ♀

body feathers
and coverts

SC

Rc

Ro

scap

briefly. Sometimes these 'large' feathers (especially in plucked corpses) can pose a problem. They are often brown in colour with paler markings but are significantly shorter than those of adults. The large calamus is an indication of a young bird's 'large' feather, as well as its overall shape, which resembles that of adults. However, feathers of young may be confused with those of adults of other smaller species, for example young Pheasants and adult Grey Partridges (Ps and Ss). The width of the calamus is then a clue, a thinner feather indicating a young bird, but this distinction is sometimes difficult to discern.

Certain species are domestic or bred in captivity (hens, guineafowl, peacocks, etc.); their 'large' feathers may be found near farms, parks, etc. The colours are sometimes bright, and they may be large in size, which can cause confusion with other groups.

Possible confusion: with birds of prey and bustards (Rs) mainly.

▷ **Body feather of a galliform with a hyporachis.**
Note the clear structure of the hyporachis, with parallel rachis and barbs.

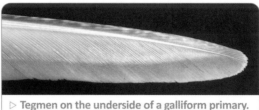

▷ **Tegmen on the underside of a galliform primary.**
Note the difference in the feather's shape compared to that of a duck or goose primary.

MEASUREMENTS OF THE 'LARGE' FEATHERS OF 'WILD' GALLIFORMS							
Espèce	Pmax	Pmin	Smax	Smin	S innner	Rmax	Rmin
Hazel Grouse (SS)	14	9	12	9	7-9	14	12
Willow Ptarmigan (SS)	18	11	12	9	7-10	14	11
Red Grouse	18 (19)	11	12	10	8-10	13	11
Rock Ptarmigan	16 (20)	11	13 (15)	10	7-10	14 (15)	11
Black Grouse	20 (25)	12	16	12	10-14	f15 / m23	11
Capercaillie	f27 / m33 (37)	f16 / m21	f21 / m24	f15 / m19	12-18	f24 / m36	f18 / m21
Red-legged Partridge	13 (15)	9	11 (12)	9 (7)	7-10	11	7
Barbary Partridge (SS)	14	11	12	10	n.d.	12	10
Grey Partridge	14 (15)	9	11	9	7-10	9 (10)	7
Common Quail	8.5	5.5	6.5	5	4.5-6.5	4.5	3
Reeves's Pheasant	27	17	18	14	n.d.	140	13
Common Pheasant	23	15	19	15	n.d.	60	13

Note: no measurements could be found for the following species: California Quail *callipepla californica*, Northern Bobwhite *Colinus virginianus*, Chukar *Alectoris chukar* and Rock Partridge *Alectoris graeca*.

Colour of 'large' feathers: very variable; remiges often brown with pale markings.

Remiges: they must support the relatively high weight of these birds. In addition, these species take flight abruptly and vertically. The remiges are therefore curved in profile and are very rigid to compensate for the strong thrust exerted when taking off and in flight. A carpal remex is sometimes present, but not always; if so, it resembles S1 but is about one-third smaller (not shown in the relative size graphs or measurement tables).

Primaries: Ps of characteristic shape and structure. The outer 6 Ps are emarginated very low, which leaves space between them when the wing is opened. This creates an air movement that sucks the wing upwards both when the bird rises and descends, facilitating sudden acceleration on take-off. The Ps therefore appear narrow but enlarged at the base. The notch can be situated at 30-50% of the feather. A tegmen is also present, quite narrow and along the rachis but clearly visible in most species, at least on the notched Ps. Be careful when studying a size graph as P10 may be the smallest of the Ps (P10 < P1 in size).

Secondaries: they are wide and rigid to support the high wing-loading in flight. Because the Ps are quite short, the size of most Ss is 75-80% Pmax. The innermost Ss are shorter.

Rectrices: their shape is variable. Confusion possible with the barred tail feathers of bustards (T > 10 cm), also with other species with a uniform tail. Except in the pheasants, the rectrices are robust and rectangular, and quite short compared to those of birds of prey.

Body feathers: some are identifiable by colour, especially those of male pheasants and of partridges (flank and belly). This group has a highly developed and structured hyporachis on the majority of its body feathers, and sometimes also on its 'large' feathers (visible rachis and parallel sides to a large extent). There is also the peculiar shape of the secondary coverts, in connection with their use for flight. As with the primary coverts, but to a lesser degree, the calamus can be quite long and slightly curved towards the tip.

Species with a square or very slightly rounded tail (Rmin > 90% Rmax): Hazel Grouse, ptarmigans, female Black Grouse, partridges, California Quail.

Species with a slightly rounded tail (Rmin > 80% Rmax): Capercaillie, Common Quail.

HAZEL GROUSE *Bonasa bonasia:* Ps and Ss brown-grey with pale markings, and with a thin pale border at the tip. Outer vane of P4-P10 buff with a hint of brown bars; sometimes white to buff without bars on P10 and P9. P1-P3 often uniform or with pale mottling towards the tip of the inner vane. Ss with a hint of pale bars on the border of the outer vane. Inner Ss more obviously barred on outer vane and also on the tip of the inner vane. Rs brown-grey with black and white vermiculations in vague bars, wide black subterminal band (of irregular margin) and white tip (reduced in young?). Central Rs more rufous and without black and white tip

Possible confusion: central Rs with owls, Ps and Ss with partridges (but well-defined bars on both vanes in Grey Partridge, and no bar in Red-legged Partridge and closely related species).

ROCK PTARMIGAN *Lagopus muta:* Ps and Ss white. Outer Ps usually marked with black on the rachis, sometimes with dark flecks towards the tip. Inner Ss brown-grey flecked with buff in summer; sometimes yellow-buff barred with black (young?). Rs (except R1) black with a small white base and usually a small white tip, sometimes grey. R1 white in winter and barred with black in summer, on a yellow-buff (female) to buff-grey (male) background.

WILLOW PTARMIGAN *Lagopus lagopus:* 'large' feathers similar to those of the Rock Ptarmigan, but slightly larger. Overall less yellow and more rufous in the plumage. Inner Ss have a browner background. Rs (except R1) black with a small white base and a more or less narrow white tip. R1 white in winter and barred with black in summer, on a reddish-brown background.

RED GROUSE *Lagopus scotica:* Ps and Ss dark brown-grey. Outer vane of the outer Ps paler, silvery-grey (not uniform). Outer Ss sometimes uniform; towards the medians buff-red mottling often appears towards the tip and on the outer vane, which can form a thin subterminal bar towards the median and inner Ss. Inner Ss have more quite irregular pale speckling (rather buff-yellow). Rs black without a white base and with a small white tip or uniform. Central Rs black and quite irregularly speckled with rufous-buff (to yellow-buff). R2 and R3 sometimes with rufous at the tip.

Ptarmigan and Red Grouse young have brown remiges barred with black, the paler parts ranging from reddish brown to yellow-buff.

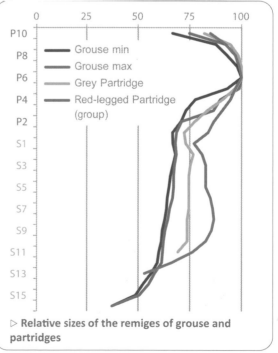

▷ **Relative sizes of the remiges of grouse and partridges**

10 cm: scale ²/₃

Al

PC

P9

flank

broken bases

P7

P1

Rock Ptarmigan

flan wint

Sm

S1

underwing-covert

scap

R2

R6

white tip

3 body feathers

Red Grouse

P10

P damaged

P8

P5

carpal remex

P2

S13

S10

S3

R2 tip

So

Pc

R1

R5 different individual

R8

Rm

Hazel Grouse

3 body feathers

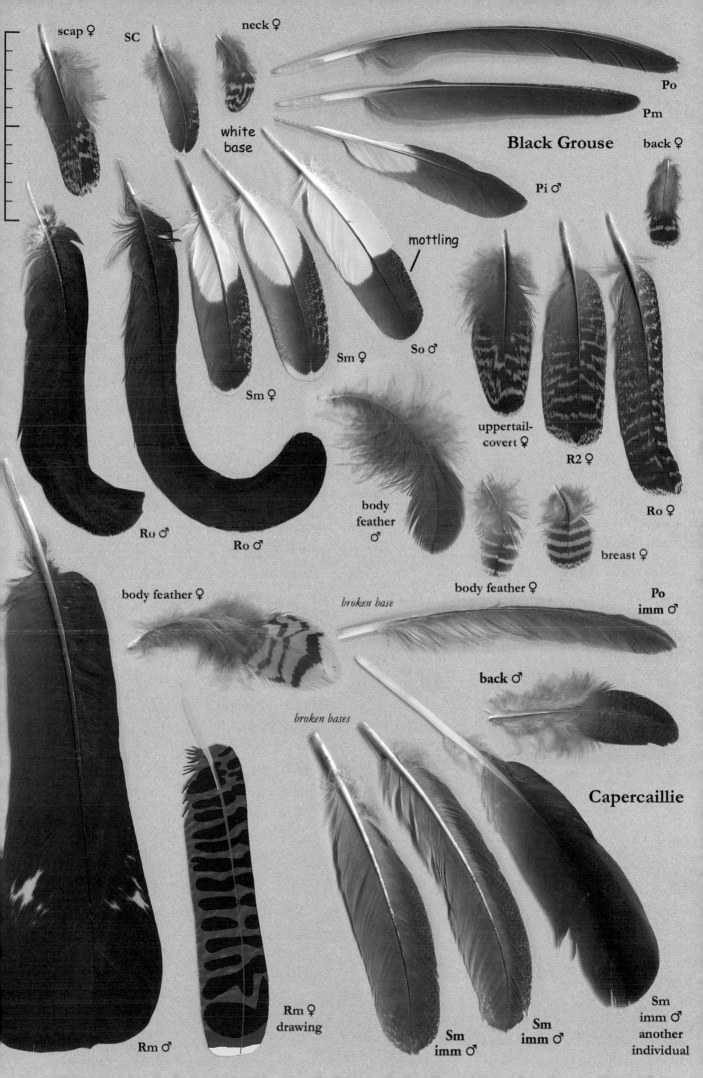

scap ♀

SC

neck ♀

white base

Po

Pm

Black Grouse

back ♀

Pi ♂

mottling

Sm ♀

So ♂

Sm ♀

uppertail-covert ♀

R2 ♀

Ro ♀

Ro ♂

Ro ♂

body feather ♂

breast ♀

body feather ♀

body feather ♀

broken base

Po imm ♂

back ♂

broken bases

Capercaillie

Rm ♂

Rm ♀ drawing

Sm imm ♂

Sm imm ♂

Sm imm ♂ another individual

BLACK GROUSE *Tetrao tetrix*

Male: Ps and Ss black-brown with a white base. P6-P10 without white but with a paler rachis. P5 with a small white base which increases towards P1 (basal half or more). Outer vane of Ps sometimes slightly marked with pale. Ss with basal half to two thirds white, the rest black (less white on the inner Ss). The black descends along the rachis and a little on the side of the two vanes. Ss with small white tip, possible lightly speckled on the edge of one or both vanes. Rs black often tinged with green-blue, central ones sometimes with a small white tip. Rs bent outwards, increasingly more well-defined towards the outer Rs; usually the 4 outer pairs hockey-stick shaped. Rs short and wide in the young.

Female: Ps and Ss similar to those of male but generally slightly browner with better developed pale markings. Well-defined white edge at the tip of the Ss, sometimes also on the Ps. Well-defined white to buff vermiculations on the Ss, especially on the outer vane. Rs black with thin, rufous wavy bars (or marks), more or less largely edged with white at the tip. Median and outer Rs slightly bent outwards.

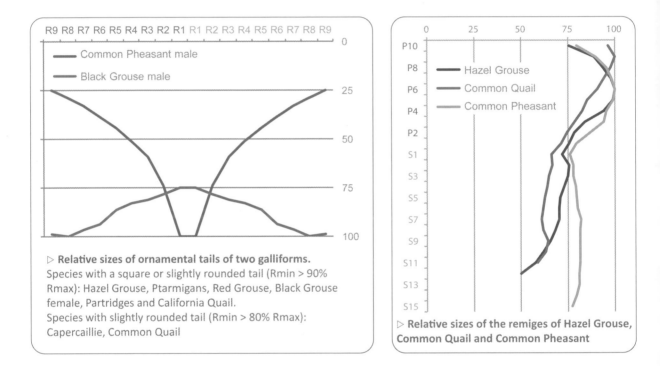

▷ **Relative sizes of ornamental tails of two galliforms.**
Species with a square or slightly rounded tail (Rmin > 90% Rmax): Hazel Grouse, Ptarmigans, Red Grouse, Black Grouse female, Partridges and California Quail.
Species with slightly rounded tail (Rmin > 80% Rmax): Capercaillie, Common Quail

▷ **Relative sizes of the remiges of Hazel Grouse, Common Quail and Common Pheasant**

CAPERCAILLIE *Tetrao urogallus*

Male: Ps and Ss brown-black. P5-P10 with a white border in the basal part (more or less extensive); possibly more or less sharp pale speckling. Ss finely speckled with light (buff-red to white) at the end and on the edge of the outer vane (often not very visible), with a (very) thin white tip, increasingly speckled towards the inner Ss. Rs wide, black, usually speckled with white over half of their length; sometimes with a small white tip. Rs sometimes spotless black. Slightly rounded tail (Rmin > 85% Rmax).

Female: Ps and Ss brown, marked with rufous, especially towards the tip and on the outer vane. Faint rufous bars on the outer vane of outer Ps becoming very irregular towards the inner Ps (few marks) and the Ss. Ss edged with buff towards the tip of the outer vane, with a thin white tip and rufous-buff speckling on the outer vane. Rs rufous with black bars, the last bar being wider and with a small white tip. Slightly rounded tail.

Juvenile (SS): Ps and Ss black-brown, Ps speckled and Ss barred with rufous.

Possible confusion: Rs of female with those of diurnal birds of prey.

CALIFORNIA QUAIL *Callipepla californica* (SS): Ps and Ss greyish-brown, vaguely tinged with rufous. One third of the distal part of the Ss bordered with buff on the outer vane. Rs bluish-grey.

No data could be found for Northern Bobwhite *Colinus virginianus*.

The Red-legged Partridge group

RED-LEGGED PARTRIDGE *Alectoris rufa:* Ps and Ss grey-brown. Ps with pale rachis, especially towards the outer Ps; outer vane browner. Buff border on the outer vane, on the narrower part of the Ps and diminishing towards the inner Ps (sometimes absent on P1 and P2). This pale border reappears, but is wider on the outer Ss, then disappears towards the median Ss. Ss very finely speckled with pale especially on the inner vane, sometimes not very visible. Inner Ss tinted rufous especially on their border and at the tip. Rs brown-grey at the base and turning dark rufous towards the tip. Central Rs brown-grey without the rufous. Flank feathers grey with rufous towards the base. The grey ends in a cream bar, then a single black stripe and a wide dark rufous tip. Neck feathers with the same colours (sometimes without rufous) but the black forms a rounded or drop-shaped mark.

ROCK PARTRIDGE *Alectoris graeca* (SS): 'large' feathers similar to those of Red-legged Partridge. Buff of the outer vane a little less extensive on the Ps and paler on the Ss. Rs with a little less rufous. Flank feathers grey, possibly with rufous towards the base. The grey is followed by a narrow black line, then a wide cream band, then a thin black stripe and a thin dark rufous tip (sometimes absent).

CHUKAR *Alectoris chukar* (SS). 'large' feathers similar to those of Red-legged Partridge. Flank feathers like those of Rock Partridge but often with a little more cream.

BARBARY PARTRIDGE *Alectoris barbara* (SS): the 'large' feathers similar to those of Red-legged Partridge. Ps with a lot of pale (buff to off-white) and the outer vane a little redder. Ss with few pale markings. R1 and R2 (R3) grey. Other Rs grey towards the base and dark rufous towards the tip, which increases in size towards the outer Rs. Flank feathers grey, possibly with rufous towards the base. The grey finishes with a thin black stripe, then a wide rufous band becoming cream-coloured, then a fine black stripe and a small dark rufous tip (sometimes absent).

Grey Partridge and Quail

GREY PARTRIDGE *Perdix perdix.* Ps and Ss brown marked with buff. Ps with thin, paler bars more or less complete becoming more broken and less regular towards the inner Ps (sometimes brown vaguely speckled with buff). Ss barred with buff, coupled with pale speckling on the borders and tip. Possibly reddish areas in brown of the outer vane. The brown background becomes black towards the tip on the inner Ss. Ss with a pale line on the rachis in the distal third, wider towards the inner Ss. Central Rs (R1–R2) vermiculated with buff-grey and black; more barred towards the base with a rufous rachis. R3 rufous at the tip (or almost totally rufous throughout). Others (R4–R9)

▷ **Grey Partridge wing**

become rufous with a small, pale tip and base spotted or barred with dark. Secondary coverts and many of the dorsal body feathers recognisable by their pattern and buff line along the rachis. Body feathers of the 'horseshoe' marking on the chest grey with black vermiculations and broad dark rufous subterminal band.

COMMON QUAIL *Coturnix coturnix:* the large feathers are moulted before migration. Very small. Pattern of the remiges similar to those of the Grey Partridge. Ps and Ss brown-grey, paler towards the base of the inner vane. Outer vane marked with buff to tawny-rufous, often in small notches or wavy bars or in irregular marks. Inner vane of P10-P7 most often uniform, sometimes like other Ps with more or less regular, well-defined pale marks, especially towards the tip. Ss paler finely barred with buff on the outer vane and at least the distal third of the inner one. Inner Ss almost entirely barred, with a pale line on the rachis on its distal third. The pale bars of the Ss are bordered with dark; the rest of the feather is more or less speckled. Rs dark brown with pale border at the tip and with more or less wavy, buff bars, narrower on the inner vane. Rs curved and squat. Tail square but outer Rs (R5, R6) a little shorter (about 80% Rmax).

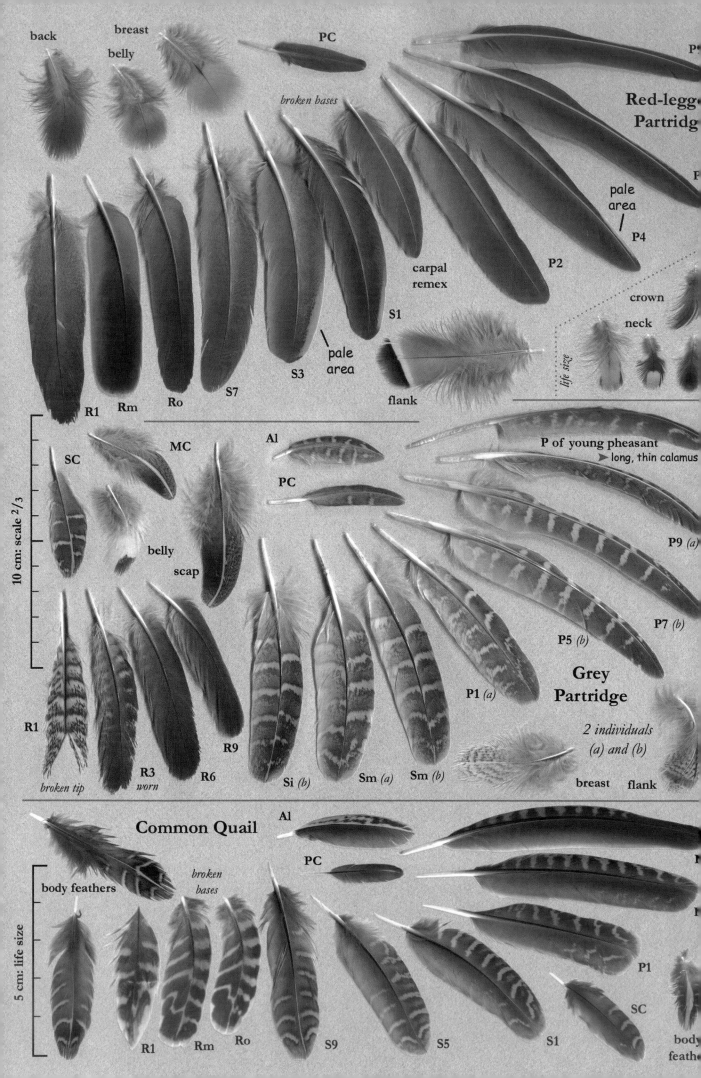

back

breast

belly

PC

Red-legge
Partridg

pale
area

P4

P2

broken bases

carpal
remex

pale
area

S1

S3

S7

flank

crown

neck

life size

R1

Rm

Ro

10 cm: scale 2/3

SC

MC

Al

PC

belly

scap

P of young pheasant

long, thin calamus

P9 *(a)*

P7 *(b)*

P5 *(b)*

P1 *(a)*

Grey
Partridge

2 individuals
(a) and (b)

R1

broken tip

R3
worn

R6

R9

Si *(b)*

Sm *(a)*

Sm *(b)*

breast

flank

Al

Common Quail

PC

5 cm: life size

body feathers

broken
bases

P1

SC

S1

body
feathe

R1

Rm

Ro

S9

S5

uppertail-covert

broken base (bb)

Al f

different individuals

Po ♀

PC ♂

SC ♀

SC ♂

Po ♂

notches ♂

Pm ♀

bars ♀

R8 ♀

R6 ♂

bb

bb

Pm ♀

R2 ♀

long, thin calamus

Si ♂

Si ♀

Si ♀

Sm ♀

So ♀

So ♂

Pi ♂

belly ♂

flank ♂

Pheasant

R1

S

P

3 'large' feathers, juvenile

mantle ♂

neck ♂

body feathers (scale 1/2)

10 cm: scale 1/2

R1 ♂

other patterns of R1 of male

neck ♂

10 cm: scale 1/3

4 body feathers ♀

neck ♂

back ♂

body feathers of male

3 coverts

Al

Cockerel (domestic)

various individuals

Reeves's Pheasant

very variable in size and colour

P

P

R ♂

P

SC

P

carpal remex

P

R ♀

Lady Amherst's Pheasant

S

undertail-coverts

sometimes metallic iridescence

body feathers

S

body feathers

hackle (neck)

Pheasants

The tail is pointed with very long central feathers and the others quite narrow. The rectrices are pointed, their width decreases steadily towards the tip. The outer ones are bent towards the inside of the tail. Their size increases rapidly towards the centre of the tail.

In males, bars are present on both vanes of the central feathers, then may fade but remain more visible on the outer vane and towards the base. The bars sometimes are reduced to more irregular marks. Females have more discreet colours.

COMMON PHEASANT *Phasianus colchicus*

Ps and Ss black-brown marked with buff bars in the female; notches and elongated marks in the male. Rs: T = 14 - 60 cm. Rs of female: black and buff bars on a brown-grey to rufous-brown background. Rs of male: pink and/or orange iridescence along the outer vane. Several central feathers brown-grey with black bars, then more elongated marks. Bars often disappear towards the outer rectrices, which are grey-buff with speckling and irregular brown-grey marks, and a rufous edge.

Body feathers of male with various colours and often a green, red, or pink metallic iridescence at the tip.

▷ **Common Pheasant wing (female)**

REEVES'S PHEASANT *Syrmaticus reevesii* (SS): Male: Rs
white to light grey barred and speckled with black, outer vane with rufous border which increases and brightens towards the outer Rs. Often some reddish-brown in several black bars of the outer vane. Outer Rs golden-red with black speckling, with the base of the inner vane white and barred. Ps brown with large reddish marks speckled with dark. Ss barred in black and white, with a wide rufous border towards the tip of the outer vane. The rufous extends onto the tip towards the inner Ss. Most dorsal body feathers are white or golden-rufous and bordered with black.

GOLDEN PHEASANT *Chrysolophus pictus* (SS): Male: Rs rufous-buff marked with thin bars of black, and pale round (central feathers) or irregular marks.

LADY AMHERST'S PHEASANT *Chrysolophus amherstiae* (SS):

Male: Rs white with black bars and vermiculations; outer vane tinted rufous. Central Rs very widened. Ps brown, pale along the rachis, outer vane bordered with white on the outer feathers. Inner Ss blue-black.

Female: Compared to the female Common Pheasant the colour of the Ps, Ss and Rs is generally more rufous (no buff) and the brown bars are much wider. Central Rs with golden-rufous background.

MEASUREMENTS OF THE 'LARGE' FEATHERS OF DOMESTIC GALLIFORMS The sample size is small for all species.						
Species	Pmax	Pmin	Smax	Smin	Rmax	Rmin
Reeves's Pheasant	27	17	18	14	140	13
Golden Pheasant	21	14	18	14	70	12
Lady Amherst's Pheasant	23	17	18	16	f60 / m105	11
Cockerel (domestic)	> 19	< 13	> 19	< 12	> 20	< 12
Red Junglefowl (domestic)	21	15	20	15	25	15
Indian Peafowl (Peacock)	41	32	37	28	56	34
Domestic Turkey	> 35	n.d.	n.d.	n.d.	> 40	n.d.
Helmeted Guineafowl	28	20	25	13	20	14

Al y

PC y

P y

Po ♂

Pm ♂

S ♀

Sm ♂

Indian Peafowl

R ♂

Si
black form

S y
broken tip

S white form

'large' feathers with longitudinal
stripes along the rachis, even
visible in the white forms

rachis x2

Indian Peafowl

LC

LC

LC

neck

'eye-spot'

back

uppertail-
covert
(small)

body feathers ♂

Domestic Turkey

damaged feathers

Po ♀

*tip
broken*

3 body feathers ♂

**Helmeted Guineafowl
(and domestic form)**

10 cm: scale ¹/₂

close-up of spots (x 1.5)

body
feathers

small
white
spots

R ♀

uppertail-covert

body feather ♀

10 cm: scale ¹/₄

Si

S (or R?)

Cocks, hens and other domesticated species

Domestic 'Chicken' *Gallus gallus domesticus*, **Red Junglefowl** *Gallus gallus*, **Green Junglefowl** *Gallus varius* etc.:
Remiges the size of those of a Grey Partridge (for small species) or a Common Pheasant, or even larger for the largest
species. Outer Rs of male quite curved. Ps and Ss very concave. 'Large' feathers of females of the wild forms brown
to brown-black, marked with pale on the outer vane. 'Large' feathers of domestic forms are of variable colour, often
rufous, or mixed with black, white and/or reddish, generally with ill-defined or irregular markings. Various ornamental
body feathers in males, of variable shape and colour.

Possible confusion: with some wild species such as Red-legged Partridge (Ps) but the Ps are often more pointed and
more curved in the *Gallus* species when viewed from above.

Indian Peafowl (Peacock) *Pavo cristatus*: confusion impossible with wild species due to the coloration and large size
of the 'large' feathers and their particular shape. Note that the rachis of the remiges is very finely streaked (coloured).
There are also albino, leucistic and melanistic varieties. The only possible confusion is with swans for white varieties,
but the absence of a tegmen on the primaries and the strong curvature of Ss are distinctive.

Male: Ps rufous, sometimes faintly marked with brown on the inner vane. Ss black to bluish-black; inner ones buff
(more reddish towards the base) barred and marbled with brown. Rs brown with pale speckling, especially on the
outer vane. Rs curved inwards, with a wide tip; very rigid because they support the 'tail fan' during display and help
to keep it spread open.

Female: Ps, Ss and Rs dark brown, with pale mottling on the outer vane.

Young male: as female but quickly takes on intermediate colours between the male and female (for example, Ps rufous
strongly marked with brown).

Recognisable body feathers include secondary coverts (like the Ss) and uppertail-coverts in particular: metallic
green from a few cm to more than a metre, the largest with a black, blue, bronze. and green 'eye-spot'.

Domestic Turkey *Meleagris gallopavo* (SS): Ps barred with black and white, bars quite ill-defined with the black often
fading into the white. Ss barred with brown and buff, inner Ss browner (barred).

Rs of the male quite finely barred brown and reddish with a broad black subterminal bar, sometimes with a
few thin bars and a white tip. Rs of female dark brown with reddish-brown mottling that can form bars; base
brown or with some large white and black bars. Subterminal black bar and wide white tip. Although the species
spreads its tail in display the Rs do not necessarily have an upturned calamus.

Body feathers generally with a black (sub)terminal bar, wide tip and iridescence. Greater coverts similar.

Helmeted Guineafowl *Numida meleagris* Numididae: confusion impossible with local wild species. All 'large' feathers
and the majority of body feathers are black-grey with fine and regular white markings: small spots or discontinuous
bars. The background is sometimes also finely speckled with white. Body feathers easily recognisable, marked in the
same way.

BIRDS OF PREY

Birds of prey can be conveniently grouped as Diurnal Birds of Prey (families Falconidae, Accipitridae and
Pandionidae), and Nocturnal Birds of Prey or Owls (families Tytonidae and Strigidae).

In these species there is often great variability between individuals, both in size and coloration. Indeed,
in many of them females are significantly larger than males. Geographical origin may also influence the size
of individuals. The species are medium to large, and the disparities between individuals are therefore more
pronounced than in small birds. The colour of the plumage often differs depending on the sex, and young birds
may have colours or patterns that differ from those of adults for several years. It can therefore be difficult to
separate several species by examining only one or a few 'large' feathers. Looking at a larger sample of feathers
may be necessary.

However, several criteria can give a clue to the feather being from one or another species. The tables in the following pages are compilations of sets of characters that can be used in determining the origin of a feather.

• FALCONIDAE: FALCONS

NP = 10 (+1) / NS = 11-14 / NR = 12

Distinctive criteria of the group: shape, colour.

This family has quite narrow, pointed wings. The tail is usually long.

Possible confusion: with nightjars, Common Cuckoo (but 'large' feathers not as thick in the Cuckoo), curlews (Ps and Ss), other small diurnal birds of prey, especially sparrowhawks (Ss and Rs) and Black-winged Kite (Ps and Ss).

'Large' feather colour: the commonest coloured marks are in the form of notches or pale ovals on Ps and Ss; sometimes bars on Ss. Rs usually have bars or notches and/or a (sub)terminal bar.

Primaries: size decreases rapidly towards the inner Ps. The hand is not fingered; few Ps are notched, and the narrowing is situated very high: notch at 75-90% on P9 and P10 and quite often a slight emargination on P8 and P9.

Secondaries: the forearm is narrower than in birds of prey that often soar or glide; the Ss are quite short compared to the larger Ps (about 50-60% of Pmax).

Rectrices: the tail is long, the Rs rectangular and elongated. Compared to feathers of other diurnal birds of prey they are narrower. The tail is rectangular (Rmin > 90% Rmax) with the outer rectrices sometimes a little shorter (R6 > 85% R1).

The species of falcon in the following tables are listed approximately according to size.

CHARACTERISTICS OF THE PRIMARIES AND SECONDARIES OF FALCONS

| Species | Pmax | Pmin | Smax | Smin | Principal colours | | Marks | | | |
					Dark colour	Pale colour	Uniformly dark	Notches/ Pale ovals	Bars	Irregular marks
Falcons										
Lesser Kestrel	21	10	11	9	brown to black	white to grey, or rufous	(P grey-brown)	PS	Si	
Common Kestrel	23	11.5	12	9.5	brown to black	white to grey, or rufous		PS	Si	
Red-footed Falcon (SS)	21.5	11.5	12	9	dark grey	white to grey (m)		PS	S	P (m imm)
Merlin	19 (20)	9	11	8	brown (f); grey to black (m)	white to grey (m); buff to rufous (f)		PS	S	
Eurasian Hobby	23 (28)	9.5	11 (13)	9	dark grey to black-brown	rufous to buff	Si	PS		
Eleonora's Falcon (SS)	n.d.	12.5	10.5	9.5	dark brown-grey	pale grey to rufous-buff	PS black-grey	PS	(S)	PS
Lanner Falcon (SS)	29	n.d.	14	n.d.	grey-brown	white to grey or buff-rufous		PS	S	PS

Species					Dark colour	Pale colour				
Saker Falcon	33	17.5	20.5	15.5	brown-grey	white to buff or rufous		PS		S
Gyrfalcon (SS)	n.d.	n.d.	n.d.	n.d.	grey to black	white, grey, or buff-grey; finely speckled with dark		PS	(PS)	PS
Peregrine Falcon	f30 (34) m25	f15 m12	f16 (19) m14	f12 m10	brown-grey to black	grey to rufous-buff		PS	Si	
Similar-looking species										
Black-winged Kite (SS)	23	13	14	9	pale to dark grey	pale grey to white	PS (iv of Ss pale)			
Common Cuckoo	19 (22)	10	12	8	brown-grey	white to rufous		PS	S	
curlews (2 species)	24	8.5	13.5	7	brown	white to pale buff		PS		
nightjars (2 species)	18 (21)	8.5	12	8	brown to black	buff to cinnamon	P	PS	PS	

CHARACTERISTICS OF THE RECTRICES OF FALCONS

Species	Rmax	Rmin	Principal colours		Marks					
			Dark colour	Pale colour	Uniformly dark	Notches pale ovals	Dark bars	Irregular marks	Wide, dark terminal or subterminal band	Tip paler
Falcons										
Lesser Kestrel	17 (19)	13	brown to black	rufous-brown (f and imm); grey (male)	m: grey		imm and f/m (iv)		X	X
Common Kestrel	20	15 (14)	brown to black	rufous-brown (f and imm); grey (male)	m: grey		imm and f/m (iv)		X	X
Red-footed Falcon (SS)	15	12.5 (11.5)	black	grey, sometimes buff to rufous (imm)	m: black-grey		X (ad: thin, imm: wide)		X	
Merlin	15.5	12.5	grey to black (m); dark brown (f)	grey (m); buff to rufous (f)		(X) wide	wide (f) to narrow (m)	(X)	X	X
Eurasian Hobby	15.5 (17)	13	grey to brown	rufous(buff)	c: brown (imm) to black-grey (ad)	X	(wide)	(X)		X
Eleonora's Falcon (SS)	17	15.5	brown-grey	rufous	c: brown-grey	(X)	X		(X)	(X)
Lanner Falcon (SS)	22.5	f19 m14	brown to black	grey to buff or rufous		(X)	wide			X
Saker Falcon	26	21	brown-grey	white to buff, or rufous	c: brown-grey	X	(X)			X
Gyrfalcon (SS)	30	18	grey to black	white, grey, or buff-grey with fine, dark mottling		(X)	thin to wide			X

Al ♂ PC ♂ P10 ♂ P9 ♂ Pm y Pi y

upper tail-covert scap ♂ MC ♂ SC ♂ scap y Pi ♂ So ♂ Sm y Sm ♂ Si ♂

R1 ♀ R4 ♂ R6 y

Common Kestrel underwing-covert ♂ **Lesser Kestrel** Po ♂ Po ♀ R ♀

uppertail-covert ♀ uppertail-covert y breast ♂ breast ♀ underwing-covert ♀

stained tip P10 ♂ P9 ♀ Pm ♀ Pi ♂ Pi ♀

Merlin

underwing-covert ♂ Ter ♂ So ♂ Al ♂ Al ♀

R1 ♀ Rm ♀ R6 ♀ Si ♀ Sm ♂ Sm ♀ So ♀ PC ♀ PC ♂ SC ♀ SC ♂

Peregrine Falcon	f21 m15 (16)	f17 (15) m14	brown-grey to black	grey to rufous-buff		X	X			X
Similar-looking species										
Common Cuckoo	19	10	brown-grey to black	white to rufous			X	X	X	X
nightjars	18 (20)	12	brown to black	yellowish to buff-grey				X	mottled background	(X)

COMMON KESTREL *Falco tinnunculus:* the Ps are black-brown with broad, pale notches on the inner vane, tinted reddish in the female, little or not at all in the male (often rather greyish). Tip sometimes with pale edge. In the young, there are often also reddish marks on the outer vane of the median and inner Ps. The Ss often have notches towards the outer feathers becoming thinner and changing to bars towards the inner ones. The outer vane of Ps may be dark or have small pale marks. That of the Ss is variable: dark (more the outer Ss), with marks, clearly barred or amply tinged with rufous (male).

The Rs have a wide black subterminal bar and off-white tip. The adult male has ash-grey Rs. Thin, incomplete black bars, sometimes indistinct, may be present on the inner vane or both vanes. The Rs of the female and young are reddish-brown with well-defined black bars. The female's outer Rs are rarely without large, black subterminal bars. Bars are usually narrower in adult females, wider in young. First- and second-year males may have clearly barred rectrices on a greyish background or thinner bars on a grey-red background.

LESSER KESTREL *Falco naumanni:* the 'large' feathers are very similar to those of the previous species for corresponding sex and age but are slightly smaller. Ps of male are sometimes brown-grey, paler towards the edge and base of the inner vane, with no obvious pale markings (only slightly paler areas).

MERLIN *Falco columbarius:* Ps of male dark grey to black; Ss grey; Ps and Ss with broad, pale grey notches, sometimes in bars on the inner vane of the Ss. Ps and Ss of female black-brown with wide reddish notches becoming thinner and then bars on the inner Ss.

▷ **Common Kestrel wing (male), upperside and underside**

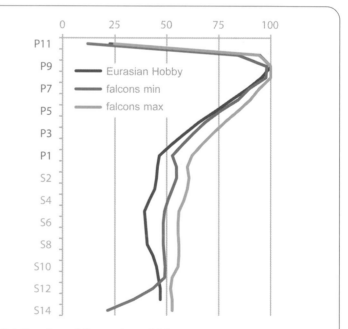

▷ **Relative sizes of the remiges of falcons**
Species included in the minima and maxima: Common Kestrel, Red-footed Falcon, Merlin, Saker and Peregrine Falcons. In the Peregrine and the Hobby the inner secondaries are quite long but smaller than the median secondaries in the other species compared here. The Hobby has a slimmer wing than the other species considered here.

Al

PC

underwing-covert

undertail-covert

body feathers

SC

SC y

P10 y

bent tip

P9

Pm y

Pm

Pi

So y

So

Sm

Si

Ter y

R6

R4 y

R1 y

underwing-covert

Eurasian Hobby

Eleonora's Falcon
young male

P9

Pm

Pi

body feather

scap

So

Sm

growing

R1

R5

R6

Po y ♂

Pm y ♂

Pi y ♂

Red-footed Falcon

R1 ♂

R5 y

R5 ♂

Rs of male grey finely speckled with black and with 5-6 dark bars (sometimes absent on the inner vane), a very wide subterminal bar and pale tip. The female's Rs are rufous-buff with 6-8 wide dark brown bars, the last (and sometimes also the penultimate) slightly wider; pale tip. The dark is often dominant; the light bars sometimes become incomplete (only around the rachis). The 'large' feathers of the young are similar to those of the female. R6 has a more pointed tip and regular bars at the tip in young; a squarer end and an often irregular pattern at the tip in females.

RED-FOOTED FALCON *Falco vespertinus:* The 'large' feathers of the male are uniform dark grey; Rs sometimes almost black. The Ps have the upperside more or less obviously silvery-grey. Females and immatures have 'large' feathers with pale markings: notches or ovals on Ps and Ss, bars on inner Ss. The Rs are grey with black bars with a paler tip, and a wide subterminal bar in females. Central Rs often have thinner, even incomplete bars. The Ps and Ss of subadult males sometimes have irregular pale marks on the inner vane.

EURASIAN HOBBY *Falco subbuteo:* Ps and Ss black-brown with rufous oval marks on the inner vane, touching slightly or not at all the edge of the vane. Pale marks sometimes very small or absent on the outer Ps. Fewer and smaller spots in the young. Rs dark brown-grey with pale tip and 8-11 rufous ovals or notches becoming semi-bars towards the outer Rs. R1 uniform with or without the pale tip. R3 and R4 viewed from below have a pale tip and much more well-defined bars in young than in adults. Moulting is suspended during migration; usually some Ps are moulted before.

ELEONORA'S FALCON *Falco eleonorae:* the remiges and central rectrices of adults are a uniform grey-brown in the dark morph. The other rectrices have rufous markings on the inner vane (notches or bars). In the light morph there are irregular or notched paler spots on the inner vane of remiges. In the immature the pale marks are more well-defined and wider, the rectrices are barred

▷ **Eurasian Hobby wing: upperside and underside**

with rufous or have notch-shaped marks (except the central feathers that are uniform).

LANNER FALCON *Falco biarmicus* (SS): Ps and Ss brown-grey with buff to rufous notches or ovals on the inner vane of the Ps and both vanes of the Ss. These pale marks are quite numerous and narrow (often 9-13 on the Ps). They lessen in intensity on the Ss and are sometimes grey and not obvious. Rs have many narrow brown-black-grey and reddish bars, with a pale tip. Young birds have greater amount of dark; on the Ps the pale notches are reduced to ovals; on the Rs the pale bars are incomplete.

PEREGRINE FALCON *Falco peregrinus:* the age and sex of individuals strongly influences the size and colour of the 'large' feathers. Males on average have 'large' feathers measuring 75-85% those of females. The marks in the young are wide, well-defined and pale (oval or buff notches on a dark background). With age the pale areas may be tinged with grey (especially male) and their margin becomes more irregular; they can be punctuated with small spots. Rs of adult have paler, greyer markings than younger adults. The Ps and Ss of young have pale edges at the tip and when fresh.

SAKER FALCON *Falco cherrug:* very large size. The brown-grey remiges have many large off-white to buff notches, decreasing in size towards the inner Ss (small ovals or vague marks). There are often 11-13 notches on the outer Ps, sometimes merged towards the base. Immatures have fewer notches and they disappear more rapidly on the Ss. The Rs are brown-grey with buff tips. The central Rs are uniform; the others have pale notches (buff to rufous or regular ovals (sometimes still pale spots on the central Rs).

GYRFALCON *Falco rusticolus* (SS): very large size. Very variable colouring. Rs usually barred, with pale notches or bars. The morphs with barred remiges are often pale with a wide, dark bar at the tip of the Ps, sometimes the Ss. Pale morph has white remiges with a hint of black bars (last bar wider, or a dark tip).

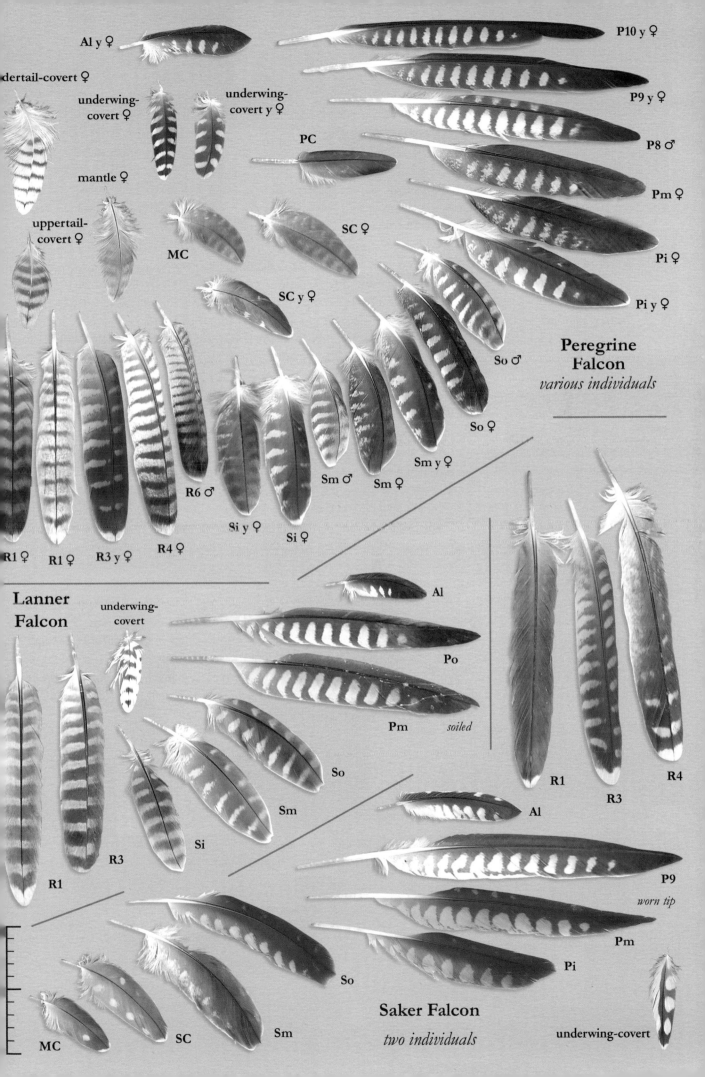

Al y ♀

undertail-covert ♀

underwing-covert ♀

underwing-covert y ♀

mantle ♀

MC

uppertail-covert ♀

SC y ♀

SC ♀

PC

P10 y ♀

P9 y ♀

P8 ♂

Pm ♀

Pi ♀

Pi y ♀

So ♂

So ♀

Sm y ♀

Sm ♂

Sm ♀

Si y ♀

Si ♀

R6 ♂

R1 ♀

R1 ♀

R3 y ♀

R4 ♀

Peregrine Falcon
various individuals

Lanner Falcon

underwing-covert

Al

Po

Pm   *soiled*

So

Sm

Si

R3

R1

R1

R3

R4

Al

P9

worn tip

Pm

Pi

So

MC

SC

Sm

Saker Falcon

two individuals

underwing-covert

• ACCIPITRIDAE AND PANDIONIDAE: BUZZARDS, EAGLES, KITES, HARRIERS, VULTURES, ETC.

In general: NP = 10 (+1) / NS = 13-14 (17) / NR = 12.

Other groups: vultures (4 species): NP = 10 / NS = > 15 / NR = 12-14 and Osprey: NS = 20.

Distinctive criteria of group: shape, size, colour and where found for some species.

Species in this family often use gliding or soaring and therefore have wide, rectangular and fingered wings. The tail is often long and/or wide and helps to provide lift during flight, with relatively rigid rectrices. Some species use flapping flight more often such as Black-winged Kite, sparrowhawks and Goshawk. Others fly at low level and have narrower wings such as harriers and Osprey.

There can be significant differences in coloration between young and adults, and sometimes between males and females. Some species may show gradual changes in plumage during the early years. Size can also vary greatly between the two sexes. All these parameters must be taken into account in the identification of isolated feathers as they multiply the diversity in coloration and size of 'large' feathers that may be found.

Note: There is sometimes a slight 'hairy' appearance to the surface on new feathers, especially towards the base, but it is much more localised and poorly developed than in owls (Tytonidae and Strigidae).

Possible confusion: according to species (colour and size), confusion with other large soaring species that have emarginated primaries (storks, cranes, pelicans, etc.).

'Large' feather colour: quite variable. Rather uniform in vultures and White-tailed Eagle (brown to black or white), most often with obvious markings in others. Rs with more or less regular bars; Ps and Ss often partially barred or with pale notches. The bars are often more distinct on the inner vane. Pale marks, sometimes irregular (for example, eagles). The marks sometimes visible on the wrist are those on the underwing-coverts and are therefore not visible on the remiges.

Primaries: apart from Black-winged Kite, the outer Ps are notched, the notch quite low on the feather (Not = 45-90%, sometimes 35-40% on outer P) and on at least 3 Ps (as far as P7). There is usually a prominent emargination on the outer vane. Logically, the indentation of a P is at the level of the emargination of the feather that follows it (more internal). These outer primaries are 'fingered' with a neat emargination and/or notch, which creates air movements which draw the bird forward. These Ps are often a little longer than those that follow.

Secondaries: the Ss are rather broad and quite similar throughout the forearm. Ss typically measure between 50% (pointed wings) and 75% (broad wings) of Pmax.

Rectrices: their proportions vary according to species, but they are quite stout. They are rather short and wide in vultures, a little more elongated in eagles, long in kites, etc. The rectrices are usually rectangular to elongate (except short-tailed vultures). They are very frequently barred (multiple bars or terminal bar), except in vultures.

Tertials: ventral feathers often have drop-shaped spots or more or less wide bars. The dorsal feathers are often brown and as they wear the tip usually becomes quite triangular.

The following table gives criteria of shape and colour for several species of diurnal birds of prey.

CHARACTERISTICS OF THE PRIMARIES AND SECONDARIES OF DIURNAL BIRDS OF PREY														
Species	Pmax	Pmin	Smax	Smin	Dark colour	Pale colour	Base of inner vane	Uniform or lightly marked (grey, brown, black, etc.)	Outer Ps bicoloured and without marks (distal area brown to black; iv white at base)	Pale notches (or ovals) on darker background)	Dark bars on paler background, complete or not.	Irregular marks	Wide dark subterminal or terminal bar(s) (rest barred or not)	Tip paler than the rest (the rest barred or not)
Black-winged Kite (SS)	23	13	14	9	grey	white	white S	PS						
Eurasian Sparrowhawk	f21 m18	f14 m12	f15 m12	f11 m9	brown	brown-grey	white to cinnamon				PS			
Levant Sparrowhawk	n.d.	n.d.	n.d.	n.d.	black-grey	grey (m); brown (f)					PS			
Northern Goshawk	f31 (35) m27	f18 m16	f23 (25) m18	f16 m14	black-brown	brown to buff	white to cinnamon				PS		S	PS imm
Montagu's Harrier female	33 (37)	17	19	13	black-brown	brown-grey					PS		P(s)	PS
Montagu's Harrier male	33 (37)	17	19	13	black	ashy-grey	white on S (m)		Po and Pm black	4 to 8	Pi S	Pi		P
Hen Harrier female	36 (37)	17	22 (25)	14	black-brown	white, cream to grey-brown					PS			
Hen Harrier male	36 (37)	17	22 (25)	14	black	ashy-grey	white	Po	5		(P) S Inc		S	
Marsh Harrier female	36 (41)	22	24	17	choco-late-brown	rufous		PS		(5 to 6: little white)			(PS) iv	
Marsh Harrier male	36 (41)	22	24	17	black-brown	grey, cinnamon	white	Po black	5 to 6			(PS)	(PS blurred)	
Honey Buzzard	39 (42)	22	27	18	black-brown	brown-grey	white			1 to 2 (3)	PS irr		PS	PS (imm)
Booted Eagle	31	21	22	16	black-brown	brown-grey	white to brown-grey			pale morph, 2 to 4	PS	PS	P	PS
Common Buzzard	37 (43)	21	26	17	black-brown	brown-grey to rufous	white			0 to 3	PS	Po	PS	
Rough-legged Buzzard	40 (45)	24	28	17	brown to black	brown-grey	white			3 to 5	PS		PS	(S)
Long-legged Buzzard	40	20	22	16	brown to black	brown-grey, rufous	white to cinnamon			2 to 5	PS		S	
Black Kite	45 (50)	21	23	18	dark brown	chestnut to buff-grey	brown; white on Po and Si			(1 or 2)	PS	Po		
Red Kite	47	23	25 (27)	20	black-brown	brown-grey to rufous	white on Po (or P) and Si			3 to 5	S	PS		
Osprey	41 (43)	23	25	20	black-brown	brown-grey	white	Po	3 and 4	PS	PS	Po		

Species					Dark colour	Pale colour								
Bonelli's Eagle (SS)	41	25	30	20	black-brown	grey-brown			2 or 3 (imm)		PS	PS	PS	
Lesser Spotted Eagle (SS)	42	n.d.	24	21	black-brown	grey-brown				PS	PS			PS imm
Egyptian Vulture	48 (50)	25	28	23	black	grey to white		P				S		
Short-toed Eagle	48	28	30	23 (16)	black-brown	grey-brown	white		2 to 5		PS			PS edge
Spotted Eagle	n.d.	n.d.	n.d.	n.d.	black-brown	grey-brown			?	(PS)	PS			PS imm
Imperial Eagle (SS)	52	35	36	26	black-brown	grey-brown				PS	PS	(PS)	PS (ad)	PiS (imm)
Spanish Imperial Eagle (SS)	n.d.	n.d.	34	n.d.	black-brown	grey-brown		(PS)			PS	PS	(PS)	PiS (imm)
Golden Eagle	55 (68)	34	39 (42)	26	black-brown	buff-grey to white		Po	imm, > 5		PS (imm especially)	PS	PS	
White-tailed Eagle	68	40	49	30	grey-brown			PS						
Lammergeier	75	40	47	36	grey-brown			PS						
Griffon Vulture	60 (68)	35	42 (48)	35	black-brown			PS						
Black Vulture	65	30	42	30	black-brown			PS						

Note: n.d.: no data available

CHARACTERISTICS OF THE RECTRICES OF DIURNAL BIRDS OF PREY

Species	Rmax	Rmin	Dark colour	Pale colour	Uniform (grey, brown, black, etc.)	Pale notch, mark or bar on dark background	Regular (x) or irregular (irr) bars	Wide terminal (t) or subterminal (s) band (rest barred or not)	Tip paler than rest (rest barred or not)	Irregular marks or pattern (no real bar)
Black-winged Kite (SS)	15	10	grey	white	white to pale grey			t fine (imm)		
Eurasian Sparrowhawk	f21 m17	f17,5 m15	black-brown	grey to brown or buff			X	ts		
Levant Sparrowhawk	n.d.	n.d.	dark grey	pale grey	central (m)		X			
Northern Goshawk	f31 m26	f26 (23) m22	black-brown	brown to buff			X		imm	(Ro)
Montagu's Harrier female	26	21	black-brown	brown-grey			wide	(s)	(X)	

Montagu's Harrier male	26	21	dark grey to rufous	pale grey to ashy	central dark grey		wide	tip darker			
Hen Harrier female	27 (29)	22	brown-grey	cream to brown-grey			wide	(s)			
Hen Harrier male	27 (29)	22	grey	ash-grey / white			incomplete	(s)	(X)		
Marsh Harrier female	29	22	chocolate-brown	rufous	brown		sometimes	(s)	X		
Marsh Harrier male	29	22	silvery-grey	white	central, outers bicoloured			(s)			
Honey Buzzard	28	23	dark brown	chestnut to buff-grey			Irr	s	imm		
Booted Eagle	23	19	brown	chestnut-grey to white			subdued	s	X	(X)	
Common Buzzard	26 (28)	21	brown	white to brown-grey or rufous			X	s ad	X	(X)	
Rough-legged Buzzard	27	23	brown to black	white to grey or rufous			X	s	X	X	
Long-legged Buzzard	29	21	brown	white to grey-brown, rufous	rufous; extremities pale		X	(s)	X		
Black Kite	30 (34)	23	dark brown	grey-brown to rufous-brown			X		(buff to ochre)		
Red Kite	40	26	brown	rufous, browner on the Ro; iv whiter towards the Ro	(central uniform rufous)		X		(blurred)		dark marks along the rachis
Osprey	26	19	brown	white to brown-grey	central grey darker at tip	X	X		X		
Bonelli's Eagle (SS)	31	26	black-brown	brown-grey			X	t ad		(X)	
Lesser Spotted Eagle (SS)	25	21	black-brown	brown-grey			X		X imm		
Egyptian Vulture	35 (37)	19		white (adult); buff to brown (immature)		X					
Short-toed Eagle	33	29 (19)	black-brown	brown-grey; white on iv at least			2 or 3	last, wider	small		
Spotted Eagle	n.d.	n.d.	black-brown	brown-grey			X		X imm	(X)	
Imperial Eagle (SS)	36	32	black-brown	brown-grey		X	X	t ad	X imm	(X)	
Spanish Imperial Eagle (SS)	31	n.d.	black-brown	brown-grey		X	X	t ad	X imm	(X)	
Golden Eagle	37 (41)	33	black-brown	grey-buff to white			X	t imm (white base)			X
White-tailed Eagle	40	24	brown	white	white with dark base (adult)			t (imm)			imm: brown with white marks
Lammergeier	56	40	brown-grey		silvery tint; darker border						
Griffon Vulture	40	30 (27)	dark brown	rufous, browner on the Ro; iv whiter towards the Ro							
Black Vulture	40	24	black-brown								

Note: n.d.: no data available

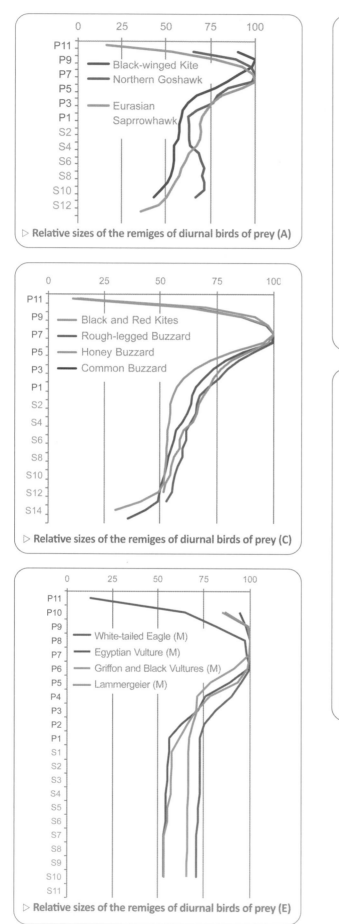

▷ **Relative sizes of the remiges of diurnal birds of prey (A)**

▷ **Relative sizes of the remiges of diurnal birds of prey (C)**

▷ **Relative sizes of the remiges of diurnal birds of prey (E)**

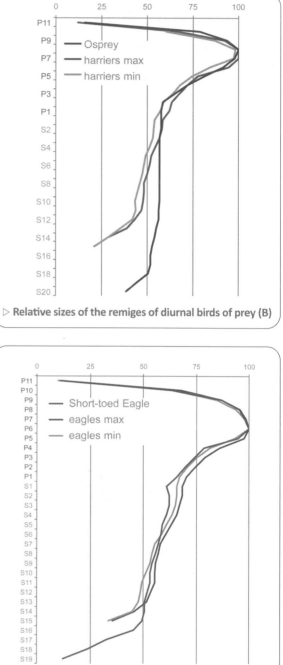

▷ **Relative sizes of the remiges of diurnal birds of prey (B)**

▷ **Relative sizes of the remiges of diurnal birds of prey (D)**

Note: for graph E, all measurements used are of type M (measurement of the feather in place on the bird)

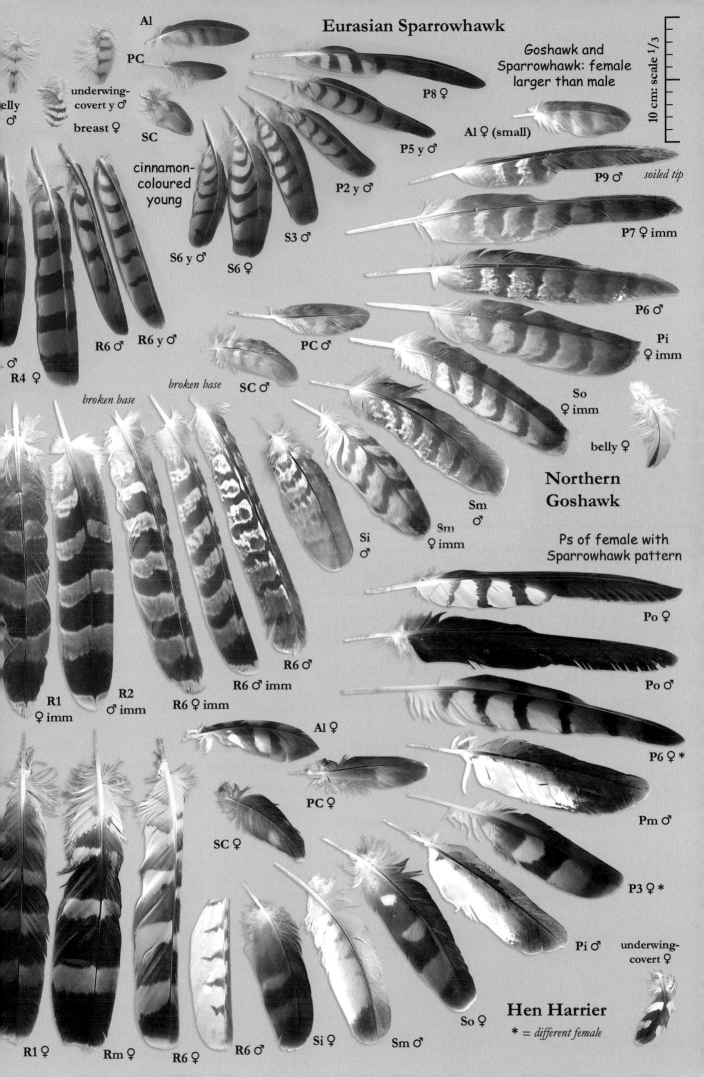

Eurasian Sparrowhawk

Al

PC

belly ♂

underwing-covert y ♂

breast ♀

SC

Goshawk and Sparrowhawk: female larger than male

10 cm: scale ⅓

P8 ♀

Al ♀ (small)

P5 y ♂

P9 ♂   *soiled tip*

cinnamon-coloured young

P2 y ♂

P7 ♀ imm

S3 ♂

S6 y ♂

S6 ♀

P6 ♂

R6 ♂

R6 y ♂

Pi ♀ imm

PC ♂

So ♀ imm

R4 ♀

broken base

SC ♂

belly ♀

broken base

broken base

Northern Goshawk

Si ♂

Sm ♂

Sm ♀ imm

Ps of female with Sparrowhawk pattern

R1 ♀ imm

R2 ♂ imm

R6 ♀ imm

R6 ♂

Po ♀

Po ♂

P6 ♀ *

Al ♀

PC ♀

Pm ♂

SC ♀

P3 ♀ *

Si ♀

Sm ♂

So ♀

Pi ♂

underwing-covert ♀

Hen Harrier

* = *different female*

R1 ♀   Rm ♀   R6 ♀   R6 ♂

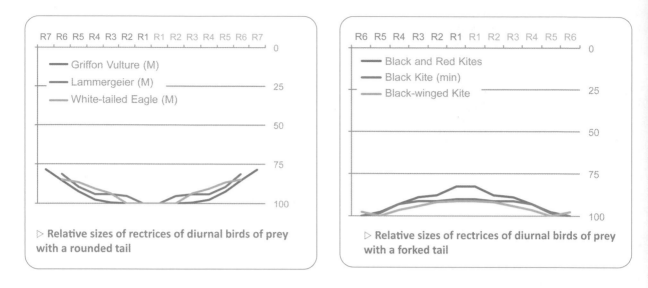

▷ Relative sizes of rectrices of diurnal birds of prey with a rounded tail

▷ Relative sizes of rectrices of diurnal birds of prey with a forked tail

BLACK-WINGED KITE *Elanus caeruleus* (SS): shape of remiges similar to those of falcons. Few Ps emarginated/notched and only at the tip. Possible confusion: of 'large' feathers with those of gulls (colour). Feathers ash-grey to dark grey, the Ps often with a small darker crescent at the tip. The Ps have a darker underside (dark grey), the darker inner P only at the tip, rarely also the inner Ss. Inner vane of Ss largely white. Tail quite short and square. Central Rs ash-grey, the others light grey to white (outer vane a little darker).

EURASIAN SPARROWHAWK *Accipiter nisus:* size of P10 (SS): 8-11 cm. Remiges brown-grey with the edge of the inner vane white to cinnamon, with curved brown bars, the terminal one wider. Altogether 5-7 bars on the Ps, 5-6 on the brown-grey Ss. Rs most often with 5 black-brown bars, the outer one wider. In the young, bars a little wider; pale area of remiges often cinnamon; all body feathers and coverts slightly tinged with cinnamon. Males have 'large' feathers measuring on average 80-85% of those of females.

▷ Eurasian Sparrowhawk wing (male)

LEVANT SPARROWHAWK *Accipiter brevipes* (SS): the 'large' feathers of young are very similar to those of the Eurasian Sparrowhawk, but the bars are narrower and more numerous (6-8 on the rectrices c.f. 5 on average). In adult males, remiges are grey to black on the outer vane, barred on a greyish background on the inner vane. Those of the adult female are browner. Outer Ps of both sexes are dark at the tip. Rs of the male barred on a grey background, central ones grey.

NORTHERN GOSHAWK *Accipiter gentilis:* size of P10 (SS): 14–19 cm. Remiges brown extensively barred with black-brown; inner vane paler towards the base (white to cinnamon). The pale areas are speckled with dark; the bars are more well-defined on the underside. The central Rs have 4 wide bars with up to 6 on the outer ones. In the young, contrast between the bars and the rest of the 'large' feathers is more marked, the last bar is often wider, and there is often a pale tip. Males have 'large' feathers measuring on average 80-85% of those of females.

MONTAGU'S HARRIER *Circus pygargus* (SS)

Male: Median and outer Ps black-brown sometimes dark grey (most often P3 to P10), with pale tip; the base of the inner vane may be grey to white; rachis pale on underside. P2 often dark grey with paler marks on the inner vane, P1 often grey with 2 (or 3) black bars on the inner vane. The black Ps/grey Ps boundary can vary

SC imm

PC imm

bent

underwing-
covert imm

Al imm

Al

PC

SC

P10 imm

P9

bent at tip

P6 imm

Pi imm

bars very
irregular in
adult

Pi

R1

R3 imm

R6

Ter
imm

Si
imm

Sm

So imm

Honey Buzzard

uppertail-
covert (d)

with few bars

SC (p)

SC (d)

P10 (p)

P9 (d)
soiled at tip

P8 (p)
soiled at tip

P6 (p) *soiled at tip*

Pi always paler

Pi

Booted Eagle

Pale morph (p) and dark morph (d)

1 (p)

R2 (d)

R5 (d)

R6 (p)

Si (p)

Sm (p)

So (d)

Al ♀

MC
♂ imm

PC ♂ imm

*broken
base*

SC ♂ imm

Po ♀

Po ♀

*soiled at
tip*

Pm ♂ imm

Pi ♂ imm

**Marsh
Harrier**

R1
♀

Rm
♂ imm

R6
♂ imm

Si ♂ imm

Sm ♀

So ♂ imm

by one or two feathers. Central Rs dark grey, the others ash-grey with 4 (5) dark grey bars, becoming reddish towards the outer Rs.

Female: Ps grey-brown with darker bars (4-5), thinner and more numerous (5-7) on the outer Ps. Ss with 3 very wide bars (often a fourth on the inner vane very near the base). The contrast between bars and background is almost absent on the inner Ss (sharper below). Rs with 5 wide dark bars.

Juvenile: pattern similar to the female, but wider bars (especially on Rs) and more rufous background, greyer base. Fewer bars on the outer Ps. The fingered parts of the Ps are dark. Ss very dark brown on both sides (bars often barely visible), with small pale tips. Young males may have remiges like those of females but on a paler grey background.

HEN HARRIER *Circus cyaneus*

Male: Outer Ps (P6-P10) black-brown with a small white base that has a punctuated outline on the inner vane. P1 to P5 silver-grey with triangular black tip; inner vane white towards the border with incomplete dark grey bars (or marks along the rachis). P5 with black outer vane, sometimes totally dark. Ss like the inner Ps, but with a darker more rounded tip, and becoming less contrasting and greyer towards the inner Ss. Central Rs grey, darker at the tip. Moving towards the outer Rs, the grey becomes paler (the outer Rs may be white); dark grey bars appear on the inner vane and then on both vanes (between R5 and R8).

Female: Ps grey-brown with darker bars (4 or 5), thinner and more numerous (5 or 6) on the outer Ps. Ss with 4 wide bars (often a fifth on the inner vane very near the base). The contrast between bars and background is almost absent on the inner Ss (sharper below). Compared to female Montagu's Harrier the bars of the remiges are thinner, and the last dark bar appears much wider than the others. Rs with 4 (5) wide brown to dark rufous bars on an off-white to cinnamon background, central Rs with greyer hues; bars often with a 'washed out' look on outer Rs (R4–R6).

Immature male: compared to the female the tone is generally greyer. The bars on Ps and Ss are a little thinner, often narrowed towards the edge. Often a pale edge towards the tip from median Ps to median Ss. The bars of the Rs are very wide, less numerous (4-6), often tinged with rufous and with dark border. The outer bar is sometimes still wide when the bird is several years old.

Juvenile: similar to the female. Underside of Ss often significantly darker than that of the Ps.

MARSH HARRIER *Circus aeruginosus*

Male, remiges:

2nd-year male (summer, autumn): 'large' feathers of young, but inner Ps dark blue-grey above with a large black tip. Sometimes other 'large' feathers like those of the female.

Male 2nd-year (winter) to 3rd-year (summer): Outer Ps (P5-P10) with more black from the tip downwards (descends to the primary coverts; dark grey), base of the inner vane white to cinnamon. Other Ps and Ss grey with pale cinnamon or whitish base and a clear dark tip. Inner Ss brown. Inner Ps sometimes with broken bars or dark marks.

Male 3rd-year and older: Outer Ps (P5-P10) black-brown with a rounded pale base on the inner vane; median and inner Ps silvery-grey, often with darker tip and mainly white inner vane. Ss silvery-grey; inner Ss largely marked with brown or entirely brown.

Male, rectrices:

Male 2nd- to 4th- or 5th-years: Rs silver-grey often with a well-defined dark subterminal mark and a rufous tint on the inner vane, sometimes barred in 2-3 year-old birds.

Old male: Rs pale silver-grey without dark mark at the tip and with white, especially on the inner vane.

Female: Ps and Ss usually chocolate brown with paler grey-brown underside than in young; the fingers of Ps and Ss are darker. Very little white at the base. Sometimes rufous marks on the inner vane. Rs rufous-brown paler than in the young, with reddish speckling on all Rs except on the uniform R1. Rarely inner Ps more rufous with darker incomplete bars; Rs rufous with incomplete dark bars.

Juvenile: Ps, Ss and Rs chocolate brown. Rs quite dark; only R8 may have rufous marks on the inner vane.

Common Buzzard

different individuals:
variable in colour and
density of barring

Al

PC

PC

derwing-
covert

flank

back

SC

P10

P7

P4

P3

P1

So

Si

Sm y

R6 y

broken base

Al

P10

PC

P8

SC

P4

So

undertail-
covert

Sm

undertail-
covert

Long-legged
Buzzard
immature

1 ad

R2 y

R4 ad

uppertail-
covert

Si

Rm ad

R1

Rm

R6

PC

SC

LC

P9

P7

P5

P1

underwing-
coverts

R1

R3

R5

Si

Sm

So

Osprey

EUROPEAN HONEY BUZZARD *Pernis apivorus:* this is the only species of raptor in the area considered here to have most 'large' feathers marked with two types of bars: large, dark bars that are small in number alternate with paler, narrower and more numerous bars. This pattern is present on remiges and rectrices (except the outer most Ps) but it varies with age. Sex has a greater influence on the density of pigments in the bars, with the males being paler.

▷ Honey Buzzard wing (missing base)

The outer Ps are dark brown with the base of the inner vane white; sometimes barred once or twice or with a mark in the palest individuals and barred like the others in dark individuals.

The 'large' feathers have a brown-grey background with brown markings; the background is more or less white depending on the individual, from the base towards the border of the inner vane. All 'large' feathers have a wide, dark terminal bar; the tip of the rectrices is paler. There are 2 or 3 wide, well-defined bars in adults, towards the base of the feather. Between these there are less contrasting thin bars.

In immatures, there are 3 or 4 dark bars; they are wider and more regularly arranged along the feather. The thin bars are less sharp than in adults, so the whole feather is generally less contrasting and darker. The tip of the 'large' feathers is often pale.

This pattern of the 'large' feathers is also found on the primary and greater coverts.

BOOTED EAGLE *Hieraaetus pennatus* (SS):

Dark morph: the pattern on the remiges resembles that of Black Kite; black-brown with partial or complete paler bars and a very small white base. The outer Ps are black-brown with paler wave-shaped marks on the inner vane (P10 = 18-19 cm). The Rs are grey-brown with a wide dark subterminal band and pale tip. The inner vane becomes white towards the base. On a very speckled background there are 6-7 barely distinguishable darker bars.

Pale morph: the marks on the remiges are more contrasting. The Rs have the same pattern but are lighter in colour, with more white on the inner vane (especially on the outer Rs).

In both morphs the inner 3 or 4 Ps are much paler than the outer ones, buff-grey with brown-grey bars.

COMMON BUZZARD *Buteo buteo:* a very variable species; some variability between individuals is also visible in the coloration of the 'large' feathers. However, there are certain criteria common to the majority of patterns. The 'large' feathers are finely barred with dark (6-8 bars on the Ps, 7-9 on the Ss, 9-11 on the Rs). The outer Ps are black-brown in the emarginated part; the rest is white, either barred, marked with a few spots, or pure white. The remiges have some white at the base of the inner vane, at least between the dark bars (see difference with the Black Kite). In adults, barred 'large' feathers have a wider terminal bar; it is absent on the Rs of young birds.

▷ Common Buzzard wing (young)

The remiges of young often have one or two fewer bars than in adults and these are wider.

The Rs show most variability (confusion possible with Long-legged Buzzard, among others). They usually have regular barring (9-11 bars), on a background ranging from brown-grey to white, through rufous or warm brown. On very pale Rs, the bars tend to fade on the inner vane and towards the base; some are barred 'normally'

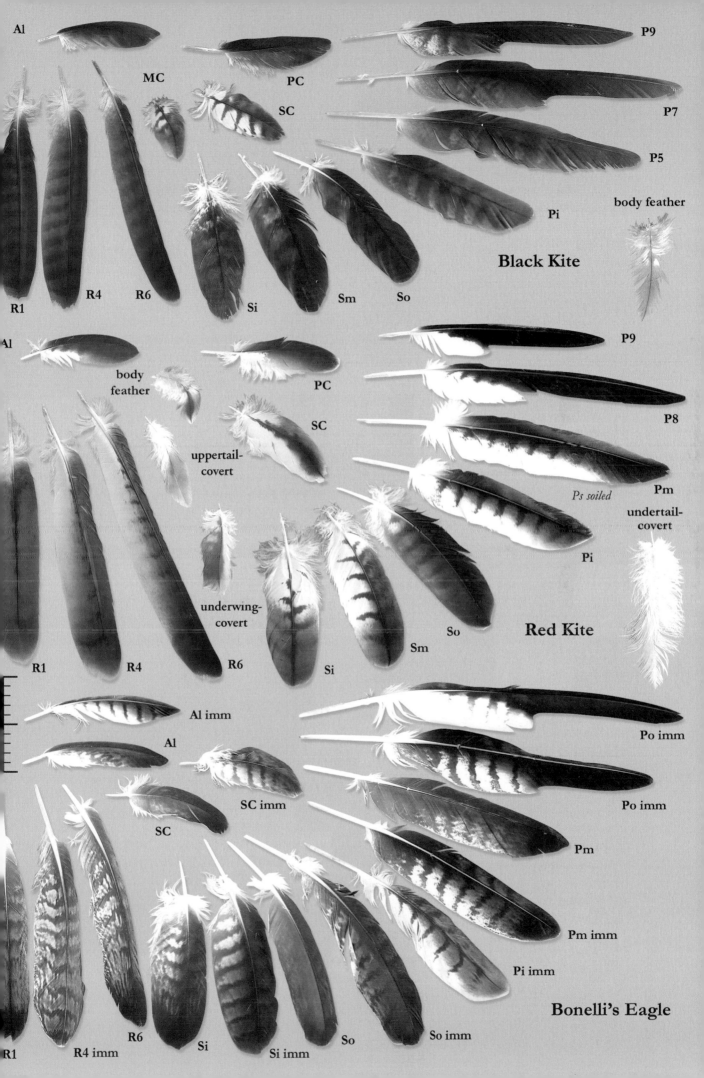

Al

MC

PC

SC

P9

P7

P5

Pi

body feather

Black Kite

R1

R4

R6

Si

Sm

So

Al

body
feather

PC

SC

uppertail-
covert

P9

P8

Pm

Ps soiled

Pi

undertail-
covert

underwing-
covert

R1

R4

R6

Si

Sm

So

Red Kite

Al imm

Al

SC imm

SC

Po imm

Po imm

Pm

Pm imm

Pi imm

R1

R4 imm

R6

Si

Si imm

So

So imm

Bonelli's Eagle

on the outer vane and white with only the tip dark on the inner vane. Sometimes the inner vane is white only along the rachis.

ROUGH-LEGGED BUZZARD *Buteo lagopus* (SS): outer Ps brown to black in the emarginated part, white on the base of the inner vane, dark area possibly paler and vaguely barred towards the base. Other Ps grey-brown to rufous with 2-3 dark bars and dark tip; more or less white on the inner vane (at least half). Bars often incomplete, pointed towards the tip. Ss like the inner Ps but with more brown on the outer vane, 2-3 incomplete bars and a broad dark tip.

Rs with at least the basal two thirds white (rarely only half). The proximal area may be grey-brown or tinged with rufous. Incomplete dark bars; 2-3 on the central feathers (sometimes 1), (3) 4-5 on the outer ones and a wide dark end with a small, paler tip. Rs of adult female without bars but always with the dark end and a paler washed-out area on the underside. In the male, the end is black and the bars contrasting. In the young, the contrast is less clear (dark brown / grey-brown and the terminal band is less well-defined.

LONG-LEGGED BUZZARD *Buteo rufinus:* outer Ps brown-black in the emarginated part, white towards the base; sometimes a rufous hue towards the tip and the outer vane (P10 measures about 20 cm). Other Ps brown-grey often marked with rufous, with incomplete dark bars around the rachis and often a lot of white on the inner vane. Ss have the same pattern; white or cinnamon on the inner vane.

Rs quite variable, often with some rufous, examples:
– pattern 1: rufous, paler to white on the inner vane towards the base;
– pattern 2: rufous with a hint of dark bars (at least 8), the outer one larger;
– pattern 3: mixture of 1 and 2 (some bars and some white);
– pattern 4: brown-grey tinged with rufous, with a suggestion of at least 9 dark bars, with white towards the base and on the inner vane (looks like those of an immature Common Buzzard);
– pattern 5: as 4 but without rufous, sometimes almost black and white, wider terminal bar.

BLACK KITE *Milvus migrans:* the remiges and rectrices are brown with dark barring. The remiges of young can be tinged with rufous on the outer vane. The outer web of the remiges is white at the base, covering only a small area on P8 and P9 and on inner Ss (at least for the nominate race), and generally speckled with brown (see Common Buzzard).

The tail is slightly forked, the central Rs measure about 80-90% of the outer Rs.

RED KITE *Milvus milvus:* the remiges are dark, barred with brown. The inner vane of at least P6-P10 is largely white towards the edge and base; the bars only in the brown areas. In pale individuals, the white may persist on all Ps but is less extensive. The background may be slightly greyer or more rufous on the Ps. The Ss are brown and barred, except the innermost ones that are largely tinged rufous. The remiges of the young are tinged with rufous, especially the inner Ps.

The tail is prominently forked; the central Rs measure about 75-85% of the outer ones. The Rs are rufous; the central ones almost uniform, becoming browner and the tip darker towards the outer ones, with the inner vane becoming whiter towards the outer Rs. More or less complete bars are usually visible; these are sometimes only dark marks along the rachis. Sometimes there is a poorly defined pale, more or less wide tip.

▷ **Red Kite wing (young)**

PC

underwing-
covert

Short-toed
Eagle

P10

Al

P7

Pi

SC

Pi

So

body feather

Sm

Si

Rc

Rm

R6

PC

SC

Golden Eagle

Pi imm

S

Rc imm

Ro
imm

Ro

Imperial Eagle

young

underwing-
covert

PC

Po

Pi

*broken
tip*

Pi pale

1 ad
le ⅛

S

Rc

Ro

SC

Osprey *Pandion haliaetus:* remiges grey-brown barred with dark; 5-7 bars on Ps, 4-6 on the Ss. Outer Ps black-brown with white base of the inner vane. Rs with 5 (6) dark brown bars and a large dark terminal bar (quite thin in young), often with pale tip. The central Rs are dark brown-grey; the bars barely visible. The others are brown-grey, white on much of the inner vane. Sometimes the Rs are white with black barring.

▷ Osprey wing (young)

Bonelli's Eagle *Aquila fasciata* (SS): remiges brown-grey with broad dark tips (absent in young birds); barred with black-brown. The bars often have an irregular margin and the light background is speckled with dark. The pale bands are sometimes reduced to pale speckling or irregular marks.

Rs buff-grey with wide, dark terminal bar; absent in young birds. Inner vane sometimes partly white. Darker bars quite thin and numerous (7-9) but sometimes incomplete or not very marked (especially on central Rs).

Immature: the remiges resemble those of Common Buzzard but are on average a little larger; the white is less extensive on the inner vane; the bars are thinner and have less regular margins. The inner Ps have a buff-grey background.

Lesser Spotted Eagle *Aquila pomarina* (SS): dark brown 'large' feathers, often with a small white base, with brown-grey barring. Rachis pale at least at the base. The bars are numerous (at least 8); fewer on outer Ps. In some dark individuals the marks are not obvious, in the form of bars or notches, or even absent (particularly outer Ps). Juvenile: Inner Ps, Ss and Rs with broad white to buff tips.

Egyptian Vulture *Neophron percnopterus:* six outer Ps fingered. Remiges black-brown. Ps and Ss have a large, white patch with blurred borders on the inner vane; they become more extensive on the inner Ss. Outer Ps sometimes almost without pale patch. Wedge-shaped tail (outer Rs measuring about 70-75% of central ones). Rs white to yellowish in adults, dark grey-brown in young.

Confusion possible with the Ps and Ss of White Stork and White Pelican.

Short-toed Eagle *Circaetus gallicus:* the colour pattern is quite 'simple'. The remiges and rectrices are brown-grey with a few dark, well-defined bars (often 3 or 4, the outer one a little wider). The Ps with notches have from 0-2 bars. The inner vane of the 'large' feathers is mainly white. The white starts from the rachis at the base and then runs diagonally towards the edge of the inner vane as far as the top of the right-hand side of the vane. On the outer Ps the white extends to the limit of the emargination; on the other Ps and Ss as far as the curvature on the edge of the vane, except on the inner Ss where it stops at the last dark bar. White edge on the median and inner Ps and the Ss when the feather is fresh. The Rs are quite wide. They have a slightly more variable pattern with much white on the outer Rs (on both vanes); usually 3 bars sometimes only 2 and little or no white on the central Rs. The last bar is wider.

Spotted Eagle *Aquila clanga* (SS): remiges black-brown with brown-grey barring (outer Ps sometimes without bars on the white base?). Rs brown-grey with black-brown barring. Dark individuals have brown 'large' feathers with barely visible pale markings.

Juvenile: Inner Ps, Ss and Rs with broad white to buff tips.

Imperial Eagle *Aquila heliaca* (SS): remiges black-brown with grey-brown bars, notches, or ovals; speckled and with an irregular margin. Vague, large darker tip. Remiges of young without dark tip, paler, with a few pale inner Ps (sometimes greyish without bars). Small pale tip on the inner Ps and the Ss.

Rs of adults with broad black-brown tip (at least the distal quarter) and small white base. The rest is brown marked with thin greyish bars with a wavy appearance. Rs of young dark brown with pale tip and small white base slightly marked with paler, often thin greyish bars.

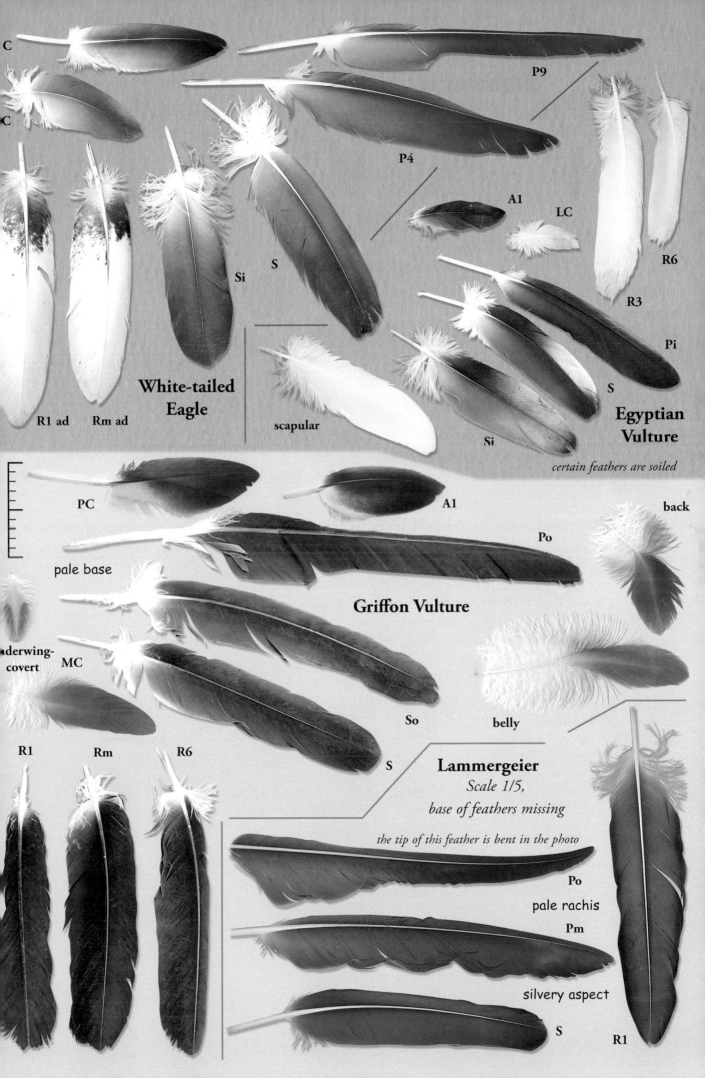

C

C

P9

P4

A1

LC

R6

R3

Pi

Si

S

**White-tailed
Eagle**

S

Si

**Egyptian
Vulture**

R1 ad   Rm ad

scapular

certain feathers are soiled

PC

A1

back

Po

pale base

Griffon Vulture

nderwing-
covert

MC

belly

R1   Rm   R6

So

S

Lammergeier

*Scale 1/5,
base of feathers missing*

the tip of this feather is bent in the photo

Po

pale rachis

Pm

silvery aspect

S

R1

GOLDEN EAGLE *Aquila chrysaetos*

Adult: black-brown 'large' feathers with a large darker tip, usually with irregular paler markings or brown, greyish or white bars. Outer Ps sometimes without markings but usually at least with pale flecks on the base of inner vane. Towards the inner Ps and on the Ss and Rs, the pale marks often line up to give the appearance of clear bars (2-4) but can be of other very different shapes.

Juvenile: base of 'large' feathers white, more extensive on the inner vane. The rest is black-brown sometimes marked with pale as in adults. The Ps often have half to two thirds of inner vane white; the Ss one third to one half. The Rs have at least the basal two thirds white, the rest black.

Immature: white gradually diminishes and large pale marks appear; the Rs retain a lot of white for a long time. The primary and greater coverts have the same pattern as that of the 'large' feathers.

WHITE-TAILED EAGLE *Haliaeetus albicilla:* the 7 outer Ps are fingered. Remiges dark grey-brown, paler towards the base. Rachis white at the base becoming dark towards the tip. Remiges of immatures often largely speckled with pale; those of adults sometimes still slightly speckled on the inner vane. Adult-type Rs white, densely speckled with brown-grey at the base (generally on fewer than one third of the vanes). Rs of immature dark brown with irregular white marks on the inner vane (often elongated) increasing in area and on the outer vane with age; Rs of subadults often only dark at the tip and base.

LAMERGEIER *Gypaetus barbatus* (SS): only 5 outer Ps clearly notched. The hand is more triangular than in other vultures. The tail is wedge-shaped and long (outer Rs measuring about 70-75% of central feathers). 'Large' feathers of adult are dark brown with bronze-coloured upperside; the rachis is white to pale brown. A slight silvery wash covers the 'large' feathers. Those of young and immatures often have irregular pale markings (grey, white, or buff) on the inner vane, the rachis is sometimes dark.

GRIFFON VULTURE *Gyps fulvus:* 7 outer Ps fingered. 'Large' feathers black with dark rachis, a little white at the base (rachis white in this part). Outer vane of Ss browner; tip of tertials grey to tawny.

BLACK VULTURE *Aegypius monachus* (SS): 7 outer Ps fingered. 'Large' feathers black-brown, almost black, with dark rachis, often very little white at the base (rachis dark as far as the base of the barbs).

• TYTONIDAE, STRIGIDAE: OWLS

NP = 10 / NS = 12-14 / NR = 12. Barn Owl: NS = 12-16

Distinctive criteria of group: shape, colour. This group can be identified by the velvety surface of the feathers (see key) and the tones/patterns of the majority of feathers.

These families use flapping flight over short distances, usually seeking prey from a perch. Some species with longer wings catch sight of their prey by flying at low-level above the ground, such as the Short-eared Owl. They often hunt at night. These birds are not very fast compared to diurnal birds of prey but their rounded wings allow them great manoeuvrability and above all, silent flight. This adaptation is visible on the feathers: the fringed edge of the outer Ps and their velvety surface ('fluffy' and very soft to the touch, on all feathers) reduces sound-producing turbulence. The border is irregular on the notched part of the inner vane. On the edge of the outer vane (on 1 or 2 outer Ps most often) there is even a comb with curved teeth. Species of this order also have fur-like feathers on the tarsus, as far as the feet for most, which reduces the amount of noise created when the feet are extracted from the plumage just before capturing prey.

Possible confusion: between the different species of owl, possibly with small diurnal birds of prey, nightjars, or other species with 'large' feathers marked with pale notches or bars. Usually, the obviously velvety surface of feathers is sufficient to identify them as belonging to a species of owl.

'Large' feather colour: bars, complete or not, more or less regularly spaced.

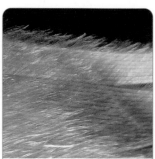

▷ **Upperside of an owl's feather.** Barbs (horizontal) support the barbules (velvety) which softens the sound produced in flight.

Main colours:
– white + terminal black marks or bars: Snowy Owl;
– rufous-yellow: Barn, Long-eared, Short-eared and Eagle Owls;
– buff-brown-grey: Tawny, Little Owls, etc.
The 'large' feathers of both sexes are generally identical and those of young differ little from those of adults.

Primaries: the Ps are quite wide, with a rounded tip (except outer Ps). Outer Ps with outer vane with fringed edge (like a comb); inner vane also a little fringed; notched high on the inner vane (Not = 60-90%).

Secondaries: the Ss are quite long (about 70-85% of Pmax in species with rounded wings).

Rectrices: the tail is short and square. The Rs resemble the Ss in their outline. However, they are distinguished by several criteria: the calamus is upturned in the Rs; the curvature of the rachis towards the inner vane is better defined on the Rs and located towards the base. The rest of the rachis is straight (totally straight in the central Rs). In the Ss, the curvature is more regular. On outer Rs, the outer vane is much narrower (never very narrow in the Ss).

Body feathers: the body feathers usually do not have regular, sharp, horizontal bars (as is the case in some diurnal birds of prey, such as Eurasian Sparrowhawk). The marks are usually longitudinal, sometimes with other transverse marks in addition to a dark centre. The marks sometimes present on the wrist are those of the underwing-coverts and are therefore not visible on the remiges (Tawny Owl, Long-eared Owl, etc.).

▷ **Fringed border of outer primary of an owl.**
Barbs with rigid tips curved towards the feather base.

COMPARISON OF THE REMIGES OF OWLS										
Species	Pmax	Pmin	Smax	Smin	Dark colour	Pale colour	Notches/ pale ovals	Bars	Tip	Irregular marks
Snowy Owl	35	23	24	20	black	white		PS		PS
Eurasian Eagle Owl	43	29	30	22	brown	buff to rufous		PS	on darker brown back- ground	Si
Great Grey Owl (SS)	41-33	25	n.d.	19	brown-grey	whitish to grey or rufous		PS, with or without speckles	on darker brown back- ground	
Ural Owl (SS)	32	23	n.d.	n.d.	brown-grey	whitish to grey or rufous		PS		
Tawny Owl	25	17	20.5	13	brown-grey	buff-grey to rufous		PS		
Barn Owl	25	14.5	15.5	12	grey to black	white and rufous		PS inc	speckled grey	
Long-eared Owl	25	14.5	16	12	black- brown	white and rufous (-yellow)		PS	on speckled grey-brown background	
Short-eared Owl	26	15.5	16	12.5	black- brown	white and rufous-yellow		PS	paler (except Po)	Po
Northern Hawk Owl (SS)	20.5	18	n.d.	n.d.	brown to black	white	PS (6-8)	(PS)		
Tengmalm's Owl (SS)	14.5	9.5	11	8	brown to black	white	PS (4-5)			
Little Owl	14	10	11	7.5	brown	buff	PS (4-6)	Si	S pale	
Eurasian Scops Owl	14	9.5	10	7.5	brown to brown-grey	buff to rufous		PS	speckled	
Eurasian Pygmy Owl (SS)	9	6.5	7	5.5	dark brown-grey	white to buff	PS (4-5)	Si	S pale tip	

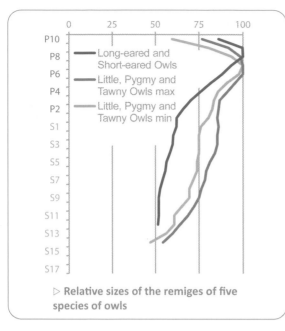

▷ Relative sizes of the remiges of five species of owls

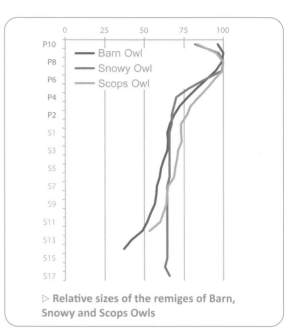

▷ Relative sizes of the remiges of Barn, Snowy and Scops Owls

COMPARISON OF THE RECTRICES OF OWLS								
Species	Rmax	Rmin	Dark colour	Pale colour	Notches, pale ovals	Bars	Other	Tip
Eurasian Eagle Owl	33	22	brown	buff to rufous		dark; fine; very wide in centrals (R1 and R2)		paler
Great Grey Owl (SS)	35	28						
Ural Owl (SS)	30	n.d.	brown-grey	whitish to grey or rufous		dark		
Tawny Owl	18 (20)	15.5	brown-grey	buff to rufous		dark	R1 with speckling; R1 uniform (rufous form)	paler in y
Barn Owl	13.5	11	grey to black	white to rufous		dark		vermicul-ated
Long-eared Owl	16	13.5	brown-grey	yellow-rufous		dark		paler in y
Short-eared Owl	17.5	14.5	brown-grey	yellow-rufous to pale-buff		dark		paler in y
Northern Hawk Owl (SS)	19	17	brown	white	X	8-10 pale and narrow (R1–R3)		(paler)
Tengmalm's Owl (SS)	11.5	9.5	brown	white	X			
Little Owl	9.5	7.5	brown	buff to rufous		wide; dark		(small pale tip)
Eurasian Scops Owl	8	6.5	brown to brown-grey	buff to rufous		speckled		
Eurasian Pygmy Owl (SS)	7	5.5	dark brown-grey	white to buff	R3–R6	4 or 5 fine; pale (R1 and R2)		(small pale tip)

SNOWY OWL *Bubo scandiacus:* this species is distinguished by the large size and colour (white, or white and black) of its 'large' and body feathers. In the female and young, there are many black bars of variable width. In adult males, there are fewer bars and 'large' and body feathers may be pure white.

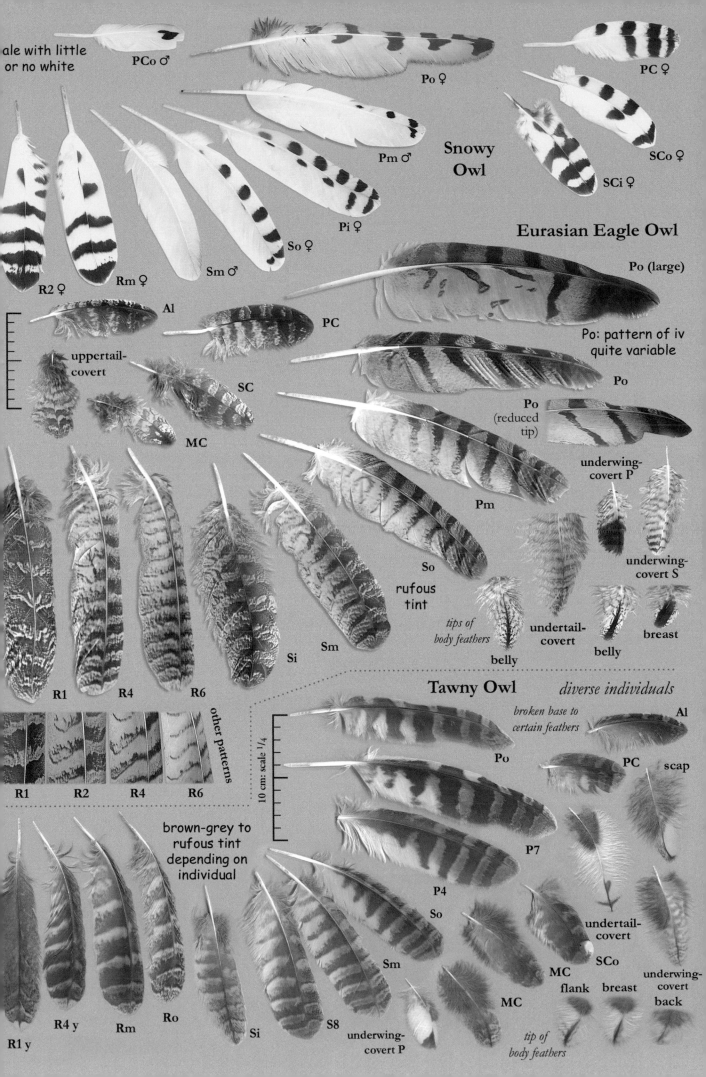

ale with little
or no white

PCo ♂

Po ♀

PC ♀

Pm ♂

SCo ♀

SCi ♀

Snowy
Owl

Pi ♀

So ♀

Sm ♂

Rm ♀

R2 ♀

Eurasian Eagle Owl

Al

PC

Po (large)

uppertail-
covert

SC

Po: pattern of iv
quite variable

MC

Po

Po
(reduced
tip)

underwing-
covert P

Pm

underwing-
covert S

So

rufous
tint

R1

R4

R6

Si

Sm

*tips of
body feathers*

belly

undertail-
covert

belly

breast

other patterns

R1

R2

R4

R6

Tawny Owl   *diverse individuals*

*broken base to
certain feathers*

Al

10 cm: scale ¹/₄

Po

PC

scap

brown-grey to
rufous tint
depending on
individual

P7

P4

So

undertail-
covert

Sm

SCo

MC
flank

breast

underwing-
covert
back

R1 y

R4 y

Rm

Ro

Si

S8

underwing-
covert P

MC

*tip of
body feathers*

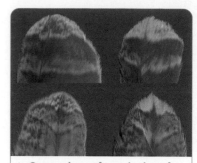

EURASIAN EAGLE OWL *Bubo bubo:* this species is distinguished by the large size of its 'large' feathers and the rufous tint often present. Overall orange-rufous with brown bars, with more dark in the background at the tips of the remiges and the central rectrices (wider bars). Inner secondaries also browner.

GREAT GREY OWL *Strix nebulosa* (SS): large 'large' feathers. Bars quite wide, often oblique on the Ps (see Eurasian Eagle Owl). Pale areas and edge of bars often speckled. No definite rufous, but possible light red hue.

URAL OWL *Strix uralensis* (SS): a little like a large Tawny Owl but with more uniform and fewer bars. The bars usually remain complete even towards the base of the 'large' feathers.

▷ **Comparison of rectrix tips of Tawny Owl.** Adult (left), young (right). Median (top) and central (bottom) rectrices.

TAWNY OWL *Strix aluco:* wide brown bars on a buff to brown background; often incomplete towards the base of the Ps and Ss. In the rufous morph, significantly more warm brown, even definite rufous, especially on the Rs, sometimes with few bars.

In young, the narrow terminal band of Ps and Ss is incomplete; the tip of the R is pale, (almost) without speckling.

BARN OWL *Tyto alba:* feathers of this species are quite easily distinguished from those of other owls: they have much white and orange, the outer Ps are not notched or emarginated, drop-shaped marks are present (see below).

Other than the usual grey colour of the fluffy base of the body feathers, the feathers have a mix of three colours (only one or two on the ventral side of body feathers): white, orange-buff, and brown-grey. Depending on the morph (pale to rufous) the orange colour is more or less dense and extensive. In the palest morph the bars are less marked, sometimes barely visible. In the dark form, the bars are wider. The bars are incomplete and wider near the rachis. The tip is often vermiculated with white and grey. The inner Ss have a rufous and white base and a vermiculated tip.

▷ **Tawny Owl wing**

▷ **Barn Owl wing**

The majority of the feathers on the upperside (wings, tail, back and head) have drop-shaped spots on a speckled background (white/grey). These spots are absent in other owls in the region considered here. In the Barn Owl this spot is often absent or reduced to a dark point on the outer Ps, and in the pale morph on all 'large' feathers (Ps, Ss and Rs). Its presence, on the other hand, can only mean that the feather is of this species.

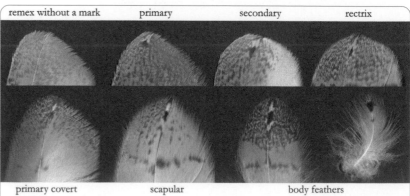

remex without a mark   primary   secondary   rectrix

primary covert   scapular   body feathers

▷ **Barn Owl: different appearances of the drop-shaped mark characteristic of this species**

facial disc

Al

PC dark

P10 pale

back

belly

SC

P8 dark

scap

P5 dark

P2

much white on the inner vane

shade and shape of bars variable

R1

R3

R4 dark

R5

R6

Si dark

Sm

So dark

Barn Owl

ear-tuft

Al

PC

obvious background colour gradient: orangish at base to brown-grey at tip

breast

MCo

SCm

P9

ack

P6

P2

belly

slight variation in number of bars

undertail-covert

R1

R4

R6

S11

S8

S2

Long-eared Owl

background quite homogeneous

sometimes brown-grey at tip

Al

PC

Rs well ontrasted

damaged

P10

bent tip

base broken

SC tip

Pm

everal dividuals

number and width of bars slightly variable

Pi

tip paler

rounded pale marks

R1

Rm

R6

Si

S10

S6

S2

Short-eared Owl

The body feathers of the ventral surface are uniform (from white to rufous depending on the morph) and have 2 small dark marks (from 0-3) along the rachis. *Possible confusion:* the body feathers of other owls have extensive marks or bars. The Snowy Owl is pure white or with broad black spots. In other species (Long-eared Owl, etc.) the flank feathers (very fluffy appearance) can be uniform or with elongated marks.

LONG-EARED OWL *Asio otus:* Ps, Ss and Rs yellowish-rufous with fine black-brown bars. Outer Ps with wider bars absent towards the base. Tip of Ps on a greyer and

▷ Long-eared Owl wing

speckled background and the base of inner vane white. Rs with speckled tip; central Rs with wider speckled bars. In young, the tip of Rs (except R1) is pale, (almost) without speckling.

SHORT-EARED OWL *Asio flammeus:* overall a paler rufous-buff than Long-eared Owl, with wider dark bars. Tips of Ps and Ss often with less grey speckling; paler tip (except outer Ps), with more obvious contrast. On the outer Ps the dark tip appears to enclose the pale spot on the inner vane. Outer Ps sometimes with diagonal bars. Few bars on the inner vane of the male's outer Ss. Rs with wider black bars. Central Rs with large dark bars and often incomplete intermediate bars.

NORTHERN HAWK OWL *Surnia ulula* (SS): Ps and Ss brown to black with 6-8 white notches or ovals. These are in fact pale bars but all around the rachis the 'background' is dark brown. Rs brown barred with white, more as notches towards the outer Rs and fine bars on the central ones; often a brown background around the rachis (except R1).

TENGMALM'S OWL *Aegolius funereus:* Ps, Ss and Rs brown-grey with white spots. Outer Ps with 4-5 ovals on the inner vane becoming notches towards the median Ps and on the Ss. Marks less numerous and clearly separated compared to those of Northern Hawk Owl. Central Rs with 4-5 pairs of white ovals becoming notches towards outer Rs.

LITTLE OWL *Athene noctua:* Ps, Ss and Rs brown with buff markings, more rufous on the outer vane. Ps and Ss with

4-5 (6) wide pale notches on the inner vane. The marks at the base sometimes merge. Inner Ss barred. Rs with 4-5 pale bars or notches and pale tip.

EURASIAN SCOPS OWL *Otus scops:* Buff to rufous Ps, Ss and Rs with brown to brown-grey barring. Pattern of 'large' feathers reminiscent of those of Tawny Owl. Brown bars on a paler background, becoming buff to cream on the vanes' borders. Mottling on the bars and background; more towards the tip. 7-9 bars on the Ps, 6-7 bars on the Ss, 5-7 bars on the Rs. Bars of Rs quite sharp on the outer feathers becoming increasingly speckled towards the central Rs. It performs a partial moult before its migration to Africa.

EURASIAN PYGMY OWL *Glaucidium passerinum:* very small 'large' feathers, brown to black-brown with pale markings. Ps and Ss brown with 4-5 white to buff notches, quite rounded and wide on the inner vane. This effectively means brown bars on a pale background, with the background becoming brown around the rachis. Pale bars on inner Ss. Rs brown with 4-5 pale notches becoming thin bars on R1 and R2 (bars often surrounded by black-brown).

▷ Little Owl wing

▷ Eurasian Scops Owl wing

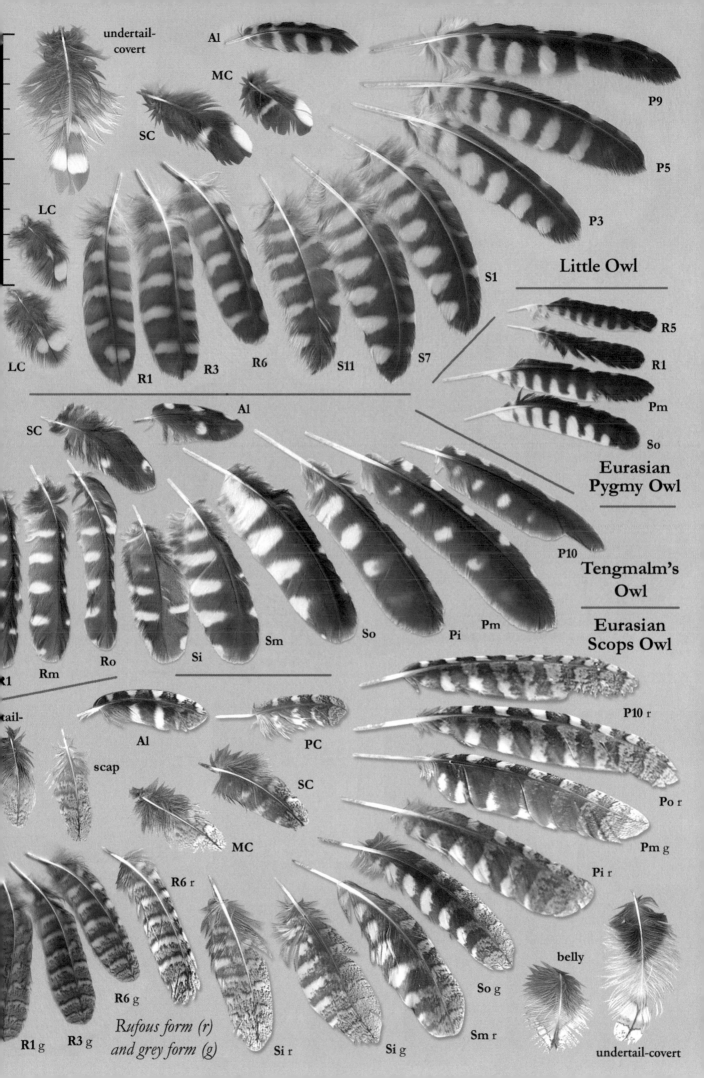

undertail-covert

Al

MC

SC

LC

LC

R1   R3   R6   S11   S7

Al

S1

P9

P5

P3

Little Owl

R5

R1

Pm

So

**Eurasian
Pygmy Owl**

SC

Al

Ro   Si   Sm   So   Pi   Pm

Rm

R1

**Tengmalm's
Owl**

**Eurasian
Scops Owl**

P10

P10 r

Po r

Pm g

Pi r

tail-

Al

PC

scap

SC

MC

R6 r

R1 g   R3 g   Si r   Si g   Sm r   So g

R6 g

belly

undertail-covert

*Rufous form (r)
and grey form (g)*

9. APPENDICES

GLOSSARY

Albinism Total absence of melanins. In truly albino individuals, the feathers are totally white, the naked parts of the body are red or pink. Partial albinism results in white feathers covering certain areas or scattered in speckles among other feathers. Other pigments may be present.

Alula Set of a few stiff and short feathers attached to the pollex, on the leading edge of the wing. The digit carrying the alula is often mistakenly referred to as the 'thumb', it is in fact the second finger of the bird (the equivalent of our index finger), fingers 3 and 4 bearing the primaries. The main function of the alula is to delay stalling when landing, and to allow the bird to slow down further without falling.

Apical At the top or at the tip.

Axillary Feather attached to the axilla (armpit) and covering the base of the wing on the underside.

Barb Structure formed of keratin, attached to the rachis, their very large number forming the vane. The barbs can be free or attached depending on the structure of their barbules. See illustration page 10.

Barbule Extension of the barb touching the neighbouring barb, which allows them to be locked together by the hooks of a barbule attaching themselves to a barbule of the next barb.

Bastard wing (n) see **Alula**.

Buff Pale, creamy-brown.

Body feather Also termed contour feather, small feather, with a down base and a coloured tip, implanted on the bird's body (back, belly, head).

Breeding plumage Plumage mainly developed for display purposes. When plumage varies in colour (or shape) through the year, breeding plumage is observed from spring until the post-breeding moult (late summer/autumn). It is obtained by moulting or abrasion of body feathers.

Calamus Shaft at the base of a feather, devoid of barbs, whose end is implanted in the skin.

Carotenoid Pigment that can colour feathers, which must be ingested by the bird. The different carotenoids (carotenes and various xanthophylls) are produced by plants and cyanobacteria. They are assimilated by the bird that consumes these micro-organisms or an animal that

has consumed them. Apart from the health benefits, their integration is responsible for orange or yellow colours in the feathers. Genetic mutations affect their integration into feathers and therefore the colour of these.

Carpal remex Small flight feather intermediate between the primaries and secondaries, fixed at the wrist level and only occurs in a few certain groups of species (pheasants, woodpeckers...).

Distal Pertains to that part of the feather (tip) farthest from the base. Opposite of proximal.

Down Small, soft and light feather, whose barbs remain unfixed between each other. The down is close to the body and is the insulating layer of the plumage. It is not visible on the bird under normal conditions (healthy bird, preened plumage).

Eclipse Eclipse plumage is worn by male ducks when moulting their remiges, while flight is almost impossible. Discreet for camouflage, this plumage resembles that of the female's worn year-round. The most conspicuous areas of plumage are much duller. This plumage concerns body feathers.

Emargination Occurs on a primary, when the width of the outer vane decreases more or less sharply. It is located almost at the same position on the feather as the possible notch on the next primary.

Fawn Reddish-brown (colour of natural leather).

Feral Describes populations or a species that has been introduced and is now breeding in the wild (examples in Europe: Canada Goose, Rose-ringed Parakeet, Feral Pigeon, etc)

Filoplume Feather resembling a hair with only a few barbs at its tip, growing amongst other feathers. Its function is probably linked to plumage maintenance and to moulting.

First- (1st-) winter A bird born in the year and observed during the first 'winter' of its life, from July/August to the following spring. The first-winter plumage appears during the first moult when the juvenile body feathers are replaced. The first-winter bird therefore carries 'large' feathers dating from its fledging and more recent body feathers (in some cases, 'large' feathers are also moulted). Depending on the colour of the body feathers, its plumage is similar or not to that of an adult.

Folded wing Measurement taken when ringing a bird or studying a specimen. It consists of measuring the folded wing of the bird, from the wrist to the end of the longest primary.

Fresh plumage Newly grown plumage not abraded. Generally, it is visible after postnuptial moulting (late summer/autumn), when 'large' feathers and body feathers are new (or at least the body feathers for birds of the year). Partial moult (body feathers only) does not lead to fully fresh plumage, since the 'large' feathers are not moulted.

Hyporachis Small fluffy feather doubling with some body or even 'large' feathers. From the single calamus, the rachis divides into two. The main part has a usual feather, the hyporachis is often downy. Also called the afterfeather. It is likely that the first feathers that appeared during the evolution of the birds were similar, their structure having subsequently been simplified. See illustration page 81.

Immature Any non-adult individual and its associated characteristics.

Juvenile A young bird that has fledged, with its first plumage and associated characteristics.

Keratin Complex set of proteins. Of various types, they are the main constituents of feathers, scales, nails, hairs and horns (reptiles, birds, mammals).

'Large' feather Includes all large feathers of the wings or tail: primaries, secondaries (including tertials) and rectrices. Collectively, also known as flight feathers (although flight feathers sometimes excludes the tail).

Leading edge Front edge of the wing and the part that cuts through the air when the bird is flying. It is the thickest part of the wing and where the feathers are most deeply implanted.

Leucism Term used to describe a wide variety of conditions which result in the partial loss of pigmentation in a bird. Melanins are present in lower concentrations than normal. The plumage may be white, pale, or patchily coloured, but not the eyes. A bird with this condition is termed leucistic.

Melanin The most common pigment produced by a bird, colouring its feathers black or brown. They also strengthen the structure of the feather. Some genetic mutations affect their production.

Melanism Excess amount of melanin resulting in darker-than-normal, even black, plumage. A bird with this condition is termed melanistic.

Moult Natural change of feathers, during which the old feathers fall out more or less gradually and are replaced by new ones.

Moulting can involve all or part of the plumage and take place over a short period (a few weeks) or throughout the year (large birds, pigeons...). Several types of moult are distinguished depending on the time of year or age of the bird: post-juvenile (first plumage after fledging), pre-breeding (before breeding), post-breeding, etc.

Non-breeding plumage Plumage observed outside the breeding period. When plumage varies within the year, non-breeding plumage is visible after the main moult, also called winter plumage (late summer/autumn), and until the following spring. It is then replaced or gradually transforms into breeding plumage by abrasion of the feathers.

Notch On a primary, it is the more or less abrupt narrowing of width of the inner vane, forming a notch in the contour of the feather. The term is also used to describe a "notch"-shaped mark on a feather, usually on its edge.

Pollex Small digit ('finger') on the wing, on which the alula feathers are attached.

Primary Large feather of the outer wing implanted on the 'hand' (usually 10 or 11 per wing).

Proximal Refers to the part near the base of the 'large' feather. Opposite of distal.

Rachis The main axis of the feather, to which the barbs are attached. See illustration page 10.

Rectrix (plural **Rectrices**) Refers to the large feathers of the tail (from 10 to 24 depending on the species).

Remex (plural **Remiges**) The 'large' feathers of the wing.

Remicle Name of the outer, often vestigial, primary.

Scapulars Feathers attached to the shoulder that cover the upper base of the wing

Secondary Large feather of the inner wing implanted on the 'arm' or ulna (often 10 or 11 per wing, sometimes more.).

Speculum Contrasting patch in a bird's plumage, which reflects light more than the plumage around it. Most commonly refers to an iridescent patch in the secondaries of some ducks.

Stalling More or less sudden loss of lift caused by the detachment of air currents along the wing. The stall may cause a fall. Its occurrence is reduced by the action of the alula.

Tegmen Characteristic structure of the primaries of Anatidae and Galliformes. Visible on the underside as a shiny area around the rachis, the tegmen is a thickening of the base of the barbs that increases the stiffness of the feather and its resistance in flight. See illustration page 80

Tertial Feather of the wing located at the 'elbow' or attached to the 'arm'. By default, designates the three innermost secondaries.

Trailing edge Rear edge of the wing, often highlighted by light or dark tips to the primaries and/or secondaries. Opposite to the leading edge. On the wing, it is the thinnest area and where some turbulence is formed during flight.

Undertail-coverts Feathers that cover the base of the underside of the rectrices (tail feathers).

Uppertail-coverts Feathers that cover the base of the upperside of the rectrices (tail feathers).

Vane Fine blade made of barbs, forming the surface of the feather. Also known as a 'web'. There are 2 vanes per feather (on each side of the rachis). The vane may be completely or partially (at the base) fluffy. See illustration page 10.

Vibrissa A stiff, hair-like feather situated at the base of the bill or around the face (also known as rictal bristles).

Wing-bar Pale (more usual) or dark band on the wing that contrasts with the overall colour of the wing, usually formed by pale tips of the wing-coverts

Wing-coverts Feathers of variable form that cover the base of the remiges. A distinction is made between underwing- and upperwing-coverts and in each case, primary coverts on the hand and secondary coverts on the forearm. The upperwing-coverts are themselves classified as greater, median and lesser coverts depending on their position between the leading edge and the trailing edge of the wing.

Wing loading Body mass of a bird compared to the surface area of its wings, expressed for example in g/cm². The smaller it is, the lighter the bird is for its wing size. The wing loading partly determines how the bird flies. The higher it is, the more muscular energy it must use to fly in flapping flight (pigeons, rails and crakes, auks, divers…) The lower it is, the more efficiently the bird can use air currents (eagles, storks, petrels...).

Wingspan Distance from the tip of one wing to the tip of the other in flight.

ANSWERS TO THE IDENTIFICATION EXERCISE IN CHAPTER 7

The plates and clues are given on pages 107–113.

• IN TOWNS AND VILLAGES

1- Collared Dove. Characteristic shape and colour.
2- Magpie. Diagnostic size and colour.
3- Feral Pigeon. White and chocolate form.
4- Feral Pigeon.
5- House Sparrow. The pale base of the outer vane is a very good clue.
6- Collared Dove. Marked contrast on underside, typical of Columbidae.
7- House Martin.
8- Common Swift. Characteristic shape for the family.
9- Magpie. An outer rectrix, therefore small.
10- Young Starling. Less contrasting than in adult; short tail.
11- House Martin. No markings and not very elongated.
12- Common Redstart or Black Redstart. Impossible to differentiate between the two in this case.
13- House Sparrow. Note, there are other very similar species.
14- White Wagtail. Size and lack of yellow eliminate Grey Wagtail; Yellow Wagtail also possible but unlikely in a town at this season.
15- Feral Pigeon. The variegated form.

• IN PARKS AND IN THE COUNTRYSIDE

21- Robin. Very small reddish tip is a useful character (not visible here); other passerines very similar.
22- Great Spotted Woodpecker. Wide, rounded white notches.
23- Mistle Thrush.
24- Jay. Some secondaries have much more blue, with dark barring.
25- Common Pheasant Female.
26- Green Woodpecker.
27- Blackbird. Male.
28- Chaffinch.
29- Linnet. Female (the white of the outer vane does not reach the rachis; hardly visible here).
30- Goldfinch. Shape of the marks is characteristic.
31- Greenfinch. Male (a lot of yellow).
32- Barn Owl. Body feather, viewed from the upperside.
33- Skylark. Shape of the tip typical of this family.
34- Common Kestrel. Probably female (it would be more rufous in the male).
35- Tawny Owl. The grey-brown form.

• ALONG LAKESIDES

41- Mallard. Speculum colour diagnostic.
42- Tufted Duck.
43- Great Crested Grebe.
44- Coot. By its diagnostic colour.
45- Black-headed Gull. Middle primary. Slender-billed Gull is very similar but eliminated due to habitat and region.
46- Cormorant. Shape, calamus and rigidity are diagnostic of the family. One of the outer rectrices (rounded tail) so size is indicative of this species.
47- Black-headed Gull. Can be confused with other gulls of a similar size.
48- Mute Swan. Could be confused with other swans (here the area and season are good indications).
49- Red-crested Pochard. Female. Tegmen, colour and size diagnostic.

• ON THE SEASHORE

61- Bar-tailed Godwit. Other waders can be very similar.
62- Eurasian Curlew. Colour indicates a curlew. Size and the feather's position on the bird are indicative of the species.
63- Tern sp. Possibly Common Tern. Shape and colour characteristic of this family.
64- Guillemot. This is the secondary, S1, but its shape is the same as the primaries in auks. Secondaries of the usual shape have a white tip in this species.
65- Turnstone. Pattern and size distinctive, but other waders are similar.
66- Oystercatcher. Pattern and size very characteristic.
67- Black-tailed Godwit. Pattern common to other waders but size and details distinctive.
68- Lesser Black-backed Gull. Shade, width of white tip and size are distinctive.
69- Black-headed Gull. Size and the absence of a white tip place it with Slender-billed Gull, though the latter can normally be eliminated by where it was found.
70- Tern sp. Possibly Arctic Tern. Shape and colour characteristic of this family. Feather shed before migration.
71- Dunlin. Shape and size indicate a *Calidris* wader. The only slightly contrasting shades and narrow, regular coloured edge associated with the bicoloured rachis are indicative of the species.

TABLE OF BIOMETRIC DATA

The double-page table below shows measurements in cm of 'large' feathers, the position of their emarginations and notches (partial data) and various other measurements (length, wingspan, folded wing and mass for all species considered here) that can help to show certain peculiarities of some 'large' feathers. See the bibliography for more details. Species are listed according to the systematic order usually adopted in more recent identification guides (but note: this order has changed markedly in the last few years).

Remarks on the measurements of 'large' feathers

The number of individuals measured varies widely. If there is a difference between the values shown in this table and the text, this table should be taken as definitive.

– Paler grey boxes: three or fewer individuals were measured; values indicated by SS (= small sample) in the text.

  – Darker grey boxes: only one individual was measured, or only a few feathers, etc. In the case of very partial data, the signs '?', '>x' '<x' indicate, respectively, assumed measurements, an observed minimum, and an observed maximum.

ENGLISH NAME	SCIENTIFIC NAME	Outer P (small)	Pmax	Pmin	Smax	Smin	Inner S (tertials)	Rmax
Mute Swan	Cygnus olor	10-12	46	32	36	27	24-29	24
Black Swan	Cygnus atratus	9-10	38	26	27	23	19-26	14
Tundra (Bewick's) Swan	Cygnus columbianus	7-8	35	22	23	21	n.d.	21
Whooper Swan	Cygnus cygnus	11-12	47	32	33	31	n.d.	22
Bean Goose	Anser fabalis	n.d.	n.d.	n.d.	n.d.	n.d.	n.d.	16
Pink-footed Goose	Anser brachyrhynchus	n.d.	n.d.	33	n.d.	n.d.	n.d.	n.d.
Greater White-fronted Goose	Anser albifrons	n.d.	n.d.	22	n.d.	n.d.	n.d.	14
Lesser White-fronted Goose	Anser erythropus	n.d.	n.d.	n.d.	n.d.	n.d.	n.d.	n.d.
Greylag Goose	Anser anser	5-8	40 (45)	23	24	20	15-23	19
Bar-headed Goose	Anser indicus	n.d.	n.d.	n.d.	n.d.	n.d.	n.d.	n.d.
Snow Goose	Anser caerulescens	n.d.	n.d.	n.d.	n.d.	n.d.	n.d.	n.d.
Canada goose	Branta canadensis	6-8	48	23	29	20	22-32	20
Barnacle Goose	Branta leucopsis	6-8	39	18	22	18	16-26	18
Brent Goose	Branta bernicla	5-7	32 (44)	14	18 (24)	12	11-16	18
Egyptian Goose	Alopochen aegyptiacus	6-9	45	24	30	27	29-33	16
Ruddy Shelduck	Tadorna ferruginea	n.d.	n.d.	n.d.	n.d.	n.d.	n.d.	n.d.
Common Shelduck	Tadorna tadorna	4-7	26	12	16	10	10-18	14
Wood Duck	Aix sponsa	n.d.	n.d.	n.d.	n.d.	n.d.	n.d.	n.d.
Mandarin Duck	Aix galericulata	4-5	21	11	13	11	?	14
Eurasian Wigeon	Anas penelope	4-5	22	11	13	9	9-15	12 (16)
Gadwall	Anas strepera	4-5	21	10	12	8	8-13	10
Eurasian Teal	Anas crecca	3-4	16 (18)	8	9	6	8-11	9 (10)
Mallard	Anas platyrhynchos	4-7	22	13	15	11	9-17	12
Domestic Mallard	Anas platyrhynchos	n.d.	> 25	12	> 20	10	8-22	20
Northern Pintail	Anas acuta	5-7	22 (25)	11	12	9	10-14	14f / 22m
Garganey	Anas querquedula	3-5	16	9	10	7	9-12	9
Northern Shoveler	Anas clypeata	4-6	19 (22)	10	11	7	9-14	13
Marbled Duck	Marmaronetta angustirostris	n.d.	n.d.	n.d.	n.d.	n.d.	n.d.	n.d.
Red-crested Pochard	Netta rufina	n.d.	19	11	13	11	11	> 9
Common Pochard	Aythya ferina	4-5	16	11	11	9	9-13	8
Ferruginous Duck	Aythya nyroca	n.d.	n.d.	n.d.	n.d.	n.d.	n.d.	n.d.
Tufted Duck	Aythya fuligula	4-5	16	9	10	7	8-11	8
Greater Scaup	Aythya marila	4-5	17	9	11	8	9-13	8
Common Eider	Somateria mollissima	3-7	23	14	15	11	(6) 10-14	14
King Eider	Somateria spectabilis	n.d.	n.d.	n.d.	n.d.	n.d.	n.d.	n.d.
Steller's Eider	Polysticta stelleri	n.d.	n.d.	n.d.	n.d.	n.d.	n.d.	n.d.
Harlequin Duck	Histrionicus histrionicus	n.d.	n.d.	n.d.	n.d.	n.d.	n.d.	n.d.
Long-tailed Duck	Clangula hyemalis	4-6	18	9	11	8	8-9	14f / 28m
Common Scoter	Melanitta nigra	n.d.	19	10	12	8	9-14	13
Velvet Scoter	Melanitta fusca	n.d.	21	11	13	9	9-14	9 (11)
Barrow's Goldeneye	Bucephala islandica	n.d.	n.d.	n.d.	n.d.	n.d.	n.d.	n.d.
Common Goldeneye	Bucephala clangula	5-7	22	10	11	8	8-11	13
Smew	Mergellus albellus	4-6	15	8	9	6	7-10	8 (10)
Red-breasted Merganser	Mergus serrator	5-7	19 (21)	10	13 (14)	8	9-13	10 (12)
Goosander	Mergus merganser	5-8	22	12	14	11	13-15	13
Ruddy Duck	Oxyura jamaicensis	n.d.	n.d.	n.d.	n.d.	n.d.	n.d.	n.d.
White-headed Duck	Oxyura leucocephala	n.d.	n.d.	n.d.	n.d.	n.d.	n.d.	n.d.
Hazel Grouse	Bonasa bonasia		14	9	12	9	7-9	14
Willow Ptarmigan	Lagopus lagopus		18	11	12	9	7-10	14
Red Grouse	Lagopus lagopus scoticus		18 (19)	11	12	10	8-10	13

– n.d.: given when no data could be collected.

– Particular attention should be paid to the measurements of outer Ps and inner Ss. Samples are often smaller than for the rest of the 'large' feathers and it is more likely that variation exists outside of the indicated range.

Remarks on interpretation of the Em and Not values

The given values refer to the numbers of the primaries in which an **emargination** and/or a **notch** are found. Single parentheses (x) indicate a less marked or less frequently observed narrowing, double parentheses ((x)) indicate that narrowing rarely occurs. Depending on the origin of the populations and their behaviour (migrant or not, for example), there may be variation within a species.

Abbreviations: ov = outer vane, iv = inner vane, Po = outer primary, Pi = inner primary.

Value 0 indicates no narrowing. Empty boxes indicate no data available.

Example reading of the table: the numbers indicate inclusive ranges, so if four successive values – 9 / 7 / 10 / 6 – are given in columns 5, 6, 7, and 8, this means that there is an emargination on feathers P9-P7 and a notch on feathers P10-P6.

Rmin	NP (number of Ps)	NS (number of Ss)	NR (number of Rs)	Em (ov) Po	Em (ov) Pi	Not (iv) Po	Not (iv) Pi	L (length)	W (wingspan)	FW (folded wing)	M (mass in grams or <kg)
14	10+1	22-28	20-24	9	7	10	8	140-160	200-240	52-58	(5.3) 6.3-16.3 kg
12	10+1	22-28	≥ 20	9	7	10	8	140	160-200		3.7-8.75 kg
16	10+1	22-28	20	9	7	10	8	115-127	170-195		5.7-6.5 kg
13	10+1	≥ 20	20	9	7	10	8	140-160	205-235		8-11 kg
15	10+1	≥ 12	≥ 16	10	9	9	8	69-88	140-174		3-4.1 kg
n.d.	10+1	≥ 12	≥ 16	10	9	9	8	64-76	137-161		2.2-2.7 kg
12	10+1	≥ 12	≥ 16	10	9	9	8	64-78	130-160		1.4-3.3 kg
n.d.	10+1	≥ 12	≥ 16	10	9	9	8	56-66	115-135		1.9-2.3 kg
13	10+1	20	(14) 18-20	10	9	9	8	74-84	149-168	38-52	1.8-5.2 kg
n.d.	10+1	≥ 12	≥ 16					68-78	140-160		2-3 kg
n.d.	10+1	≥ 12	≥ 16					65-75	132-165		2.7 kg
16	10+1	≥ 11	16	9	7	10	8	90-105	160-185		4.3-5 kg
11	10+1	≥ 12	16	10	9	9	8	58-70	120-142		1.4-2.4 kg
9	10+1	≥ 11	14 ?	10	9	9	8	55-62	105-117		1.1-1.7 kg
10	10+1	20 ?	14 ?	9	8	10	9	63-73	134-154		1.5-2.25 kg
n.d.	10+1	20 ?	14 ?					58-70	110-135		625-1500
10 (8.5)	10+1	20	14	9		10	9	55-65	100-120	28-35	(560) 690-1500 (1710)
n.d.	10+1	≥ 14	≥ 14					44-51	68-74		480-865
10	10+1	≥ 13	16	9		10		41-49	65-75		425-695
9	10+1	(12) 16-18	14-18 (20)	9		10		42-50	78-87	22.5-28	500-1090
8	10+1	(12) 16-18	14-18 (20)					46-56	78-90	23.3-28.4	500-1100 (1300)
6	10+1	(12) 16-18	14-18 (20)	9		10		34-38	53-59	16.6-20	(163) 200-450 (500)
8	10+1	16-18	14-18 (20)	9	(8)	10	(9)	50-60	81-95	24.5-29.8	720-1580
8	10+1	16-18	14-18 (20)								
8	10+1	(12) 16-18	14-18 (20)	9		10		51-62 (+10)	79-87	23.6-28.9	400-1300
7	10+1	(12) 16-18	14-18 (20)					37-47	59-67	18.2-21.1	250-450 (600)
8	10+1	(12) 16-18	14-18 (20)	9		10		44-52	73-82	21.3-25.1	(300) 470-800 (1000)
n.d.	10+1	≥ 12	≥ 14					39-42	63-70		450-590
n.d.	10+1	(12) 16-18	14-18 (20)					53-57	65-90	23.7-27.3	830-1400
5	10+1	(12) 16-18	14-18 (20)	9		10		42-49	67-75	18.5-22.6	(467) 715-1300
n.d.	10+1	≥ 12	≥ 14					38-42	60-67		650-800
6	10+1	(12) 16-18	14	9		10		40-47	65-72	18.5-21.5	(335) 500-1070 (1100)
6	10+1	≥ 12	≥ 14					42-51	71-80		700-1100
8	10+1	≥ 18	14	9		10		60-70	95-105		1200-2800
n.d.	10+1	≥ 12	≥ 14					55-63	87-100		1500-2010
n.d.	10+1	≥ 12	≥ 14					42-48	68-77		860
n.d.	10+1	≥ 12	≥ 14					38-45	63-70		500-730
8	10+1	≥ 12	≥ 14	9				39-47 (+15)	73-82		600-900
6	10+1	≥ 12	≥ 14					44-54	70-84		1300-1500
6	10+1	≥ 12	≥ 14					51-58	79-97		f1100-1250 / m 1200-2000
n.d.	10+1	≥ 12	≥ 14					42-53	67-82		1000-1305
5	10+1	≥ 12	16	9		10		40-48	62-77		f700-950 / m820-1150
6	10+1	≥ 12	≥ 14					38-44	56-69		500-800
7	10+1	≥ 12	18					52-58	69-82		1000-1250
10	10+1	≥ 12	20					58-68	78-94		1050-1650
n.d.	10+1	≥ 12	≥ 14					35-43			350-800
n.d.	10+1	≥ 12	≥ 14					43-48			700-900
12	10	≥ 12	16	9	4	10	6	34-39			300-450
11	10	≥ 15	16					35-43	55-65		370-750
11	10	≥ 15	16	9	5	(9)	(6)	33-42	55-66		550-750

ENGLISH NAME	SCIENTIFIC NAME	Outer P (small)	Pmax	Pmin	Smax	Smin	Inner S (tertials)	Rmax
Rock Ptarmigan	*Lagopus mutus*		16 (20)	11	13 (15)	10	7-10	14 (15)
Black Grouse	*Tetrao tetrix*		20 (25)	12	16	12	10-14	f15 / m23
Capercaillie	*Tetrao urogallus*		f27 / m33 (37)	f16 / m21	f21 / m24	f15 / m19	12-18	f24 / m36
California Quail	*Callipepla californica*		n.d.	n.d.	n.d.	n.d.	n.d.	n.d.
Northern Bobwhite	*Colinus virginianus*		n.d.	n.d.	n.d.	n.d.	n.d.	n.d.
Chukar	*Alectoris chukar*		n.d.	n.d.	n.d.	n.d.	n.d.	n.d.
Rock Partridge	*Alectoris graeca*		n.d.	12	n.d.	n.d.	n.d.	n.d.
Red-legged Partridge	*Alectoris rufa*		13 (15)	9	11 (12)	9 (7)	7-10	11
Barbary Partridge	*Alectoris barbara*		14	11	12	10	n.d.	12
Grey Partridge	*Perdix perdix*		14 (15)	9	11	9	7-10	9 (10)
Common Quail	*Coturnix coturnix*		8.5	5.5	6.5	5	4.5-6.5	4.5
Reeves's Pheasant	*Syrmaticus reevesii*		27	17	18	14	n.d.	140
Common Pheasant	*Phasianus colchicus*		23	15	19	15	n.d.	60
Golden Pheasant	*Chrysolophus pictus*		21	14	18	14	n.d.	70
Lady Amherst's Pheasant	*Chrysolophus amherstiae*		23	17	18	16	n.d.	f60 / m105
Domestic cockerel	*Gallus gallus*	n.d.	> 19	< 13	> 19	< 12	n.d.	> 20
Green Junglefowl	*Gallus varius*	n.d.	n.d.	n.d.	n.d.	n.d.	n.d.	n.d.
Red Junglefowl	*Gallus gallus*	4-5	21	15	20	15	9-15	25
Indian Peafowl	*Pavo cristatus*	26	41	32	37	28	n.d.	56
Domestic turkey	*Meleagris gallopavo*	n.d.	> 35	n.d.	n.d.	n.d.	n.d.	> 40
Helmeted Guineafowl	*Numida meleagris*	7-9	28	20	25	13	n.d.	20
Red-throated Diver	*Gavia stellata*	5-6.5	21 (23)	11	13	8	8-10	8
Black-throated Diver	*Gavia arctica*	n.d.	22 (23)	11	13	9	9-10	8
Great Northern Diver	*Gavia immer*	6-7.5	30	16	16 (21)	11	n.d.	9
White-billed Diver	*Gavia adamsii*	n.d.	n.d.	n.d.	n.d.	n.d.	n.d.	n.d.
Little Grebe	*Tachybaptus ruficollis*	0.5-1	8 (8.5)	6	7.5	5.5	5-6	
Great Crested Grebe	*Podiceps cristatus*	4-5.5	14 (18)·	10	11	8	8-12	
Red-necked Grebe	*Podiceps grisegena*	4-5	14 (18)	9	11	7.5	7.5-11.5	
Slavonian Grebe	*Podiceps auritus*	3-4	11 (14)	7.5	8.5	6.5	6.5-7.5	
Black-necked Grebe	*Podiceps nigricollis*	2-3	10.5 (14)	7	7.5	5.5	5.5-6.5	
Northern Fulmar	*Fulmarus glacialis*	6.5-7.5	26	11.5	12.5	9	n.d.	15
Cory's Shearwater	*Calonectris diomedea*	n.d.	26	11.5	12.5	10	12-13.5	15
Great Shearwater	*Puffinus gravis*	5.5-6.5	25 (28.5)	12	n.d.	n.d.	n.d.	13
Sooty Shearwater	*Puffinus griseus*	n.d.	n.d.	n.d.	n.d.	n.d.	n.d.	n.d.
Manx Shearwater	*Puffinus puffinus*	5-6	18.5	9	9.5	7	n.d.	10 (11)
Yelkouan/Balearic Shearwater	*Puffinus yelkouan*	n.d.	n.d.	n.d.	n.d.	n.d.	n.d.	n.d.
Wilson's Storm Petrel	*Oceanites oceanicus*	n.d.	n.d.	n.d.	n.d.	n.d.	n.d.	n.d.
European Storm Petrel	*Hydrobates pelagicus*	n.d.	10.5	5	5	4	n.d.	6.5
Leach's Storm Petrel	*Oceanodroma leucorhoa*	n.d.	13.5	6.5	6.5	6	4.5-5.5	9.5
Madeiran Storm Petrel	*Oceanodroma castro*	n.d.	n.d.	n.d.	n.d.	n.d.	n.d.	n.d.
Northern Gannet	*Morus bassanus*	6-8	37	18	19	14	13-15	29
Great Cormorant	*Phalacrocorax carbo*	6-7	27 (31)	19	20	16	15-16	20 (23)
European Shag	*Phalacrocorax aristotelis*	6-7	21	17	18	13	10-13	17
Pygmy Cormorant	*Phalacrocorax pygmaeus*	n.d.	16	12	14	11	6-12	15 ?
White Pelican	*Pelecanus onocrotalus*	n.d.	50 (60)	33	38	30	30-36	22
Dalmatian Pelican	*Pelecanus crispus*	n.d.	50	32	38	30	30-34	28
Eurasian Bittern	*Botaurus stellaris*	4-6	30	20	22	17	13.5-18	12
Little Bittern	*Ixobrychus minutus*	n.d.	12	9	9	8	7.5-9	6
Black-crowned Night Heron	*Nycticorax nycticorax*	6-8	26	19	20	14	15-18 ?	13
Squacco Heron	*Ardeola ralloides*	n.d.	n.d.	n.d.	n.d.	n.d.	n.d.	12
Cattle Egret	*Bubulcus ibis*	n.d.	22	14	16 (17)	12	10-13	11
Little Egret	*Egretta garzetta*	n.d.	24	17	19	14	n.d.	13.5
Great Egret	*Casmerodius albus*	6-7	35 (38)	25 (22)	28 (30)	21	?-26	20
Grey Heron	*Ardea cinerea*	9-12	37	25	28	20	15-25	20
Purple Heron	*Ardea purpurea*	8-11	34	24	26	19	15-20	15 (20)
Black Stork	*Ciconia nigra*	8-10	50 (54)	30	33	26	26-29	26 (32)
White Stork	*Ciconia ciconia*	7-10	49	28	31	26	22-29	27
Glossy Ibis	*Plegadis falcinellus*	5-7	24	15	17	14	n.d.	13
Sacred Ibis	*Threskiornis aethiopicus*	n.d.	30	24	28	20	n.d.	n.d.
Eurasian Spoonbill	*Platalea leucorodia*	5-6	31	21	22	17	17-21	15
Greater Flamingo	*Phoenicopterus roseus*		33	20	22	16	15-26	16
European Honey Buzzard	*Pernis apivorus*	4.5-6	39 (42)	22	27	18	9-19	28
Black-winged Kite	*Elanus caeruleus*	n.d.	23	13	14	9	6-10	15
Black Kite	*Milvus migrans*	5-7	45 (50)	21	23	18	16-19	30 (34)
Red Kite	*Milvus milvus*	4-6	47	23	25 (27)	20	16-21	40
White-tailed Eagle	*Haliaeetus albicilla*	7-10	68	40	49	30	25-32	40

Rmin	NP (number of Ps)	NS (number of Ss)	NR (number of Rs)	Em (ov) Po	Em (ov) Pi	Not (iv) Po	Not (iv) Pi	L (length)	W (wingspan)	FW (folded wing)	M (mass in grams or kg)
11	10	18-19	16	9	5	(9)	(6)	31-38	55-66		430-750
11	10	≥ 14	18					f40-45 / m49-58			f280-380 / m400-450
f18 / m21	10	18-19	18-24	10	4	(10)	(5)	f54-63 / m74-90			f1.5-2 kg / m3.4-3.6 kg
n.d.	10	> 10	≥ 12					27	38-43		
n.d.	10	> 10	≥ 12					25	35-40		
n.d.	10	15	14 (16)					32-35			
11	10	≥ 14	≥ 14					33-36			500-750
7	10	15	14 (16-18)					32-35		14.7-17.5	(320) 390-560
10	10	≥ 14	≥ 14					32-35			
7	10	15	14-18	9	4	10	6	28-37		14.7-17	310-455
3	10	14	10-12	9	8	10		16-18		10.1-11.9	(65) 70-155
13	10	16	18-20					m < 210			
13	10	16	≥ 18					f55-70 / m70-90			f0.9-1.1 / m1-1.4 kg
12	10	16	≥ 18					f60-80 / m90-105			
11	10	16	≥ 18					f60-80 / m105-120			
< 12	10+1	≥ 15	≥ 14					env 25-70			env 0.7-5 kg
n.d.	10+1	≥ 15	≥ 14					70			450-795
15	10+1	≥ 15	≥ 14					41-78			300-700
34	≥ 10	≥ 14	≥ 16					90-230	130-160		4-6 kg
n.d.	≥ 10	≥ 14	≥ 16					76-125			2.7-6.3 kg
14	10+1	≥ 15	≥ 14					58-68			
6	10+1	18-23	12	0		(10)	9	55-67	91-110		f 1500 / m 1750
5	10+1	18-23	12	0		(10)	9	63-75	100-127		
7	10+1	18-23	12	0		(10)	9	73-88	122-148		
n.d.	10+1	18-23	12					77-91	137-152		
	11+1	17-22	reduced	10	9 (8)	11	10	23-29		9.5-10.8	100-236 (315)
	11+1	17-22	reduced	10	9	11	10	46-51	59-73	16.8-21.5	568-1300 (1490)
	11+1	17-22	reduced	10	9	11	9	40-46			700-900
	11+1	17-22	reduced	11	9	11	9	31-38	46-55		375-450
	11+1	17-22	reduced	10	(9)	11	(10)	28-34		12.4-13.9	213-450
10	10+1	20	14	0	0	0	0	43-52	101-117		700-900
12	10+1	≈ 20	12	0	0	0	0	45-56	112-126	28.2-39.3	(390) 514-1150
11	10+1	≥ 14	12	0	0	0	0	43-51	105-122		
n.d.	10+1	≥ 14	12	0	0	0	0	40-50	93-106		
8	10+1	22	12	0	0	0	0	30-35	72-82		350-450
n.d.	10+1	≥ 14	12	0	0	0	0	34-39	78-90		350-450
n.d.	10+1	≥ 14	12	0	0	0	0	16-19	38-42		
5.5	10+1	≈ 14	12	0	0	0	0	15-16	37-41	11.6-12.9	(18) 20-30 (39)
6.5	10+1	≥ 14	12	0	0	0	0	18-21	43-48		40-50
n.d.	10+1	≥ 14	12	0	0	0	0	19-21	43-46		
12 (11)	10+1	≥ 25	12	10	(9-8)	(10)	8	85-97	170-192		2800-3200
14	10+1	≈ 22	14-16	9	7	10	8	77-94	121-149		2000-2500
12	10+1	18-19	12	9	7	10	8	68-78	95-110	24-28.9	1360-2300
9 ?	10+1	> 18	12	9	7	10	8	45-55	75-90		
15	10+1	30	24	9	8 (6)	10	6	140-175	245-295		
20	10+1	30 ?	24 ?					160-180	270-320		
8	10+1	15-20	10	0	0	0	0	69-80	100-130	27.8-37.7	(390) 661-2060
5	10+1	15-20	12	0	0	0	0	33-38	49-58	12.7-17	(64) 80-150 (180)
10	10+1	15-21	12	9	7	10	8	58-65	90-100		550
8	10+1	≥ 15	12	(9)	(6)			40-49	71-80		250-300
8	10+1	≥ 15	12	(9)	(6)	10	9 (8)	45-52	80-90		300-400
9	10+1	15-20	12	(9)	(6)	10	9 (8)	55-65	88-106	24.5-33	(220) 330-550 (710)
16	10+1	≥ 15	12	10	(9-8)	(9)	(7)	85-100	143-169		
15	10+1	15-20	12	9	7 (6)	10	8 (7)	84-102	155-175	42.5-49	(600) 1020-2075 (2300)
12	10+1	15-20	12	9	7	10	8	70-90	107-143	33.7-39.5	525-1345
20	11+1	22	12	10	6	11	7	90-105	175-202		3000
18	11+1	22	12	10	6	11	7	95-110	183-217	53-63	2610-4400
10	10+1	≥ 14	12	9	8	10	9	55-65	88-105		500-800
n.d.	10+1	≥ 14	12	9	7	10	7	89	112-124		1250-1500
12	10+1	≥ 17	12	9	7	10	7	80-93	120-135		1200-1700
11	11	≥ 25	14	10	9	11	8	120-145	140-170		2500-3500
23	10+1	≥ 13	12	9	6	10	6	52-59	113-135		600-1000
10	10+1	≥ 12	12	9	8	10	9 (8)	31-36	71-88		
23	10+1	≥ 13	12	10	6	9	5	48-58	130-155		650-950
26	10+1	14	12	9	6	10	6	61-72	140-165		f950-1300 / m750-1050
24	10+1	≥ 17	12	9	4	10	5	76-94	190-240		f4.1-7 kg / m3.1-5.5 kg

ENGLISH NAME	SCIENTIFIC NAME	Outer P (small)	Pmax	Pmin	Smax	Smin	Inner S (tertials)	Rmax
Lammergeier	*Gypaetus barbatus*		75	40	47	36	32-38	56
Egyptian Vulture	*Neophron percnopterus*		48 (50)	25	28	23	21-25 ?	35 (37)
Griffon Vulture	*Gyps fulvus*		60 (68)	35	42 (48)	35	25-44	40
Eurasian Black Vulture	*Aegypius monachus*		65	30	42	30	25-40	40
Short-toaed Eagle	*Circaetus gallicus*	5-6	48	28	30	23 (16)	16-24	33
Western Marsh Harrier	*Circus aeruginosus*	4.5-6	36 (41)	22	24	17	12-18	29
Hen Harrier	*Circus cyaneus*	4-5.5	36 (37)	17	22 (25)	14	12-16	27 (29)
Montagu's Harrier	*Circus pygargus*	n.d.	33 (37)	17	19	13	(6) 11-16	26
Northern Goshwak	*Accipiter gentilis*	1-2	f31 (35) / m27	f18 / m16	f23 (25) / m18	f16 / m14	f12-17 / m8-15	f31 / m26
Eurasian Sparrowhawk	*Accipiter nisus*	?	f21 / m18	f14 / m12	f15 / m12	f11 / m9	5-11	f21 / m17
Levant Sparrowhawk	*Accipiter brevipes*	n.d.	n.d.	n.d.	n.d.	n.d.	n.d.	n.d.
Common Buzzard	*Buteo buteo*	5-7	37 (43)	21	26	17	10-20	26 (28)
Long-legged Buzzard	*Buteo rufinus*	n.d.	40	20	22	16	13-17	29
Rough-legged Buzzard	*Buteo lagopus*	4-6	40 (45)	24	28	17	(5) 11-22	27
Greater Spotted Eagle	*Aquila clanga*	n.d.	n.d.	n.d.	n.d.	n.d.	n.d.	n.d.
Lesser Spotted Eagle	*Aquila pomarina*	4-4.5	42	27	27	20	13-17	26
Booted Eagle	*Aquila pennata*	18-19	31	21	22	16	13-16	23
Golden Eagle	*Aquila chrysaetos*	6-8	55 (68)	34	39 (42)	26	20-30	37 (41)
Bonelli's Eagle	*Aquila fasciata*	n.d.	41	25	30	20	14-23	31
Eastern Imperial Eagle	*Aquila heliaca*	5-6	52	35	36	26	(10) 15-25	36
Spanish Imperial Eagle	*Aquila adalberti*	n.d.	n.d.	n.d.	n.d.	34	n.d.	31
Osprey	*Pandion haliaetus*	6-7	41 (43)	23	25	20	15-20	26
Lesser Kestrel	*Falco naumanni*	n.d.	21	10	11	9	7-10	17 (18)
Common Kestrel	*Falco tinnunculus*	4-5	23	11	12	9	7-11	20
Red-footed Falcon	*Falco vespertinus*	n.d.	22	11	12	9	?	15
Merlin	*Falco columbarius*	3-4.5	19 (20)	9	11	8	8-9	16
Eurasian Hobby	*Falco subbuteo*	4-6	23 (28)	9	11 (13)	9	6-11	16 (17)
Eleonora's Falcon	*Falco eleonorae*	2.5-3.5	n.d.	12	11	9	n.d.	17
Lanner Falcon	*Falco biarmicus*	n.d.	29	n.d.	14	n.d.	13	23
Saker Falcon	*Falco cherrug*	3-4	33	17	21	15	10-17	26
Gyrfalcon	*Falco rusticolus*	n.d.	n.d.	n.d.	n.d.	n.d.	n.d.	30
Peregrine Falcon	*Falco peregrinus*	2.5-4	f30 (34) / m25	f15 / m12	f16 (19) / m14	f12 / m10	f7-13 (15) / m5-11	f21 / m15 (16)
Common Buttonquail	*Turnix sylvaticus*	n.d.	n.d.	n.d.	n.d.	n.d.	n.d.	n.d.
Water Rail	*Rallus aquaticus*	7-8	10.5	7	8.5	7	5.5-10	7
Spotted Crake	*Porzana porzana*	7-7.5	9.5	7	7.5	6	5-8.5	6
Little Crake	*Porzana parva*	n.d.	9	6	6.5	5	4-6	6
Baillon's Crake	*Porzana pusilla*	5-5.5	7.5	5	5.5	4	4.5-6	?
Corncrake	*Crex crex*	7.5-10 (12)	12 (14)	7.5	9	6.5	6.5-10	7.5
Common Moorhen	*Gallinula chloropus*	1-2.5	15	10	11	8.5	6-10.5	9
Purple Swamphen	*Porphyrio porphyrio*	15-16	20 (25)	16	16.5 (20)	14.5	11-16	9.5
Eurasian Coot	*Fulica atra*	2-3	16.5	10.5	12.5	9	8-12	8
Red-knobbed Coot	*Fulica cristata*	n.d.	20	14 ?	16 ?	12 ?	n.d.	10 ?
Common Crane	*Grus grus*	4-6 ?	45	26	32	26	30-36	24
Little Bustard	*Tetrax tetrax*	14-15.5	18	12.5	13.5	11.5	13-16	12
Great Bustard	*Otis tarda*	n.d.	61	33	43	21	23-30	29
Eurasian Oystercatcher	*Haematopus ostralegus*	5-5.5	20.5	11	12	9.5	11.5-14	13
Black-winged Stilt	*Himantopus himantopus*	4-4.5	20	10.5	11	9	11-13	9.5
Pied Avocet	*Recurvirostra avosetta*	3.5-4	18	10	10.5	9	9-12	10.5
Stone Curlew	*Burhinus oedicnemus*	4.5-5	20	5	12	8.5	12-14	14
Collared Pratincole	*Glareola pratincola*	3.5	18	7	8	5.5	4.5-6.5	11.5
Black-winged Pratincole	*Glareola nordmanni*	n.d.	19 ?	7 ?	8.5 ?	5.5 ?	n.d.	9 ?
Little Ringed Plover	*Charadrius dubius*	n.d.	9.5	5	6	4.5	5.5-7	7
Common Ringed Plover	*Charadrius hiaticula*	2-3	11	5	7	4.5	5.5-7.5	8
Kentish Plover	*Charadrius alexandrinus*	n.d.	10	5	6.5	4.5	5-8.5	7
Eurasian Dotterel	*Charadrius morinellus*	n.d.	13	5.5	7.5	5.5	5.5-11.5	8.5
European Golden Plover	*Pluvialis apricaria*	4-4.5	15.5	7.5	9	7.5	7.5-10	9.5
Grey Plover	*Pluvialis squatarola*	4-4.5	16	8	9	6.5	9-11.5	9.5
Spur-winged Plover	*Hoplopterus spinosus*	n.d.	20 ?	13 ?	13.5 ?	9 ?	10-11	13 ?
Northern Lapwing	*Vanellus vanellus*	3.5-5	20	13	13.5	9	7.5-14	13
Red Knot	*Calidris canutus*	3	13.5	6.5	7.5	6	5.4-9.1	7.5
Sanderling	*Calidris alba*	2	10.5	4.5	6	4.5	4.5-8	6

Rmin	NP (number of Ps)	NS (number of Ss)	NR (number of Rs)	Em (ov) Po	Em (ov) Pi	Not (iv) Po	Not (iv) Pi	L (length)	W (wingspan)	FW (folded wing)	M (mass in grams or kg)
40	10	≥ 17	12	((10	1))	10	6	105-125	235-275		5-7 kg
19	10	≥ 15	14	9	5	10	5	55-65	155-170		1.6-2.2 kg
30 (27)	10	≥ 21	14	9	4	10	4	95-110	230-270		f8-11 kg / m7.5-10.5 kg
24	10	≥ 17	12	9	4	10	5	100-115	250-295		7-12.5 kg
29 (19)	10+1	≥ 15	12	9	6	10	6 (5)	62-69	162-178		1.1-2 kg
22	10+1	13-14 (17)	12	9	6	10	7 (6)	43-55	115-140	35-43.2	f500-1030 / m400-750
22	10+1	13-14 (17)	12	9	6 (5)	10	7 (6)	42-55	97-118	32.3-39.5	f300-700 / m300-400
21	10+1	13-14 (17)	12	9	7 (6)	10	8 (7)	39-50	96-116	32-41	f300-450 / m200-300
f26 (23) / m22	10+1	13-14 (17)	12	9	5	10	6	f58-64 / m49-56	f108-120 / m90-105	28-37	f800-2200 / m415-1100
f17 / m15	10+1	13-14 (17)	12	9	6 (5)	10	6	f35-41 / m29-34	f67-80 / m58-65	18.6-26	f185-360 (430) / m105-200
n.d.	10+1	≥ 12	12					30-37	63-76		
21	10+1	13-14 (17)	12	9	5 (4)	10	7 (6)	46-58	110-132	35-43	380-1400, f heavier; vulpinus lighter
21	10+1	≥ 14	12	9	6 (5)	10	6	50-62	130-155		
23	10+1	≥ 14	12	9	6 (5)	10	7 (6)	49-59	123-148		f950-1300 / m600-950
n.d.	10+1	≥ 14	12					59-69	153-177		
21	10+1	≥ 14	12	9	6 (5)	10	6	55-65	143-168		
19	10+1	≥ 14	12	9	6 (5)	10	6	42-51	110-135		f850-1250 / m500-800
33	10+1	≥ 17	12	9	4	10	5	80-93	190-225		f3.8-6.7 kg / m2.8-4.5 kg
26	10+1	≥ 14	12	9	6	10	6 (5)	65-72	145-165		1.5-2.5 kg
32	10+1	≥ 17	12	9	5	10	6 (5)	70-83	175-205		f3.1-4 kg / m2.4-2.7 kg
n.d.	10+1	≥ 14	12	9	5	10	6 (5)	72-85	180-210		f3.1-4 kg / m2.4-2.7 kg
19	10+1	20	12	9	6	10	7	52-60	152-167		1.2-2 kg
13	10+1	11-14	12	9	(8)	10	(9)	27-33	63-72		90-210
15 (14)	10+1	11-14	12	9	(8)	10	(9)	31-37	68-78	20-27.5	(135) 155-300 (360)
12 (11)	10+1	≥ 13	12	10	9	(9)		28-34	65-76		
12	10+1	11-14	12	9	8	10	9	26-33	55-69	19.1-23.8	125-300
13	10+1	11-14	12	9	(8)	10	(9)	29-35	70-84	23.5-28.6	130-340
15	10+1	≥ 11	12	?	?	?	?	36-42	87-104		350-450
19 / m14	10+1	≥ 11	12	9	8	10	9 (8)	43-50	95-105		f700-900 / m500-600
21	10+1	≥ 11	12	9	?	10		47-55	105-129		
18	10+1	≥ 11	12					53-63	109-134		f1400-2100 / m800-1300
f17 (15) / m14	10+1	14	12	9	((8))	10	((9))	38-51	m89-100 / f104-113		f900-1300 / m600-750
n.d.	10 ?	> 10 ?	> 12 ?					15-17			
4	10	12-20	12 (16)	0	0	0	0	23-26		9.9-13.2	(74) 80-190
4	10	10-20	12 (16)	0	0	0	0	19-25		10.9-13.1	60-115
4.5	10	10-20	12 (16)	0	0	0	0	17-19		9.6-11.7	(36) 40-60 (72)
?	10	10-20	10-12 (16)	0	0	0	0	16-18		8.5-10	33-50
5	10	(10) 13-14 (20)	12 (16)	0	0	0	0	22-25		12.7-15.8	120-235
6	10+1	(10) 12-14 (20)	12 (16)	0	0	0	0	27-31		15.9-19.8	(190) 240-420 (493)
5.5	10 ?	> 11	10-12	0	0	0	0	45-50			500-1000
5	10+1	(10) 17-20	(12) 14 (16)	0	0	0	0	36-42		19.3-23.5	475-1100 (1250)
6 ?	10+1	(10) 17-20	(12) 14 (16)	0	0	0	0	39-44			500-900
21	10+1	16+7	12					96-119	180-222		4-7 kg
10	10	> 15	20					40-45	83-91		600-950
21	10	> 15	20	9	5	10	6	f75-85 / m90-105	170-240		f3.5-5 kg / m8-16 kg
9.5	10+1	18-20	12	0	0	0	0	39-46	72-83	23.4-28.5	300-810
8	10+1	≈ 15	12	0	0	0	0	33-40	67-83	20.6-23.1 (25.5)	140-289
9.5	10+1	≈ 15	12	0	0	0	0	42-46	67-77	20.6-24	228-435
9	10+1	≈ 15	12 (14)	0	0	0	0	38-45	76-88	22-26	290-535
5.5	10+1	> 10	12	0	0	0	0	23-28	60-70	17.1-20.3	70-90
5.5 ?	10+1	> 10	12	0	0	0	0	23-28	60-70	18-21.6	
5.5	10+1	≈ 15	12	0	0	0	0	14-18	32-35	10.5-12.4	(26) 32-49 (53)
5.5	10+1	≈ 15	12	0	0	0	0	17-20	35-41	12-14.4	(35) 40-74 (84)
5.5	10+1	≈ 15	12	0	0	0	0	15-17.5	42-45	10-11.8 (12.3)	(27) 39-60 (69)
7	10+1	> 10	12	0	0	0	0	20-24	57-64	14.3-16.3	90-145
7	10+1	≈ 15	12	0	0	0	0	25-29	53-59	17-20.3	(140) 160-280 (312)
7.5	10+1	≈ 15	12	0	0	0	0	26-30	56-63	17.8-21.5	(118) 170-280 (345)
9.5	10+1	≈ 15	12	?		9	8	25-28	70-80	19-22	
10	10+1	≈ 15	12	(9	6)	10	8 (7)	28-31	67-72	21.4-24.3	(104) 150-290 (330)
6	10+1	15	12	0	0	0	0	23-26	47-53	15.4-18.1	65-215
5	10+1	15	12	0	0	0	0	18-21	36-39	11.6-13.6	(37) 43-85 (102)

ENGLISH NAME	SCIENTIFIC NAME	Outer P (small)	Pmax	Pmin	Smax	Smin	Inner S (tertials)	Rmax
Little Stint	Calidris minuta	2.5-3	8.5	4.5	5	4	4.5-6	5
Temminck's Stint	Calidris temminckii	n.d.	8.5 ?	4.5 ?	5 ?	4 ?	n.d.	5 ?
Curlew Sandpiper	Calidris ferruginea	n.d.	11	5	6.5	4.5	7-8	6
Purple Sandpiper	Calidris maritima	n.d.	11	5	6.5	4	5-7	6
Dunlin	Calidris alpina	3-3.5	9.5	4.5	5.5	4.5	5-7.5	6
Broad-billed Sandpiper	Limicola falcinellus	n.d.	n.d.	n.d.	n.d.	n.d.	n.d.	n.d.
Ruff	Philomachus pugnax	2.5-3	14.5	7	8.5	6.5	7.5-10.5	8.5
Jack Snipe	Lymnocryptes minimus	3.5-4	9	6	7.5	6	5.5-7.5	6.5
Common Snipe	Gallinago gallinago	4-4.5	11	6	7.5	6	5.5-9	7 (9)
Great Snipe	Gallinago media	n.d.	11.5	6	7.5	6	4.5-9	6.5
Eurasian Woodcock	Scolopax rusticola	2.5-3.5	15.5	10.5	12	10	7.5-11	10.5
Black-tailed Godwit	Limosa limosa	5 ?	18	8.5	10	7.5	7.5-12	10
Bar-tailed Godwit	Limosa lapponica	2-2.5	17.5	8.5	9.5	7	6.5-11	10
Whimbrel	Numenius phaeopus	4-4.5	19.5	8.5	10.5	7	9-13.5	12
Eurasian Curlew	Numenius arquata	4.5-5	24	13	13.5	11	11-17	14
Spotted Redshank	Tringa erythropus	n.d.	13.5	7	8	6.5	< 10.5	8.5
Common Redshank	Tringa totanus	2-3	13.5	7.5	8	6.5	4.5-10.5	8
Marsh Sandpiper	Tringa stagnatilis	n.d.	13	6.5	7.5	5	< 10.5	7
Common Greenshank	Tringa nebularia	n.d.	15.5	6	8.5	5.5	8-11.5	9
Green Sandpiper	Tringa ochropus	2.5-3	13.5	7	8	5.5	7.5-9	7.5
Wood sandpiper	Tringa glareola	n.d.	11.5	5.5	6.5	5	4-6.5	6.5
Common Sandpiper	Actitis hypoleucos	n.d.	9.5	5	6	4.5	5-8	7
Ruddy Turnstone	Arenaria interpres	3-3.5	13.5	6.5	7.5	5.5	< 9	8.5
Red-necked Phalarope	Phalaropus lobatus	n.d.	n.d.	n.d.	n.d.	n.d.	n.d.	n.d.
Grey Phalarope	Phalaropus fulicarius	2.5	10.5	5	6	4.5	6-8	7
Pomarine Skua	Stercorarius pomarinus	7.5-9	27 (30)	13	15	13	11-14	18
Arctic Skua	Stercorarius parasiticus	5-6.5	31	12	14	11	12-14	17
Long-tailed Skua	Stercorarius longicaudus	6-7.5	32	13	n.d.	n.d.	n.d.	33
Great Skua	Stercorarius skua	n.d.	31 (34)	18	20	16	n.d.	18
Mediterranean Gull	Larus melanocephalus	3.5-8	30	12	15	10	11-13.5	14
Little Gull	Hydrocoloeus minutus	2.5-4	(10.5) 18.5 (21)	8.5 (5)	10 (11.5)	7.5	7.5-10.5	11
Black-headed Gull	Chroicocephalus ridibundus	3-7	25	11	13	10	(6.5) 8-13	13
Slender-billed Gull	Chroicocephalus genei	5-9	25 (30)	12	15	12	9-14	14
Audouin's Gull	Larus audouinii	n.d.	31	16	16	13	13-17	16
Common Gull	Larus canus	7.5-9	29	13	14	9	12-14	16
Yellow-legged Gull	Larus michahellis	n.d.	35	18	19	15 (12)	14-18	20
Great Black-backed Gull	Larus marinus	9-11	50	22	24	18	19-22	23
Lesser Black-backed Gull	Larus fuscus	8-10	33	16	18	12	11-17	18
Herring Gull	Larus argentatus	9-11	34	18	20	13	12-18	19
Caspian Gull	Larus cachinnans	n.d.	n.d.	n.d.	n.d.	n.d.	n.d.	n.d.
Glaucous Gull	Larus hyperboreus	9-11	40	24	24	17	15-19	22
Iceland Gull	Larus glaucoides	5-7	32	18	19	17	15-17	18
Black-legged Kittiwake	Rissa tridactyla	7-8	26	12	14	10	9-14	15
Little Tern	Sternula albifrons	3-4	14 (16.5)	5.5	6.5	4.5	4.5-5	9
Gull-billed Tern	Gelochelidon nilotica	5-6.5	28	10.5	11	9.5	9.5-10	14.5
Caspian Tern	Hydroprogne caspia	n.d.	32	14	15	11	?	17.5
Whiskered Tern	Chlidonias hybrida	4-5	21	8.5	10	8	7-8.5	10
Black Tern	Chlidonias niger	n.d.	> 13	10	n.d.	n.d.	n.d.	11
White-winged Tern	Chlidonias leucopterus	n.d.	n.d.	n.d.	n.d.	n.d.	n.d.	8.5
Sandwich Tern	Sterna sandvicensis	5.5-6.5	23 (30)	9	9	7.5	6.5-10	14 (18)
Common Tern	Sterna hirundo	n.d.	22	8	8.5	7	7-8.5	17 (18.5)
Arctic Tern	Sterna paradisaea	4.5-5.5	22	8	8.5	7	7-8.5	20
Roseate Tern	Sterna dougallii	3.5	20	7	7	5	6-7	n.d.
Common Guillemot	Uria aalge	3-4	13.5 (15)	8	9	5	4-5	7
Brünnich's Guillemot	Uria lomvia	n.d.	n.d.	n.d.	n.d.	n.d.	n.d.	n.d.
Razorbill	Alca torda	4-4.5	15	7	9	5	4-5	9.5
Black Guillemot	Cepphus grylle	3.5-4.5	14.5 (16)	9	10	6.5	4.5-6.5	8
Little Auk	Alle alle	2.5-3	10	5.5	7	4.5	3.5-5	5
Atlantic Puffin	Fratercula arctica	3.5-4.5	13	6.5	8	5	4-6	7
Black-bellied Sandgrouse	Pterocles orientalis	n.d.	16.5	10	n.d.	n.d.	n.d.	n.d.
Pin-tailed Sandgrouse	Pterocles alchata	3-4	17	8.5	8.5	7.5	7.5-10	18
Rock Dove	Columba livia		20	12	13	8	7.5-8.5	14
Feral Pigeon	Columba livia dom		20.5	11	13	8	7.5-8.5	14
Stock Dove	Columba oenas		19.5	11	12	9	7.5-10	14
Wood Pigeon	Columba palumbus		21.5	13.5	14.5	9.5	7.5-11	19
Eurasian Collared Dove	Streptopelia decaocto		15	9	10	9	7-9.5	16
European Turtle Dove	Streptopelia turtur		15	7.5	10	8	7-8.5	14
Laughing Dove	Streptopelia senegalensis		15 ?	7.5 ?	10 ?	8 ?	7-8.5 ?	14 ?
Rose-ringed Parakeet	Psittacula krameri		18	10	10	5.5		23

Rmin	NP (number of Ps)	NS (number of Ss)	NR (number of Rs)	Em (ov) Po	Em (ov) Pi	Not (iv) Po	Not (iv) Pi	L (length)	W (wingspan)	FW (folded wing)	M (mass in grams or kg)
4	10+1	15	12	0	0	0	0	12-15.5	27-30	9.1-11.1	16-40 (44)
4 ?	10+1	15	12	0	0	0	0	13-15	28-31	9.4-10.5	15-36
5	10+1	15	12	0	0	0	0	18-23	38-41	12.1-14.1	(32) 43-90 (103)
5	10+1	15	12	0	0	0	0	19-22	37-44	12-14.2	48-105 (112)
4	10+1	15	12	0	0	0	0	16-22	32-40	10.4-12.4 (13.1)	(27) 33-60 (82)
n.d.	10+1	15	12	0	0	0	0	15-18	34-37	10-11.5	30-45
5.5	10+1	15	12	0	0	0	0	f20-26 / m26-32	46-60	13.2-21 (f20-25 / m26-32)	f(62) 70-165 (175 / m(121) 135-280 (299)
5	10+1	15	12	0	0	0	0	17-20	33-36	10.3-12.2	33-86 (106)
5	10+1	10+5-6	(12) 14-16 (18)	0	0	0	0	23-28	39-45	12.6-14.7	(72) 80-140 (169)
5.5	10+1	≈ 15	14-18	0	0	0	0	26-30	43-50	13-15.8	(92) 140-225 (265)
7.5	10+1	≈ 15	12	0	0	0	0	33-38	55-65	18.2-21.8	131-420
7.5	10+1	≥ 15	12	0	0	0	0	36-44	63-74	16.8-24	(198) 245-390 (535)
7	10+1	18	12	0	0	0	0	35-41	62-72	(19) 20-23.6	(151) 230-455
10	10+1	19	12	0	0	0	0	37-46	78-88	(21.4) 22.3-27.8	270-550 (595)
11.5	10+1	20-21	12	0	0	0	0	48-60	89-106	(26.8) 27.3-33.3	540-1300
7	10+1	≈ 15	12	0	0	0	0	29-33	48-52	15.8-18	(121) 130-210 (230)
6	10+1	≈ 15	12	0	0	0	0	24-29	47-53	14.6-18.4	85-155 (184)
6	10+1	≈ 15	12	0	0	0	0	22-25	39-44	12.8-14.8	
8	10+1	17	12	0	0	0	0	30-34	55-62	17.7-20.2	(102) 130-270
6	10+1	~ 15	12	0	0	0	0	20-24	39-44	13.5-15.5	(53) 60-90 (112)
5	10+1	15	12	0	0	0	0	18-21	35-40	11.5-14	(41) 44-90 (101)
5	10+1	≈ 15	12	0	0	0	0	18-21	32-35	9.9-12.4	(25) 34-60 (88)
6	10+1	≈ 15	12	0	0	0	0	21-25.5	43-49	14.4-16.7	(72) 80-150 (201)
n.d.	10+1	≈ 15	12	0	0	0	0	17-19	30-34	10.2-11.8	25-50
5.5	10+1	≈ 15	12	0	0	0	0	20-22	36-41	12.2-14.3	37-77
15	10+1	> 18	12					42-50	110-125		
12	10+1	> 18	12					37-44	100-118		380-600
12	10+1	> 18	12					35-41	88-112		250-450
15 (14)	10+1	> 18	12					50-58	125-140		1200-1650
11	10+1	≈ 20	12					37-40	94-102	29.4-32.2	232-382
8	10+1	≈ 21	12					24-28	62-69		90-150
11	10+1	20-21	12					35-39	86-99	27.1-33.8	190-365 (400)
12	10+1	20-21	12					37-42	90-102		250-350
14	10+1	> 19	12					44-25	117-128		550-800
13	10+1	20	12					40-46	99-115	31.9-41	284-505 (582)
17	10+1	21	12					52-58	120-140	39.5-48.5	750-1500
18	10+1	23-25	12					61-74	144-166	44-52.3	1033-2272
15	10+1	22	12					48-56	117-134	37.7-46.2	545-1045
16	10+1	21	12					54-60	123-148	38-48	690-1495
n.d.	10+1	> 19	12					55-60	138-147		
18	10+1	> 19	12					63-68	138-158		1200-2000
17	10+1	> 19	12					52-60	123-139		
13	10+1	≈ 20	12					37-42	93-105	25.8-33.6	(210) 305-525
4	10+1	16	12	0	0	0	0	21-25	41-47	16.3-18.9	(40) 44-64 (72)
9	10+1	≥ 15	12	0	0	0	0	35-42	76-86		
12	10+1	≥ 15	12	0	0	0	0	48-55	96-111		
7	10+1	≈ 15	12	0	0	0	0	24-28	57-63	22.7-26	78-102
7	10+1	≥ 15	12	0	0	0	0	22-26	56-62		
7	10+1	≥ 15	12	0	0	0	0	20-24	50-56		
7.5	10+1	≈ 20	12	0	0	0	0	37-43	85-97	29.3-32.6	(198) 215-275
7.5	10+1	19-20	12	0	0	0	0	34-37	70-80	24.3-29.5	80-165 (175)
7.5	10+1	≥ 15	12	0	0	0	0	33-39	66-77		
n.d.	10+1	≥ 17	12	0	0	0	0	33-36	67-76		
5	10+1	≥ 21	12	0	0	0	0	38-46	61-73	18.7-22.5	490-1285 (1450)
n.d.	10+1	≥ 21	12	0	0	0	0	44-44	64-75		
6.5	10+1	21	12	0	0	0	0	38-43	60-69	18.1-21.7	372-890
5	10+1	≥ 20	12	0	0	0	0	32-38	49-59		340-390
4	10+1	≥ 16-17	12	0	0	0	0	19-21	34-38		
5	10+1	≥ 15	12 (16 ?)	0	0	0	0	28-34	50-60		320-550
n.d.	10+1	≈ 18	16-18	0	0	0	0	30-35			400-550
6.5	10+1	≈ 18	16-18	0	0	0	0	28-32	55-63		200-280
11	10	(10) 11-12 (15)	12	9	(8)	10	(9)	30-35	62-68		240-300
12	10	(10) 11-12 (15)	12	9	(8)	10	(9)	29-35	60-68		
10	10	(10) 11-12 (15)	12	9	(8)	10	(9)	28-32	60-66	20.6-23.6	(242) 250-350 (400)
15.5	10	(10) 11-12 (15)	12	9	(8)	10	(9)	38-43	68-77	21.5-27	193-620
12	10	(10) 11-12 (15)	12	9	(8)	10	(9)	31-34	48-56	16.3-19.1	134-249 (264)
10	10	(10) 11-12 (15)	12	9	(8)	10	(9)	25-27	49-55	16.7-18.7	(85) 110-225 (240)
10 ?	10	(10) 11-12 (15)	12	?				23-26	40-45		
8	10	9-11	12					(27) 39-43	42-48		

ENGLISH NAME	SCIENTIFIC NAME	Outer P (small)	Pmax	Pmin	Smax	Smin	Inner S (tertials)	Rmax
Budgerigar	Melopsittacus undulatus		9	4	5	3.5	2.5-3.5	11
Great Spotted Cuckoo	Clamator glandarius	8-10	20 (22)	10	11 (13)	7	5-8	23
Common Cuckoo	Cuculus canorus	10-12	19 (22)	10	12	8	4-9	19
Barn Owl	Tyto alba	22	25	14	16	12	8-13	14
Eurasian Scops Owl	Otus scops	10-12	14	9	10	7	6-9	8
Snowy Owl	Nyctea scandiaca	29	35	23	24	20	n.d.	34
Eurasian Eagle Owl	Bubo bubo	25-27	43	29	30	22	19-26	33
Northern Hawk Owl	Surnia ulula	n.d.	21	18	n.d.	n.d.	n.d.	19
Eurasian Pygmy Owl	Glaucidium passerinum	5	9	6	7	5	5-6	7
Little Owl	Athene noctua	10-11	14	10	11	7	6-9	10
Tawny Owl	Strix aluco	14	25	17	21	13	9-14	18 (20)
Ural Owl	Strix uralensis	n.d.	32	23	n.d.	n.d.	n.d.	30
Great Grey Owl	Strix nebulosa	n.d.	35 (41 ?)	25	n.d.	19	n.d.	35
Long-eared Owl	Asio otus	21-22	25	14	16	12	10-13	16
Short-eared Owl	Asio flammeus	21-22	26	15	16	12	7-13	18
Tengmalm's Owl	Aegolius funereus	8.5	15	9	11	8	6-8	12
European Nightjar	Caprimulgus europaeus		18	9	11	8	6-8	16
Red-necked Nightjar	Caprimulgus ruficollis		18 (21)	8	12	8	6-8	18 (20)
Common Swift	Apus apus		15	4.5	5	4	3-3.5	9.5
Pallid Swift	Apus pallidus		15	4.5	5.5	4	3-3.5	9
Alpine Swift	Apus melba		22	5	6.5	4	3.5-4	10
Common Kingfisher	Alcedo atthis		7.5	5	6.5	4	2-4	4.5 (6)
European Bee-eater	Merops apiaster	2.5-4	13	7.5	8	6	3.5-6	14
European Roller	Coracias garrulus		19	11	12	9	4.5-10	17
Eurasian Hoopoe	Upupa epops	5.5-9	14 (17)	9.5	13	9	6-10.5	12.5
Eurasian Wryneck	Jynx torquilla	1.5-2	8.5	5.5	7.5	5	3-6	8
Grey-headed Woodpecker	Picus canus	n.d.	13	9	14	8	n.d.	12
Green Woodpecker	Picus viridis	4-4.5	15	10	12	8	5-8.5	12
Black Woodpecker	Dryocopus martius	7.5-8	21 (23)	13.5	17 (21)	12	8-12	21
Great Spotted Woodpecker	Dendrocopos major	3-4.5	13 (15)	8	10	7.5	4-7.5	10 (11.5)
Syrian Woodpecker	Dendrocopos syriacus	n.d.	11.5	9	10	8	n.d.	n.d.
Middle Spotted Woodpecker	Dendrocopos medius	3-3.5	11.5	8.5	9	6.5	4-6	9.5
White-backed Woodpecker	Dendrocopos leucotos	3.5-4	14	10	10.5	8.5	(2.5) 4-8	11.5
Lesser Spotted Woodpecker	Dendrocopos minor	2-3	9.5	6	7	5	2-5	7
Three-toed Woodpecker	Picoides tridactylus	n.d.	12	7	8.5	7	4-7.5	9
Dupont's Lark	Chersophilus duponti	1.4-2	8.6	6.4	7	5.4	3-6.7	7.3
Calandra Lark	Melanocorypha calandra	1.1-2.2	11.2	7.8 (7j)	7.8	6.3	3.6-7.6	7.3
Greater Short-toed Lark	Calandrella brachydactyla	1.2-1.6	7.6	5	5.7	5.3	4.4-7.1	6.4
Lesser Short-toed Lark	Calandrella rufescens	0.9-1.2	8.3	5.2	6.1	4.7	2.4-6.2	7.2
Crested Lark	Galerida cristata	1.7-2.2	9.5	6.4	7.8	5.8	3.4-8	7.5
Thekla's Lark	Galerida theklae	2.1-2.5	9	6.3	7.4 (7.7)	6.7	3.5-6.9	6.8
Woodlark	Lullula arborea	1.5-2.4	7.2	6	6.5	5.8	3-5.8	5.9
Eurasian Skylark	Alauda arvensis	1.6-2.3	9.9	6.5 (5.4j)	6.9	5	3.1-7.3	7.9
Horned Lark	Eremophila alpestris	1-1.4	9.3	6.3 (5.6j)	6.5	5.7 (4.8j)	3.4-6.2	7.9
Common Sand Martin	Riparia riparia	0.5-1.5	11	5	5.5	4	3-4.5	6
Eurasian Crag Martin	Ptyonoprogne rupestris	0.5-1	11.5	5	5.5	5	3-5	7
Barn Swallow	Hirundo rustica	0.5-1	12	4.5	6	4	2.5-4.5	12 (13)
Common House Martin	Delichon urbicum	0.5-1.5	11	4.5	5.5	4	2-4	7
Red-rumped Swallow	Cecropis daurica	0.5-1	12	5	6	3	2-3.5	11
Tawny Pipit	Anthus campestris	n.d.	8	6.5	6.7	6	5.1-6.7	7.8
Tree Pipit	Anthus trivialis	2.1	7.8	5.4	6.6	5.4	5.4-7.0	7.4
Meadow Pipit	Anthus pratensis	1	7.9	5.6	6.5	5.1	3.4-6.4	7
Red-throated Pipit	Anthus cervinus	0.6-0.9	7.7	5.3	5.7	5.2	3.4-6.8	7.3
Water Pipit	Anthus spinoletta	1.7	8.9	6.8	7.1	5.8	?-7.6	8.0
Rock Pipit	Anthus petrosus	0.7	8.1	6.1	6.7	5.6	3.3-6.8	7.7
Yellow Wagtail	Motacilla flava	1.1	7.5	5.2	6.0	4.9	3.6-6.5	8.3
Grey Wagtail	Motacilla cinerea	1.6	7.7 (8.2)	5.0	5.2	4.7	3.0-6.4	11.7
Pied / White Wagtail	Motacilla alba	1.1-1.3	8.1	5.4	5.8	5.2	3.8-7.1	10
Bohemian Waxwing	Bombycilla garrulus	2.4	10.8	6.7	6.7	5.5	3.4-5.6	7.4
White-throated Dipper	Cinclus cinclus	2.5-3.7	7.6 (9.5)	6.0	6.2	4.2	3.4-5.4	5.9
Eurasian Wren	Troglodytes troglodytes	1.8-1.9	4.2	3.2	4	2.9	1.2-3.1	3.7
Dunnock	Prunella modularis	1.1-1.6	6.3	4.7	5.6	4.3	2.2-4.6	6.8
Alpine Accentor	Prunella collaris	1.5-2.5	10	7	7.2	5.4	3.3-5.9	8
Rufous Bush Robin	Cercotrichas galactotes	2-3	7.3 (8.2)	5.9	6.2 (7)	5.1	3.1-5.5 (6.2)	8
European Robin	Erithacus rubecula	2.4-2.6	6.6	4.9	5.7	4.2	2.5-4.5	6.5
Thrush Nightingale	Luscinia luscinia	1.2-1.6	7.5	5.9	6	4.6	2.8-4.9	7.5
Common Nightingale	Luscinia megarhynchos	1.9-2.7	7.5 (8.5)	5.5	5.9	4.3	2.7-4.9	7.4
Bluethroat	Luscinia svecica	1.8-2.4	6.6 (7.3)	5	5.4	4.2	3-5.7	6.3
Red-flanked Bluetail	Tarsiger cyanurus	2.3-2.4	7.1	5.4	5.5	4	1.9-4.2	6.4
Black Redstart	Phoenicurus ochruros	2.3-2.7	7.8	6	6.3	4.4	3.1-5.1	6.8
Common Redstart	Phoenicurus phoenicurus	1.8-2.2	7.3 (8.2)	5.4	5.6	4.2	2.6-4.6	6.5
Whinchat	Saxicola rubetra	2.1-2.3	6.5 (7.3)	5.2	5.3	3.7	3.1-4.5	5.3

Rmin	NP (number of Ps)	NS (number of Ss)	NR (number of Rs)	Em (ov) Po	Em (ov) Pi	Not (iv) Po	Not (iv) Pi	L (length)	W (wingspar)	FW (folded wing)	M (mass in grams or kg)
4	10	9	12								
10	10	11	10	0	0	10	7	35-39			140-170
10	10	9	10	0	0	0	0	32-36	54-60	19-24.1	(67) 70-143 (160)
11	10	12+4	12	0	0	0	0	33-39	80-95	27-31.5	(187) 240-370 (460)
6	10	13	12	9	8 (7)	10	8	19-21	47-54	14.2-16.8	(54) 75-95 (135)
23	10	≥ 12	12	9	6	10	6	53-65	125-150		1500-2400
22	10	≥ 12	12	9	6	10	6	59-73	138-170		1500-3000
17	10	≥ 12	12	?				35-43	69-82		250-380
5	10	≥ 12	12	9	6	10	7	15-19	32-39		50-80
7	10	14	12	9	6	10	6	23-28	50-57	14.5-18.1	(99) 112-226 (265)
15	10	13-14	12	9	6	10	7 (6)	37-43	81-96	23-30.3	215 (295)-715 (780)
n.d.	10	14	12	9	6	10	7	50-59	103-124		
28	10	≥ 12	12	9	6 ?	10	7 ?	59-68	128-148		
13	10	12-14	12	9		10	9	31-37	86-98	27.4-32	(151) 210-370 (435)
14	10	14	12	9	(8)	10		33-40	95-105	29.8-33.4	260-425
9	10	≥ 12	12	9	8	10	9	22-27	50-62	16.2-18.8	(87) 90-236
12	10	13	10	9	8	10	9	24-28	52-59	17.3-20.8	53-108
15	10	13	10	9	8	10	9	30-34	60-65	19.4-21.4	60-75
4.5	10	6+3 (4)	10					17-19	40-44	15.7-18.7	(26) 27-56
5	10	8+2	10					16-18	39-44	15.6-18.4	32-50
5	10	10	10					20-23	51-58		80-120
3.5	10	(11) 12 (14)	12 (14 ?)	0	0	(10)		17-20		7.4-8.2	(31) 34-46 (55)
9	9+1	14	12	0	0	7	1	25-29	36-40	13.6-16	(36) 44-78
12	10	> 9	12	9	7	10	8	29-32	52-58	18.9-21.1	110-160 (189)
9	9+1	10+1	10	8	5	9	6	25-29	44-48	13.1-15.6	(39) 47-87 (98)
6	9+1	10-12	10+2	8	7	9	7	16-18	25-27	8.2-9.7	(22) 30-45 (54)
7	9+1	10-12	10+2					25-32	38-40	13.5-15.5	(98) 125-165
8	9+1	10-12	10+2	8	5	10	5	30-36	40-51	15.4-17.2	(138) 180-220 (250)
11.5	9+1	10-12	10+2	7	4	10	4	40-57	64-73	22.5-26	(200) 300-370 (460)
6.5	9+1	10-12	10+2	8	5	(9)	5	22-26	34-44	12.4-15	67-102
n.d.	9+1	10-12	10+2					22-25	34-40		55-90
6	9+1	10-12	10+2					19-22	33-34	12-13.8	46-85
8.5	9+1	10-12	10+2					24-28	38-40		100-115
4	9+1	10-12	10+2					14-17	24-29	8.4-10	16-26
6	9+1	10-12	10+2					21-24	32-35		54-75
5	9+1	9 (10)	12					16-18			
5.4	9+1	9	12	8	6	9	6	17-20		11.1-14.1	40-75
5.6	9+1	9	12					14-16		8.2-10.2	(16) 18-35
5.1 (4.1)	9+1	9	12					13-15			
5.9	9+1	9	12					17-19		9.4-11.4	30-55
5.8	9+1	9	12					15-17			
5.1	9+1	9 (10)	12					13-15		8.7-10.4	20-34 (39)
6	9+1	9	12	8	6			16-18		9.3-12.6	25-51 (56)
6.5 (5.2j)	9+1	9	12					16-19			
4	9+1	9	12	0	0	0	0	12-13		9.5-12	(8) 11-17 (21)
5	9+1	9	12	0	0	0	0	14-15		12.1-14.4	15-35
4.5	9+1	9	12	0	0	0	0	17-21		11-14	(10) 15-25 (28)
4	9+1	9	12	0	0	0	0	13-15		9.5-12.3	9-23 (26)
4.5	9+1	9	12	0	0	0	0	14-19		11.1-13	15-30
6.7	9+1	9	12					15-18		8.2-10.2	(16) 21-32 (40)
5.9	9+1	9	12					14-16		7.9-9.9	(15) 17-32 (40)
5.6	9+1	9	12	8	6			14-16		7.3-8.8	13-23 (26)
5.8	9+1	9	12					14-15		7.7-9.2	15-24 (28)
6.4	9+1	9	12					15-17		8.1-9.8	(17) 20-30 (34)
6	9+1	9	12					15-17		8-9.9	18-33
7.1	9+1	9	12					15-16		7.3-9	(10) 12-23 (28)
9.5	9+1	9	12	8	7 (6)			17-20		7.8-9.2	13-23 (25)
8.4	9+1	9	12	8	6			16-19		8-9.7	16-27 (30)
6.2	9+1	9	12	8	7			18-21		11.1-12.5	(34) 45-83 (85)
5	9+1	9	12	8	6			17-20		8-10.5	45-85
2.7	9+1	9	12	8	5			9-11		4.6-5.6 (6.1)	6-15 (20 Islande)
5.9	9+1	9	12	8	5			13-15		6.2-7.9	(14) 16-24 (26)
6.7	9+1	9	12	8	5			15-18		9.3-11.3	32-53
6.3	9+1	9	12	8	5	9	6	15-17			
5.8	9+1	9	12	8	5			12-14		6.5-7.8	12-21 (26)
6.2	9+1	9	12					15-17			
6.2 (5.5)	9+1	9	12	8	7			15-17		7.9-9.4	(11) 17-25 (36)
4.7	9+1	9	12	8	6			13-14		6.2-8.2	(11) 13-26
5.6	9+1	9	12					13-14			
5.8	9+1	9	12					13-15		7.9-9.1	12-20 (25)
4.8	9+1	9	12	8	6	9	7	13-15		7.3-8.9	(9) 11-21 (25)
4.6	9+1	9	12	8	6			12-14		6.8-8.3	(11) 13-19 (27)

ENGLISH NAME	SCIENTIFIC NAME	Outer P (small)	Pmax	Pmin	Smax	Smin	Inner S (tertials)	Rmax
Common Stonechat	Saxicola rubicola	2-2.4 (2.8)	5.7 (6)	4.6	5	3.6	2.5-4.3	5.1
Isabelline Wheatear	Oenanthe isabellina	1.7-2.2	8.4	6	6.6	5	3.1-5.6	6.9
Northern Wheatear	Oenanthe oenanthe	1.7-2.2	8.8	6.5 (6)	6.4	5	3.1-5.6	7
Pied Wheatear	Oenanthe pleschanka	1.7-2.4	8.1	5.9	6.4	4.7	2.6-4.9	7
Black-eared Wheatear	Oenanthe hispanica	2-2.4	8.2	6.1	6.4	4.7	3.2-4.9	7.2
Black Wheatear	Oenanthe leucura	2.3-2.6 (3.7)	8.3	6.7	7.1	5.8	3.5-5.9	8 (8.2)
Common Rock Thrush	Monticola saxatilis	1.4-1.6	10.7	7.3	7.5	5.8	3.9-6.5	7.2
Blue Rock Thrush	Monticola solitarius	2-2.6	11	8	8.2	6.5	3.5-7	9.3
Ring Ouzel	Turdus torquatus	2.4-3.2	12.8	9.1	9.5	7.2	4.8-7.8	12.5
Common Blackbird	Turdus merula	2.8-3.5	11.4 (13)	8.7	9.6	7.5	4.4-7.8	11.8 (13)
Fieldfare	Turdus pilaris	2.4-3	12.4 (14)	9.2 (8.5)	9.3 (10.5)	7.5	4.8-8	12.3 (14)
Song Thrush	Turdus philomelos	2-2.4	10.2 (11.5)	7.5	7.9 (8.5)	6.5	4-6.7	9.8 (10.3)
Redwing	Turdus iliacus	1.5-2	10.5 (12)	7.4	7.7	6.3	3.7-6.5	9.1 (11)
Mistle Thrush	Turdus viscivorus	2.5	13.4 (14)	9.6 (8.5)	9.5 (10)	7.6	4.8-7.8	12.8 (13.3)
Cetti's Warbler	Cettia cetti	2.2-2.6	5.6	4.4	5.4	3.5	1.9-4.2	7
Zitting Cisticola	Cisticola juncidis	1.7-2	4.8	3.6	4.5	3.2	2.4-3.9	5
Common Grasshopper Warbler	Locustella naevia	1.3-1.6	5.6	4.9	4.6	3.8	2.3-4.1	6.2
River Warbler	Locustella fluviatilis	1.2-1.4	6	4.9	4.9	3.8	2.4-4.3	6.2
Savi's Warbler	Locustella luscinioides	1.1-1.5	5.8	4.7	4.7	3.6	2.5-4.1	6.8
Moustached Warbler	Acrocephalus melanopogon	1.8-2	5.5	3.9	4.8	3.2	1.7-3.9	5.4
Aquatic Warbler	Acrocephalus paludicola	1.1-1.4	5.6	4.1	4.5	3.4	2.3-4.1	5.4
Sedge Warbler	Acrocephalus schoenobaenus	1.1-1.3	5.7 (6.8)	4.3	4.6 (5.5)	3.5	2.2-4	5.6
Paddyfield Warbler	Acrocephalus agricola	1.3-1.6	5.4	4.2	4.6	3.4	2.1-3.7	5.7
Blyth's Reed Warbler	Acrocephalus dumetorum	1.1-1.3	5.8	4.5	4.8	3.5	2.3-4	5.8
Marsh Warbler	Acrocephalus palustris	1.1-1.3	6	4.5	4.8	3.6	2-4	5.8
Eurasian Reed Warbler	Acrocephalus scirpaceus	1.2-1.5 (2)	5.8 (6.3)	4.5	5	3.5	2-4.1	5.8 (6.3)
Great Reed Warbler	Acrocephalus arundinaceus	1.8-2.2	8.3 (9.2)	6.1	7	4.8	3-5	8.6
Eastern Olivaceous Warbler	Hippolais pallida	1.5-1.8	5.7	4.6	4.7	3.5	2-3.7	5.6
Western Olivaceous Warbler	Hippolais opaca	1.8-2.1	6.2	5.2	5.3	4.2	2.3-4.3	6.3
Booted Warbler	Hippolais caligata	1.4-1.7	5.3	4.5	4.7	3.4	2.1-3.7	5
Olive-tree Warbler	Hippolais olivetorum	1.1-1.3	7.6	5.3	5.6	4.1	2.4-5.1	7.6
Icterine Warbler	Hippolais icterina	1.4-1.6	7.2 (7.7)	5.1	5.7	4.2	2.3-4.5	6.5 (7.6)
Melodious Warbler	Hipolais polyglotta	1.6-2	5.9	4.4	5.1	3.5	1.8-4.1	6
Blackcap	Sylvia atricapilla	1.7-2.1	6.5 (7.5)	5.3	5.3	4.5	2.9-4.6	6.8
Garden Warbler	Sylvia borin	1.3-1.5	6.6	4.9	5.1	3.7	2.5-4.6	6.2
Barred Warbler	Sylvia nisoria	1.2-1.9	7.5 (8.6)	5.8	6.2	4.2	2.5-5.6	8.2
Lesser Whitethroat	Sylvia curruca	1.4-2	6.3 (6.6)	5.3	5.4	4.4	2.8-4.4	6.6
Western Orphean Warbler	Sylvia hortensis	2.1-2.4	6.8	5.7	5.7	4.7	2.4-4.9	7.5
Common Whitethroat	Sylvia communis	0.8-1.3	6.5	5.2	5.7	4.3	2.7-4.7	7.1
Spectacled Warbler	Sylvia conspicillata	1.3-1.7	5	4.2	4.5	3.5	2.1-4	5.8
Dartford Warbler	Sylvia undata	1.5-1.8	4.8	3.8	4.5	3.2	2-3.5	6.9
Marmora's Warbler	Sylvia sarda	1.2-1.9	5.3	4.5	4.7	3.6	2.1-4	6.9
Rüppell's Warbler	Sylvia rueppelli	1.2-1.7	6.2	5	5.3	3.9	2.4-2.5	6.8
Subalpine Warbler	Sylvia cantillans	1.1-1.6	5.6	4.2	4.6	3.5	2.1-4.1	6.4
Sardinian Warbler	Sylvia melanocephala	1.5-2.2	5.3	4.3	4.8	3.8	2.4-4	6.6
Greenish Warbler	Phylloscopus trochiloides	1.6-1.7	5.2	4.1	4.3	3.2	1.8-3.4	4.8
Arctic Warbler	Phylloscopus borealis	1-1.6	5.7	4.6	4.6	3.7	1.9-3.7	4.8 (5.1)
Yellow-browed Warbler	Phylloscopus inornatus	1.4-1.6	4.9	3.6	4.2	3	1.7-3.2	4.3
Western Bonelli's Warbler	Phylloscopus bonelli	1.4-1.7	5.9	4.2	4.9	3.3	2-3.9	5.6
Eastern Bonelli's Warbler	Phylloscopus orientalis	1.5-1.8	5.3	4.3	4.5	3.3	1.9-3.5	5
Wood Warbler	Phylloscopus sibilatrix	1.1-1.4	6.5	4.6	5	3.7	2-3.8	5.6
Common Chiffchaff	Phylloscopus collybita	1.5-1.9	5.4	4.1	4.5	3.5	1.9-3.6	5.3
Iberian Chiffchaff	Phylloscopus ibericus	n.d.	n.d.	n.d.	n.d.	n.d.	n.d.	n.d.
Willow Warbler	Phylloscopus trochilus	1.5-1.9	5.7	4.5	4.7	3.5	2.1-3.8	5.6
Goldcrest	Regulus regulus	1-1.6	5	4	4.2	3	2.3-3.3	4.4
Firecrest	Regulus ignicapilla	1.5-2.2	5 (5.6)	3.9	4.3	2.8	1.5-3	4.7
Spotted Flycatcher	Muscicapa striata	2-2.2	7.8	5.6	5.5	4.5	2.8-4.6	6.8
Red-breasted Flycatcher	Ficedula parva	1.7-2.1	6.3	4.9	5	3.7	2.2-4.1	6.2
Semicollared Flycatcher	Ficedula semitorquata	1.9-2.3	7.4	5.3	5.4	4.3	2.4-4.6	6
Collared Flycatcher	Ficedula albicollis	1.7-2.2	7.4	5.3	5.4	4.1	2.4-4.6	5.9
Pied Flycatcher	Ficedula hypoleuca	1.9-2.6	7.2	5.3	5.2	3.7	2.2-4.3	6.1
Bearded Reedling	Panurus biarmicus	0.9-2.1	5.8	4	4.8	3.9 (3.2j)	2.14.4	9.6
Long-tailed Tit	Aegithalos caudatus	1.9 (2.4)	5.8 (6.7)	4	4.8	3.6	1.8-3.7	9.6
Marsh Tit	Poecile palustris	2-2.4	6	4.8	4.9	3.5	2.1-4.1	6.1
Sombre tit	Parus lugubris	2.8-2.9	6.7	5.7	6	4.6	2.5-4.8	7.1
Willow Tit	Poecile montanus	1.9-2.2	6.1	4.6 (4)	5.3	3.5	1.9-4.1	6.4
Siberian Tit	Parus cinctus	2.4-2.6	6.3	5.3	5.5	4	1.9-4.1	7.3
Crested Tit	Lophophanes cristatus	2-2.2	6.4	4.9 (4.2)	5.1	3.5	2-3.8	6
Coal Tit	Periparus ater	1.8-2.2	6.1	4.7	5.1	3.5	1.9-4	5.2
European Blue Tit	Cyanistes caeruleus	1.9	6.1	4.9	5.1	3.9	2.5-4	5.8
Azure Tit	Cyanistes cyanus	2-2.3	6.3	5.2	5.8	4	2.3-4.4	6.9
Great Tit	Parus major	2.1-2.2	7	4.9	6.1	4.1	2.5-4.5	7.2

Rmin	NP (number of Ps)	NS (number of Ss)	NR (number of Rs)	Em (ov) Po	Em (ov) Pi	Not (iv) Po	Not (iv) Pi	L (length)	W (wingspan)	FW (folded wing)	M (mass in grams or kg)
4.6	9+1	9	12	8	5			11-13		6.1-7.2	(11) 13-19 (21)
5.5	9+1	9	12					15-17			
5.5	9+1	9	12					14-17		8.5-11	(17) 19-47
6	9+1	9	12					14-17			
5.7	9+1	9	12					13-16		8.1-9.7	12-22 (26)
7.4	9+1	9	12	8	5			16-18			
6.2	9+1	9	12					17-20		11.2-13.1	(34) 40-65 (72)
8.2 (7.5)	9+1	9	12	8	6	9	7	21-23			
8	9+1	9	12	8	6			24-27		13-15.2	82-138
10.2	9+1	9	12	8	5			23-29		11.2-14	69-125 (149)
10.5 (10)	9+1	9	12	8	6			22-27		13-15.9	(68) 83-141 (160)
8	9+1	9	12	8	6			20-22		11.1-12.8	(41) 58-107 (112)
7.9 (7)	9+1	9	12	8	6			19-23		11.1-13.3	(43) 50-90
8.8 (8)	9+1	9	12	8	6 (5)			26-29		14.2-16.6	94-170
4.9	9+1	9	10					13-14		5-6.8	8-19
2.9	9+1	9	12	8	4	9	6	10-11		4.4-5.6	6.5-13
4.1	9+1	9	12	8 ((7))	9			12-14		5.7-6.9	9-17 (20)
5 (4.2)	9+1	9	12					14-16			
1.9	9+1	9	12					13-15		6.2-7.7	11-18 (33)
4.3	9+1	9	12					12-14		5.2-6.4	(7) 9-14 (17)
3.8	9+1	9	12					11-13		5.6-6.8	9-15 (20)
4.4	9+1	9	12	8		9		11-13		5.9-7.4	(7) 9-17 (22)
4.7	9+1	9	12	8	6			12-14		5.1-6.3	7-14
4.8	9+1	9	12	8	7			12-14		5.8-6.7	8-15 (25)
4.9	9+1	9	12	8		9		13-15		6.3-7.6	(7) 10-15 (20)
4.7	9+1	9	12	8	(7)	9		12-14		5.9-7.3	7-16 (22)
6.9	9+1	9	12	8	((7))			16-20		8.2-10.5	(14) 24-40 (52)
4.8	9+1	9	12					12-14			
5.2	9+1	9	12					12-14			
4.5	9+1	9	12					11-13			
6.2	9+1	9	12					16-18			
5.1	9+1	9	12	8	7 (6)			12-14		7.1-8.7	10-18 (23)
5	9+1	9	12	8	6 (5)	9	7	12-13		6.5-7.2	(8) 10-14 (17)
6.5	9+1	9	12	8	6			13-15		6.5-8.1	(8) 12-25 (32)
5.7	9+1	9	12					13-15		6.7-9	(10) 15-24 (37)
7.1	9+1	9	12	9	7	9	(8)	15-17		8-9.6	(17) 21-30 (40)
5.5	9+1	9	12	8	6			11-14		5.9-7.2	7-15 (20)
6.4	9+1	9	12					15-16		7.7-8.6	14-31
5.8	9+1	9	12					13-15		6.4-8.1	(8) 12-19 (25)
5	9+1	9	12					12-13		5-6.1	7-13
5.2	9+1	9	12	8	6	9		13-14		4.8-5.8	7-11
5.4	9+1	9	12					13-14		5-6.2	8.5-13
5.8	9+1	9	12					12-14			
5.1	9+1	9	12					12-13		5.1-6.8	(7) 9-14 (19)
5.4	9+1	9	12	8	6	9		13-14		5.1-6.7	(7) 9-15 (20)
4.6	9+1	9	12	8	6 (5)			9-11		5.5-6.8	5-10
4.4	9+1	9	12	8	6			11-13			8-15
3.9	9+1	9	12	8	5			9-11		5-6.1	5-10
4.7	9+1	9	12	8	6 (5)			10-12		5.6-7	(5) 7-12
4.3	9+1	9	12	8	6 ((5))					6-7.3	6-12
4.9 (4.4)	9+1	9	12	8	7 (6)			11-13		6.8-8.3	(6) 8-12 (15)
4.6	9+1	9	12	8	5			10-12		4.9-6.9	5-13
n.d.	9+1	9	12	8	5			10-12		5.4-6.5	6-11
4.6	9+1	9	12	8	6			11-13		6-7.5	(5) 7-13 (16)
3.6	9+1	9	12					8-10		5-6	4-8
3.9	9+1	9	12					9-10		4.9-6.2	4-7
5.7	9+1	9	12	8	7 (6)			13-15		7.5-9.6	(10) 13-23 (28)
5	9+1	9	12					11-12		6.3-7.3	7-11 (14)
5.5 (5.2)	9+1	9	12					12-14		7.5-8.6	12-17 (28)
5	9+1	9	12					12-14		7.5-9	(9) 12-16 (19)
5.2	9+1	9	12	5	6			12-14		7.1-8.7	9-15 (22)
4.3 (4j)	9+1	9	12	8	5			14-16		5.1-6.7	(9) 11-19 (21)
4	9+1	9	12	8	5			13-15		5.4-7	6-11
5.2	9+1	9	12	8	4			11-13		5.7-7	9-13 (15)
6.3	9+1	9	12					13-14			
4.8	9+1	9	12					12-13		5.5-7.2	8-14
6.2	9+1	9	12					12-14			
4.9	9+1	9	12					10-12		5.7-7	9-13 (16)
4.6	9+1	9	12	8	5			10-12		5.4-6.8	8-13
5	9+1	9	12	8	4			10-12		5.7-7.3	(7) 9-15
6.1	9+1	9	12	8	4			12-13			
5.3	9+1	9	12	8	4			13-15		7-8.3	14-22 (25)

ENGLISH NAME	SCIENTIFIC NAME	Outer P (small)	Pmax	Pmin	Smax	Smin	Inner S (tertials)	Rmax
Corsican Nuthatch	Sitta whiteheadi	1.9-2.1	6.3	5.3	5.4	4.3	2.4-4.3	4.3
Eurasian Nuthatch	Sitta europaea	1.9-2.3	7.8	5.8	6.5	4.7	2.7-5.3	5.5
Western Rock Nuthatch	Sitta neumayer	2.7-2.9	7.1	6.5	6.4	5.1	3.1-5.4	5.3
Wallcreeper	Tichodroma muraria	3.1-3.6	9.6	7.2	8.2	5.2	3.4-6	6.4
Eurasian Treecreeper	Certhia familiaris	2.1-2.8	6 (6.5)	4.1	5.1	3.7	2.6-4	6.7
Short-toed Treecreeper	Certhia brachydactyla	1.9-2.4	6	4.5	5	3.8	2.5-4.0	6.7
Eurasian Penduline Tit	Remiz pendulinus	0.8-1.2	5 (5.5)	4.1	4.3	3.2	1.9-3.5	5.4
Eurasian Golden Oriole	Oriolus oriolus	5.5-6.5	13.5	8.3	9	6.7 (6)	(3.6) 4.4-7.3	10
Red-backed Shrike	Lanius collurio	2.1-3.2	8.1 (9.3)	5.9	6.5	5	3.3-5.4	8.9 (9.3)
Lesser Grey Shrike	Lanius minor	2.3-3.1	10.3	6.9	7.4	5.6	1.7-6.5	10
Great Grey Shrike	Lanius excubitor	3.7-5.5	10.4 (11)	7.7	7.8 (9)	6.2	4.2-6.5	12.2
Iberian Grey Shrike	Lanius meridionalis	3.8-4.6	10	7	8.1	6.1	4.2-6.7	11
Woodchat Shrike	Lanius senator	2.5-3	8.5	6.3	6.5	4.5	3-5.7	8.8
Masked Shrike	Lanius nubicus	1.7-2.9	8	6.2	6.5	5.3	3.4-5.5	9.6
Eurasian Jay	Garrulus glandarius	6-8	18 (20)	12 (11)	16 (17)	9	4-10	17 (18)
Siberian Jay	Perisoreus infaustus	5-6.5	14	9	12	8	4.5-10	15
Azure-winged Magpie	Cyanopica (cyana) cooki	7-9	16	8	14	8	7-10	25
Common Magpie	Pica pica	6-7	18 (19)	11	15	11	7-12	27
Spotted Nutcracker	Nucifraga caryocatactes	7-8	17 (18)	11	14	10	7-11	14
Alpine Chough	Pyrrhocorax graculus	11-12	24	15	16	12	5-12	19
Red-billed Chough	Pyrrhocorax pyrrhocorax	13-15	27 (31)	17	18	14	6-14	17
Western Jackdaw	Corvus monedula	10-12	20 (28)	13	16	11	6-12	17
Rook	Corvus frugilegus	13-16	29 (35)	18	21	15	8-15	20 (23)
Carrion Crow	Corvus corone corone	12-16	31	17	21	12	8-14	21
Hooded Crow	Corvus corone cornix	13-15	32	18	21	13	8-14	20 (22)
Common Raven	Corvus corax	24-29	47	26	26 (28)	16	12-17	30
Common Starling	Sturnus vulgaris	1.6-2.2	11.3	7.5	7.8	6.4	4-6.7	7.3
Spotless Starling	Sturnus unicolor	1.4-2.5	11.6	7.4	8.2	6.1	4.1-6.8	7.9
Rose-coloured Starling	Sturnus roseus	1.4-2	11.1	7.5	7.7	5.8	3.9-6.7	8.1
House Sparrow	Passer domesticus	1.4	6.6	5.2	5.5	4.4	2.7-4.9	6.1
Italian Sparrow	Passer italiae	1.1-1.5	6.9	4.8	5.7	4	2.3-5.2	6.4
Spanish Sparrow	Passer hispaniolensis	0.8-1.8	6.8	4.8	5.3	4.9 (4j)	2.9-5.2	6.2
Eurasian Tree Sparrow	Passer montanus	0.8-1.5	6.4 (6.8)	4.7	4.9 (5.2)	4.2	2.4-4.7	5.8
Rock Sparrow	Petronia petronia	1.2-1.5	8.4	5.6	5.9	4.9	3.2-5.6	6.1
White-winged Snowfinch	Montifringilla nivalis	1.4-1.8	10.9	6.2	6.8	5.3	3.5-6.3	8.3
Common Chaffinch	Fringilla coelebs	0.5-1.4	8.1	5.6	6.3	4.6	2.8-5.1	7.8
Brambling	Fringilla montifringilla	1.1-1.3 (2.1)	8.1 (9.4)	5.7	5.8 (6.5)	4.7	2.9-5.3	7.5
European Serin	Serinus serinus	0.7-0.8 (1.7)	6.6 (7.2)	4.4	5.2	3.7	2.4-4.8	5.8
Citril Finch	Serinus citrinella	n.d.	6.6	5.1	5	4.1	2.6-4.1	6.1
Corsican Finch	Serinus corsicana	n.d.	n.d.	n.d.	n.d.	n.d.	n.d.	5.7
European Greenfinch	Chloris chloris	1.1-1.3	7.7	5.3	5.8	4.6	3-5.2	6.4
European Goldfinch	Carduelis carduelis	1.1-1.2	7	4.9	5.1	4.4	2.5-4.7	5.7
Eurasian Siskin	Spinus spinus	1.7-1.9	6.6 (7.1)	4.4	4.8	3.7	2.5-4.5	5.5
Common Linnet	Linaria cannabina	0.8-1.1 (1.6)	7	4.7	5	4	2.2-4.5	6.2
Twite	Linaria flavirostris	0.5-1.5	6.9	4.8	5.1	4.1	2.2-4.5	7.1
Common / Lesser Redpoll	Acanthis flammea / cabaret	0.6-1 (2)	6.6 (7.5)	4.7	4.8	3.7	2.2-4.2	6.5
Arctic Redpoll	Acanthis hornemanni	0.7-0.9	7 (7.6)	4.8	5.2	3.8	2.4-4.6	6.8 (7.5)
Two-barred Crossbill	Loxia leucoptera	0.8-1.3	8.3	5.2	5.6	4.6	2.8-5	7.2
Common Crossbill	Loxia curvirostra	0.9-1.1	8.5	5.3	5.7	4.8	3.2-5.3	7.3
Scottish Crossbill	Loxia scotica	1-1.1	8.4	5.7	5.5	5	3.1-5.1	6.8
Parrot Crossbill	Loxia pytyopsittacus	1.1-1.2 (2)	9.6	5.9	6.8	5.1	4.7-6.3	8
Trumpeter Finch	Bucanetes githagineus	0.9	7.6	4.9	5.2	4.4	2.8-5.2	5.9
Common Rosefinch	Carpodacus erythrinus	0.7-0.8	7.3	5.5	5.6	4.8	2.9-5.2	6.4
Pine Grosbeak	Pinicola enucleator	0.9-1.1 (2.2)	10 (12)	7	8.2	6	3.5-6.3	10.6
Eurasian Bullfinch	Pyrrhula pyrrhula	0.7-1.1	8.4	5.8	6.4	4.8	2.6-5.3	7.5
Hawfinch	Coccothraustes coccothraustes	1.2-2.6 ?	9.2 (10.5)	6.3	7 (7.3)	5.4	3.5-6	6.5
Lapland Bunting	Calcarius lapponicus	0.5-0.9	8.3	5.4	6	4.8	3.1-5.4	7.1
Snow Bunting	Plectrophenax nivalis	0.9-1.4	9.8 (10.7)	5.8	6.7 (7.3)	5	3.4-6.1	7.9 (8.3)
Yellowhammer	Emberiza citrinella	0.7-0.8	8	5.8	6.7	5.4	2.7-6.2	8.5
Cirl Bunting	Emberiza cirlus	1.8	7.5	5.8	6.2	5	3.3-5.3	8.7
Rock Bunting	Emberiza cia	0.8	7.2	5.9	6.1	5.1	2.5-5.5	8.3
Ortolan Bunting	Emberiza hortulana	0.7-0.8	7.9	5.6	6.2	5.5	3.2-6.1	7.7
Cretzschmar's Bunting	Emberiza caesia	0.7-0.8	7.5	5.6	6	5.5	3.2-5.8	7.7
Rustic Bunting	Emberiza rustica	0.8	6.7 (7.1)	5.1	5.5	4.5	2.8-4.9	6.8
Little Bunting	Emberiza pusilla	1	6.4	4.9	5.2	4.4	2.8-4.8	6.6
Yellow-breasted Bunting	Emberiza aureola	0.6-0.8	6.5	5.4	5.4	5.2	3.2-5.5	6.6
Reed Bunting	Emberiza schoeniclus	0.9	7	5.9	6	5.3	3.2-5.4	7.6
Black-headed Bunting	Emberiza melanocephala	0.7-0.8	7.5	5.9	6.2	5.3	3.1-6	8.2
Corn Bunting	Emberiza calandra	1.3	8.7	6.3	6.6 (7.3)	5.9	4.5-6	7.7 (8.4)

Rmin	NP (number of Ps)	NS (number of Ss)	NR (number of Rs)	Em (ov) Po	Em (ov) Pi	Not (iv) Po	Not (iv) Pi	L (length)	W (wingspan)	FW (folded wing)	M (mass in grams or kg)
3.9	9+1	9	12					11-12			
4.5	9+1	9	12	8	6			12-15		8-9.4	15-28
4.5	9+1	9	12					14-16			
5.2	9+1	9	12	8	5	(9)	(6)	15-17			
4.7	9+1	9	12	7	5			12-14		5.8-7.1	6-11 (13)
4.1	9+1	9	12					12-13		5.6-6.8	(6) 8-11
4.2	9+1	9	12	8	5	9	6	10-12		5-6.1	7-13
8.2	9+1	9-10	12	8	7			22-25		13.5-16.1	(42) 46-87 (98)
6.9	9+1	9	12	8	7			16-18		8.2-10.2	(20) 25-35 (45)
7.5	9+1	9	12					19-21			
9.5	9+1	9	12	8		6		22--26		10.4-12.3	(48) 60-81 (90)
8.5	9+1	9	12					22-26		10.2-11.7	48-70 (93)
7 (6.5)	9+1	9	12	8	6	9	7	17-19		9.1-10.6	(20) 28-53 (60)
7.4	9+1	9	12					17-19			
12	9+1	10	12	7	4 (3)			32-35	54-58	16-20.1	(110) 135-197 (230)
11.5	9+1	10	12	8	3 (2)	9	5	26-29			
9	9+1	10	12	8	5			31-35			
11	9+1	10	12	8	4			40-51		16.2-21.9	(133) 142-290 (300)
11	9+1	10 ?	12	(8)	(6)	10	6	32-35	49-53	16-20	(124) 140-190 (200)
16	9+1	10 ?	12	9	6 (5)	10	7 (6)	36-39	65-74	24.2-28.9	160-285 (310)
16	9+1	10 ?	12	9	5	10	6	37-41	68-80	24.5-34.4	(230) 253-390
9	9+1	10	12	9	6			30-34	64-73	20.5-25.5	(140) 174-288 (300)
16	9+1	11	12	9	6			41-49	81-94	26.4-34.7	(292) 340-560 (670)
15	9+1	11	12	9	5			44-51	84-100	28.3-38.5	350-660
16	9+1	11	12	9	5			44-51	84-100		
20	9+1	11	12	9	6			54-67	115-130		
6.3	9+1	9	12	9	8			19-22		11.6-14.1	(41) 60-97 (115)
5.8	9+1	9	12					19-22		11.7-14.3	70-115
6.6	9+1	9	12					19-22			
5.6	9+1	9	12	8	6			14-16		6.9-8.9	22-35 (39)
5	9+1	9	12					14-16			
5.4 (4.4j)	9+1	9	12					14-16			
5	9+1	9	12	8	6			12-14		6-7.8	(16) 18-25 (29)
4.8	9+1	9	12					15-17		8.9-10.3	(23) 26-39
6.8	9+1	9	12					16-19			
6.1	9+1	9	12	8	5			14-16		7.6-9.9	(14) 17-30
5.5	9+1	9	12	8	6			14-16		8-10	17-33 (40)
4.3	9+1	9	12	8	6			11-12		6.5-7.8	(8) 11-14 (17)
4.4	9+1	9	12					11-13		7.1-8.4	10-15
4.7	9+1	9	12					11-13		6.8-7.7	(11-12)
4.7	9+1	9	12	8	6			14-16		7.9-9.8	(18) 22-37 (40)
4.7	9+1	9	12	8	6			12-14		6.7-8.9	10-22
3.9	9+1	9	12	8	7 (6)			11-13		6.7-8	10-15 (18)
4.6	9+1	9	12	8	6			12-14		7.3-8.6	14-26
4.6	9+1	9	12	8	6			12-14			
4.8	9+1	9	12	8	6			11-14		6.4-8.6	9-24
5	9+1	9	12					12-14			
5.7	9+1	9	12					14-16		8.4-9.9	24-38
5.7	9+1	9	12	8		6		15-17		8.6-10.7	(27) 33-48 (62)
5.8	9+1	9	12					15-17		9-10.5	33-45 (50)
6.1	9+1	9	12	0	0			16-18		9.5-11.1	(37) 47-59 (69)
4.6	9+1	9	12					11-13			
5.5	9+1	9	12					13-15		7.6-8.9	(16) 18-27 (31)
8.6	9+1	9	12	8	5	(10)	(7)	19-22			
6.3	9+1	9	12	8	5			15-18		7.6-10.1	16-39 (42)
5.3	9+1	9	12	8	6	5	2	16-18		9.4-11.2	(35) 46-72
6.1 (5)	9+1	9	12					14-16			
5.9	9+1	9	12	8	7			15-18			
6.7	9+1	9	12	8		5		15-17		7.6-10.1	18-38
7.1	9+1	9	12	8		4		15-17		7.1-8.6	20-29
7.4	9+1	9	12					15-17		7.1-8.7	(18) 21-29
6.7	9+1	9	12					15-17		7.7-9.7	(16) 19-28 (37)
6.6	9+1	9	12					14-16			
5.6	9+1	9	12					13-15			
5.3	9+1	9	12					12-14		6.4-7.8	12-20
5.8	9+1	9	12					14-16			
6	9+1	9	12	8	5			13-16		7.4-8.5	(12) 14-25
6.7	9+1	9	12					15-18			
6.8	9+1	9	12	5	6			16-19		8.2-10.9	38-58

TABLE OF NUMERICAL DATA USED IN THE GRAPHS OF RELATIVE 'LARGE' FEATHER SIZE

These data come from the measurements taken from a single individual of each species. If several species are included in a group (e.g. crows and Rook) a single individual representing each species from the group was measured; minimum and maximum measurements are given. A slight variation in the curves in the graphs may therefore be found depending on the individual measured. The values are given as a percentage of the largest feather in the category concerned (remiges and rectrices).

PASSERINES

SPECIES OR GROUP	P10	P9	P8	P7	P6	P5	P4	P3	P2	P1	S1	S2	S3	S4	S5	S6	S7	S8	S9	S10	S11	R6	R5	R4	R3	R2	R1
Passerines type A max	47	93	99	100	100	100	97	92	91	90	88	88	87	86	83	81	74	65	50								
Passerines type A min	23	74	93	96	99	94	91	88	87	84	82	81	78	76	73	70	63	50	35								
Passerines type A(b) min	12	80	95	97	97	89	82	80	78	75	75	74	73	71	68	65	62	53	39								
Passerines type B max	29	88	100	100	98	94	88	86	83	80	75	76	73	73	72	62	60	59	46								
Passerines type B min	9	88	98	97	90	82	78	75	72	68	67	66	63	63	62	61	60	59	37								
Passerines type A(b) max	36	95	100	100	100	100	99	89	87	84	83	82	79	78	74	70	68	60	47								
magpies, jays, (max)	44	73	92	98	98	100	100	99	97	92	87	86	84	82	76	71	75	66	53	44							
magpies, jays, (min)	38	70	86	94	96	97	100	97	94	90	84	83	81	77	75	72	67	61	43	42							
Spotted Nutcracker	43	75	87	96	98	100	100	93	87	84	81	78	77	75	72	70	68	65	57	45							
crows and Rook (max)	53	85	96	100	100	95	89	82	77	75	73	70	69	66	65	62	62	58	50	34	31						
crows and Rook (min)	51	80	95	99	99	89	82	76	72	69	65	63	62	60	59	57	56	54	46	33	29						
Common Raven	65	92	100	100	96	84	74	69	64	59	59	59	58	55	55	53	50	46	41	38	30						
Common Magpie																						56	65	71	77	86	100
Common Raven																						75	84	88	93	97	100
Siberian Jay																						84	92	95	97	99	100
Common Blackbird	25	86	96	98	100	98	92	88	87	85	84	83	81	79	75	71	68	58	43								
Blue Rock Thrush and Thrushes	20	91	99	100	97	89	84	82	80	78	76	75	73	71	68	66	63	54	42								
Common Rock Thrush and Ring Ouzel	23	97	100	97	91	87	84	82	79	76	72	70	69	67	67	64	64	56	40								
Mistle Thrush																						74	87	95	99	100	100
others (min)																						93	94	97	99	97	92
others (max)																						99	100	100	100	100	98
Common Starling	15	97	100	95	91	86	82	80	77	73	69	67	66	65	62	61	59	49	40								
Eurasian Golden Oriole	44	89	100	100	94	87	82	79	76	73	70	68	66	63	61	59	58	54	44								
Bohemian Waxwing	11	95	100	99	92	86	83	79	75	72	66	65	63	60	58	57	55	49	40	31							
Wallcreeper	37	76	91	96	99	100	96	89	79	76	71	72	82	80	80	76	71	66	55	41							
Woodlark	20	95	99	100	96	100	85	85	83	80	79	77	78	78	77	74	77	66	49	30							
Lesser Short-toed Lark	15	96	99	99	100	88	85	81	78	76	75	75	78	78	77	74	64	49	22								
Calandra, Greater Short-toed & Horned Larks and Skylark	11	96	100	100	99	93	82	79	76	75	68	68	67	67	66	67	74	65	49	34							
Crested, Thekla & Dupont's Larks max	30	93	99	99	100	100	90	89	76	71	70	82	85	83	82	84	84	71	56	30							
Crested, Thekla & Dupont's Larks min	16	92	96	98	98	99	86	82	80	75	71	68	75	77	79	77	64	50	34								
White-throated Dipper	34	94	96	100	100	100	92	90	89	87	83	81	79	77	76	75	70	61	50	30							
Eurasian Wren	49	83	94	99	99	100	96	94	94	94	91	91	91	86	75	70	57	44	29								
treecreepers max	11	97	99	100	100	91	84	82	81	81	81	80	79	79	76	72	59	39				81	89	92	95	97	100
treecreepers min	8	94	96	99	98	88	81	79	77	74	74	74	74	74	71	69	66	54	36			75	84	86	89	91	100
pipits max	11	97	99	100	100	91	84	82	81	81	79	79	80	79	76	76	62	46									
pipits min	8	94	96	98	98	88	81	79	77	74	74	74	74	74	71	74	71	65	46								
wagtails	12	96	99	99	100	87	82	79	77	75	73	72	72	71	70	72	69	50									
European Goldfinch	31	81	95	99	99	88	83	94	91	88	88	88	86	83	79	75	72	59	39			100	100	98	94	90	84
Common Linnet	23	86	96	100	98	91	84	82	79	77	75	73	71	69	66	54	36					100	91	88	85	80	78
European Serin	20	92	100	100	94	82	80	82	83	85	80	77	73	72	71	67	65	58	46								
Long-tailed Tit	29	84	89	94	96	96	96	94	93	91	90	89	84	71	63	54	42					49	75	86	98	100	95
Bearded Reedling	33	90	90	97	100	100	96	95	94	93	91	89	83	77	71	63	54	46				53	67	81	90	98	100
Common/Iberian Chiffchaffs	23	86	96	100	98	93	88	86	84	82	79	79	77	75	73	71	69	59	39								
Arctic Warbler	29	89	97	100	100	96	94	90	84	79	77	75	72	69	68	66	63	54	36								
Wood Warbler	20	92	100	100	98	94	89	85	83	82	80	77	75	72	71	69	65	58	46	34							
other Phylloscopus max	35	88	100	100	100	98	94	92	90	88	87	83	80	77	73	69	67	55	40								
other Phylloscopus min	29	81	96	98	98	92	86	81	80	77	75	73	71	69	65	63	55	46	35								
Booted Warbler	33	90	97	100	100	100	96	95	94	93	91	89	83	77	71	63	54	46									
Icterine & Olive-tree Warblers	21	94	100	99	99	96	89	86	83	80	77	75	72	69	66	64	63	61	54	42							
other Hippolais max	32	89	97	100	100	93	91	90	86	89	86	85	81	78	75	73	69	61	45								
other Hippolais min	29	84	96	98	98	91	89	88	85	84	80	80	77	75	73	67	65	63	56	41							
Great Reed & Marsh Warblers	22	97	100	98	94	92	90	88	85	83	79	78	78	79	73	67	65	58	45								
other reed warblers max	30	94	100	100	99	96	94	93	90	90	87	85	84	82	79	74	71	69	58	46							
other reed warblers min	23	86	98	98	100	94	92	90	88	89	83	83	81	79	75	71	69	57	44								
Cetti's Warbler	46	79	93	98	95	100	98	98	96	95	93	93	91	88	82	73	69	57	39								
Zitting Cisticola	45	86	96	99	100	100	95	95	94	95	95	95	91	93	93	82	77	73	59								

SPECIES OR GROUP	P10	P9	P8	P7	P6	P5	P4	P3	P2	P1	S1	S2	S3	S4	S5	S6	S7	S8	S9	S10	S11	R6	R5	R4	R3	R2	R1
Cetti's Warbler																							86	94	96	99	100
Zitting Cisticola	36	84	99	99	100	100	100	98	97	95	91	91	88	86	83	77	70	58	43			67	83	90	94	98	100
Moustached Warbler	30	99	100	100	99	95	93	91	90	88	83	82	82	79	75	73	73	70	54								
'grasshopper' and 'sedge' warblers max	20	95	100	96	93	91	88	87	85	83	77	77	75	72	70	68	66	55	40								
'grasshopper' and 'sedge' warblers min																											
Locustella max																						84	93	97	99	100	100
Locustella min																						77	89	95	98	99	98
Moustached Warbler																						86	91	95	96	98	100
Sedge Warbler																						90	96	98	100	100	98
Aquatic Warbler																						81	90	92	94	97	100
Hirundines (max)	20	100	100	96	88	82	75	69	63	57	51	50	49	49	48	47	44	39	30			100	100	99	98	95	92
Hirundines (min)	9	97	98	92	86	79	72	65	58	53	48	46	44	43	42	41	39	34	24			100	99	93	90	87	86
Eurasian Crag Martin																											
Sand Martin																											
House Martin, Red-rumped (y) & Barn Swallows (y)																						100	94	79	72	68	66
Red-rumped Swallow ad																						100	83	72	65	60	57
Barn Swallow m ad																						100	61	53	48	45	43

NEAR-PARRERINES

SPECIES OR GROUP	P11	P10	P9	P8	P7	P6	P5	P4	P3	P2	P1	S1	S2	S3	S4	S5	S6	S7	S8	S9	S10	S11	S12	S13	S14	S15	S16	S17	R6	R5	R4	R3	R2	R1
Alpine Swift		97	100	95	85	73	61	52	42	34	30	26	27	28	28	27	28	26	20	18										100	94	84	75	62
Common Swift		97	100	95	87	77	67	58	49	42	36	31	32	32	32	31	32	30	23	21										100	90	77	66	60
Black Woodpecker		28	67	88	95	98	100	98	93	83	78	78	73	70	70	68	67	67	65	57	44	33												
Eurasian Wryneck		19	91	100	100	96	94	91	90	88	86	85	83	82	81	79	79	79	71	58	44													
Green Woodpecker		34	80	97	100	99	95	92	89	89	86	83	83	81	80	80	77	75	69	61	49													
'pied' woodpeckers min		23	78	98	98	99	99	96	87	78	76	76	74	73	77	70	67	64	53	39														
'pied' woodpeckers max		30	80	96	100	100	99	99	84	80	79	78	75	74	73	71	69	66	65	51	40													
woodpeckers max																													43	84	89	94	95	100
woodpeckers min																													26	69	80	87	93	100
Eurasian Wryneck																													43	96	99	100	100	99
Eurasian Hoopoe		35	79	95	98	100	100	95	88	86	84	84	85	83	80	78	75	72	66	58	41													
Common Kingfisher		89	95	100	98	98	97	95	94	92	89	89	86	85	82	80	78	77	74	71	68	60	46	35										
European Roller		95	99	100	97	91	87	82	78	75	73	71	70	69	68	66	65	65	63	65	60	56	47	38	28									
European Bee-eater		19	100	99	93	88	77	75	74	71	67	63	61	60	60	58	56	55	55	55	57	59	53	44	33				86	89	89	89	90	100
cuckoos max		77	95	100	100	95	87	80	75	71	67	62	62	59	57	57	59	62	55	47	41	30												
cuckoos min		48	87	98	95	88	78	70	64	56	53	45	45	45	47	49	50	45	40	37														
Great Spotted Cuckoo																																		
Common Cuckoo																																		
nightjars min		87	96	100	84	75	67	62	59	55	53	52	52	50	50	50	50	50	50	50	47													
nightjars max		95	99	100	87	80	69	65	60	58	56	54	54	54	54	54	53	52	52	54	52	48	40	38										
Collared Dove		92	99	100	94	89	82	77	73	70	70	70	70	70	69	68	66	63	61	56	47	39												
other species min		85	99	96	89	84	77	71	68	65	61	61	61	59	57	56	55	53	51	51	47	41	39	36										
other species max		96	100	100	94	88	83	79	75	71	68	65	64	63	61	61	59	58	57	56	54	50	41	39										
Turtle Dove																													84	96	97	99	100	98
other species: min																													90	94	93	91	87	86
Pin-tailed Sandgrouse	20	99	100	100	94	88	83	76	69	64	57	50	50	50	51	52	52	52	52	53	57	59	61	63	62	61	51	44						

AQUATIC BIRDS OF THE COAST AND WETLANDS: PRIMARIES AND SECONDARIES

SPECIES OR GROUP	P12	P11	P10	P9	P8	P7	P6	P5	P4	P3	P2	P1	S1	S2	S3	S4	S5	S6	S7	S8	S9	S10	S11	S12	S13	S14	S15	S16	S17	S18	S19	S20	S21	S22	S23	S24	S25	S26
gulls max		31	100	100	98	93	88	82	74	70	64	59	57	57	57	57	56	55	54	54	54	54	54	54	54	54	54	54	54	54	52	50						
gulls min		17	96	99	98	86	78	70	65	52	47	47	47	46	46	46	46	45	44	43	43	44	39	44	43	44	42	43	40	38	40	40						
Black Tern		23	100	96	91	84	76	68	60	52	47	45	44	44	42	42	42	42	41	41	39	39	39	39	36	36	36	36	40				37	28				
terns max		23	100	99	92	83	79	69	64	56	49	43	39	38	38	37	37	37	37	37	37	37	37	37	38	38	38	39	40	42	43	43	37					
terns min		19	100	94	87	79	70	61	54	48	42	38	35	35	35	35	35	35	35	35	34	34	32	32	32	32	32	32	33	33								
Pomarine Skua		30	100	99	95	90	83	77	71	64	58	53	53	52	52	52	52	53	54	54	55	55	55	55	55	55	55	53	53	49	43							
auks min		29	91	95	98	94	90	86	80	74	67	60	54	52	49	49	49	49	49	49	49	49	48	46	45	43	42	41	40	38	38	36	33						
auks max		36	99	100	100	99	97	95	90	86	80	74	66	62	59	57	58	58	57	57	56	54	53	51	49	47	45	43	42	40	37	36						
Calidris sandpipers min			96	100	95	87	81	74	67	62	53	53	50	50	49	49	49	49	51	53	56	57	62	65	68	59	43											
Calidris sandpipers max			100	100	97	95	87	83	75	68	65	60	56	52	56	55	53	52	51	52	54	54	57	57	78	59	43											
snipes min			93	99	98	93	89	84	79	75	71	67	63	63	63	63	63	63	64	64	65	73	77	76	78	59	43											
snipes max			99	100	100	97	93	88	85	79	76	72	68	68	68	67	68	70	72	73	75	85	87	85	70	58												
plovers, ringed plovers, Turnstone min			98	98	92	86	80	73	66	61	55	48	46	46	45	45	43	43	44	46	45	48	50	67	74	59	55											
plovers, ringed plovers, Turnstone max			100	100	100	93	85	79	72	66	60	55	55	53	53	52	52	52	54	56	60	63	73	86	82	74	60											
Curlew, Stone Curlew, Stilt min			95	99	96	91	85	78	73	67	62	57	52	52	51	50	49	48	47	47	48	49	51	53	56	59	63	59	64	47								
Curlew, Stone Curlew, Stilt max			100	100	100	95	87	83	75	68	65	60	56	56	56	55	53	52	47	52	54	54	55	57	61	66	76	77	64	47								
Tringa sandpipers, Avocet min			96	100	97	90	86	80	73	67	63	59	56	56	55	54	52	52	51	51	52	53	56	59	67	63	63	52										
Tringa sandpipers, Avocet max			99	100	100	96	93	86	81	75	68	64	62	62	63	62	60	60	61	61	63	63	69	79	80	78	73	71	66									
godwits, Oystercatcher min			95	98	91	85	78	70	64	61	58	54	52	51	50	50	49	50	52	52	54	55	56	57	61	64	66	71	64	57								
godwits, Oystercatcher max			100	100	99	95	81	84	79	72	66	62	58	58	57	56	56	56	57	58	60	62	64	69	79	74	74	71	64	57								
Grey Phalarope			100	99	96	92	81	74	67	62	57	53	53	50	50	50	49	49	49	48	47	52	62	75	62													
Eurasian Woodcock		98	98	100	100	96	93	88	85	80	77	75	74	75	74	72	72	72	71	72	71	71	72	72	67	60	52											
Northern Lapwing		85	85	95	98	99	100	99	98	94	88	79	66	66	61	58	56	55	55	57	57	58	59	62	67	70	67	59	50									
Collared Pratincole	21	99	99	100	95	88	81	74	69	62	55	49	45	44	44	42	41	41	40	41	43	44	44															
Procellariiformes minimum	19	88	95	100	97	92	85	77	68	60	52	47	45	42	41	41	41	41	41	40	40	39	37	37	35	35	39	39	39	39	39	39	35	35				
Procellariiformes maximum	30	99	97	100	96	93	88	83	77	67	62	55	50	50	47	47	48	47	47	48	49	47	47	47	47	47	47	47	47	47	47	47	45	44				
Northern Gannet	20	86	100	100	97	94	92	90	86	83	80	81	81	80	78	77	75	75	75	73	72	71	71	71	70	72	73	72	71	70	67	64	62	55	51	51	52	52
cormorants min	30	86	86	95	99	95	90	87	80	75	73	70	70	68	66	65	63	62	61	61	61	61	61	62	62	65	66	67	68	69	64	67						
cormorants max	32	95	95	99	100	100	97	92	90	87	85	82	87	86	86	85	83	83	82	81	79	76	75	73	73	72	72	70	68	66	64	67						
divers min	24	93	93	98	98	92	86	82	78	73	69	64	64	64	63	63	60	59	57	54	52	52	51	51	50	50	49	49	44	44	40	40	40	40	39			
divers max	29	100	100	100	99	95	94	87	78	76	73	71	67	67	67	69	69	69	67	67	66	66	63	60	60	60	60	56	53	53	51	51	45	45	45	45		
Little Grebe	11	88	96	100	98	96	94	93	92	91	89	89	84	84	83	81	80	78	75	72	75	74	74															
Black-necked Grebe	21	94	99	100	97	93	90	87	82	76	74	70	67	66	65	65	63	61	55	56	63	62	63	62	62	64	67	63	56									
other grebe species min	22	90	97	100	96	93	90	87	82	76	74	71	67	66	66	71	71	69	68	67	67	66	66	66	65	65	65	65	65	65	64	61	55					
other grebe species max	22	94	100	100	97	94	92	90	86	85	83	81	80	80	78	77	75	75	75	73	72	71	71	71	70	72	73	72	71	70	67	64	62	55				
Mute Swan	22	85	85	96	100	100	92	86	80	75	73	71	70	68	66	65	63	62	61	61	61	61	61	62	64	66	66	67	68	69	64	67						
Bean Goose	25	94	95	99	100	100	96	91	78	74	65	63	63	62	61	60	58	57	57	57	56	56	57	55	58	59	63	61	52	66	64	67						
Common Shelduck	19	96	100	100	100	96	91	89	78	76	72	66	62	59	61	60	62	61	57	60	56	56	59	65	65	61	61	52										
other dabbling ducks max	20	97	97	100	99	94	89	83	77	71	66	61	53	54	54	54	55	55	60	54	55	58	63	70	70	60	45	52	53	53	51	51	55					
other dabbling ducks min	20	95	95	100	98	92	87	79	72	67	61	56	49	50	50	50	50	50	50	51	54	54	58	60	55	48	41											
Mallard	26	95	95	99	95	91	88	85	80	75	70	65	60	60	59	58	56	56	56	56	57	59	63	69	77	64	44											
diving ducks (group 1) max	28	97	96	100	96	93	90	87	85	71	69	64	61	60	60	59	58	58	58	58	58	60	67	69	66	60		63	65	65	64	61	55					
diving ducks (group 1) min	28	93	88	91	97	94	86	80	74	67	63	61	61	58	57	54	54	54	57	57	59	53	57	61	61	42		60	65	70	67	64	62	55				
Common Eider	29	94	94	100	96	94	92	90	81	70	66	67	60	66	66	65	63	61	60	60	61	59	63	62	64	66	66	67	68	69	67	67						
Common Pochard		96	96	100	100	93	90	86	74	69	65	62	62	59	59	58	58	58	60	60	56	60	63	66	75	76	69	60	52	47	36							
Common Scoter		88	88	100	100	91	86	80	72	67	61	66	66	65	66	65	62	61	60	60	59	59	61	67	76	77												
Ardeidae max	28	90	96	99	99	100	100	96	91	83	79	84	82	82	80	80	80	79	78	81	83	81	80	79	73	76	69	60	43									
Ardeidae min	19	87	94	95	99	97	91	86	81	76	72	69	75	75	75	76	75	74	75	76	76	73	76	73	73	68	66	50	60									
Little Bittern	22	97	97	100	99	99	93	91	87	84	81	79	77	74	75	74	74	74	73	75	78	73	76	73	73	67	57											

SPECIES OR GROUP	P12	P11	P10	P9	P8	P7	P6	P5	P4	P3	P2	P1	S1	S2	S3	S4	S5	S6	S7	S8	S9	S10	S11	S12	S13	S14	S15	S16	S17	S18	S19	S20	S21	S22	S23	S24	S25	S26
White Stork	16	79	90	100	100	99	95	82	72	67	64	59	50	50	50	50	49	49	49	49	49	49	48	49	50	51	52	52	49									
Common Crane		10	91	97	98	100	99	94	73	68	66	65	55	66	65	65	65	66	66	65	66	66	66	66	66	66	79	81	80									
Glossy Ibis		22	96	100	99	95	90	86	80	75	71	67	56	66	66	65	65	65	65	65	65	65	64	63	61	61												
Greater Flamingo			95	99	100	93	87	81	74	71	67	65	61	61	61	60	59	58	56	54	52	52	52	53	53	61	61	61										
Eurasian Spoonbill			87	95	99	100	95	88	84	81	77	74	73	72	71	70	73	68	68	68	68	68	69	70	71	70	69	67	63									
Water Rail			73	90	97	100	100	93	92	88	86	83	80	79	77	75	75	73	74	80	84	74	82	59	48													
Corncrake			83	98	100	99	97	94	91	87	83	78	72	71	69	67	65	65	66	71	83	91	64	70	58													
Purple Swamphen			75	100	100	96	96	91	89	82	77	71	67	67	67	68	69	69	74	75	76	74	64	54	47	27												
Eurasian Coot		15	85	91	98	100	99	97	94	89	86	81	79	77	74	72	70	70	74	75	75	75	79	81	85	85	78	72	67	67	66	65	60	58				
Common Moorhen		13	81	98	100	100	99	96	92	89	85	79	76	74	72	69	67	67	69	75	77	73	65	53	44													
Spotted Crake			81	100	100	99	98	96	93	88	82	77	75	73	70	68	67	66	67	75	89	86	65	54														

AQUATIC BIRDS OF THE COAST AND WETLANDS: RECTRICES

SPECIES OR GROUP	R1	R2	R3	R4	R5	R6	R7	R8	R9	R10
Black and White-winged Terns	84	85	88	91	96	100				
Caspian and Gull-billed Tern	73	74	82	90	96	100				
Little Tern ad	60	62	66	74	83	100				
Arctic Tern ad	41	44	49	57	68	100				
Sandwich and Little Terns juv	71	74	79	86	94	100				
Arctic Tern juv	64	66	69	79	91	100				
Common Tern	51	53	59	66	80	100				
Arctic Tern ad	41	44	49	57	68	100				
Pomarine Skua	135	103	103	101	100	100				
Arctic Skua	155	104	102	100	100	101				
Long-tailed Skua	235	104	104	100	101	101				
Great Skua	110	107	105	103	102	100				
Razorbill min	100	94	89	85	81	76				
Razorbill max	100	95	94	92	88	85				
Other auks min	89	95	98	97	95	89				
Eurasian Woodcock	100	98	98	95	90	89				
Great Snipe	100	98	97	94	92	88				
Jack Snipe	100	94	87	85	83	81				
Grey Phalarope	100	98	92	91	88	83				
Common Sandpiper	98	98	100	98	95	84				
Stone Curlew	100	98	98	90	83	74				
Collared Pratincole	54	55	92	96	96	100				
Leach's Petrel	74	77	79	85	94	100				
other petrel species: minimum	94	97	97	97	97	91				
Northern Gannet min	100	88	81	75	67	43				
Northern Gannet max	100	92	87	80	73	54				
Great Cormorant	98	100	98	96	93	89				
European Shag	100	99	96	94	88	75				
Black-throated Diver	97	99	100	100	93	83				
Anatidae with rounded/wedge-shaped tails max	100	100	98	98	95	91	88	79	67	64
Anatidae with rounded/wedge-shaped tails min	96	95	91	89	86	79	73	71	65	61
Common Scoter	100	95	89	84	78	72	64	57		
Long-tailed Duck male	343	214	140	124	105	102	100			
Northern Pintail male	200	136	116	113	112	110	106	100		
Eurasian Wigeon male	122	107	106	105	103	102	100			
White Stork	100	99	96	88	81	75				
Black Stork	100	100	99	98	94	89				
Purple Swamphen	100	95	89	83	74					
Common Moorhen	100	100	99	98	93	80				
Spotted Crake	100	96	94	90	85	77				

OTHER GROUPS

SPECIES OR GROUP	P11	P10	P9	P8	P7	P6	P5	P4	P3	P2	P1	S1	S2	S3	S4	S5	S6	S7	S8	S9	S10	S11	S12	S13	S14	S15	S16	S17	S18	S19	S20	R9	R8	R7	R6	R5	R4	R3	R2	R1
grouse min		67	88	93	98	100	93	78	73	71	67	67	68	67	66	65	64	64	63	62	61	60	59	56	52	49	37													
grouse max		75	92	97	100	100	99	90	87	71	69	68	69	68	67	67	67	65	64	63	62	62	61	58	55	50	37													
Grey Partridge		82	95	99	100	100	97	91	84	78	72	73	77	75	75	75	75	75	74	74	74	69																		
'red-legged' partridges		85	93	97	99	100	99	95	91	87	84	77	81	75	83	83	85	85	87	86	83	77	66	53																
Hazel Grouse		75	89	94	98	100	99	94	84	78	75	72	75	75	73	71	70	70	68	66	62	58	50																	
Common Quail		96	100	98	94	91	85	83	78	74	71	66	67	65	64	63	62	61	62	64	63	59																		
Common Pheasant		80	90	95	98	100	99	95	94	87	79	76	78	78	79	80	82	80	81	81	81	81	81	81	79	77													
Black Grouse male																																25	29	33	39	44	51	59	74	100
Common Pheasant male																																99	100	97	94	86	83	81	79	75
Great Bustard	23	91	99	99	100	97	93	77	70	66	65	68	68	68	68	69	69	68	68																					
Eurasian Hobby		95	100	95	89	81	73	65	59	52	46	45	45	44	41	39	40	40	41	43	45	46	47	47																
falcons min	12	84	98	98	89	84	76	68	63	57	53	55	55	53	51	49	48	48	48	49	49	49	43	34	22															
falcons max	25	94	94	100	94	90	84	78	72	69	62	60	61	60	58	56	55	55	56	56	55	53	52	53																
Black-winged Kite		90	100	99	91	85	78	67	63	60	59	58	58	57	55	55	54	54	53	51	47	43																		
Northern Goshawk (M measurements)	16	66	90	98	100	98	85	79	76	67	63	63	63	63	64	69	70	71	70	71	71	67																		
Eurasian Sparrowhawk	12	53	77	93	99	100	93	81	76	73	71	69	70	69	67	63	61	59	57	55	53	50	46	36																
harriers max	12	67	93	100	100	95	78	72	68	64	62	59	58	57	55	52	51	50	48	48	48	47	44	39	29	21														
harriers min	10	63	85	94	97	84	75	68	64	59	57	54	54	53	51	49	48	47	46	45	43	44	42	35	29	21														
Osprey	16	79	84	100	98	92	81	72	65	59	58	58	58	57	57	57	57	57	57	57	57	56	56	55	54	53	52	51	44	38										
Common Buzzard	14	62	86	97	100	100	91	85	80	76	71	67	66	66	64	62	60	60	60	58	57	56	55	53																
Rough-legged Buzzard	13	64	86	97	100	99	87	79	73	69	66	64	64	62	60	57	56	55	54	53	52	51	50	49	43	36														
European Honey Buzzard	13	70	92	98	100	98	95	81	77	73	72	69	67	66	64	60	59	59	56	55	54	53	51																	
Red & Black Kites	11	62	86	98	100	95	82	72	65	60	57	56	55	55	54	53	53	53	53	52	52	51	50	41																
eagles max	11	68	87	96	99	100	98	86	79	75	70	68	68	67	66	63	61	58	57	55	55	55	52	52	45															
eagles min	10	63	85	94	98	100	93	82	75	71	67	66	66	65	63	61	59	55	54	53	51	49	49	48	34															
Short-toed Eagle	11	64	87	95	98	100	95	79	74	69	66	61	62	62	61	59	58	57	56	54	53	53	52	51	49	45	32	23	12											
White-tailed Eagle (M)	13	65	81	97	98	100	95	90	84	75	73	73	73	73	73	72	71	71	71	71	71	71																		
Egyptian Vulture (M)		95	98	100	100	99	89	76	72	63	56	56	55	55	54	54	54	53	53	53	53																			
Griffon & Black Vultures (M)		85	99	100	100	99	82	78	72	69	67	67	67	67	67	66	66	66	66	66	66																			
Lammergeier (M)		87	99	100	100	92	79	71	71	66	62	58	58	57	57	55	55	55	53	53	53	53	51																	
Griffon Vulture (M)																																79			85	92	97	99	100	100
Lammergeier (M)																																		81	81	90	94	94	95	100
White-tailed Eagle (M)																																			85	87	91	93	100	100
Red & Black Kites																																			100	98	93	89	88	83
Black Kite (min)																																			100	98	93	92	92	90
Black-winged Kite																																			98	100	96	94	92	91
Little, Pygmy & Tawny Owls min		60	80	93	100	98	91	86	84	83	80	76	75	75	75	74	74	72	70	70	65	61	61	56	47															
Little, Pygmy & Tawny Owls max		77	92	99	100	100	96	91	87	86	85	86	85	85	84	81	79	78	76	75	72	69	64	60	54															
Long-eared and Short-eared Owls		86	100	100	94	88	82	76	70	66	62	62	60	60	58	56	54	54	53	52	52	52	52																	
Barn owl		97	100	99	95	89	83	77	72	69	66	64	65	64	62	60	58	60	57	57	54	52	49	42	37															
Snowy Owl		82	97	100	99	88	78	70	69	67	67	66	66	66	66	66	64	64	64	64	64	64	64	64	64	63	66													
Eurasian Scops Owl		83	96	100	100	94	90	84	79	77	73	73	73	71	70	69	65	65	64	64	62	60	53																	

TABLE OF ILLUSTRATION CREDITS AND ORIGIN OF FEATHERS IN CHAPTER 8 'SPECIES DESCRIPTIONS'

Sources: C.F. = author / S.U. = S. Uriot / M.K. = M. Klemann
Origin: P = plucked prey / M = moulted / C = corpse / ni = not indicated
For the photos of wings: if not otherwise indicated, the scale bar is 10 cm long.

ENGLISH NAME	C.F.	S.U.	M.K.	ORIGIN	SCALE
Alpine Accentor	x	x		C	1
Alpine Chough		x		ni	1/3
Alpine Swift		x		M	2/3
Arctic Skua	x		x	C	1/2
Arctic Tern	x	x	x	C	1/2
Atlantic Puffin		x		C	2/3
Azure Tit	x			C	1
Azure-winged Magpie		x		C	1/2
Bar-tailed Godwit	x			C	1/2
Barn Owl	x	x		C	1/3
Barn Swallow	x			CP	2/3
Barnacle Goose	x	x		M	1/3
Bearded Reedling	x	x	x	CM	1
Black Grouse	x		x	M	1/2
Black Guillemot		x		C	2/3
Black Kite	x	x		C, ni	1/4
Black Redstart	x	x		C	1
Black Stork		x		MC	1/4
Black Stork			x	C	1/4
Black Woodpecker	x			P	1/2
Black-bellied Sandgrouse		x		ni	sans
Black-crowned Night Heron	x			C	1/3
Black-eared Wheatear		x		C	1
Black-headed Bunting			x	C	1
Black-headed Gull	x			MC	1/3
Black-necked Grebe		x		C	2/3
Black-tailed Godwit	x			M	1/2
Black-throated Diver		x		CM	1/3
Black-winged Stilt		x		C	1/2
Blackbird	x			P	2/3
Blackcap	x			C	1
Blue Rock Thrush	x			C	2/3
Blue Tit	x			CM	1

ENGLISH NAME	C.F.	S.U.	M.K.	ORIGIN	SCALE
Bluethroat		x		C	1
Bohemian Waxwing		x	x	C	2/3
Bonelli's Eagle		x		ni	1/4
Bonelli's Warbler		x		C	1
Booted Eagle		x		ni	1/4
Brambling	x	x		CP	1
Brent Goose	x			MC	1/3
Budgerigar	x			M	1/2
Calandra Lark			x	C	2/3
Canada Goose	x			M	1/3
Capercaillie	x		x	MC	1/2
Carrion Crow	x			M	1/3
Caspian Tern	x			M	1/2
Cattle Egret		x		C	1/2
Cetti's Warbler		x		C	1
Cirl Bunting		x		C	1
Citril Finch		x		C	1
Coal Tit		x		C	1
Collared Dove	x			PCM	1/2
Collared Pratincole			x	C	1/2
Common Crossbill		x		C	1
Common Buzzard	x			MC	1/4
Common Chaffinch	x			PM	1
Common Chiffchaff		x		C	1
Common Crane	x	x		PCM	1/4
Common Cuckoo	x	x	x	PC	1/2
Common Eider		x	x	C	1/2
Common Goldeneye			x	C	1/2
Common Guillemot		x		C	2/3
Common Gull	x		x	MC	1/3
Common Kestrel	x	x		MC	1/2
Common Kingfisher	x		x	C	1
Common Linnet	x			PC	1

ENGLISH NAME	C.F.	S.U.	M.K.	ORIGIN	SCALE
Common Magpie	x			M	1/2
Common Moorhen	x			P	1/2
Common Nightingale		x		C	1
Common Pheasant	x			MPC	1/3
Common Pochard	x	x		MC	1/2
Common Quail		x	x	C	1
Common Raven	x			M	1/3
Common Redshank	x	x		MC	2/3
Common Redstart			x	C	1
Common Rock Thrush	x			C	2/3
Common Rosefinch			x	C	1
Common Sandpiper	x			P	2/3
Common Scoter	x	x		C	1/2
Common Shelduck	x	x	x	MC	1/2
Common Skylark		x	x	C	2/3
Common Snipe	x			C	2/3
Common Starling	x			CPM	2/3
Common Stonechat		x		C	1
Common Swift	x			C	2/3
Common Tern	x	x	x	CM	1/2
Common Whitethroat		x	x	C	1
Corn Bunting		x		C	1
Corncrake		x	x	C ni	2/3
Cory's Shearwater		x		C	1/2
Crag Martin		x		ni	2/3
Crested Lark		x	x	C	2/3
Crested tit		x		C	1
Curlew Sandpiper		x	x	C	1
Dartford Warbler		x		C	1
Domestic duck	x			M	1/2
Domestic hen	x			M	1/3
Domestic turkey	x			MC	1/4
Dunlin	x			MC	1

ENGLISH NAME	C.F.	S.U.	M.K.	ORIGIN	SCALE
Dunnock	x	x		P	1
Egyptian Vulture		x		ni	1/4
Elenora's Falcon		x		C	1/2
Eurasian Eagle Owl		x	x	CM	1/4
Eurasian Bittern			x	C	1/3
Eurasian Bullfinch	x	x		C	1
Eurasian Coot	x			CM	1/2
Eurasian Curlew	x			M	1/2
Eurasian Golden Oriole	x	x		C M	2/3
Eurasian Hobby		x		C	1/2
Eurasian Hoopoe	x			P	2/3
Eurasian Jay	x			PM	1/2
Eurasian Nuthatch		x		C	1
Eurasian Oystercatcher	x			M	1/2
Eurasian Pygmy Owl		x		C	2/3
Eurasian Reed Warbler		x		C	1
Eurasian Scops Owl	x	x		C	2/3
Eurasian Siskin	x	x		CP	1
Eurasian Sparrowhawk	x			MC	1/3
Eurasian Spoonbill	x		x	C	1/3
Eurasian Teal	x	x		MCP	1/2
Eurasian Treecreeper		x		C	1
Eurasian Wigeon	x	x	x	C	1/2
Eurasian Wren		x		C	1
Eurasian Wryneck		x		C	1
European Bee-eater		x		C	2/3
European Goldfinch	x			CP	1
European Greenfinch	x			CPM	1
European Honey Buzzard		x		ni	1/4
European Nightjar	x	x		C	1/2
European Robin		x		C	1
European Roller	x	x		CM	2/3
European Serin		x		C	1
European Shag	x			M	1/3
Feral Pigeon	x			CPM	1/2
Fieldfare	x			CM	2/3
Firecrest	x	x		C	1

ENGLISH NAME	C.F.	S.U.	M.K.	ORIGIN	SCALE
Gadwall	x		x	C	1/2
Garden Warbler		x		C	1
Glaucous Gull		x	x	C	1/3
Goldcrest		x		C	1
Golden Eagle		x		M	1/4
Golden Plover		x		C	2/3
Goosander	x			M	1/2
Great Black-backed Gull	x	x		MC	1/3
Great Bustard		x	x	MC	1/4
Great Cormorant	x	x		MC	1/3
Great Crested Grebe	x			C	2/3
Great Egret		x	x	CM	1/2
Great Grey Shrike		x		C	2/3
Great Northern Diver		x		CM	1/3
Great Skua	x			M	1/2
Great Spotted Cuckoo		x		C	1/2
Great Spotted Woodpecker	x			CPM	2/3
Great Tit	x			CM	1
Greater Flamingo	x	x	x	MPC	1/4
Greater Scaup		x		C	1/2
Greater Short-toed Lark		x		C	2/3
Green Sandpiper	x	x	x	MC	2/3
Green Woodpecker	x			PMC	2/3
Grey Heron	x			M	1/3
Grey Partridge	x			PC	2/3
Grey Plover	x			M	2/3
Grey Wagtail		x		C	1
Greylag Goose	x	x		MC	1/3
Griffon Vulture	x	x		M, ni	1/4
Hawfinch		x		C	1
Hazel Grouse	x		x	M	2/3
Helmeted Guineafowl	x			M	1/2
Hen Harrier		x		ni	1/3
Herring Gull	x	x		MC	1/3
Hooded Crow		x		C	1/3
Horned Lark			x	C	2/3
House Martin	x			C	2/3

ENGLISH NAME	C.F.	S.U.	M.K.	ORIGIN	SCALE
House Sparrow	x			P	1
Iberian Grey Shrike			x	C	2/3
Icterine Warbler		x		C	1
Imperial Eagle		x		ni	1/4
Indian Peafowl	x			M	1/4
Jack Snipe	x			C	2/3
Kentish Plover			x	C	1
Kittiwake	x		x	CM	1/3
Lady Amherst's Pheasant	x			M	1/3
Lammergeier		x		ni	1/5
Lanner Falcon		x		CM	1/3
Leach's Storm Petrel	x	x		C	1/2
Lesser Black-backed Gull	x	x	x	MC	1/3
Lesser Kestrel		x		M	1/2
Lesser Redpoll			x	C	1
Lesser Short-toed Lark			x	C	2/3
Lesser Spotted Woodpecker	x	x		C ni	2/3
Lesser Whitethroat			x	C	1
Lessser Grey Shrike			x	C	2/3
Little Auk			x	C	2/3
Little Bittern		x		C	1/2
Little Bustard		x	x	MC	1/2
Little Egret	x			MC	1/2
Little Grebe		x		C	2/3
Little Gull			x	C	1/3
Little Owl	x	x		C	2/3
Little Ringed Plover		x	x	C	1
Little Stint		x		C	1
Little Tern		x	x	C	1/2
Long-eared Owl	x			C	1/3
Long-legged Buzzard		x		ni	1/4
Long-tailed Duck			x	C	1/2
Long-tailed Skua			x	C	1/2
Long-tailed Tit		x		C	1
Mallard	x	x	x	MCP	1/2
Marsh Harrier		x		ni	1/4
Marsh Tit			x	C	1

ENGLISH NAME	C.F.	S.U.	M.K.	ORIGIN	SCALE
Marsh Warbler			x	C	1
Masked Shrike		x		C	2/3
Meadow Pipit		x		C	1
Melodious Warbler		x		C	1
Merlin		x	x	C	1/2
Middle Spotted Woodpecker			x	C	2/3
Mistle Thrush	x			P	2/3
Mute Swan	x			C	1/4
Northern Fulmar	x			C	1/2
Northern Gannet	x			MC	1/3
Northern Goshawk		x	x	ni	1/3
Northern Lapwing	x	x		MCP	1/2
Northern Pintail	x	x	x	C	1/2
Northern Shoveler	x		x	C	1/2
Northern Wheatear		x		C	1
Ortolan Bunting		x	x	C	1
Osprey		x		ni	1/4
Pallid Swift		x		C	2/3
Parrot Crossbill			x	C	1
Peregrine falcon		x	x	C	1/3
Pied Avocet	x	x	x	M	1/2
Pied Flycatcher		x		C	1
Pin-tailed Sandgrouse		x		ni	sans
Purple Heron		x		C	1/3
Purple Swamphen	x			C	1/2
Razorbill		x		C	2/3
Red Grouse			x	C	2/3
Red Kite	x	x		C, ni	1/4
Red Knot		x		C	2/3
Red-backed Shrike		x		C	2/3
Red-billed Chough		x		ni	1/3
Red-breasted Merganser			x	C	1/2
Red-crested Pochard	x			M	1/2
Red-footed Falcon		x	x	C	1/2
Red-legged Partridge	x		x	C	2/3
Red-necked Grebe		x		C	2/3
Red-necked Nightjar		x	x	C	1/2
Red-throated Diver		x		CM	1/3

ENGLISH NAME	C.F.	S.U.	M.K.	ORIGIN	SCALE
Redwing	x			PC	2/3
Reed Bunting	x			P	1
Reeves's Pheasant	x			M	1/3
Ring Ouzel	x			P	2/3
Ringed Plover		x	x	C	2/3
Rock Bunting		x		C	1
Rock Ptarmigan	x			PC	2/3
Rock Sparrow		x		ni	1
Rook	x			M	1/3
Rose-ringed Parakeet	x			P	1/2
Ruddy Turnstone	x			M	2/3
Ruff		x		C	2/3
Rufous Bush Robin			x	C	1
Rustic Bunting			x	C	1
Sacred Ibis	x			M	1/3
Saker Falcon		x	x	CM	1/3
Sand Martin	x		x	C	2/3
Sanderling	x		x	MC	1
Sandwich Tern			x	C	1/2
Sardinian Warbler		x		C	1
Sardinian Warbler	x			M	1/3
Savi's Warbler		x		C	1
Sedge Warbler			x	C	1
Short-eared Owl		x	x	C	1/3
Short-toed Eagle		x	x	ni	1/4
Siberian Jay	x			C	1/2
Slavonian Grebe		x		C	2/3
Snow Bunting		x	x	CP	1
Snowy Owl		x	x	CM	1/4
Song Thrush	x			P	2/3
Spanish Sparrow			x	C	1
Spotless Starling			x	C	2/3
Spotted Crake		x		C	2/3
Spotted Flycatcher			x	C	1
Spotted Nutcracker	x			M	1/2
Spotted Redshank	x	x		MC	2/3
Spur-winged Lapwing		x		C	1/2
Stock Dove	x			P	1/2

ENGLISH NAME	C.F.	S.U.	M.K.	ORIGIN	SCALE
Stone Curlew		x		C	1/2
Subalpine Warbler		x		C	1
Syrian Woodpecker		x		ni	2/3
Tawny Owl	x			CM	1/3
Tawny Pipit		x		C	1
Tengmalm's Owl		x		C	2/3
Thekla's Lark		x		C	2/3
Tree Pipit		x		C	1
Tree Sparrow	x	x		P	1
Tufted Duck	x			M	1/2
Turtle Dove	x	x		PC	1/2
Twite			x	C	1
Two-barred Crossbill			x	C	1
Velvet Scoter	x			M	1/2
Wallcreeper	x			C	2/3
Water Pipit		x		C	1
Water Rail		x		C	2/3
Western Jackdaw	x			M	1/3
Western Orphean Warbler		x		ni	1
Whimbrel	x			C	1/2
Whinchat			x	C	1
White Pelican	x			M	1/4
White Stork		x	x	MC	1/4
White-tailed Eagle		x		ni	1/4
White-throated Dipper		x		C	1
White-winged Tern			x	C	1/2
White/Pied Wagtail		x		C	1
Willow Warbler		x		C	1
Wood Pigeon	x			PCM	1/2
Wood Warbler		x		C	1
Woodchat Shrike		x		C	2/3
Woodcock	x			C	1/2
Woodlark		x		C	2/3
Yellow Wagtail		x		C	1
Yellow-legged Gull	x	x		MC	1/3
Yellowhammer	x			CP	1
Zitting Cisticola		x		C	1

REFERENCES

Feather illustrations

Bezzel E. (2004) *Plumes des oiseaux d'Europe.* Artémis.

Brown R., Ferguson J., Lawrence M. & Lees D. (2003) *Tracks & Signs of the Birds of Britain and Europe (Second Edition).* Helm, London.

Burnie D. (2002) *Le nid, l'œuf et l'oiseau.* Gallimard, Paris.

Demongin L. (2016) *Identification guide to Birds in the Hand.* (privately published)

Fraigneau C. (2007) *Reconnaître facilement les plumes.* Delachaux et Niestlé, Lonay.

Hartmann, G. (ed.) (in press) *Atlas of Feathers for Western Palearctic Birds.* Concise Edition: Passerines. World Feather Atlas Foundation.

Species Guides (illustrations)

Beaman M. & Madge S. (1998) *The Handbook of Bird Identification: for Europe and the Western Palearctic.* Helm, London

Jonsson L. (1992) *Birds of Europe with North Africa and the Middle East.* Helm, London

Svensson L., Mullarney K. & Zetterström D. (2009) *Collins Bird Guide.* Second Edition. HarperCollins, London

Taylor D. & Message S. (2005) *Waders of Europe, Asia and North America.* Helm, London

Species Guides (photos)

Svensson L. & Delin H. (1993) *Photographic Guide to the Birds of Britain and Europe.* Hamlyn, London.

Hume, R., Still, R., Swash, A. & Harrop, H. (2021) *Europe's Birds: An Identification Guide.* Princeton University Press.

Jiguet, F. & Audevard, A. (2017) *Birds of Europe, North Africa and the Middle East: A Photographic Guide.* Princeton University Press

Shirihai, H. & Svensson, L. (2018) *Handbook of Western Palearctic Birds: Passerines Vols 1 & 2.* Helm, London

Websites consulted during the writing of this book:

In English

http://www.featherguide.org

http://www.michelklemann.nl

In French

http://www.alulawebsite.com

In French (English and German)

http://www.federn.org

In German

http://www.federbestimmung.de

http://www.ornithos.de/Ornithos/Feather_Collection

http://www.gefiederkunde.de

In Italian

http://www.cuneobirding.it

ADDITIONS AND CORRECTIONS

This work is the result of the observation of thousands of feathers whose species of origin is considered proven. However, it is possible that identification errors may have been made, due to the author's inability to verify the provenance of certain feathers (illustrations viewed on the internet, images courtesy of contributors, etc.).

Therefore, if the reader of this present work detects a real error in identification, he or she is strongly requested to share their thoughts with the author (ipoeo2017@yahoo.fr).

In addition, each reader is encouraged to provide additional information on descriptions, illustrations, or identification criteria, particularly for species with a small sample size (mentioned in the text or in the biometrics table).

ACKNOWLEDGEMENTS

I should like to thank my parents who have hosted a large part of my feather collection and other bird material for many years. My thanks also go to all my family for their support during the development of this guide, both moral and logistical. I thank all the people who helped me, from near and far, to complete this work by allowing me to examine live birds, by providing me with illustrations, by lending me their collections, or by helping me to identify mine, especially: Fabien M., Hervé R., Laurent C., Ma'lys H., Mathieu S., Michel K., Stéphane M., Sylvain U., Theo V., Vincent R., Virginie N. My thanks also to the team at Delachaux and Niestlé who worked on this project and also to my English publishers, Bloomsbury Publishing. And, finally, thank you to the birds for revealing some of their mysteries to the curious beings that we are, but above all for keeping many hidden.

INDEX OF ENGLISH NAMES

Only the relevant chapters 6 and 8 are indexed.
The page numbers in italics refer to photographic plates.

LIST OF SCIENTIFIC/ENGLISH NAMES

Calandrella rufescens	Lesser Short-toed Lark	*Dendrocopos medius*	Middle Spotted Woodpecker	*Hippolais opaca*	Western Olivaceous Warbler
Calcarius lapponicus	Lapland Bunting	*Dendrocopos minor*	Lesser Spotted Woodpecker	*Hippolais pallida*	Eastern Olivaceous Warbler
Calidris alba	Sanderling				
Calidris alpina	Dunlin	*Dendrocopos syriacus*	Syrian Woodpecker	*Hirundo daurica*	Red-rumped Swallow
Calidris canutus	Red Knot	*Dryocopus martius*	Black Woodpecker	*Hirundo rupestris*	Eurasian Crag Martin
Calidris ferruginea	Curlew Sandpiper	*Egretta garzetta*	Little Egret	*Hirundo rustica*	Barn Swallow
Calidris maritima	Purple Sandpiper	*Elanus caeruleus*	Black-winged Kite	*Histrionicus histrionicus*	Harlequin Duck
Calidris minuta	Little Stint	*Emberiza aureola*	Yellow-breasted Bunting	*Hydrobates pelagicus*	European Storm Petrel
Calidris temminckii	Temminck's Stint			*Hydrocoloeus minutus*	Little Gull
Callipepla californica	California Quail	*Emberiza caesia*	Cretzschmar's Bunting	*Hydroprogne caspia*	Caspian Tern
Calonectris diomedea	Cory's Shearwater	*Emberiza calandra*	Corn Bunting	*Ixobrychus minutus*	Little Bittern
Caprimulgus europaeus	European Nightjar	*Emberiza cia*	Rock Bunting	*Jynx torquilla*	Eurasian Wryneck
Caprimulgus ruficollis	Red-necked Nightjar	*Emberiza cirlus*	Cirl Bunting	*Lagopus lagopus scoticus*	Red Grouse
Carduelis carduelis	European Goldfinch	*Emberiza citrinella*	Yellowhammer	*Lagopus lagopus*	Willow Ptarmigan
Carpodacus erythrinus	Common Rosefinch	*Emberiza hortulana*	Ortolan Bunting	*Lagopus mutus*	Rock Ptarmigan
Casmerodius albus	Great Egret	*Emberiza melanocephala*	Black-headed Bunting	*Lanius collurio*	Red-backed Shrike
Cecropis daurica	Red-rumped Swallow	*Emberiza pusilla*	Little Bunting	*Lanius excubitor*	Great Grey Shrike
Cepphus grylle	Black Guillemot	*Emberiza rustica*	Rustic Bunting	*Lanius meridionalis*	Iberian Grey Shrike
Cercotrichas galactotes	Rufous Bush Robin	*Emberiza schoeniclus*	Reed Bunting	*Lanius minor*	Lesser Grey Shrike
Certhia brachydactyla	Short-toed Treecreeper	*Eremophila alpestri*	Horned Lark	*Lanius nubicus*	Masked Shrike
Certhia familiaris	Eurasian Treecreeper	*Erithacus rubecula*	European Robin	*Lanius senator*	Woodchat Shrike
Cettia cetti	Cetti's Warbler	*Falco biarmicus*	Lanner Falcon	*Larus argentatus*	Herring Gull
Charadrius alexandrinus	Kentish Plover	*Falco cherrug*	Saker Falcon	*Larus audouinii*	Audouin's Gull
Charadrius dubius	Little Ringed Plover	*Falco columbarius*	Merlin	*Larus cachinnans*	Caspian Gull
Charadrius hiaticula	Common Ringed Plover	*Falco eleonorae*	Eleonora's falcon	*Larus canus*	Common Gull
Charadrius morinellus	Eurasian Dotterel	*Falco naumanni*	Lesser Kestrel	*Larus fuscus*	Lesser Black-backed Gull
Chersophilus duponti	Dupont's Lark	*Falco peregrinus*	Peregrine Falcon		
Chlidonias hybrida	Whiskered Tern	*Falco rusticolus*	Gyr falcon	*Larus genei*	Slender-billed Gull
Chlidonias leucopterus	White-winged Tern	*Falco subbuteo*	Eurasian Hobby	*Larus glaucoides*	Iceland Gull
Chlidonias niger	Black Tern	*Falco tinnunculus*	Common Kestrel	*Larus hyperboreus*	Glaucous Gull
Chloris chloris	European Greenfinch	*Falco vespertinus*	Red-footed Falcon	*Larus marinus*	Great Black-backed Gull
Chroicocephalus genei	Slender-billed Gull	*Ficedula albicollis*	Collared Flycatcher	*Larus melanocephalus*	Mediterranean Gull
Chrysolophus amherstiae	Lady Amherst's Pheasant	*Ficedula hypoleuca*	Pied Flycatcher	*Larus michahellis*	Yellow-legged Gull
Chrysolophus pictus	Golden Pheasant	*Ficedula parva*	Red-breasted Flycatcher	*Larus minutus*	Little Gull
Ciconia ciconia	White Stork			*Larus ridibundus*	Black-headed Gull
Ciconia nigra	Black Stork	*Ficedula semitorquata*	Semicollared Flycatcher	*Larus sabini*	Sabine's Gull
Cinclus cinclus	White-throated Dipper	*Fratercula arctica*	Atlantic Puffin	*Limicola falcinellus*	Broad-billed Sandpiper
Circactus gallicus	Short-toed Eagle	*Fringilla coelebs*	Chaffinch	*Limosa lapponica*	Bar-tailed Godwit
Circus aeruginosus	Western Marsh Harrier	*Fringilla montifringilla*	Brambling	*Limosa limosa*	Black-tailed Godwit
Circus cyaneus	Hen Harrier	*Fulica atra*	Eurasian Coot	*Linaria cannabina*	Common Linnet
Circus pygargus	Montagu's Harrier	*Fulica cristata*	Red-knobbed Coot	*Linaria flavirostris*	Twite
Cisticola juncidis	Zitting Cisticola	*Fulmarus glacialis*	Northern Fulmar	*Locustella fluviatilis*	River Warbler
Clamator glandarius	Great Spotted Cuckoo	*Galerida cristata*	Crested Lark	*Locustella luscinioides*	Savi's Warbler
Clangula hyemalis	Long-tailed Duck	*Galerida theklae*	Thekla's Lark	*Locustella naevia*	Common Grasshopper Warbler
Coccothraustes coccothraustes	Hawfinch	*Gallinago gallinago*	Common Snipe		
Colinus virginianus	Northern Bobwhite	*Gallinago media*	Great Snipe	*Lophophanes cristatus*	Crested Tit
Columba livia dom	Feral Pigeon	*Gallinula chloropus*	Common Moorhen	*Loxia curvirostra*	Common Crossbill
Columba livia	Rock Dove	*Gallus gallus*	Red Junglefowl	*Loxia leucoptera*	Two-barred Crossbill
Columba oenas	Stock Dove	*Gallus gallus domesticus*	Domestic chicken	*Loxia pytyopsittacus*	Parrot Crossbill
Columba palumbus	Wood Pigeon	*Gallus varius*	Green Junglefowl	*Loxia scotica*	Scottish Crossbill
Coracias garrulus	European Roller	*Garrulus glandarius*	Eurasian Jay	*Lullula arborea*	Woodlark
Corvus corax	Common Raven	*Gavia adamsii*	White-billed Diver	*Luscinia luscinia*	Thrush Nightingale
Corvus cornix	Hooded Crow	*Gavia arctica*	Black-throated Diver	*Luscinia megarhynchos*	Common Nightingale
Corvus corone	Carrion Crow	*Gavia immer*	Great Northern Diver	*Luscinia svecica*	Bluethroat
Corvus frugilegus	Rook	*Gavia stellata*	Red-throated Diver	*Lymnocryptes minimus*	Jack Snipe
Corvus monedula	Western Jackdaw	*Gelochelidon nilotica*	Gull-billed tern	*Marmaronetta angustirostris*	Marbled Duck
Coturnix coturnix	Common Quail	*Glareola nordmanni*	Black-winged Pratincole	*Melanitta fusca*	Velvet Scoter
Crex crex	Corncrake			*Melanitta nigra*	Common Scoter
Cuculus canorus	Common Cuckoo	*Glareola pratincola*	Collared Pratincole	*Melanocorypha calandra*	Calandra Lark
Cyanistes caeruleus	European Blue Tit	*Glaucidium passerinum*	Eurasian Pygmy Owl	*Meleagris gallopavo*	Domestic turkey
Cyanistes cyanus	Azure Tit	*Grus grus*	Common Crane	*Melopsittacus undulatus*	Budgerigar
Cyanopica cooki	Iberian Magpie	*Gypaetus barbatus*	Lammergeier	*Mergellus albellus*	Smew
Cyanopica cyanus	Azure-winged Magpie	*Gyps fulvus*	Griffon Vulture	*Mergus merganser*	Goosander
Cygnus atratus	Black Swan	*Haematopus ostralegus*	Eurasian Oystercatcher	*Mergus serrator*	Red-breasted Merganser
Cygnus columbianus	Bewick's Swan	*Haliaeetus albicilla*	White-tailed Eagle		
Cygnus cygnus	Whooper Swan	*Hieraaetus fasciatus*	Bonelli's Eagle	*Merops apiaster*	Bee-eater
Cygnus olor	Mute Swan	*Hieraaetus pennatus*	Booted Eagle	*Milvus migrans*	Black Kite
Delichon urbicum	Common House Martin	*Himantopus himantopus*	Black-winged Stilt	*Milvus milvus*	Red Kite
Dendrocopos leucotos	White-backed Woodpecker	*Hipolais polyglotta*	Melodious Warbler	*Monticola saxatilis*	Rock Thrush
		Hippolais caligata	Booted Warbler	*Monticola solitarius*	Blue Rock Thrush
Dendrocopos major	Great Spotted Woodpecker	*Hippolais icterina*	Icterine Warbler	*Montifringilla nivalis*	Snowfinch
		Hippolais olivetorum	Olive-tree Warbler	*Morus bassanus*	Northern Gannet

Motacilla alba	White Wagtail	Picoides tridactylus	Three-toed Woodpecker	Sternula albifrons	Little Tern
Motacilla alba yarrellii	Pied Wagtail	Picus canus	Grey-headed Woodpecker	Streptopelia decaocto	Eurasian Collared Dove
Motacilla cinerea	Grey Wagtail			Streptopelia senegalensis	Laughing Dove
Motacilla flava	Yellow Wagtail	Picus viridis	Green Woodpecker	Streptopelia turtur	European Turtle Dove
Muscicapa striata	Spotted Flycatcher	Pinicola enucleator	Pine Grosbeak	Strix aluco	Tawny Owl
Neophron percnopterus	Egyptian Vulture	Platalea leucorodia	Eurasian Spoonbill	Strix nebulosa	Great Grey Owl
Netta rufina	Red-crested Pochard	Plectrophenax nivalis	Snow Bunting	Strix uralensis	Ural Owl
Nucifraga caryocatactes	Spotted Nutcracker	Plegadis falcinellus	Glossy Ibis	Sturnus roseus	Rose-coloured Starling
Numenius arquata	Eurasian Curlew	Pluvialis apricaria	European Golden Plover		
Numenius phaeopus	Whimbrel			Sturnus unicolor	Spotless Starling
Numida meleagris	Helmeted Guineafowl	Pluvialis squatarola	Grey Plover	Sturnus vulgaris	Common Starling
Nyctea scandiaca	Snowy Owl	Podiceps auritus	Slavonian Grebe	Surnia ulula	Northern Hawk Owl
Nycticorax nycticorax	Black-crowned Night Heron	Podiceps cristatus	Great Crested Grebe	Sylvia atricapilla	Blackcap
		Podiceps grisegena	Red-necked Grebe	Sylvia borin	Garden Warbler
Oceanites oceanicus	Wilson's Storm Petrel	Podiceps nigricollis	Black-necked Grebe	Sylvia cantillans	Subalpine Warbler
Oceanodroma castro	Madeiran Storm Petrel	Poecile montanus	Willow Tit	Sylvia communis	Common Whitethroat
Oceanodroma leucorhoa	Leach's Storm Petrel	Poecile palustris	Marsh Tit	Sylvia conspicillata	Spectacled Warbler
Oenanthe hispanica	Black-eared Wheatear	Polysticta stelleri	Steller's Eider	Sylvia curruca	Lesser Whitethrot
Oenanthe isabellina	Isabelline Wheatear	Porphyrio porphyrio	Purple Swamphen	Sylvia hortensis	Western Orphean Warbler
Oenanthe leucura	Black Wheatear	Porzana parva	Little Crake		
Oenanthe oenanthe	Northern Wheatear	Porzana porzana	Spotted Crake	Sylvia melanocephala	Sardinian Warbler
Oenanthe pleschanka	Pied Wheatear	Porzana pusilla	Baillon's Crake	Sylvia nisoria	Barred Warbler
Oriolus oriolus	Eurasian Golden Oriole	Prunella collaris	Alpine Accentor	Sylvia rueppelli	Rüppell's Warbler
Otis tarda	Great Bustard	Prunella modularis	Dunnock	Sylvia sarda	Marmora's Warbler
Otus scops	Eurasian Scops Owl	Psittacula krameri	Rose-ringed Parakeet	Sylvia undata	Dartford Warbler
Oxyura jamaicensis	Ruddy Duck	Pterocles alchata	Pin-tailed Sandgrouse	Syrmaticus reevesii	Reeves's Pheasant
Oxyura leucocephala	White-headed Duck	Pterocles orientalis	Black-bellied Sandgrouse	Tachybaptus ruficollis	Little Grebe
Pandion haliaetus	Osprey			Tadorna ferruginea	Ruddy Shelduck
Panurus biarmicus	Bearded Reedling	Ptyonoprogne rupestris	Eurasian Crag Martin	Tadorna tadorna	Common Shelduck
Parus cinctus	Siberian Tit	Puffinus gravis	Great Shearwater	Tarsiger cyanurus	Red-flanked Bluetail
Parus lugubris	Sombre Tit	Puffinus griseus	Sooty Shearwater	Tetrao tetrix	Black Grouse
Parus major	Great Tit	Puffinus puffinus	Manx Shearwater	Tetrao urogallus	Capercaillie
Passer domesticus	House Sparrow	Puffinus yelkouan	Yelkouan/Balearic Shearwater	Tetrax tetrax	Little Bustard
Passer hispaniolensis	Spanish Sparrow			Threskiornis aethiopicus	Sacred Ibis
Passer italiae	Italian Sparrow	Pyrrhocorax graculus	Alpine Chough	Tichodroma muraria	Wallcreeper
Passer montanus	Eurasian Tree Sparrow	Pyrrhocorax pyrrhocorax	Red-billed Chough	Tringa erythropus	Spotted Redshank
Pavo cristatus	Indian Peafowl	Pyrrhula pyrrhula	Eurasian Bullfinch	Tringa glareola	Wood Sandpiper
Pelecanus crispus	Dalmatian Pelican	Rallus aquaticus	Water Rail	Tringa nebularia	Common Greenshank
Pelecanus onocrotalus	White Pelican	Recurvirostra avosetta	Avocet	Tringa ochropus	Green Sandpiper
Perdix perdix	Grey Partridge	Regulus ignicapilla	Firecrest	Tringa stagnatilis	Marsh Sandpiper
Periparus ater	Coal Tit	Regulus regulus	Goldcrest	Tringa totanus	Common Redshank
Perisoreus infaustus	Siberian Jay	Remiz pendulinus	Penduline Tit	Troglodytes troglodytes	Eurasian Wren
Pernis apivorus	European Honey Buzzard	Riparia riparia	Sand Martin	Turdus iliacus	Redwing
		Rissa tridactyla	Kittiwake	Turdus merula	Common Blackbird
Petronia petronia	Rock Sparrow	Saxicola rubetra	Whinchat	Turdus philomelos	Song Thrush
Phalacrocorax aristotelis	European Shag	Saxicola rubicola	Common Stonechat	Turdus pilaris	Fieldfare
Phalacrocorax carbo	Great Cormorant	Saxicola torquatus	African Stonechat	Turdus torquatus	Ring Ouzel
Phalacrocorax pygmaeus	Pygmy Cormorant	Scolopax rusticola	Eurasian Woodcock	Turdus viscivorus	Mistle Thrush
Phalaropus fulicarius	Grey Phalarope	Serinus citrinella	Citril Finch	Turnix sylvaticus	Common Buttonquail
Phalaropus lobatus	Red-necked Phalarope	Serinus corsicana	Corsican Finch	Tyto alba	Barn Owl
Phasianus colchicus	Common Pheasant	Serinus serinus	European Serin	Upupa epops	Eurasian Hoopoe
Philomachus pugnax	Ruff	Sitta europaea	Eurasian Nuthatch	Uria aalge	Common Guillemot
Phoenicopterus roseus	Greater Flamingo	Sitta neumayer	Western Rock Nuthatch	Uria lomvia	Brünnich's Guillemot
Phoenicopterus ruber	American Flamingo	Sitta whiteheadi	Corsican Nuthatch	Vanellus spinosus	Spur-winged Lapwing
Phoenicurus ochruros	Black Redstart	Somateria mollissima	Common Eider	Vanellus vanellus	Northern Lapwing
Phoenicurus phoenicurus	Common Redstart	Somateria spectabilis	King Eider		
Phylloscopus bonelli	Western Bonelli's Warbler	Spinus spinus	Eurasian Siskin		
		Stercorarius longicaudus	Long-tailed Skua		
Phylloscopus borealis	Arctic Warbler	Stercorarius parasiticus	Arctic Skua		
Phylloscopus collybita	Common Chiffchaff	Stercorarius pomarinus	Pomarine Skua		
Phylloscopus ibericus	Iberian Chiffchaff	Stercorarius skua	Great Skua		
Phylloscopus inornatus	Yellow-browed Warbler	Sterna caspia	Caspian Tern		
Phylloscopus orientalis	Eastern Bonelli's Warbler	Sterna dougallii	Roseate Tern		
		Sterna hirundo	Common Tern		
Phylloscopus sibilatrix	Wood Warbler	Sterna nilotica	Gull-billed Tern		
Phylloscopus trochiloides	Greenish Warbler	Sterna paradisaea	Arctic Tern		
Phylloscopus trochilus	Willow Warbler	Sterna sandvicensis	Sandwich Tern		
Pica pica	Common Magpie				